PROGRESS IN BIOMASS AND BIOENERGY PRODUCTION

Edited by **S. Shahid Shaukat**

Progress in Biomass and Bioenergy Production
Edited by S. Shahid Shaukat

Published by InTech

Janeza Trdine 9, 51000 Rijeka, Croatia

Publishing Process Manager Niksa Mandic

Technical Editor Teodora Smiljanic

Cover Designer InTech Design Team

Additional hard copies can be obtained from orders@intechopen.com

Progress in Biomass and Bioenergy Production, Edited by S. Shahid Shaukat
p. cm.
ISBN 978-953-307-491-7

For EU product safety concerns:
IN TECH d.o.o., Prolaz Marije Krucifikse Kozulić 3, 51000 Rijeka, Croatia,
info@intechopen.com or visit our website at intechopen.com.

Meet the editor

S. Shahid Shaukat is Professor at the Institute of Environmental Studies, University of Karachi, Pakistan. He worked as teaching assistant and Tutorial Incharge at the University of Wales, U.K. and The University of Western Ontario, Canada. He has been teaching Ecology, Biostatistics and Bioinformatics in various departments, including, Agriculture, Environmental Studies, Marine Biology and Microbiology. He also teaches Mathematical Ecology in the Master of Industrial and Business Mathematics Program in the Department of Mathematics, University of Karachi. He attended a Postgraduate Forestry course at the University of Wales, Bangor and was awarded a Master's degree by research. Dr. Shaukat earned his Ph.D. from the University of Western Ontario, Canada. His major research interests include Mathematical Ecology, Biostatistics, Bioinformatics, Renewable Energy sources, Remote Sensing and GIS. He has authored a book titled "Essentials of Mathematical Ecology". Dr. Shaukat has published around 300 research papers in journals of international repute.

Contents

Preface

The fossil fuels that are principally used to provide energy today are in limited quantity, they are diminishing at an alarming rate, and their worldwide supplies will eventually be exhausted. Fossil fuels provide approximately 60 percent of the world's global electric power. Carbon dioxide levels in the atmosphere will continue to rise unless other cleaner sources of energy are explored. Biomass has the potential to become one of the major global primary energy source in the years to come. Biomass is the source of bioenergy which is produced by burning biomass or biomass fuels and provides cleanest energy matrix. Biomass, currently the most important source of energy, is organic matter which can be in the form of leaves, wood pieces, grasses, twigs, seeds and all other forms that plants and animals can assume whether living or recently dead. Often biomass has to be converted to usable fuel. This book addresses the challenges encountered in providing biomass and bioenergy. The book explores some of the fundamental aspects of biomass in the context of energy, which include: biomass types, biomass production system, biomass characteristics, recalcitrance, and biomass conversion technologies. The natural resistance of plant cell walls to microbial and enzymatic breakdown together is known as biomass recalcitrance. This characteristic of plant contributes to increased cost of lignocellulose conversion. Some of the articles included here address this issue. Besides exploring the topics of biomass and bioenergy, the book also deals with such diverse topics as biosorption, waste water treatment, fuel production including ethanol and hydrogen, and bio-economics.

The book is divided into seven sections which contain different number of chapters. Section I includes papers on Gasification and pyrolysis. The first Chapter by Jernej Mele presents a cold-flow model of FICFB biomass gasification process and its scale-up to industrial pilot plant. In Chapter 2, B. Fakhim and B. Farhanieh focus on Second Law analysis of bubbling fluidized bed gasification. Chapter 3 written by Milan Hrabovsky elucidates some new results on the production of syngas through thermal plasma technique, using gasification as well as pyrolysis. Chapter 4 authored by Jiri Jenista provides a numerical investigation of hybrid-stabilzed argon-water electric arc used for biomass gasification.

The Section II of the book covers biomass production and includes two chapters. In Chapter 5 Janis Abolins and Janis Gravitis present a simple analytical model for remote assessment of the dynamics of biomass accumulation. H. Viana, D. Lopes and J. Aranha, in Chapter 6 suggest a methodology for assessment of forest above ground biomass and dynamics using remote sensing and geostatistical modelling.

Section III which contains three chapters deals with Metal Biosorption and Reduction. Chapter 7 by J. F. Cardenas-Gonzalez and I. Acosta-Rodriguez describe a technique of removal of hexavalent chromium using a strain of the fungus *Paecilomyces* sp. Chapter 8 presents a comprehensive review of biosorption of metals by R.C. Oliveira and C. Palmieri which includes general features of the biosorption phenomenon as well as potential applications for environmental and technological processes. Chapter 9 authored by Zhu Guocai examines reduction of manganese ores using biomass as reductant. Section IV that deals with Wastewater treatment contains two chapters. Chapter 10 by Nima Badkoubi and H. Jazayeri-Rad attempts to investigate the parameters of wastewater treatment plant using extended Kalman filters (EKF) and some constrained methods. In Chapter 11 Dr. P. Vega discussed different control strategies for wastewater treatment. Section V, a large section, devoted to Characterization of biomass, pre-treatment, recovery and recalcitrance, comprises of seven chapters. Chapter 12 written by Yufu Xu, Xianguo Hu, Wendong Li and Yinyan Shi provides an elaborated review on Preparation and Characterization of Bio-oil from biomass. The investigation on bio-oils led to the conclusion that the bio-oils present bright prospects as an alternative renewable energy source instead of the popular fossil fuels. In Chapter 13 S. Adena-Elena focuses on Combined microwave-acid pretreatment of the biomass. Chapter 14 by Olfa Bouzaiane investigates the relationships of C, N and DNA content of municipal solid waste during the composting process. In Chapter 15 Hale Sütcü characterizes activated carbon produced from Oleaster stones. In Chapter 16 by A. Wijanarko, the effect of substituted urea and ammonia in the growth medium on the lipid content of *Chlorella* is investigated.

Chapter 17 by E. Fumoto, T. Tago and T. Masuda focuses on the recovery of ammonia and ketones from biomass waste. Recovery of ammonia is achieved through adsorption while that of ketones through catalytic cracking process. Chapter 18 written by O. Senneca characterizes biomass as nonconventional fuels by thermal techniques and presents a comprehensive protocol for the same. Section VI contains articles on Fuel production: ethanol and hydrogen. In Chapter 19 V.P. Obade, D.E. Clay, C.G. Carlson, K. Dalsted, B. Wylie, C. Ren and S.A. Clay provide the Principles and Applications of using remote sensing of nonharvested crop residue cover. In Chapter 20 Elisabeth Schröder discusses activated carbon production from waste biomass. In Chapter 21 M. Sveinsdottir, M.A. Sigurbjornsdottir and J. Orlygsson deal with the production of ethanol and hydrogen using thermophilic bateria from sugars and complex biomass. Harun Razif and M.K. Danquah in Chapter 22 focus on the analysis of process configuration for bioethanol production from microalgal biomass. Chapter 23 by R. Heard and C.R. Ozansoy reviews the Microbial conversion of biomass concentrating on microbial fuel cells.

Section VII contains one Chapter on Bio-economics. Chapter 24 written by D.V. Filatova and M. Grzywaczewski presents structural and parametric synthesis of bio-economic models using stochastic differential equations. Estimation procedures involved Monte Carlo simulation. The strength of the book rests more or less on all the contributions, my sincere thanks are due to all the authors for providing their in depth individual studies or comprehensive overviews of their research areas and the state-of-art in their fields and meeting the various deadlines.

I would like to express my gratitude to the faculty members of the Institute of Environmental Studies, University of Karachi and to postgraduate students and Prof. Dr. Moinuddin Ahmed (Foreign Faculty) of Ecological Research Laboratory, Federal Urdu University, Karachi, for some useful discussions and moral support. Finally, I would like to thank Ms Ana Pantar, Publishing Process Manager and Mr. Niksa Mandić, Publishing Process Manager, InTech Open Access Publisher, Croatia for bearing with me with delays and being generously helpful throughout the process of putting this book together.

May 2011

Dr. S. Shahid Shaukat
Institute of Environmental Studies
University of Karachi, Karachi
Pakistan

Part 1

Gasification and Pyrolysis

1

Scale-Up of a Cold Flow Model of FICFB Biomass Gasification Process to an Industrial Pilot Plant – Example of Dynamic Similarity

Jernej Mele
Faculty of mechanical engineering/Bosio d.o.o.
Slovenia

1. Introduction

In this chapter we are introducing the research of particles hydrodynamics in a cold flow model of **F**ast **I**nternal **C**irculating **F**luidized **B**ed (FICFB) biomass gasification process and its scale-up to industrial pilot plant. A laboratory unit has been made for the purposes of experimental research. The laboratory unit is three times smaller than the later pilot plant. For a reliable observation of the flow process, similar flow conditions must be created in the laboratory unit and the pilot plant. The results of the laboratory model will be similar to those of the actual device if geometry, flow and Reynolds numbers are the same. Therefore, there is no need to bring a full-scale gasificator into the laboratory and actually test it. This is an example of "dynamic similarity".

FICFB biomass gasification is a process for producing high caloric synthesis gas (syngas) from solid Hydrocarbons. The basic idea is to separate syngas from flue gas, and due to the separation we have a gasification zone for endothermic reactions and a riser for exothermic reactions. The bed material circulates between these two zones and serves as a heat carrier and a catalyst.

While researching the 250kW fluidized bed gasification pilot plant certain questions concerning particle dynamics in gas flows control arose. There is a zone where fluidized bed conditions are made with superheated steam, pneumatic transport with hot air and a pair of secondary gas inlets of CO_2. These particle flows are difficult to describe with mathematical models. This is the main reason why the three-times smaller cold-flow laboratory unit has been made. The hydrodynamics of particles will be studied in the air flow at arbitrary conditions. Flow conditions in the laboratory unit and pilot plant must be similar for a reliable evaluation of the process in the pilot plant.

2. Laboratory unit

The laboratory unit is a device three times smaller than the pilot plant. Its main purpose is to simulate the hydrodynamic process of FICFB gasification in a cold flow. It is made from stainless steel and in the case of the parts that are of greatest interest to the present study is made of glass, so that the particle behaviour may be observed. Fig. 1 shows a model of laboratory unit. Its main elements are:

- Reactor (A),
- Riser (B),
- Cyclone (C),
- Siphon (D),
- Chute (E),
- Gas distributor (J_1 and J_2),
- Auxiliary inlets (I_1 and I_2).

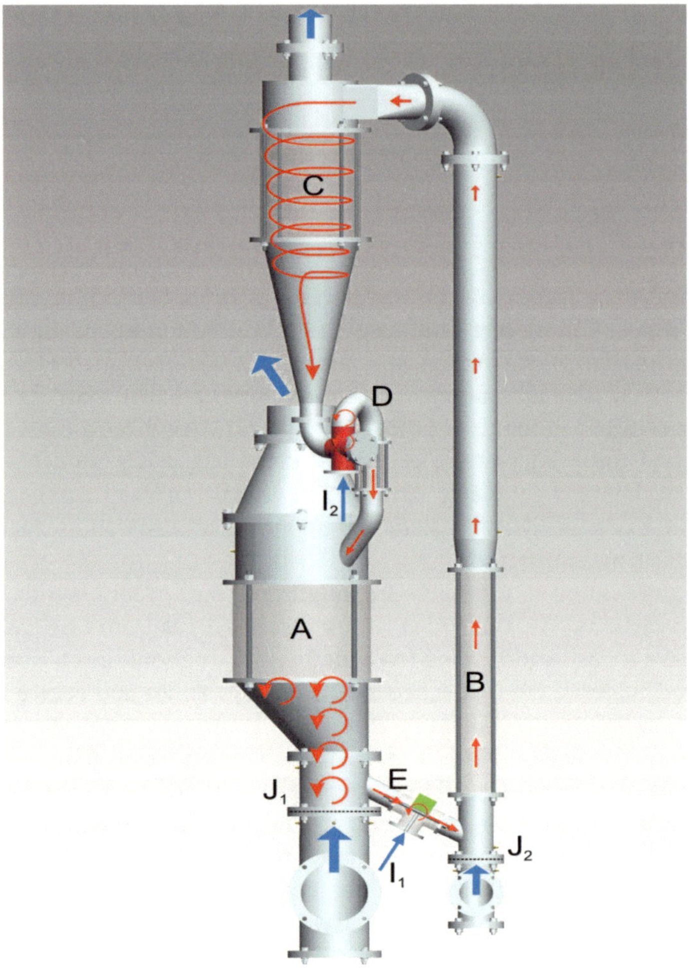

Fig. 1. 3D model of laboratory unit

Firstly, let us look at the process. There are two gas distributors at the bottom of the reactor and riser, through which air is blown vertically. The pneumatic transport of the particles takes place in the riser, where they are separated from the air flow in cyclone and finally gathered in siphon. The second auxiliary inlet acts to fluidize the gathered particles and transport them to the reactor. Here, the fluidized bed is created with the upward blowing air. From here, the particles are transported to the riser through the chute and the speed of transportation is regulated by means of the first auxiliary inlet.

	Laboratory unit	Pilot plant
$D_{gas,1}$ [mm]	100	300
$D_{gas,2}$ [mm]	190	600
D_{comb} [mm]	50	150
H_{comb} [mm]	1500	4500

Table 1. Main dimensions of laboratory unit and pilot plant

We are primarily interested in how to establish a stationary and self-sustainable process. In the laboratory unit there are glass parts through which the process in course can be directly observed. However, in the hot flow model we will not be able to see what happens inside the pilot plant, and therefore our control system must be able to initiate the process, keep it in a stationary state and halt it on the basis of measured data such as relative pressure and flow velocities. For this mater, our laboratory unit consists of 7 pressure and 2 flow velocity measuring points. Fig. 2 details the positions of the pressure places.

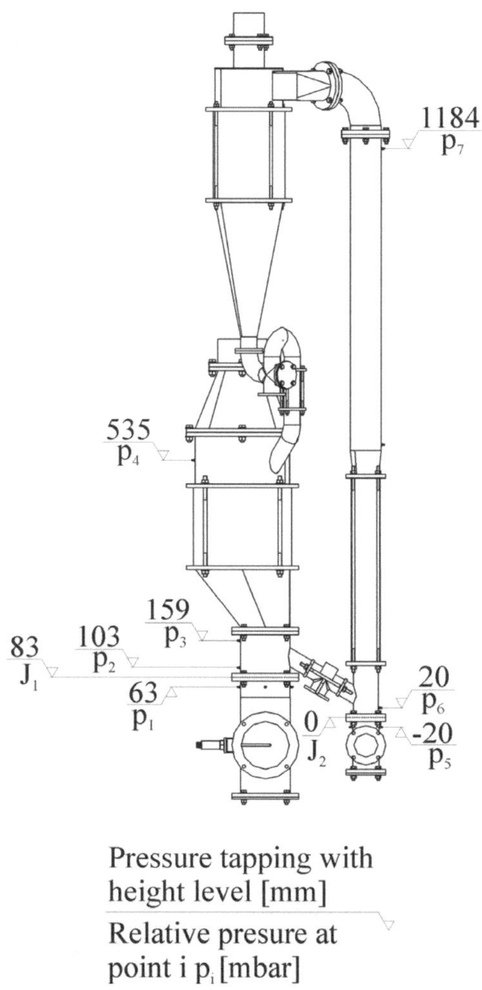

Fig. 2. Openings for the measuring of relative pressure

Trough experiments on the laboratory unit the effectiveness of elements will be studied so as to enable the correction and improvement of any construction flaws they contain. Fig. 3 shows the laboratory unit that will be used for studying the flow process. There are 7 places for pressure, 2 for temperature and 2 for gas flow measurements. For the proper operation of our solid flow system it is vital that the particles are maintained in dynamic suspension as settling down the particles can clog both the measuring openings and injection nozzles. Thus it is essential to design such systems with special care. All measurements involving the risk of clogging the measuring opening must be taken outside the solid flow zone if possible - gas flow velocity measurements with the Pitot tube must be taken in the gas pipeline before gas enters thru distributor. It is highly desirable for all measuring openings to be small and positioned rectangular to the direction of flow (Nicastro & Glicksman, 1982).

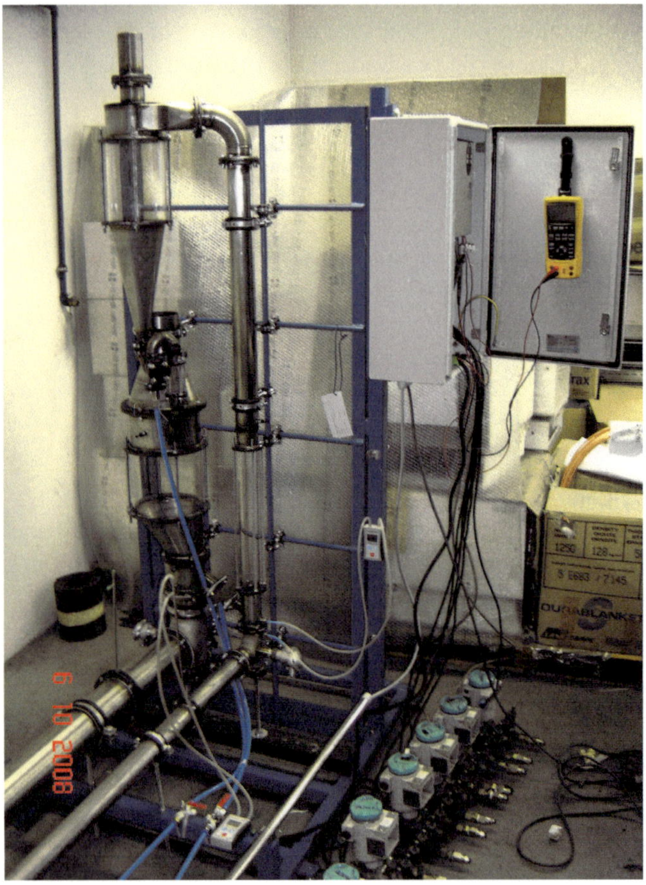

Fig. 3. Laboratory unit

2.1 Distributor

For the distributor 3 metal nets with openings of 225 μm have been used, with ceramic wool of 8mm placed in between as shown in fig. 4. We tried to achieve a sufficient pressure drop as to attain equal flow through the openings. According to Agarwal recommendation (Kunii & Levenspiel, 1991; Nicastro & Glicksman, 1982), the pressure drop across distributors must be 10 % of the pressure drop across the bed, with a minimum of 35 mm H_2O. With this we are in approximate agreement. At higher pressure drops across the distributor we get more particulate or smooth fluidization with less channelling, slugging and fluctuation in density. The pressure drop across the distributor is shown in fig. 5.

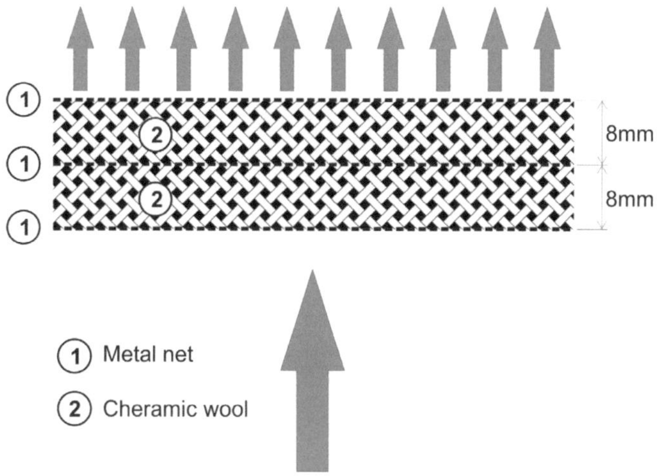

Fig. 4. Distributor structure

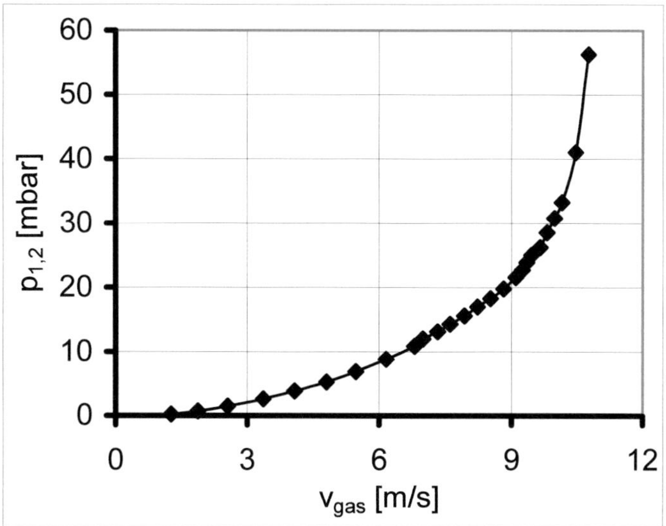

Fig. 5. Pressure drop across the distributor with blowing of air

2.2 Cyclone

In our case, the cyclone separator is placed behind the riser to separate the particles from the air flow. It has to be able to separate particles larger than 50 μm. For these conditions these particles are considered large as cyclones are often used for the removal of particles of about 10 μm diameter or larger from air streams. Our model is shown in fig. 6.

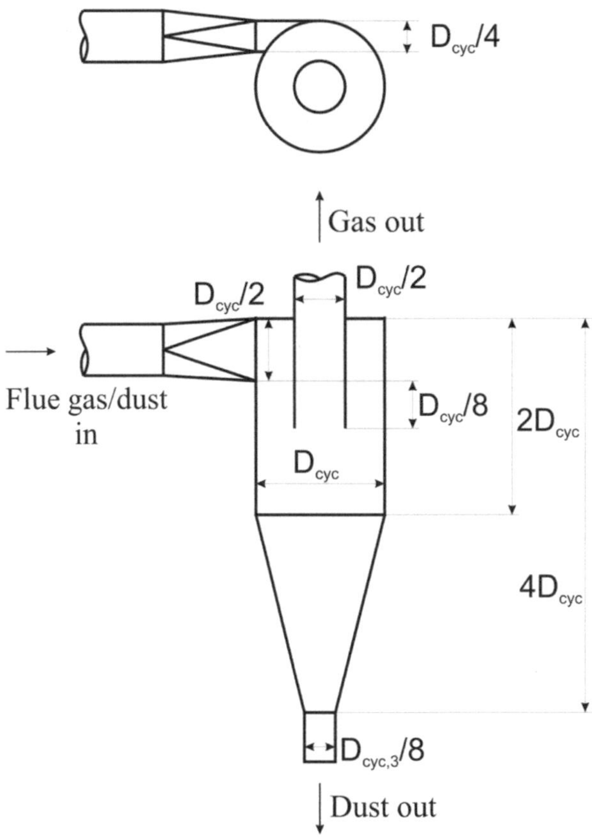

Fig. 6. The characteristic dimensions of cyclone

We dimensioned our cyclone according to Perry (Perry, 1988). $D_{p,50}$ is the particle size at which 50 % of solids of a given size are collected by the cyclone.

$$D_{p,50} = \sqrt{\frac{9 \cdot \eta_g \cdot B_{cyc}}{\pi \cdot N_s \cdot v_g \cdot (\rho_p - \rho_g)}} \tag{1}$$

By rearranging the equation (1), we obtain the following expression:

$$B_{cyc} = \frac{D_{p,50}{}^2 \cdot \pi \cdot N_s \cdot v_g \cdot (\rho_p - \rho_g)}{9 \cdot \eta_g} \tag{2}$$

The width of the cyclone entering the opening and the characteristic diameter are correlated by the following expression:

$$B_{cyc} = \frac{D_{cyc}}{4} \tag{3}$$

The diameter of our cyclone is 150 mm. Smaller particles, which are not separated in cyclone are being collected in a filter placed on the cyclone gas exit.

3. Basic equations for describing the fluidized state and similarity of flows

3.1 Reynolds number

The goal herein is to compare flows in the laboratory unit to those in the pilot plant. In order for the two flows to be similar they must have the same geometry and equal Reynolds numbers. When comparing fluid behaviour at homologous points in a model and a full-scale flow, the following holds:

Re(laboratory unit) = Re(Scale-up pilot plant)

The Reynolds number of particles can be determined by the following equation (Kunii & Levenspiel, 1991):

$$\mathrm{Re}_p = \frac{D_p \cdot v_g \cdot \rho_g}{\eta_g} \tag{4}$$

For achieving the required similarity, the following conditions must be also fulfilled:

$$\frac{p}{\rho_g \cdot v_g^{\,2}} = \frac{p_{g,ar}}{\rho_{g,ar} \cdot v_{g,ar}^{\,2}} \tag{5}$$

3.2 Minimal fluidizing velocity

The fluidization state starts when the drag force of by upward moving gas equals the weight of the particles (Oman, 2005)

$$F_{g\_p} = \frac{1}{2} \cdot C_x \cdot A_p \cdot \rho_p \cdot v_g^{\,2} \tag{6}$$

or

$$\Delta p \cdot A_t = \left(A_t \cdot L_{mf}\right)\left(1-\varepsilon_{mf}\right)\left[\left(\rho_s - \rho_g\right)\frac{g}{g_c}\right] \tag{7}$$

By rearranging equation (7), for minimum fluidizing conditions we find the following expression (Kunii & Levenspiel, 1991),

$$\frac{\Delta p_{mf}}{L_{mf}} = \left(1-\varepsilon_{mf}\right)\left(\rho_s - \rho_g\right)\frac{g}{g_c} \tag{8}$$

Voidage in fluidized bed ε_{mf} is larger than in the packed bed and it can be estimated experimentally from a random ladling sample. For small particles and low Reynolds numbers the viscous energy losses predominate and the equation simplifies to (Kunii & Levenspiel, 1991):

$$v_{mf} = \frac{\left(\Phi_s \cdot D_p\right)^2}{150} \cdot \frac{\rho_p - \rho_g}{\eta_g} \cdot g \cdot \frac{\varepsilon_{mf}^{\,2}}{\left(1-\varepsilon_{mf}\right)} \tag{9}$$

for $\mathrm{Re}_p < 20$

For large particles only the kinetic energy losses need to be considered:

$$v_{mf} = \sqrt{\frac{\Phi_s \cdot D_p}{1{,}75} \cdot \frac{(\rho_p - \rho_g)}{\eta_g} \cdot g \cdot \varepsilon_{mf}^{\,3}} \qquad (10)$$

for $Re_p > 1000$.
If Φ_S and ε_{mf} are unknown, the following modifications suggested by Wen and Yu (Kunii & Levenspiel, 1991) are used:

$$\frac{1-\varepsilon_{mf}}{\Phi_S^{\,2} \cdot \varepsilon_{mf}^{\,2}} \cong 11 \qquad (11)$$

$$\frac{1}{\Phi_S \cdot \varepsilon_{mf}^{\,3}} \cong 14 \qquad (12)$$

Equations (9) and (10) can now be simplified to:

$$v_{mf} = \frac{D_p^{\,2} \cdot (\rho_p - \rho_g) \cdot g}{1650 \cdot \eta_g} \qquad (13)$$

for $Re_p < 20$

$$v_{mf} = \sqrt{\frac{D_p \cdot g \cdot (\rho_p - \rho_g)}{24{,}5 \cdot \rho_g}} \qquad (14)$$

for $Re_p > 1000$.

3.3 Terminal velocity

The upper limit of gas flow rate is approximated by the terminal (free fall) velocity of the particles, which can be estimated from the fluid mechanics (Kunii & Levenspiel, 1991):

$$v_t = \sqrt{\frac{4 \cdot g \cdot D_p \cdot (\rho_p - \rho_g)}{3 \cdot \rho_g \cdot C_x}} \qquad (15)$$

There are spherical and non-spherical particle shapes in the bed and each of them has a different C_x value. If we combine equations (4) and (15) we get the velocity independent group:

$$C_x \, Re_p^{\,2} = \frac{4 \cdot g \cdot D_p^{\,3} \cdot \rho_g \cdot (\rho_p - \rho_g)}{3 \cdot \eta_g^{\,2}} \qquad (16)$$

An alternative way of finding v_t for spherical particles uses analytical expressions for the drag coefficient C_x (Kunii & Levenspiel, 1991).

$$C_x = \frac{24}{Re_p} \quad \text{for} \quad Re_p < 0{,}4 \qquad (17)$$

$$C_x = \frac{10}{\sqrt{Re_p}} \quad \text{for} \quad 0{,}4 < Re_p < 500 \tag{18}$$

$$C_x = 0{,}43 \quad \text{for} \quad 500 < Re_p < 200000 \tag{19}$$

But still no simple expression can represent the experimental findings for the entire range of Reynolds numbers, so by replacing these values C_x in equation (16) we obtain:

$$v_t = \frac{(\rho_p - \rho_g) \cdot g \cdot D_p^{\ 2}}{18 \cdot \eta_g} \tag{20}$$

for $Re_p < 0{,}4$

$$v_t = \sqrt[3]{\frac{4}{225} \cdot \frac{(\rho_p - \rho_g)^2 \cdot g^2}{\eta_g \cdot \rho_g}} \cdot D_p \tag{21}$$

for $0{,}4 < Re_p < 500$
and

$$v_t = \sqrt{\frac{3{,}1 \cdot D_p \cdot g \cdot (\rho_p - \rho_g)}{\rho_g}} \tag{22}$$

for $500 < Re_p < 200000$.

3.4 Determining density

In the pilot plant we will have multiple gas mixtures at different temperatures due to chemical reactions. For our calculations the density for these mixtures will be determined by the following equation (Oman et al., 2006):

$$\rho_g = \left(\sum_i \frac{w_i}{\rho_i} \right)^{-1} \tag{23}$$

To calculate the density of the gas mixture at an arbitrary temperature and an arbitrary pressure the density under normal condition must be calculated according to equation (24), with the obtained value being converted to density at the required parameters:

$$\rho_{g,ar} = \rho_g \cdot \frac{p_{g,ar}}{p_n} \cdot \frac{T_n}{T_{g,ar}} \tag{24}$$

3.5 Pressure drops

With increased gas velocity of the small solid particles across the bed a characteristic state occurs. Pressure drop starts to increase, reaching its maximum value Δp_{mf} at minimum fluidization velocity v_{mf}. At this point only part of the bed is fluidized. When the bed is fully fluidized (at v_{mff}), the pressure drop is reduced to Δp_{mff} and is almost constant until gas reaches terminal velocity. If the velocity is still increasing, the particles start transporting pneumatically and pressure drop reduces rapidly to 0. By rearranging equation (8), we obtain the following expression (Kunii & Levenspiel, 1991):

$$\Delta p_{mf} = (1-\varepsilon_{mf})(\rho_s - \rho_g)\cdot g \cdot L_{mf} \tag{25}$$

The expression can also be extended to the fully fluidized state (Kaewklum & Kuprianov, 2008):

$$\Delta p_{mff} = (1-\varepsilon_{mff})(\rho_s - \rho_g)\cdot g \cdot L_{mff} \tag{26}$$

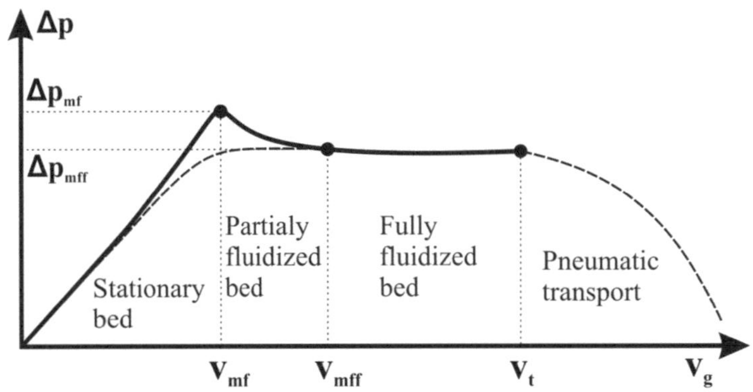

Fig. 7. The change in pressure drop relative to gas velocity for Not-too-Small Uniformly Sized Particles (Kunii & Levenspiel, 1991)

A somewhat different differential pressure characteristic occurs with a wide size distribution of particles, which are usually present in industrial processes. When the gas velocity increases through the bed of solids, the smaller particles start to fluidize and slip into the void spaces between the larger particles, while the larger particles remain stationary (Kunii & Levenspiel, 1991) (see Fig. 8). However, after a full fluidization of bed material (v_g>v_{mff}), with increasing air velocity, pressure drop mainly remains constant.

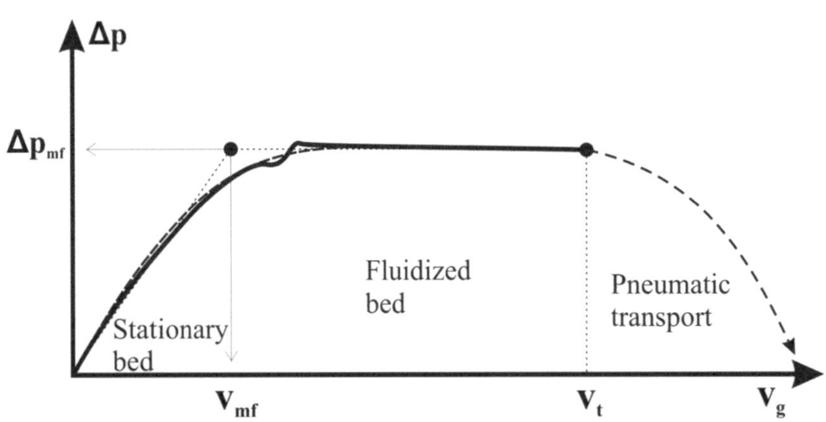

Fig. 8. The change in pressure drop relative to gas velocity for Wide Size Distribution of Particles (Kunii & Levenspiel, 1991)

3.6 Mass flows and conservation of mass

For a regular flow process we have to ensure proper gas flows at the inlets. Through defining minimal fluidizing and terminal velocities, we can estimate the mass flow of the air reactor and riser by applying the following relations:

$$\varphi_{m\_g} = \rho_g \cdot v_g \cdot \frac{\pi \cdot D_{tube}^{2}}{4} \tag{27}$$

$$\varphi_{V\_g} = \frac{\varphi_{m\_g}}{\rho_g} \tag{28}$$

It is extremely important to ensure that there are no mass losses between the ventilator and the reactor. It can be assumed:

$$v_g \cdot \frac{D_{gas,2}^{2} \cdot \pi}{4} = v_{g,ref} \cdot \frac{D_{tube}^{2} \cdot \pi}{4} \tag{29}$$

4. Calculation analyses

On the basis of the previously-mentioned equations, we can make an estimation of flow conditions in the reactor and riser. We have made a tabular comparison of physical properties between the laboratory unit and pilot plant in tables 2 and 3. The comparison is based on the established equality of Reynolds numbers. As mentioned in chapter 3.1. "In order for two flows to be similar they must have the same geometry and equal Reynolds numbers". In the laboratory unit, flows will be made with upward-blowing air at room temperature whereas in the pilot plant the fluid bed will be made with inlet of superheated steam and pneumatic transport with hot air blowing at 550 °C.

	Reactor	
	Laboratory unit	Pilot plant
Gas	Air	Steam / Syngas
T [°C]	30	550 / 800
D_p [μm]	200	600
ρ_p [kg/m³]	8250	3025
ρ_g [kg/m³]	1,204	0,288 / 0,192
η_g [Pas]	1,8·10-5	3,1·10-5 / 4,6·10-5
$v_{Re<20}$ [m/s]	0,11	0,21 / 0,14
$v_{Re>1000}$ [m/s]	0,75	1,58 / 1,95
Φ_m [kg/h]	6,4	158,9
Φ_V [m³/h]	5,4	548,5
Re_p	9,8	9,0 /4,9

Table 2. Physical properties of gas in Reactor

In the meantime endothermic chemical reactions of pyrolisys, a water-gas-shift reaction will take place in the reactor while exothermic combustion occurs in the riser. Flue gases will have a the temperature of around 1000 °C on exiting the combustor and syngas a temperature of approximately 800 °C at the reactor's point of exit. Gases in the pilot plant will have lower densities and higher viscosities than the air in the laboratory unit. The bed material will be Olivine with D_p = 600 µm. In order to establish similar conditions, we have to use smaller and denser particles. We have chosen brass particles with D_p = 200 µm . Simulation will also be tested with quartz sand and olivine.

	Riser	
	Laboratory unit	Pilot plant
Gas	Air	Air / Flue gas
T_g [°C]	30	550 / 1000
D_P [µm]	200	600
ρ_P [kg/m^3]	8250	3025
ρ_g [kg/m^3]	1,204	0,61/0,294
η_g [Pas]	1,8·10-5	3,8·10-5/4,7·10-5
$v_{Re<0.4}$ [m/s]	10,1	15,7/12,6
$v_{0.4<Re<500}$ [m/s]	3,6	5,3/6,2
$v_{500<Re<200000}$ [m/s]	6,6	9,6/13,7
Φ_m [kg/h]	47,7	154,2
Φ_V [m^3/h]	39,6	524,5
Re_p	46,6	50,8 / 23,3

Table 3. Physical properties of gas in Riser

On the basis of studied flow velocities, mass flows, as well as pressure drops through air distributors and fluid beds at different points of the laboratory unit, we may anticipate the similar results in the pilot plant.

5. Experimental work

Firstly, we have to establish the fluidized bed in the reactor. The particles will fill the chute and the lower part of the riser. The chute is installed at the bottom of the reactor and riser and has an inclination angle. The fluidizing of the particles in the chute will then be started, along with the simultaneous initialization of the pneumatic transport of the particles. When sufficient material has been gathered in the siphon, the particles must be transported back to the reactor with the help of the first auxiliary inlet. The particles are now at their starting point. We must achieve a pressure at the bottom of the fluidized bed p_2 which is larger than that at the point where the chute connects to the riser p_6. The gas flow direction will be from the reactor to the riser, pushing the particles in the desired direction. At the top of the fluidized bed we have pressure p_4 which has to be lower than p_7, so the particles can now travel back to the reactor. But there has to be enough material in the siphon at all times in order to prevent the mixing of gases between the zones. Therefore, the siphon has to serve as seal gap for gases but not for material. The more gas goes through the siphon the lower the caloric value of the gas will be. Experiments will show how pressures are distributed across the system. Fig. 9 shows which measured pressures are of greatest interest for our purposes.

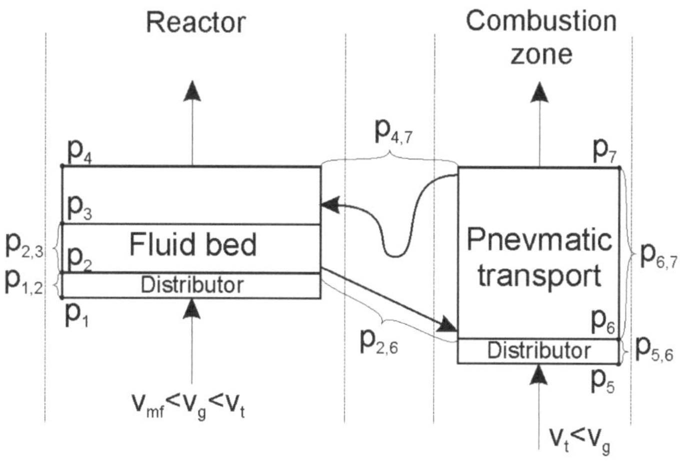

Fig. 9. Measuring scheme

By way of example, we will look at the experiment with quartz sand. The size of the particles used for simulation is shown in fig. 13. The particles have an average diameter of about 200 µm. A series of measurements were made and pressure drops at different bed heights taken. Fig. 10 represents a comparison of pressure drop across the bed in the reactor with the gas velocity for different bed heights.

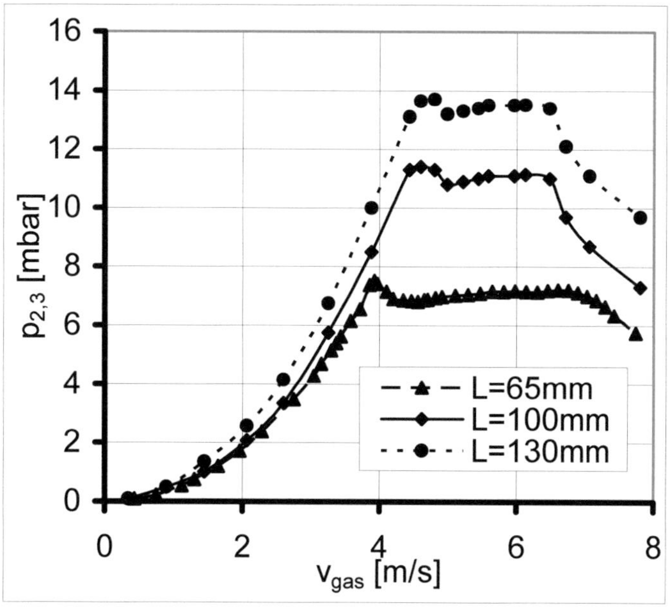

Fig. 10. Pressure drops over fluidized bed

In lower beds less aggregative bubbling occurs and results closer to calculated values are obtained. Nevertheless, still there is a lot of deviation between them. In addition, there is some leakage of gas from the reactor through chute to the riser and as the Pitot tubes are placed in front of gas entering each zone those velocities do not represent the real situation, although the mass flow of air blown through unit is quite as predicted. However, gas velocity is almost impossible to measure within the laboratory unit because attempts to do so would inevitably lead to bed material clogging the measure openings in the device. Having said that, our assessment and purpose is to define and achieve a stationary process on the basis of the measuring system. The measured quantities are presented in table 4.

Symbol	Value	[unit]
p_1	34.4	mbar
p_2	11.3	mbar
p_3	0.2	mbar
p_4	0.1	mbar
p_5	6.2	mbar
p_6	3.9	mbar
p_7	3.2	mbar
v_{gas}	5.1	m/s
v_{comb}	9	m/s

Table 4. Measurements results

Comparisons of error between calculations and experimental results of pressure drops are presented in fig. 11 and 12. Through the application of the mathematical models we find that pressure drops can be predicted within a 20 % error margin. For example let us compare results between calculated and experimental values of pressure drop across 100 mm bed of quartz sand at minimum fluidization conditions. Calculating pressure drop according to equation 23 gives us 12.5 mbar, where physical properties are as follows: $\rho_p = 2650\ kg/m^3$, $\rho_g = 1.204\ kg/m^3$, $\varepsilon_{mf} = 0.55$, $L_{mf} = 110$ mm ang $g = 9{,}81\ m/s^2$. Bed height increases for 10 mm and so does consecutive voidage. A series of measurements gives us the average value for pressure drop which is $p_{2,3} = 11.4$ mbar. As follows from this, the error of our prediction was 8.8 %.

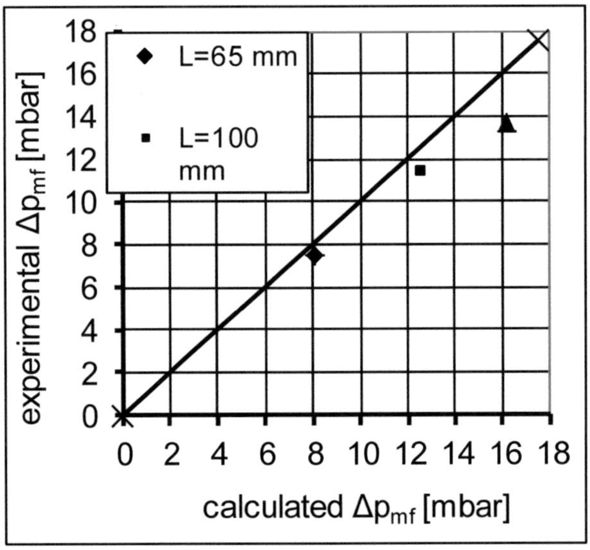

Fig. 11. The comparison of experimental and calculated Δp_{mf} for 200 μm quartz sand

For calculating pressure drops across fully fluidized bed we use equation 26. The only difference comes with a little higher bed and voidage, which remain almost constant with increasing gas velocity to terminal velocity. So if we consider that L_{mff} = 115 mm and ε_{mff} = 0.62 than pressure drop equals 11.4 mbar. With the comparison to the experimental value, which is 10.8 mbar, a 5.2 % error of prediction occurs. Error highly increases in aggregative and slugging mode of fluidization.

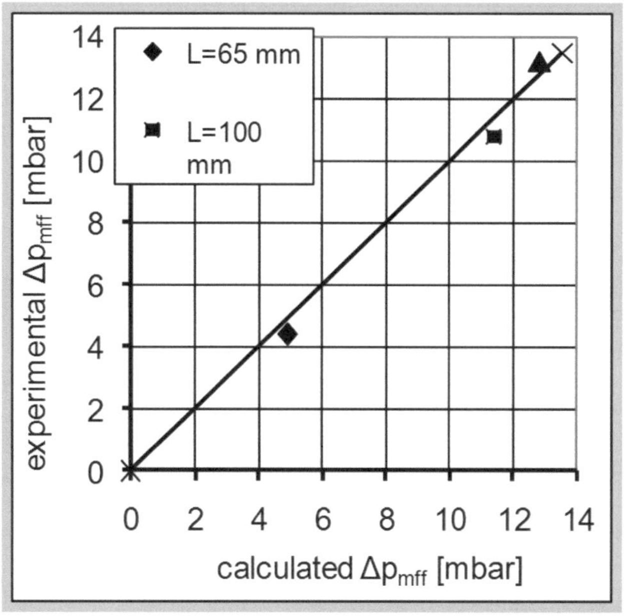

Fig. 12. Comparison of experimental and calculated Δp_{mff} for 200 μm quartz sand

Relative pressures were measured at a stationary state. One of the experiments was made when testing the process with quartz sand where the average particle diameter was about 200 μm. The stationary bed height in the reactor was 100 mm and the mass of sand used at simulation was 4.25 kg. When minimum fluidization conditions were obtained, the bed height increased by approximately 15 mm. A series of repeated measuring were carried out

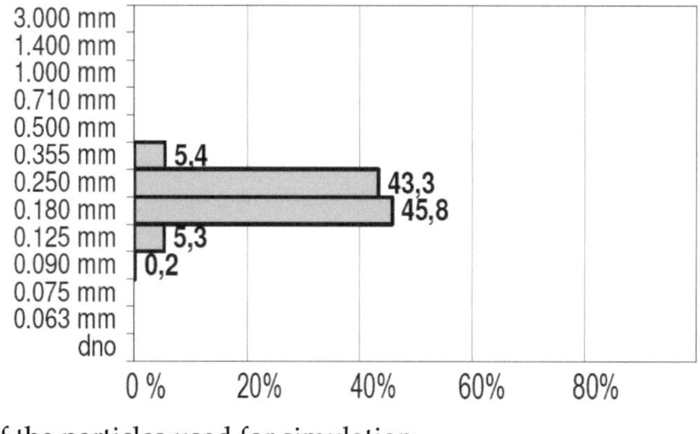

Fig. 13. The size of the particles used for simulation

and the average relative pressure at the bottom of the fluid bed was p_2 = 11.3 mbar, with p_3 = 0.2 mbar the average value at the top. As follows from this, the pressure drop across fluidized bed was $p_{2,3}$ = 11.1 mbar. Air flow had an average temperature of 25 °C. Inlet gas velocity was about 5.1 m/s in the reactor and 9 m/s in the riser. We found a higher gas velocity for fluidization than calculated, due to a certain amount of air passing through the chute to the riser. This also provides the explanation as to why the measured terminal velocity in the riser was a little lower than anticipated, as the loss of air from the reactor helped increase the air speed in the riser - resulting in the aforementioned lower value.

6. Comparison to the previously used methods

Modern gasification is occurring in fluidized beds. Its advantage is using most fuels (wood, peat and coal) including agriculture "waste" such as straw, corn stover and manure. It has a potential to use municipal waste, such as garbage, it is quicker in response and it has shorter start up time. It lends itself to complete combustion applications which would allow it to use liquid wastes, such as used engine oil, non-recyclable plastics, junk mail & old shoes and garbage for the generation of heat. However, there is a problem of complex design. Still, nowadays most research efforts are being made on fluidization bed technology.

We tested a system very similar to the one tested by G. Löffler, S. Kaiser, K. Bosch, H. Hofbauer (Kaiser S. et al., 2003) with a minor difference. Our reactor had an eccentric diffuser which proved not to be a successful idea (Mele, J. et al., 2010). That is why in future research we are planning to test a reactor with a conical bed similar to those used by Kaewklum and Kuprianov. Our mathematical model is based on the derivation of Ergun's equation (Kunii & Levenspiel, 1991). L. Glicksman pointed out that for designing an accurate scale model of a given bed all of the independent non-dimensional parameters must be identical, such as considering the case where fluidized bed is operated at an elevated temperature of flue gas or at arbitrary conditions with air (Glicksman, 1982). Our work is also based on attaining similar non-dimensional parameters such as Reynolds and Euler numbers. The Freude number based on the minimum velocity, (v_{mf}/dp_g) has been proposed as the parameter to characterize the boundary between particulate and aggregative fluidization and the Archimedes number has been used to correlate a wide array of phenomena (Zabrodsky, 1966).

7. Conclusions

By observing the CFB processes in a three-times smaller laboratory unit with air flow the size and density of particles has been determined. The preferred option was to use brass powder with an average particle diameter of 200 µm. The assumption of particle flow similarity is based on a direct comparison of Reynolds numbers. In this case the Re_p are 9.8 and 9.0 in reactor and 46.6 and 50.8 in the riser. There is a 10 % difference between Re_p in both cases. Chemical reactions cause variations in temperature, density, and dynamic viscosity all of which affect Re_p. If we compare Re_p 9.8 and 4.9 at the reactor exit 46.6 and 23.3 at the top of the riser exit, we can see that Re_p changes by 50 % and the similarity at this point is actually questioned. By way of example, the experiment carried out with quartz sand was presented. When the process is stabilized and a smooth circulation is established, then pressure drops are as follows: $p_{2,3}$ = 11.2 mbar, $p_{6,7}$ = 0.7 mbar, $p_{2,6}$ = 7.4 mbar and $p_{4,7}$ = -3.1 mbar. This result set can be characterized as $p_2 > p_6$ and $p_4 < p_7$. Pressures are as expected and gas flows are in the appropriate directions. Through the application of the mathematical models we have, pressure drops can be predicted to within a 20% error margin. The experiments highlighted one major problem, namely that the cylindrical tube and asymmetric enlargement of the tube didn't prove to be a successful construction for the reactor. With beds higher than 13 cm fluidized beds are in aggregative or bubbling fluidization states. In turn, at bed heights over 30 cm even a slugging state is attained. The solution at this point is a conical bed design in accordance with Kaewklum and Kuprianov, 2008.

8. Symbols

A_p	Cross-section of particle	[m²]
A_t	Tube cross-section	[m²]
B_{cyc}	Width of rectangular cyclone inlet dutch	[m]
C_x	Drag coefficient	
D_{cyc}	Characteristic cyclone diameter	[m]
D_{comb}	Riser diameter	[mm]
$D_{gas,1}$	Diameter of reactor upper segment	[mm]
$D_{gas,2}$	Diameter of reactor lower segment	[mm]
D_p	Diameter of particle	[μm]
$D_{p,50}$	Particle diameter at which 50% of particles are collected by cyclone	[μm]
D_{tube}	Inside tube diameter	[mm]
F_{g_p}	Gravity of particle	[N]
g	Gravity acceleration [9,81 m/s²]	
g_c	Conversion factor [9,81gm m/s² wt]	
H_{comb}	Riser height	[mm]
i	Natural number	
j	Natural number	
L	Stationary bed height	[m]
L_{mf}	Bed height at minimum fluidization condition	[m]
L_{mff}	Bed height at minimum fully fluidized state	[m]
N_s	Number of turns made by gas stream in a cyclone separator	
p	Pressure	[Pa]
$p_{g,ar}$	Pressure at arbitrary conditions	[Pa]
p_i	Relative pressure in point i	[Pa]
$p_{i,j}$	Differential pressure between points i and j	[Pa]
p_j	Relative pressure in point j	[Pa]
p_n	Pressure at normal conditions	[Pa]
Re_p	Particle Reynolds number	
$T_{g,ar}$	Temperature at arbitrary conditions	[°C]
T_n	Temperature at normal conditions	[°C]
v_{comb}	Gas velocity in riser	[m/s]
v_g	Gas velocity	[m/s]
$v_{g,ref}$	Gas velocity measured with pitot tube or orifice in tube before gas entering reactor	[m/s]
v_{gas}	Gas velocity in gasification zone	[m/s]
v_{mf}	Minimal fluidization velocity	[m/s]
v_{mff}	Minimal velocity of full fluidization	[m/s]
v_t	Terminal velocity	[m/s]
Δp	differential pressure	[Pa]
Δp_{mf}	differential pressure at minimum fluidization	[Pa]
Δp_{mff}	differential pressure at full fluidization	[Pa]

ε	Bed voidage	
ε_{mf}	Bed voidage at minimum fluidization	
ε_{mff}	Bed voidage at full fluidization	
η_g	Dynamical viscosity of gas	[Pa·s]
$\eta_{g,ar}$	Dynamical viscosity of gas at arbitrary conditions	[Pa·s]
η_n	Dynamical viscosity of gas at normal conditions	[Pa·s]
ρ_g	Density of gas	[kg/m^3]
ρ_p	Density of particle	[kg/m^3]
Φ_m	Mass flow	[kg/h]
Φ_{m_g}	Mass flow of gas	[kg/h]
Φ_V	Volume flow	[m^3/h]
Φ_{V_g}	Volume flow of gas	[m^3/h]

9. Acknowledgements

Investing in your future
OPERATION PART FINANCED BY THE EUROPEAN UNION
European Social Fund

10. References

Glicksman, L. R. (1982). Scaling Relationships For Fluidized Beds, Chemical engineering science, 39, 1373-1384

Kaewklum, R. & Kuprianov, V. I. (2008). *Theoretical And Experimental Study On Hydrodynamic Characteristic Of Fluidization In Air-Sand Conical Beds*, Chemical Engineering Science 63 1471-1479

Kaiser, S.; Löffler, G.; Bosch, K.; Hofbauer, H. (2003). *Hydrodynamics of a Dual Fluidized Bed Gasifier - Part Ii: Simulation of Solid Circulation Rate, Pressure Loop and Stability*, Chemical Engineering Science, 58, 4215 – 4223

Kunii, D. & Levenspiel, O.; (1991). *Fluidization Engineering - Second edition*, John Wiley & Sons, inc.,

Löffler G., Kaiser S., Bosch K., Hofbauer H. (2003). *Hydrodynamics of a Dual Fluidized - Bed Gasifier - Part I : Simulation of a Riser With Gas Injection and Diffuser*, Chemical Engineering Science, 58, 4197 – 4213

Mele, J.; Oman, J.; Krope, J. (Jan. 2010). *Scale-up of a cold flow model of FICFB biomass gasification process to an industrial pilot plant - hydrodynamics of particles*, WSEAS transactions on fluid mechanics, vol. 5, iss. 1, str. 15-24.

Nicastro, M. T. & Glicksman, L. R. (1982). *Experimental Verification of Scaling Relationships for Fluidized Beds*, Chemical engineering science, 39, 1373-1384

Oman J. (2005), *Generatorji Toplote*, University in Ljubljana, Faculty of mechanical engineering, Ljubljana,

Oman, J.; Senegačnik, A.; Mirandola, A. (2006). *Air, Fuels and Flue Gases: Physical Properties and Combustion Constants*, Edizioni Librerita Progeto, Padova, Italy

Perry, R. H. (1988). *Perry's Chemical Engineers Handbook (6th ed.)*, New York: McGraw Hill International Ed.

Zabrodsky, S. S. (1966). *Hydrodynamics And Heat Transfer In Fluidized Beds*, The MIT press, Cambrige

2

Second Law Analysis of Bubbling Fluidized Bed Gasifier for Biomass Gasification

B. Fakhim and B. Farhanieh
School of Mechanical Engineering,
Division of Energy Conversion, Sharif University of Technology, Tehran,
Iran

1. Introduction

The management of refused derived fuel (RDF) is one of the most significant problems especially for developing countries. Technologies to convert biomass energy already exist as well. Gasification through a bubbling fluidized bed gasifier (BFBG) is discussed in this context. A BFBG is able to deal with wide variety of fuels due to the presence of inert bed material, in which bubbles mix turbulently under buoyancy force from a fluidizing agent like air or oxygen [1]. Under such violent bed conditions biomass waste particles are able to react fully to release volatiles as a result from high solids contact rate. Gases are released from the biomass particles and can then be used for producing electricity. In the literature there are several investigations on gasification processes from the thermodynamic point of view. Altafini and Mirandola [2] presented a coal gasification model by means of chemical equilibrium, minimizing the Gibbs free energy. The authors studied the effect of the ultimate analysis and the gasifying agents/fuel ratio on the equilibrium temperature (adiabatic case) in order to obtain the producer gas composition and the conversion efficiency. They concluded that the equilibrium model fits the real process well. Similar conclusions for biomass gasification are presented by the same authors [3], simulating the gasifying process in a downdraft gasifier, where the object of study was the effect of the biomass moisture content on the final gas composition assuming chemical equilibrium. Lapuerta et al. [4] predicted the product gas composition as a function of the fuel/ air ratio by means of an equilibrium model. A kinetic model was used to establish the freezing temperature, which is used for equilibrium calculations in combination with the adiabatic flame temperature. The biomass gasification process was modeled by Zainal et al. [5] based on thermodynamic equilibrium. They analysed the influence of the moisture content and reaction temperature on the product gas composition and its calorific value. Ruggiero and Manfrida [6] emphasized the potential of the equilibrium model considering the Gibbs free energy. This proceeding can be used under different operating conditions for predicting producer gas composition and the corresponding heating value.

Many studies on the modeling of coal gasifers, in general, and coal gasification in bubbling fluidized beds, in particular, can be found in the literature. Nevertheless, thermodynamic modeling of the biomass gasification in bubbling fluidized beds has not been amply addressed. A few articles on the modeling of biomass bubbling fluidized bed gasifiers

(BBFBGs) can be found in the literature. In modeling the biomass gasification (with air) in bubbling fluidized beds (BFBG), Belleville and Capart [7] developed an empirical model which was successfully applied to the biomass gasifier of Creusot Loire in Clamecy (France). Fan and Walawender [8] and Van den Aarsen [9] reported two of the pioneering models, which are well known today; Corella et al. [10] modeled some non-stationary states of BFBBGs; Bilodeau et al. [11] considered axial variations of temperature and concentration and applied their results to a 50 kg/h pilot gasifier; Jiang and Morey [12,13] introduced new concepts in this modeling, especially related to the freeboard and the fuel feed rate; Hamel and Krumm [14] provided interesting axial profiles of temperature, although their work was mainly focused on gasification of coal and did not give many details of their model; Mansaray et al. [15,16] presented two models using the ASPEN PLUS process simulator.
In this work the equilibrium modeling of BFBG has been applied for the biomass waste gasification. The model employs equilibrium constants of all constituent reactions, in addition, the effect of the fuel/air ratio, moisture content of the fuel and gasifying temperature on the mole fraction of product gases of RDF gasification and corresponding higher heating value of it. Moreover, the exergetic efficiency and cold gas efficiency of the BFBG has been evaluated.

2. The model of the BFBG

2.1 Energy analysis

The idealized fluidized bed gasifier model is used with the following assumptions:
(i) The chemical equilibrium between gasifier products is reached, (ii) the ashes are not considered and (iii) heat losses in the gasifier are neglected.
The global gasification reaction can be written as follows:

$$\begin{aligned} & C_aH_bO_cN_dS_e + wH_2O + m(O_2 + 3.76N_2) \rightarrow n_1H_2 + n_2CO + n_3CO_2 \\ & + n_4H_2O + n_5CH_4 + n_6N_2 + n_7H_2S \end{aligned} \tag{1}$$

In which the $C_aH_bO_cS_dN_e$ is the substitution fuel formula which can be calculated by the ultimate analysis of the fuel and the mass fractions of the carbon, hydrogen, oxygen, nitrogen and sulphur. "m" and "w" are the molar quantity of air entering the gasifier and moisture molar fraction in the fuel, respectively. The variable "m" corresponds to the molar quantity of air used during the gasifying process which is entering the BFBG at the temperature of 120°C and the pressure of 45 bar and depends on the gasification relative fuel/air ratio and the stoichiometric fuel/air ratio relating to the biomass waste as a fuel[17]

$$m = 1/F_{rg}F_{st} \tag{2}$$

And w is determined from the moisture content of the fuel

$$w = M_{BM}\phi / M_{H_2O}(1-\phi) \tag{3}$$

On the right-hand side, n_i are the numbers of mole of the species i that are unknown.
In a fluidized bed gasifier, nearly the entire sulfur in the feed is converted to H2S, which must be effectively removed to ensure that the sulfur content of the final gas is within

acceptable limits. In the case of fluidized bed gasifiers, limestone can be fed into the gasifier along with coal to capture most of the H_2S produced within the bed itself. The limestone ($CaCO_3$) calcines inside the gasifier to produce lime (CaO), which in turn is converted to calcium sulfide (CaS) upon reaction with the H_2S inside the gasifier.

$$CaCO_3 \rightarrow CaO + CO_2 \tag{4}$$

$$CaO + H_2S \rightarrow CaS + H_2O \tag{5}$$

The substitution fuel formula $C_aH_bO_cS_dN_e$ can be calculated Starting from the ultimate analysis of the biomass waste and the mass fractions of the carbon, hydrogen, oxygen, nitrogen and sulphur (C, H, O, N, S), assuming a= 1, with the following expressions:

$$b = \frac{HM_C}{CM_H}, c = \frac{OM_C}{CM_O}, d = \frac{NM_C}{CM_N}, e = \frac{SM_C}{CM_S} \tag{6}$$

Ultimate analysis of the biomass waste (RDF) used in this model is shown in Table 1.

Waste Fuel	C%	H%	O%	N%	S%	Ash	HHV(MJ/Kg)
RDF	44.7	6.21	38.6	0.69	0.00	10.4	19.495

Table 1. Ultimate analysis of RDF (dry basis, weight Percentage) [18]

From the substitution fuel formula, the specific molecular weight of the biomass waste, the molar quantity of water per mole of biomass waste, the stoichiometric fuel/air ratio and the formation enthalpy of the biomass waste can be calculated.
Now for calculating the molar quantity of the product gases 7 equations are needed:
From the molar biomass waste composition $C_aH_bO_cS_dN_e$ and the molar moisture quantity, the atomic balances for C, H, O, N and S are obtained, respectively

$$\begin{aligned} C &: a = n_2 + n_3 + n_5 \\ H &: b + 2w = 2n_1 + 2n_4 + 4n_5 \\ O &: c + w + 2m = 2n_2 + n_3 + n_4 \\ N &: d + 2m \times 3.76 = 2n_6 \\ S &: e = n_7 \end{aligned} \tag{7}$$

There are now only 5 equations to calculate 7variables. To solve the system, two other equations should be added. From the first assumption, two equations in equilibrium can be used. Chemical equilibrium is usually explained either by minimization of Gibbs free energy

or by using an equilibrium constant. To minimize the Gibbs free energy, constrained optimization methods are often used which requires a realizing of complex mathematical theories. For that reason, the present thermodynamic model is developed based on the equilibrium constant. Therefore, the remaining two equations were obtained from the equilibrium constant of the reactions occurring in the gasification zone as shown below:
In the reduction zone of the gasifier, hydrogen is reduced to methane by carbon (methanation reaction).

$$C + 2H_2 \leftrightarrow CH_4 \tag{8}$$

Methane formation is preferred especially when the gasification products are to be used as a feedstock for other chemical process. It is also preferred in IGCC applications due to methane's high heating value.
The equilibrium constant K_1 relates the partial pressures of the reaction as follows:

$$k_1 = \frac{(P_{CH_4} / P_{total})}{(P_{H_2} / P_{total})} \tag{9}$$

Or as a function of the molar composition, assuming the behavior of the product gas to be ideal,

$$k_1 = \frac{n_5 \times n_{total}}{n_1^2} \tag{10}$$

The second reaction, also known as the water gas shift reaction, describes the equilibrium between CO and H2 in the presence of water

$$CO + H_2O \leftrightarrow CO_2 + H_2 \tag{11}$$

The heating value of hydrogen is higher than that of carbon monoxide. Therefore, the reduction of steam by carbon monoxide to produce hydrogen is a highly desirable reaction. The corresponding equilibrium K2 constant is obtained as follows:

$$k_2 = \frac{(P_{CO_2} / P_{total})(P_{H_2} / P_{total})}{(P_{CO} / P_{total})(P_{H_2O} / P_{total})} \tag{12}$$

Or as a function of the molar composition of the gas

$$k_2 = \frac{n_1 n_3}{n_2 n_4} \tag{13}$$

The values of the equilibrium constants K1 and K2 are calculated from the Gibbs free energy

$$K_p = \exp\left(-\Delta G_T^0 / R_u T\right) \tag{14}$$

Where ΔG_T^0 is the difference of the Gibbs free energy between the products and the reactants:

$$\Delta G_T^0 = \Delta H^0 - T\Delta S^0 \tag{15}$$

Substituting the Gibbs free energy in Eqs. (5) and (8), the equilibrium constants are obtained as

$$K_1 = \exp\left(-\left(G_{T,CH_4}^0 - 2G_{T,H_2}^0\right)/R_u T\right) \tag{16}$$

$$K_2 = \exp\left(-\left(G_{T,H_2}^0 + G_{T,CO_2}^0 - G_{T,CO}^0 - G_{T,H_2O}^0\right)/R_u T\right) \tag{17}$$

With

$$G_{T,i}^0 = \Delta\tilde{h}_{f,298}^0 + \int_{298}^{T} \overline{C}_p(T)\,dT - Ts^0 \tag{18}$$

Where $\overline{C}_p(T)$ is the specific heat at constant pressure in (J/mol K) and is a function of temperature. It can be defined by empirical equation below.

$$\overline{C}_p(T) = A + BT + CT^2 + DT^3$$

In which the coefficients are obtained from the table 2

	$\overline{C}_p(T) = A + BT + CT^2 + DT^3$ (J/mol K)			
compound	A	$B\times10^2$	$C\times10^5$	$D\times10^8$
H_2	29.062	-0.82	0.199	0.0
O_2	25.594	13.251	-0.421	0.0
CO	26.537	7.683	-0.1172	0.0
CO_2	26.748	42.258	-1.425	0.0
CH_4	25.36	1.687	7.131	-4.084

Table 2. Heat capacity of an ideal gas[19]

Gasifying temperature

For calculating K_1 and K_2, the temperature in the gasification or reduction zone must be known. It should be noted that in bubbling fluidized bed the bed, temperature will be in the range of 900-1200°K by which the equilibrium constants will be calculated.

Enthalpy definition

After defining the corresponding equations, Because of nonlinear nature of some of the equations the Newton-Raphson method has been used to calculate the values n_1-n_7.
The enthalpy of the product gas is

$$\bar{h} = \sum_{i=prod} x_i \left(\bar{h}^0_{f,i} + \Delta\bar{h}_{T,i} \right) \tag{19}$$

where xi is mole fraction of species i in the ideal gas mixture and $\bar{h}^0_f$ is the enthalpy of formation and $\Delta\bar{h}_T$ represents the enthalpy difference between any given state and at reference state. It can be approximated by

$$\Delta\bar{h}_T = \int_{298}^{T} \bar{C}_p(t)dT \tag{20}$$

Table 3 shows some the value of $\bar{h}^0_f$ for some gas components.

Compound	$\bar{h}^0_f$ (kJ/mol)
H_2	0.0
O_2	0.0
CO	-110.52
CO_2	-393.51
CH_4	-74.85
$H_2O(l)$	-285.84
H_2S	-20.501[21]
SO_2	-296.833[21]

Table 3. Enthalpy of formation at the reference state [20]

It should be noted that enthalpy of formation for solid fuel can be calculated as:

$$h_{f,bm} = HHV_{db} + \frac{1}{M_{bm}} \sum_{i=prod} v_i \tilde{h}_{f,i} \tag{21}$$

Where $\left(\bar{h}^0_f\right)_k$ is the enthalpy of formation of the product k under the complete combustion of the solid and $\overline{HHV}$ is the higher heating value of the solid fuel.
Heat of formation of any biomass waste material can be calculated with good accuracy from the following equation[22]:

$$\Delta H_C = \overline{HHV}(KJ / Kmol) = 0.2326(146.58C + 56.878H - 51.53O - 6.58A + 29.45) \tag{22}$$

Where C, H, O and A are the mass fractions of carbon, hydrogen, oxygen and Ash, respectively in the dry biomass waste.

2.2 Exergy analysis

The entropy of ideal gas is represented by:

$$S = \overline{S}^0 + \int_{T_o}^{T} \frac{\overline{C}_p}{T} dT - R \ln \frac{P}{P_o} \tag{23}$$

Where P is the pressure of the bubbling fluidized bed gasifier, and $\overline{S}^0$ is entropy at reference state. Table 4 shows some components $\overline{S}^0$

Compound	$\overline{S}^0$ (J/molK)
H_2	130.59
O_2	205.03
CO	197.91
CO_2	213.64
CH_4	186.19
$H_2O(l)$	69.94
H_2S	205.757[21]
SO_2	284.094[21]

Table 4. Entropy at the reference state(at T_{ref}=298.15K(25^{0}C),p_{ref}=1 bar) [20]

The exergy of the product gas is comprised of two components: Exergy chemical exergy $\left(E^{CH}\right)$ and physical exergy $\left(E^{PH}\right)$.Total exergy of the product gas is given as

$$E_{pg} = E^{PH} + E^{CH} \tag{24}$$

The physical exergy is the maximum theoretical work obtainable as the system(here the product gas) passes from its initial state where the temperature is the gasifying temperature and the pressure equals the gasifier pressure to the restricted dead state where the temperature is T_0 and the pressure is P_0 and is given by the expression

$$E^{PH} = (H - H_o) - T_o(S - S_0) \tag{25}$$

The physical exergy of gas mixture per mole is derived from the conventional linear mixing rule

$$\overline{e}^{PH} = \sum x_i \overline{e}_i^{PH} \tag{26}$$

The chemical exergy is the maximum theoretical useful work obtainable as the system passes from the restricted dead state to the dead state where it is in complete equilibrium with the environment.
And chemical exergy of gas mixture is given by

$$\overline{e}^{CH} = \sum_i x_i \overline{\varepsilon}^{CH}_{0,i} + RT_0 \sum_i x_i \ln x_i \tag{27}$$

Where $\overline{\varepsilon}^{CH}_{o,i}$ is the standard chemical exergy of a pure chemical compound i which is available in Table 5 for some gas components.

Substance	$\overline{\varepsilon}^{CH}_{0,i}\ (kJ\ /\ kmol)$
H_2	238490
CO	275430
CO_2	20140
$H_2O(g)$	11710
CH_4	836510
N_2	720
H_2S	812000[21]
SO_2	313.4[21]

Table 5. Standard chemical exergy of some substances at 298.15K and p_0[21]

Special considerations apply for the gasifying products when evaluating the chemical and physical exergy. When a product gas mixture is brought to P_0, T_0, some consideration would occur: At 25°C, 1 atm, the mixture consists of H_2, CO, CO_2, CH_4, N_2, together with saturated water vapor in equilibrium with saturated liquid. So it would be required to calculate the new composition at the dead state including the saturated liquid. Then the $\overline{h}_o$ and $\overline{s}_o$ values required to evaluate the physical exergy and the product gas mole fraction at the dead state essential for evaluating the chemical exergy can be calculated.
The exergy components and the total exergy for the moisture content of the fuel is obtained

$$E^{PH}_{mois} = w\left[\overline{h} - \overline{h}^{0}_{f,liq} - T_0(\overline{s} - \overline{s}^{0}_{H_2O_{(l)}})\right] \tag{28}$$

$$E^{CH}_{mois} = w \times \overline{\varepsilon}^{CH}_{0,H_2O_{(L)}} \tag{29}$$

$$E_{mois} = E^{CH}_{mois} + E^{PH}_{mois} \tag{30}$$

Exergy for the fluidizing air entering the fluidized bed is defined with molar analysis of 0.21% O_2 and 0.79% N_2 with the pressure of 45 bar and the temperature of $373\,^{0}K$, by using equations 25 and 26

$$E_{air} = E_{air}^{CH} + E_{air}^{PH} \tag{31}$$

For a biomass waste the chemical exergy is obtained as follows

$$\varepsilon_{0,biomass} = \beta HHV_{biomass} \tag{32}$$

The factor β is the ratio of the chemical exergy to the HHV of the organic fraction of biomass waste. This factor is calculated with the following correlation [18]:

$$\beta = \frac{1.0412 + 0.216(Z_H / Z_C) - 0.2499 Z_O / Z_C \left[1 + 0.7884 Z_H / Z_C\right] + 0.045 Z_N / Z_C}{1 - 0.3035 Z_O / Z_C} \tag{33}$$

Z_O, Z_C, Z_H and Z_N are the weight fractions of oxygen, Carbon, Hydrogen and Nitrogen, respectively in the biomass waste.
Therefore the total exergy of the biomass waste as a fuel can be defined:

$$E_{fuel} = \overline{\varepsilon}_{0,biomass} \times n_{fuel} \tag{34}$$

2.3 Heating value and efficiencies

2.3.1 Heating value

The heating value of the producer gas can be obtained as the sum of the products of the molar fractions of each of the energetic gases (CO, H_2 and CH_4) with its corresponding heating value (Table 6).

gas	HHV (MJ/kg mol)	LHV (MJ/kg mol)
CO	282.99	282.99
H_2	285.84	241.83
CH_4	890.36	802.34
H_2S	562.59	518.59

Table 6. Heating value of combustible gases

2.3.1 Evaluation of the efficiency

It is assumed that the fluidized bed gasifier operates as adiabatic and pseudo-homogeneous reactor at atmospheric pressure.
Gasification entails partial oxidation of the feedstock, so chemical energy of biomass waste is converted into chemical and thermal energy of product gas.
The first law thermodynamic or cold gas efficiency can be defined as the relation between the energy leaving the gasifier i.e. the energy content of the producer gas, and the energy

entering the gasifier, i.e. the biomass waste and moisture. We assume the gas leaves the process at the reference temperature (25 °C), loosing the energy corresponding to its sensible enthalpy, and define the cold gas efficiency η_{Cg} as

$$\eta_{Cg} = \frac{HHV_{gas}}{HHV_{biomass}} \tag{35}$$

Where HHV_{gas} and $HHV_{biomass}$ are the net heats of combustion (lower heating values) of gas and biomass waste, respectively.

The exergetic efficiency may be defined as the ratio between chemical exergy as well as physical exergy of product gas and the total exergy of the entering streams i.e. the biomass waste and the moisture and fluidizing air.

$$\eta_{Ex} = \frac{\dot{E}_{out}}{\dot{E}_{in}} = \frac{\dot{E}_{pg}}{\dot{E}_{air} + \dot{E}_{mois} + \dot{E}_{fuel}} \tag{36}$$

In this work variations of the exergy efficiency, cold gas efficiency and product gas concentration will be investigated as a function of temperature, gasifying fuel/air ratio (F_{rg}), and moisture content of the fuel (φ).

3. Results and discussion

3.1 Validation of the model

The model presented in this article has been compared to the experimental work for the wood particles presented by Narvaez et al. [23]. By way of illustration the predicted HHV producer gas by the model and the results from the experiments are presented in Figure 1.

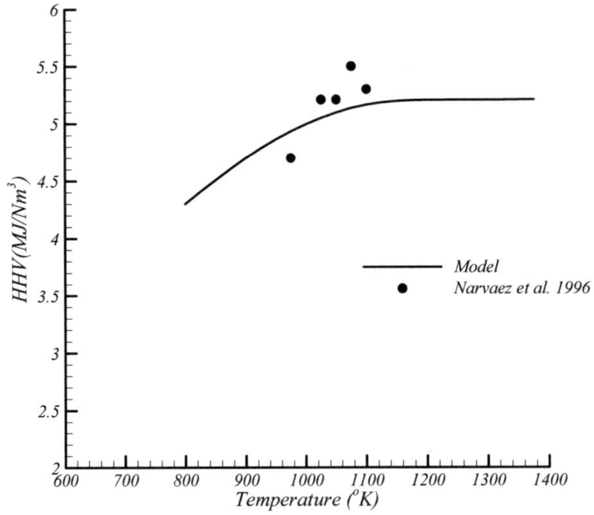

Fig. 1. Higher heating values of product gas at different temperatures for wood particles

3.2 Sensitivity analyses

The effect of F_{rg} on product gas composition and higher heating value for RDF gasification is presented in Figure 2. An increase in F_{rg} brings about an increase in the concentration of H_2 and CO and a substantial decrease in CO_2 concentration in dry gas product. This is because of the decreasing role of the char combustion in the bed compared to its gasification reaction, which results in higher concentration of combustible gases and lower CO_2.

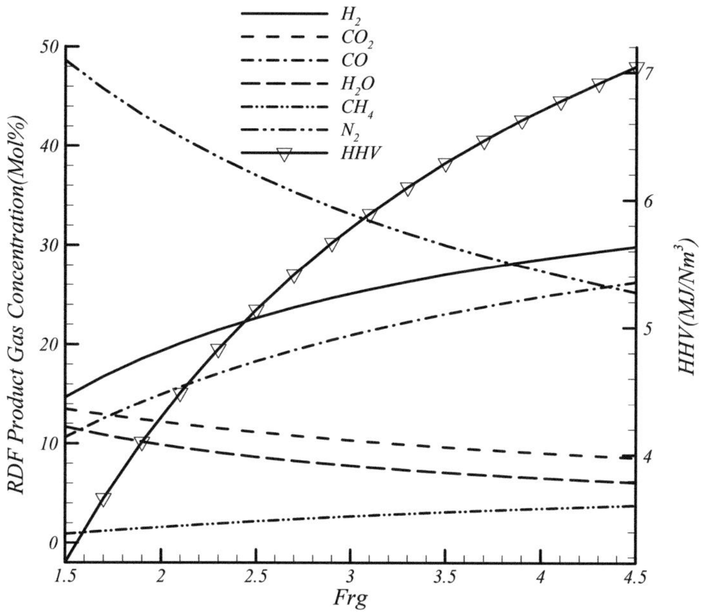

Fig. 2. Concentration of product gases and higher heating value at different F_{rg} values and T_{bed} = 1100°K.

The effect of moisture content of the fuel on product gas composition and higher heating value for RDF gasification is presented in Figure 3. As shown in the figure, an increase in moisture content brings about an increase in the concentration of H_2 and CH_4 and decrease in the concentration of CO. This is because of the increasing role of the moisture content of the fuel and effect of the methanation reaction (equation8) and the water-gas shift reaction (equation11) in which the molar concentration of the CO decreases because of the reaction with H_2O and production of H_2, and resulting an increase in the molar quantity of CH_4. Therefore the higher heating value will decrease as the moisture content increases.

The effect of gasifying temperature on product gas composition is shown in Figure 4. The figure shows that an increase in temperature brings about an increase in the concentration of H_2 and CO of RDF. This is because of the increasing role of the temperature in the equilibrium constants (16), (17) in which the equilibrium constant is dependent on the BFBG temperature, so an increase in temperature causes more production of combustible gases. The higher heating value in this temperature range at the constant Frg is to some extent constant that is valid according to experimental works [22].

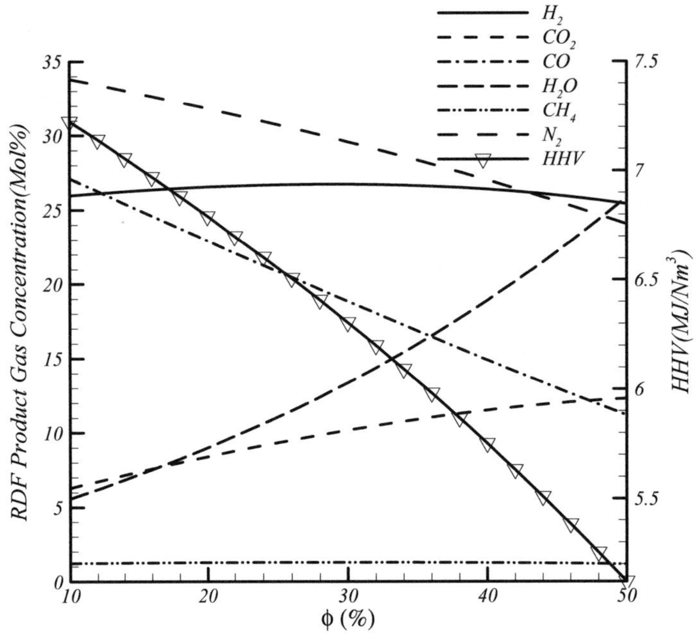

Fig. 3. Concentration of product gases and higher heating value at different moisture content of the fuel at T_{bed} = 1100°K and F_{rg}=3.

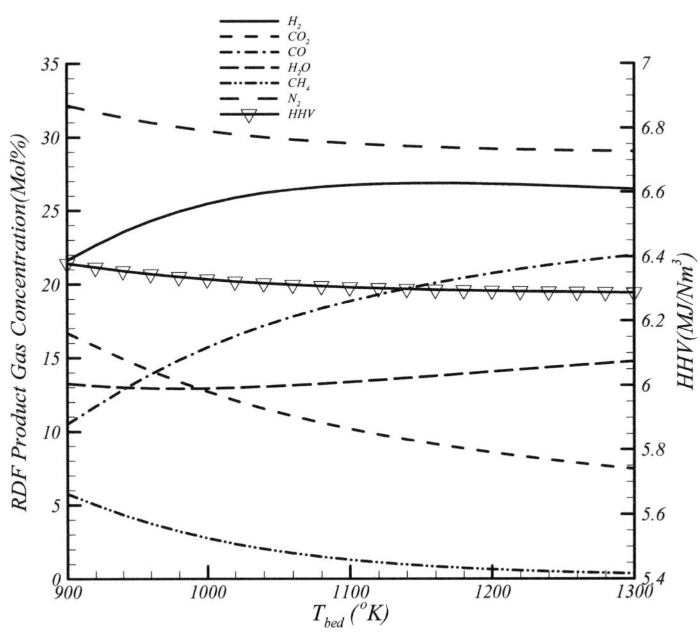

Fig. 4. Concentration of product gases and higher heating value at various gasifying temperatures at F_{rg}=3

The effect of F_{rg} with moisture content of the fuel on exergetic efficiency and cold gas efficiency for RDF gasification are presented (by line & flood contour type) in Figures 5, 6. It is shown that the exergetic efficiency of BFBG increases with rising fuel/air ratio because when less air is admitted to the process, the variations in mole fractions of product gases will influence the exergy of the product in comparison to exergy of the fuel. Higher moisture content will increase the exergetic efficiency because of its considerable effect on enthalpy of the product gases (figure5). An increase in F_{rg}, as discussed before, brings about an increase in the concentration of combustible gases and higher heating value which yields an increase in cold gas efficiency and an increase in moisture content of the fuel, as discussed before, causes decrease in the concentration of combustible gases and higher heating value which yields a decrease in the cold gas efficiency (figure6).

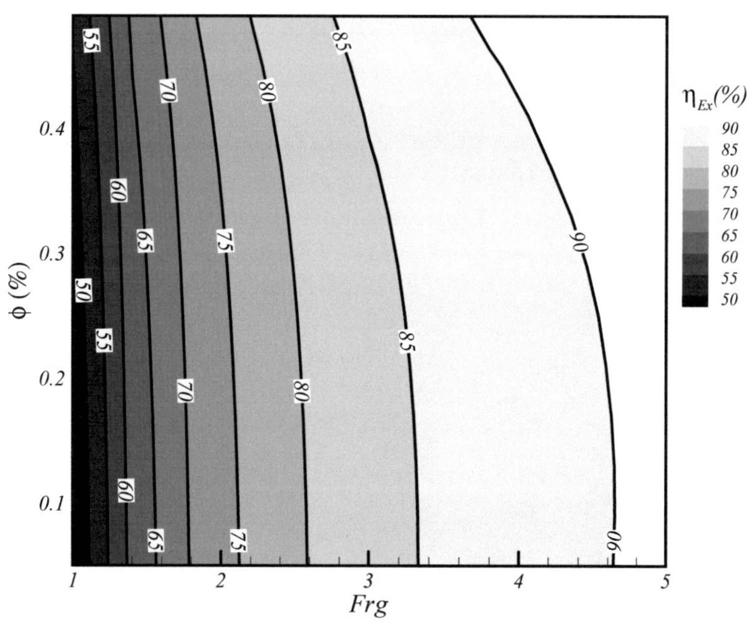

Fig. 5. Exergetic efficiency of the gasifying process as a function of the gasifying relative fuel/air ratio and the moisture content

The effect of F_{rg} and the bed temperature on exergetic efficiency and cold gas efficiency for RDF gasification are presented (by line & flood contour type) in Figures 7, 8. It is shown that the exergetic efficiency of BFG increases with rising fuel/air ratio as discussed for figures 5 and6. Higher temperature will increase the exergetic efficiency because of its considerable effect on enthalpy of the product gases (figure7). An increase in bed temperature, as discussed for figure 4, brings about an increase in the concentration of combustible gases and higher heating value which yields an increase in cold gas efficiency (figure8)

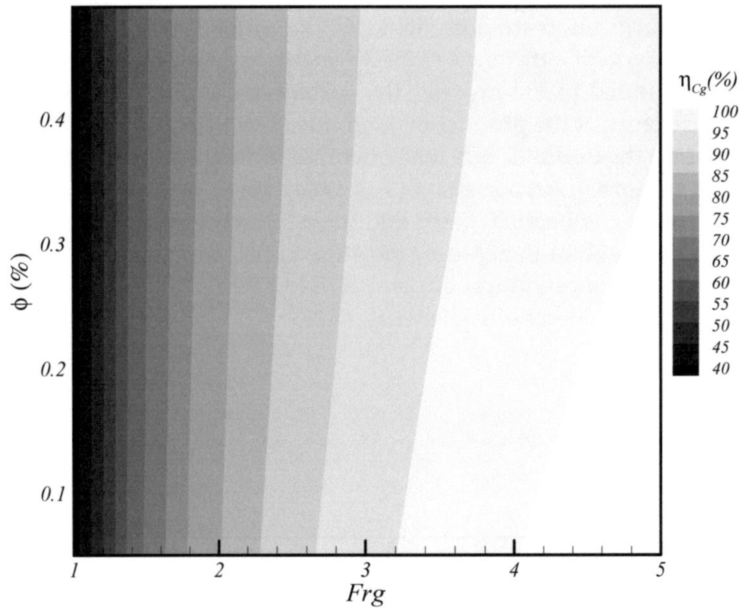

Fig. 6. Cold gas efficiency efficiency of the FBG as a function of the gasifying relative fuel/air ratio and the moisture content

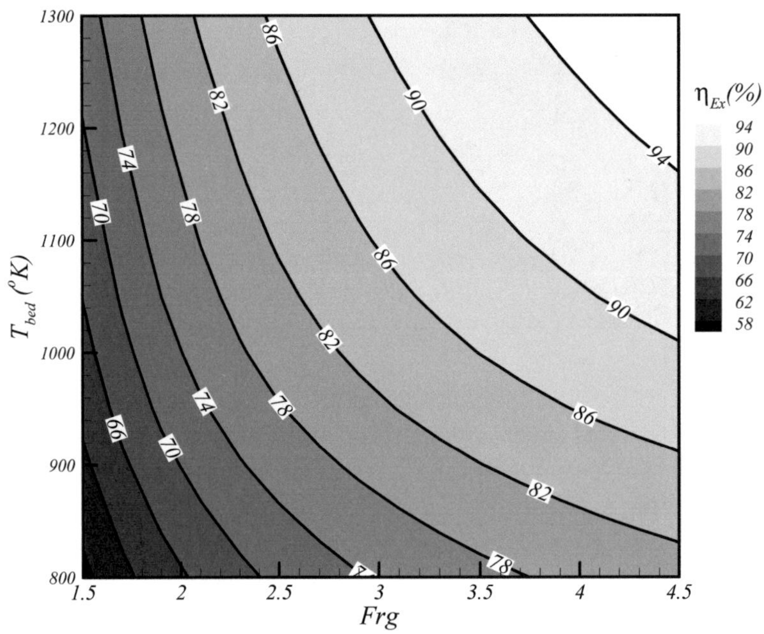

Fig. 7. Exergetic efficiency of the gasifying process as a function of the gasifying relative fuel/air ratio and the gasifying temperature

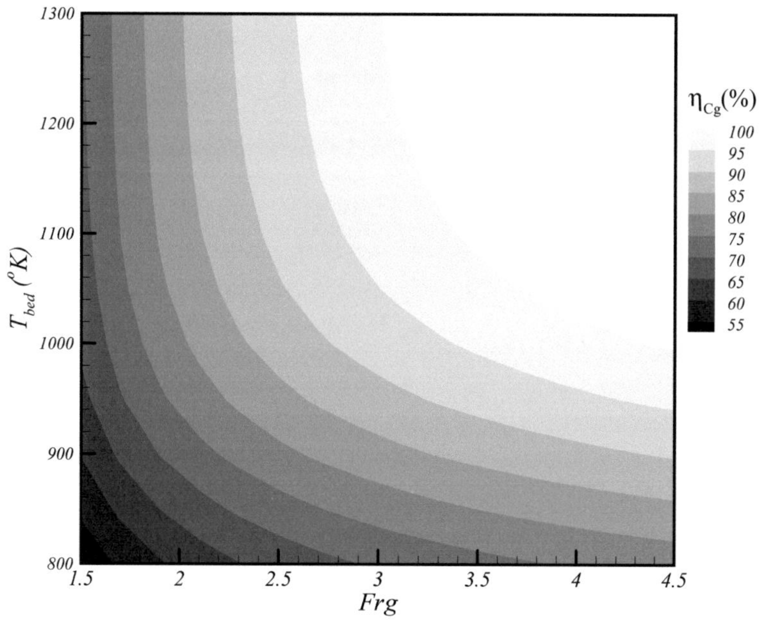

Fig. 8. Cold gas efficiency of the FBG as a function of the gasifying relative fuel/air ratio and the bed temperature

4. Conclusion

An equilibrium model was developed for the biomass waste gasification in the bubbling fluidized bed waste gasification. It was shown that higher moisture would decrease the product gas higher heating value as well as cold gas efficiency while increase the exergetic efficiency. Moreover, It was concluded that higher temperature and higher F_{rg} would increase both the product gas higher heating value, cold gas efficiency and the exergetic efficiency.

5. Nomenclature

C	mass fraction of carbon
H	mass fraction of hydrogen
F_{rg}	gasification relative fuel/air ratio
F_{st}	stoichiometric biomass waste/air ratio
M	molecular weight (kg/mol)
M_{BM}	biomass waste molecular weight (kg/mol)

N	mass fraction of nitrojen
m	molar quantity of air
w	molar quantity of biomass waste moisture content
E_{pg}	Product gas total Exergy
E^{PH}	physical Exergy
E^{CH}	chemical Ecergy
ΔG_T^0	gibbs free Energy((kJ/mol)
O	mass fraction of oxygen
HHV_{db}	higher heating value in dry base
P	pressure
S_0	standard Entropy(KJ/mol K)
S	mass fraction of sulphur
T	temperature

Greek symbols

ϕ	moisture content of the biomass waste fuel
η_{Ex}	Gasifier exergetic efficiency
η_{Cg}	Cold gas efficiency

6. References

[1] Basu, P., Combustion and gasification in fluidized beds, Taylor & Fransis, 2006

[2] Altafini CR, Mirandola A, A chemical equilibrium model of the coal gasification process based on the minimization of the Gibbs free energy. In: Flowers 97, Florence, Italy, 1997.

[3] Altafini CR, Wander PR, Barreto RM. Prediction of working parameters of a wood waste gasifier through an equilibrium model. Energy Convers Manage 2003;44:2763–77.

[4] Lapuerta M, Herna´ndez J, Tinaut FV, Horillo A. Thermochemical behaviour of producer gas from gasification of lignocellulosic biomass in SI engines. SAE Paper series 2001-01-3586, 2001.

[5] Zainal ZA, Ali R, Lean CH, Seetharamu KN, Prediction of performance of a downdraft gasifier using equilibrium modeling for different biomass materials.

Energy Convers Manage 2001;42:1499–515. [6] Ruggiero M, Manfrida G. An equilibrium model for biomass gasification processes. Renew Energy 1999;16:1106–9.

[7] P. Belleville, R. Capart, A model for predicting outlet gas concentration from a wood gasifier, in: A.V. Bridgwater (Ed.), Thermochemical Processing of Biomass, Butterworks, London, UK, 1983,pp. 217- 228.

[8] C.C. Chang, L.T. Fan, W.P. Walawender, Dynamic modeling of biomass gasification in a fluidized bed, AICHE Symposium Series 80 (234) (1984) 80–90.

[9] F.G. Van den Aarsen, Fluidized bed wood gasifier performance and modeling. PhD thesis, Univ. of Twente (NL), 1985.

[10] J. Corella, F.J. Alday, J. Herguido, A model for the non-stationary states of a commercial fluidized bed air gasifier of biomass, in: G. Grassi, et al., (Eds.), Biomass for Energy and Industry, vol. 2, Elsevier, London, 1990, pp. 2804- 2809.

[11] J.-F. Bilodeau, N. The´rien, P. Proulx, S. Czernik, E. Chornet, A mathematical model of fluidized bed biomass gasification, The Canadian Journal of Chemical Engineering 71 (1993) 549.

[12] H. Jiang, R.V. Morey, Pyrolysis of corncobs at fluidization, Biomass and Bioenergy 3 (1992) 81.

[13] H. Jiang, R.V. Morey, A numerical model of fluidized bed biomass gasifier, Biomass and Bioenergy 3 (1992) 431.

[14] S. Hamel, W. Krumm, Modelling and simulation of bubbling fluidized bed gasification reactors, Proceed. Of 3rd European Conference on Fluidization, Toulouse, France, (2000), 29-31 May, 2000, pp. 519-527.

[15] K.G. Mansaray, A.M. Al-Taweel, A.E. Ghaly, F. Hamdullahpur, V.I. Ugursal, Mathematical modeling of a fluidized bed rice husk gasifier: Part I—Model development, Energy Sources 22 (2000) 83.

[16]] K.G. Mansaray, A.M. Al-Taweel, A.E. Ghaly, F. Hamdullahpur, V.I. Ugursal, Mathematical modeling of a fluidized bed rice husk gasifier: Part II—Model sensitivity, Energy Sources 22 (2000) 167.

[17]Andre´s Melgar , Juan F. Pe´rez , Hannes Laget , Alfonso Horillo, Thermochemical equilibrium modelling of a gasifying process, Energy Conversion and Management 48 (2007) 59-67

[18] Krzysztof J. Ptasinski, Mark J. Prins, Anke Pierik, Exergetic evaluation of biomass gasification , Energy, Vol. 32, pp.568-574, 2007

[19] S. Jarungthammachote, A. Dutta, Thermodynamic equilibrium model and second law analysis of a downdraft waste gasifier, Energy 32 (2007) 1660-1669

[20] Yunus Cengel, Boles, Thermodynamics An Engineering Approach, 1999

[21] Adrian Bijan, G. Tsatsaronis, Thermal Design and Optimization, 1996 by John Wiley & Sons.

[22] Channiwala SA, Parikh PP., A unified correlation for estimating HHV of solid, liquid and gaseous fuels, Fuel, Vol. 81, pp.1051-63, 2007

[23] Narvaez I, Orio A, Aznar MP, Corella J.,Biomass gasification with air in an atmospheric bubbling fluidized bed. Effect of six operationa variables on the quality of the produced raw gas, Ind Eng Chem Re,Vol 35, pp.2110-2120, 1996.

[24] Bubbling fluidized bed biomass gasification—Performance, process findings and energy analysis, Renewable Energy 33 (2008) 2339-2343

3

Thermal Plasma Gasification of Biomass

Milan Hrabovsky
Institute of Plasma Physics ASCR
Czech Republic

1. Introduction

Since the 1980s applications of thermal plasmas experienced an important increase. In the 1990s fundamental research led to great progress in the understanding of the basic phenomena involved, and to a renewed interest in applying thermal plasmas to material processing and waste treatment. The application of plasma torches for environmental purposes is a relatively new process. Thermal plasma offers unique capability of carrying extremely high energy by small amount of plasma and ensures high heat transfer rates to treated materials. All materials can be decomposed if they are brought into contact with plasma.

Generators of thermal plasma (plasma torches) operate simultaneously as a plasma-chemical and a thermal apparatus. The electrical energy of the torches goes into the plasma which transfers its energy to the substances to be treated, thereby triggering a dual simultaneous reaction process in the plasmachemical reactor: the organic compounds are thermally decomposed into their constituent elements (*syngas* with more complete conversion of carbon into gas phase than in incinerators), and the inorganic materials are melted and converted into a dense, inert, non-leachable vitrified slag, that does not require controlled disposal. Therefore, it can be viewed as a totally closed treatment system.

While decomposition of waste and dangerous materials in thermal plasmas has been intensively studied in the last decade and industrial scale systems for treatment of various types of waste has been installed, plasma gasification of biomass is newly appearing application. For this application, the principal goal of the technology is production of fuel gases, principally mixture of carbon monoxide and hydrogen, called syngas. Thermal plasma offers possibility of decomposition of biomass by pure pyrolysis in the absence of oxygen, or with steichiometric amount of oxygen (gasification) to produce high quality syngas, with high content of carbon monoxide and hydrogen and minimum presence of other components. As production of fuel gas is the main goal of the technology, an energy balance of the process is thus much more important than in case of waste treatment, where the principal goal is material decomposition.

Gasification is a process by which either a solid or liquid carbonaceous material, containing mostly chemically bound carbon, hydrogen, oxygen, is reacted with air or oxygen. The reactions provide sufficient exothermic energy to produce a primary gaseous product containing mostly CO, H_2, CO_2, H_2O(g), and small content of higher hydrocarbons. Heat from external sources is usually supplied into the reactor to control the process and the reaction temperature, but most of heat for realization of the reaction usually comes from calorific value of biomass. In case when thermal decomposition of biomass is realized under the action of externally supplied heat and without any oxidant, we speak about pyrolysis. Especially pyrolysis is particularly well adapted to the valorization of lignocellulosic products such as wood or straw with good control of parameters of the process (gas composition, formation of tar) to maximize the production of hydrogen or syngas.

Principal problem of common biomass gasification technologies, based on the reaction between a heated carbon source with limited amounts of oxygen, consists in production of tar, which is formed from complex molecules of hydrocarbons created during the process. The gas from low temperature gasification typically contains only 50% of the energy in syngas components CO and H_2, while the remainder is contained in CH_4 and higher aromatic hydrocarbons [Boerrigter 2005]. Moreover, the syngas is diluted by CO_2 produced by partial oxidation process. The possibility of control of syngas composition in classical technologies is limited. The necessity of production of clean syngas with controlled composition leads to technologies based on external energy supply for material gasification.

Plasma pyrolysis and gasification for production of syngas is an alternative to conventional methods of biomass treatment. Plasma is a medium with the highest energy content and thus substantial lower plasma flow rates are needed to supply sufficient energy compared with other media used for this purpose. This results in minimum contamination and dilution of produced syngas by plasma gas and easy control of syngas composition. The process acts also as energy storage - electrical energy is transferred into plasma energy and then stored in produced syngas. The main advantages are better control of composition of produced gas, higher calorific value of the gas and reduction of unwanted contaminants like tar, CO_2, CH_4 and higher hydrocarbons. The other advantage of plasma is wide choice of treated materials. As energy for the process is supplied by plasma and chemical reactions in the reaction products are not primary source of energy, the process can be applied for wide choice of organic materials and biomass. For evaluation of technical and economic feasibility of plasma treatment these advantages must be taken into account together with higher energy consumption of plasma technology.

Plasma treatment offers better control of process temperature, higher process rates, lower reaction volume and especially optimum composition of produced syngas. Plasma pyrolysis or gasification exploits the thermochemical properties of plasma. The particles kinetic energy in the form of heat is used for decomposing biomass. In addition, the presence of charged and excited species renders the plasma environment highly reactive which can catalyses homogeneous and heterogeneous chemical reactions. The main advantage of plasma follows from much higher enthalpy and temperature of plasmas compared to the ones of gases used in conventional methods. Thus, substantially lower plasma flow rates can carry sufficient energy for the process and composition of produced syngas is not much influenced by plasma gas composition and moreover, substantially less energy is consumed to heating of plasma to reaction temperature. These advantages are especially noticeable in case of plasma generated in water stabilized plasma torches, which are characterized by extremely high enthalpies and low plasma mass flow rates. Compared to non-plasma methods the advantages of plasma gasification can be summarized as follows:

- Energy for gasification is supplied by plasma rather than energy liberated from combustion and thus it is independent of the treated substances, providing flexibility,

fast process control, and more options in process chemistry. Broad range of biomass feedstock, incl. biodegradable fraction of waste, can be thus gasified.
- No combustion gases generated in conventional autothermal reactors are produced.
- The temperature in the reactor can be easily controlled by control of plasma power and material feed rate.
- As sufficiently high temperatures and homogeneous temperature distribution can be easily maintained in the whole reactor volume, production of higher hydrocarbons, tars and other complex molecules is substantially reduced.
- High energy density and high heat transfer efficiency can be achieved, allowing shorter residence times and large throughputs.
- Highly reactive environment and easy control of composition of reaction products.
- Low thermal inertia and easy feedback control.
- Much lower plasma gas input per unit heating power than the gas flow of classical reactors and thus lower energy loss corresponding to the energy necessary for heating of plasma to reaction temperature; also lower amount of gases diluting produced syngas.
- Smaller plants than for conventional reactors due to high energy densities, lower gas flows, and volume reduction.

2. Thermodynamic analysis of plasma gasification of biomass

Principally all carbon and hydrogen atoms from biomass can be used for syngas production if biomass is heated to sufficiently high temperature. Maximum biomass to syngas conversion efficiency is achieved if all carbon is oxidized to CO. As most of biomass materials contain more carbon atoms than oxygen atoms, some oxygen has to be added to gasify all carbon. This can be done by addition of oxygen, air, steam or CO_2.
For the energy balance analysis, following three processes are taken into account:

a. Gasification with addition of steichiometric amount of O_2

$$biomass + \frac{(n_C - n_O)}{2} O_2 \Rightarrow n_C CO + n_{H_2} H_2 \tag{1}$$

b. Gasification with steichiometric amount of steam

$$biomass + (n_C - n_O) H_2O \Rightarrow n_C CO + (n_{H2} + n_C - n_O) H_2 \tag{2}$$

c. Gasification with steichiometric amount of CO_2

$$biomass + (n_C - n_O) CO_2 \Rightarrow (2n_C - n_O) CO + n_{H_2} H_2 \tag{3}$$

where $n_C = c/M_C$, $n_{H2} = h/2M_H$ and $n_O = o/M_O$ are molar concentrations of carbon, hydrogen and oxygen in biomass with mass fractions of carbon, hydrogen and oxygen equal to *c*, *h* and *o*, respectively.
The power balance of the process can be written as

$$\Delta Q_r = \eta W_{torch} - P_{react}(T_r) - Q_{gas}^{out}(T_r) - Q_{sol}^{out}(T_r) - Q_{pl}^{out}(T_{pl} \geq T_r) \tag{4}$$

where ΔQ_r is power available for gasification, η is efficiency and W_{torch} power of the torch, P_{react} power loss to the reactor wall, Q_{gas}, Q_{sol} and Q_{pl} are power losses carried out of the reactor by produced gases, solids and plasma, respectively. These power losses are dependent on temperature in the reactor T_r and the temperature of plasma gas leaving the reactor $T_{pl.}$ The temperature T_{pl} is equal to T_r if complete mixing and heat transfer from plasma to treated material during residence time in the reactor is ensured.

The terms on the right hand side of equation (4) are determined by the torch power and its efficiency, by the construction of plasma reactor and by the reaction temperature T_r. The reaction temperature needed for biomass gasification can be determined from calculation of temperature dependence of composition of products of reactions (1) - (3). Due to high temperature in plasma reactor, we can assume that all reaction components in plasma reactor are in thermodynamic equilibrium. Under this assumption we can calculate the composition of reaction products from thermodynamic computations.

Fig. 1 presents the temperature dependence of composition of system containing mass fractions of carbon, hydrogen and oxygen corresponding to fir wood. The equilibrium composition of this heterogeneous system was calculated using the method described in [Coufal 1994], the input data for calculations of standard reaction enthalpy and standard thermodynamic functions of system components were taken from database [Coufal 2005]. It can be seen that wood is decomposed into hydrogen, carbon monoxide and solid carbon with small amount of other components at temperatures above 1 200 K. The presence of solid carbon, which in gasification reactors leads to formation of char, can be suppressed by addition of gas containing oxygen. To maintain high concentrations of CO and H_2 in the produced gas, it is advantages to use oxygen, carbon dioxide or steam as oxidizing media.

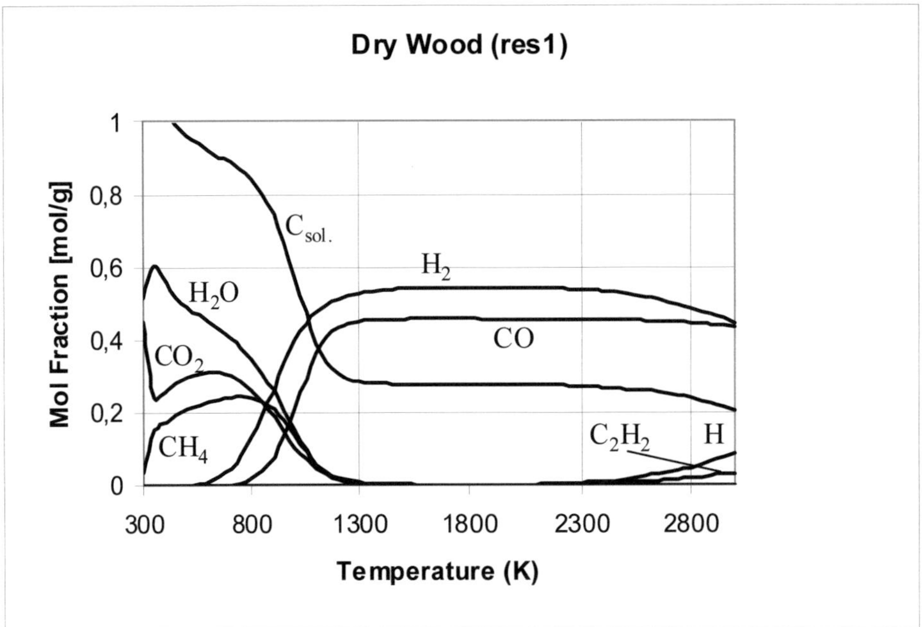

Fig. 1. Composition of products of wood pyrolysis. The mass ratios of components in wood: carbon $c = 0.511$, hydrogen $h = 0.064$, oxygen $o = 0{,}425$

Fig. 2 shows the composition of products of gasification of wood with addition of CO_2 and oxygen. The components considered in the computation correspond to the experimental conditions described in the paragraph 4. Mixture of oxygen and carbon dioxide was used as an oxidizing medium for gasification of fir wood, the atmosphere in the reactor contained also steam plasma with small amount of argon supplied into the reactor by plasma torch. It can be seen that composition of produced syngas changed substantially due to the addition of other components. It can be seen in Figs. 1 and 2 that optimum composition of syngas with high concentration of H_2 and CO is reached at temperatures higher than 1200 K. As all power losses specified in equation (1) are increasing with temperature T_r, it will be optimal for energy balance to keep reactor temperature close to temperature 1 200 K.

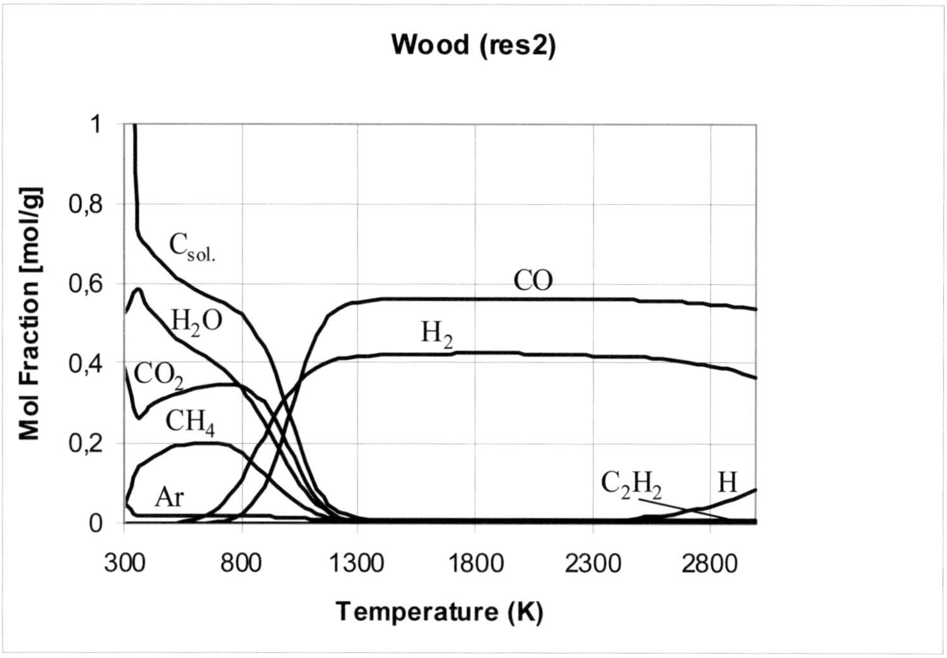

Fig. 2. Composition of products of wood gasification. The mass ratios of components in wood: $c = 0.511$, $h = 0.064$, $o = 0{,}425$. Wood 47 kg/h, humidity 6.5%, argon 13.55 slm, water plasma 18 g/min, CO_2 115 slm, O_2 30 slm.

The power available for material gasification ΔQ_r determines maximum possible feed rate of the material. We calculate energy Δh_r needed for realization of reactions represented by equations (1), (2), (3). As values representing heat of formation of biomass are not generally known, we calculate the heat Δh_r from known heats of combustion of cellulosic materials using the scheme in Fig. 3. The scheme in Fig. 3 corresponds to the reaction (1). The heat of gasification, i.e. production of syngas with composition $n_c\, CO + n_{H2}H_2$, is calculated as the difference of heat of combustion $\Delta h_{c,\,net}$ and heat of combustion of syngas $\Delta h_{c,\,syng}$

$$\Delta h_{gas} = \Delta h_{c,net} - \Delta h_{c,syng} \tag{5}$$

The heat of combustion of cellulosic materials can be calculated from the equation [Dietenberger 2002]

$$\Delta h_{c,net} = 13.23\ r_0\ [kJ\ /\ g] \tag{6}$$

where r_0 is external oxygen mass fraction needed for complete combustion

$$r_0 = (8/3)c + 8h - o \tag{7}$$

The heat of combustion of syngas (LHV) produced by complete gasification of wood can be expressed as

$$\Delta h_{c,syng} = n_C \left(\Delta_f H^o(CO_2) - \Delta_f H^o(CO)\right) + n_{H_2}\, \Delta_f H^o(H_2O) \tag{8}$$

where $\Delta_f H^o$ are heats of formation of individual molar components.

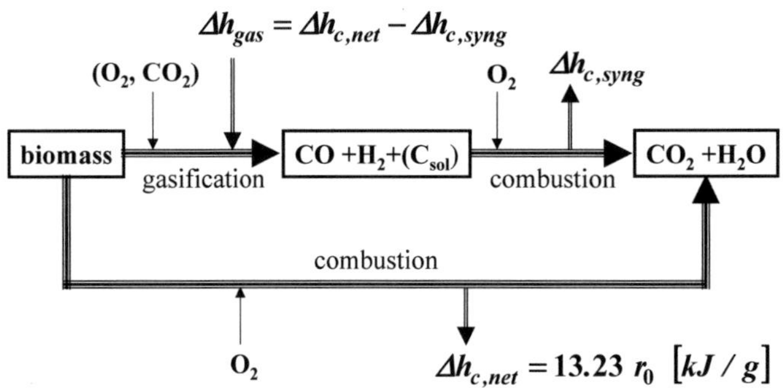

Fig. 3. Scheme of reactions for determination of reaction heat for biomass gasification.

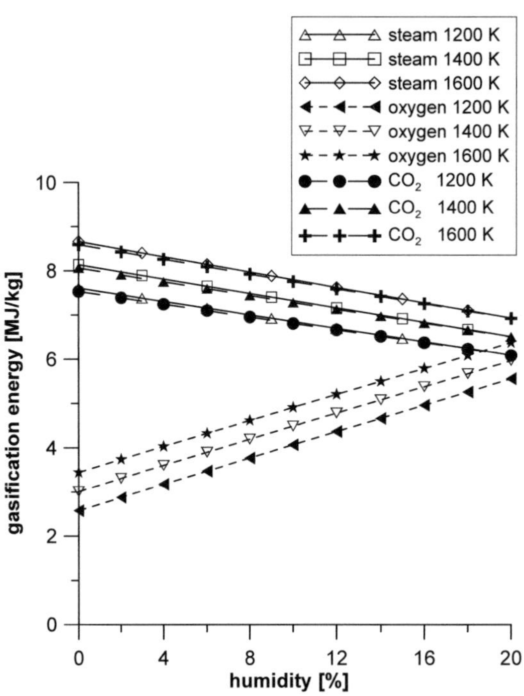

Fig. 4. Energy for gasification of wood for oxygen, steam and CO_2 process. Mass ratios of components in wood: $c = 0.511$, $h = 0.064$, $o = 0{,}425$.

In case of reactions (2) and (3), the reaction heat Δh_r includes also heat of dissociation of H_2O and CO_2. For humid biomass also heat of dissociation of water in the biomass must be taken into account. As reaction can be realized at temperature T_r, the total heat for gasification is given by the sum

$$\Delta h_r = \Delta h_{gas} + \Delta H \quad (9)$$

where Δh_{gas} is given by (4) and ΔH is heat needed for heating of components on the right hand side of equations (1-3) from standard temperature to the reaction temperature T_r.

In Fig. 4 the total external energy needed for gasification of wood by processes (1) - (3) is plotted in dependence on material humidity for three reaction temperatures T_r. The humidity of wood is given by the weight percentage of water in the material. The energy for process with addition of steam (2) and CO_2 (3) are almost the same, which is related to little difference in dissociation energies of water and CO_2.

The ratio of energy obtained by combustion of syngas (LHV) to the energy needed for its production is plotted in Fig. 5 against wood humidity. It can be seen that LHV of syngas produced from gasification of dry wood is up to 8 times higher than the heat spent for its production, for wood with 10% humidity this ratio is 5. If we consider torch efficiency 0.6 and the sum of power losses defined in (4) about 13% of the torch power, which corresponds to the experimental values measured in gasification experiments described in the paragraph 4, the ratio of LHV of syngas to total energy needed for its production is 3.6 for dry wood and 2.35 for wood with 10% humidity for reaction temperature 1300 K. As all power losses in the equation (4) for ΔQ_r are losses to the cooling water of the torch and the reactor, the real power gain after recuperation of heat in a cooling system could be even higher. If these numbers are compared with the conventional autothermal reactors, where only very low

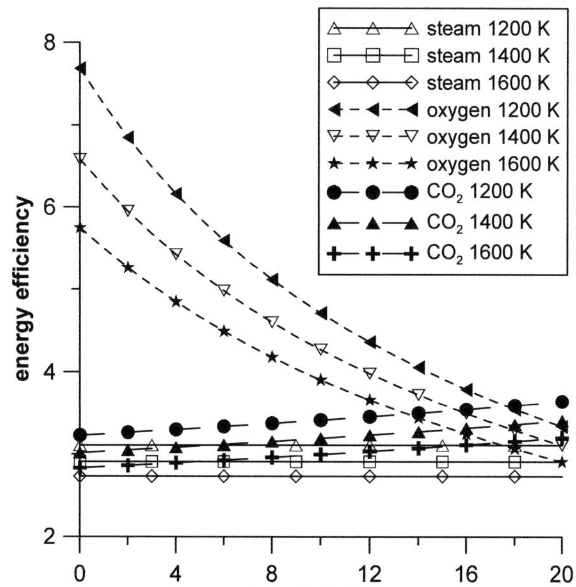

Fig. 5. Energy efficiency of gasification of wood for oxygen, steam and CO_2 processes. Mass ratios of components in wood: $c = 0.511$, $h = 0.064$, $o = 0{,}425$.

power is supplied to ignite the process of partial combustion, the energy gain in plasma systems is smaller. However, the LHV of produced syngas for autothermal reactors is usually between 35% and 60% of its theoretical value, and moreover, quality of produced syngas is low especially due to the production of tars and other contaminants.

The substantial advantage of plasma treatment is in reduction of mass flow rate of gasifying medium compared to the flow rate of gases used for non-plasma gasification. Thus, in case of plasma gasification, the produced syngas is less diluted by gas supplied into the reactor and has higher heating value. Also the power losses connected with the heating of gasifying medium to the reaction temperature are reduced. The ratio of mass of plasma, or gas at lower temperatures, needed for supply of energy Δh_r for complete gasification of wood, to the mass of wood, is plotted in dependence on gasifying medium temperature in Fig. 6 for nitrogen, oxygen and steam. The curves were calculated from thermodynamic equilibrium enthalpies of three gases [Boulos 1994, Krenek 2008] and from the total energy of gasification determined above. For temperatures lower than 3000 K the ratio is close to 1. Thus, for gasification with hot air (the amount of needed gas will be close to the values for nitrogen), almost half of the weight of produced syngas is air and thus syngas is diluted by high percentage of nitrogen (approximately 39% of weight of produced syngas). For comparison with plasma systems: for steam plasma with input temperature 16 000 K, which is the temperature corresponding to experiments described in paragraph 4, this ratio is less than 0.02 and thus almost undiluted syngas is produced.

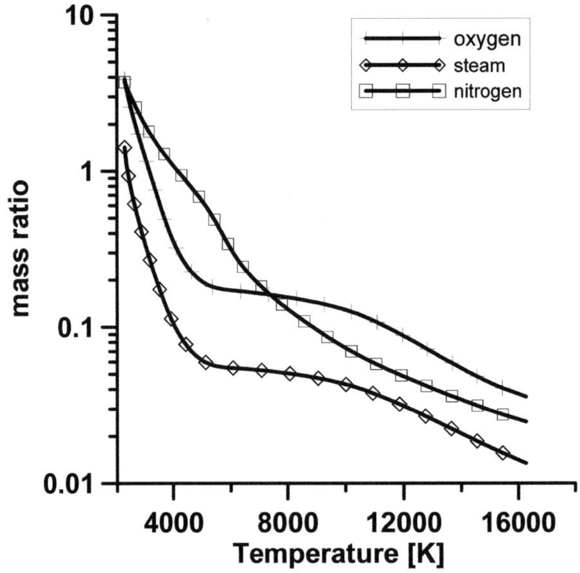

Fig. 6. Ratio of mass of gas carrying energy Δh_r for complete gasification of wood, to the mass of wood, in dependence on gas temperature.

3. Kinetics of gasification

Exact theoretical description of the plasma gasification process should be based on fluid dynamic model of plasma-material interaction, model of heat transfer to the material, its heating and volatilization as well as on description of kinetics of chemical reactions in the

reactor. We will describe here simple model of gasification kinetics based on solution of Arhenius equation describing volatilization of material at given temperature together with equations describing heat and mass transfer between reactor atmosphere and surface of material.
The rate of biomass volatilization is commonly described by Arhenius equation

$$\dot{m} = A \exp(-E / RT_s) \tag{10}$$

which determines dependence of volatilization rate m on temperature of material surface T_s. Frequency factor A and activation energy E were determined for various biomass materials, R is universal gas constant.

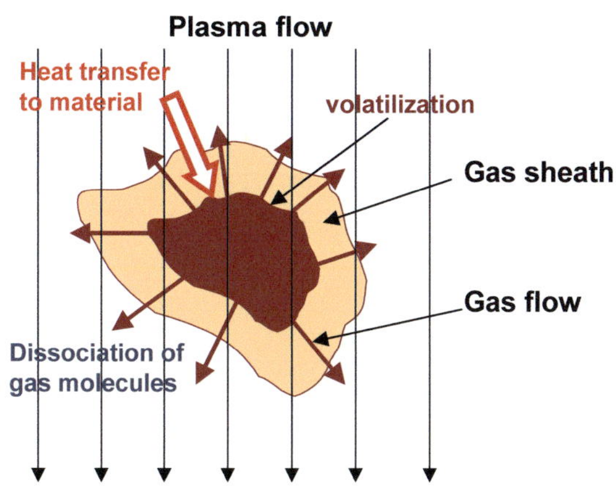

Fig. 7. Particle of gasified material with sheath of gas produced by volatilization.

In plasma gasification the heat flux to the gasified particles and thus the volatilization rate are extremely high and the sheath of gas, produced by gasification of particle, is formed. The conditions around particle of material are schematically shown in Fig. 7. The heat transfer through the sheath substantially influences the gasification rate. For description of the heat transfer we use the film model [Bird 2002]. Heat flux through the sheath created around spherical particle with surface temperature T_s is given by the equation

$$q_0 = \frac{\dot{m}C_p(T_r - T_s)}{e^{-\frac{\dot{m}C_p}{h}} - 1} \tag{11}$$

where $\dot{m}(kg / s.m^2)$ is volatilization rate, T_r is temperature in the reactor out of the sheath, C_p is specific heat and h heat transfer coefficient, both corresponding to local conditions in the sheath. We will approximate the heat transfer coefficient by relation for heat transfer to the sphere in flowing fluid [Bird 2002]

$$h = \frac{k.Nu}{D} = \frac{k}{D}\left(2 + 0.6\mathrm{Re}^{1/2}\mathrm{Pr}^{1/2}\right) \tag{12}$$

where Nu is Nusselt number, Re Reynolds number and Pr Prandtl number, D is diameter of the sphere and k thermal conductivity within the sheath.

The relation between volatilization rate and the heat flux is given by the energy balance equation

$$q_0 = \dot{m}.\Delta h_{gas} \tag{13}$$

where Δh_{gas} is energy needed for gasification, which is given by equation (5).
From (12) and (13) we get following relation between mass gasification rate and difference of temperatures in the reactor and at the particle surface:

$$\dot{m} = \frac{h}{C_p}\ln\left(\frac{C_p}{\Delta h_{gas}}(T_r - T_s) - 1\right) \tag{14}$$

By solving equations (11) and (14) we can obtain dependence of volatilization rate and surface temperature T_s on temperature in the reactor T_r and the particle diameter D. Calculated dependences are presented in Figs. 8 and 9 for various diameters of spherical particles. The computations were made for input parameters characteristic for wood (c= 0.511, h=0.064, o=0.425, A=7.7.10^6 s^{-1}, E=1.11.10^5 J/mol) and sheath values of transport and thermodynamic coefficients k, h, C_p corresponding to a mixture of hydrogen and CO with volume ratio 1:1 for zero relative velocity between particles and surrounding gas and averaged sheath temperature

$$T_{sheath} = \frac{(T_r - T_s)}{2} \tag{15}$$

Representation of wood gasification kinetics by one set of parameters A, E is simplification. Unfortunately the values corresponding to high gasification rates in plasma can not be found in the publications and thus we used values representing gasification of lignin [Miller 1997].

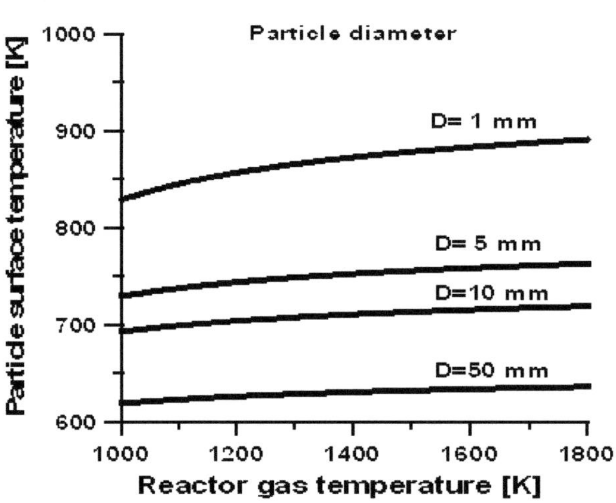

Fig. 8. Surface temperature of wood particles in dependence on reactor temperature for various particle diameters

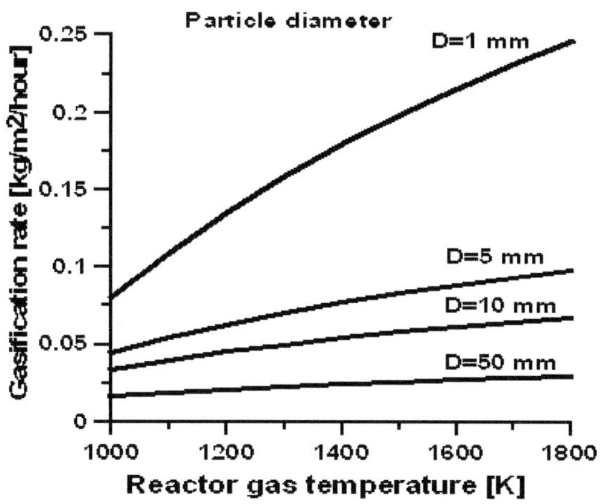

Fig. 9. Gasification rate of wood particles in dependence on reactor temperature for various particle diameters

It can be seen that the particle diameter substantially influences both the surface temperature and the gasification rate. Increase of the diameter results in reduction of heat transfer to the particle due to more intensive shielding of the particle by gas sheath formed from volatilized material. From the dependence of process rate on the size of particles a relation between throughput and minimum volume of the reactor can be estimated. The relation between total volume of particles of given diameter and gasification rate can be calculated from the equations (10) - (14). In Fig. 10 the ratio of total volume of particles to material throughput is plotted in dependence on reactor temperature for several particle diameters. A minimum reactor volume needed for given material throughput can be determined from these dependences assuming that reactor volume should be several times

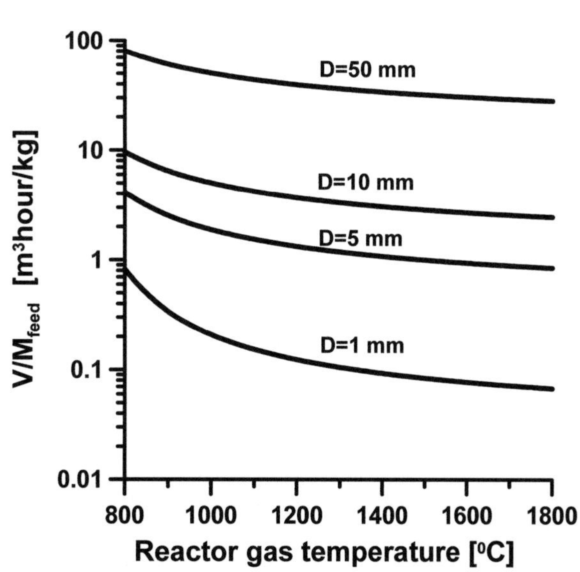

Fig. 10. Ratio of volume occupied by particles to total gasification rate

higher than volume occupied by particles to ensure good heat transfer to the particles. It can be seen from the Fig. 10 that needed volume of reactor rapidly increases with the size of the particles. The increase of the reactor volume leads to the increase of power loss $P_{react}(T_r)$ in equation (4). Optimal reactor volume can be determined on the basis of analysis of relations between process rate and power loss for given size of the particles.

4. Gasification of organic materials in steam plasma

Plasma gasification of biomass was studied in the recent years in several papers [Tang 2005, Brothier 2007, Hrabovsky 2006, Tu Wen Kai 2008, Tang 2005, Xiun 2005]. Up to now only laboratory scale experimental investigations of plasma biomass gasification have been performed. Production of syngas from wood in plasma generated in ac air plasma torches was studied in [Rutberg 2004]. In these experiments plasma with high flow rates and enthalpy not higher then 8 MJ/kg was used. The high flow rate of plasma ensures good mixing of plasma with treated material and a uniform temperature distribution in the reactor. However, the produced syngas contains plasma gas components, usually nitrogen and oxygen if air or nitrogen are used as plasma gases [Rutberg 2004, Zasypkin 2001]. The usage of mixtures of inert gas with hydrogen [Zhao 2001, Zhao 2003] eliminates this disadvantage but it increases the cost. In [Kezelis 2004] biomass was gasified in steam plasma, the usage of produced syngas as plasma gas in a special plasma torch is planned in [Brothier 2007]. This chapter presents the experimental results obtained in medium scale thermal plasma gasification reactor equipped by the gas-water dc plasma torch with arc power up to 160 kW.

4.1 Plasma gasification reactor

The experiments were performed on plasma reactor PLASGAS equipped by plasma torch with a dc arc stabilized by combination of argon flow and water vortex. The scheme of the experimental system is shown in Fig. 11. The torch power could be adjusted in the range of 90 - 160 kW. Power loss to the reactor walls was reduced by the inner lining of the reactor, which was made of special refractory ceramics with the thickness of 400 mm. The wall temperature 1100^0 to 1400^oC could be regulated by the torch power and feeding rate of the material. Inner volume of the reactor was 0.22 m^3. All parts of the reactor chamber were water-cooled and calorimetric measurements on cooling circuits were made. The material container was equipped with a continuous screw conveyer with controlled material feeding rate. Treated material was supplied into the reactor and was fed into plasma jet in the position about 30 cm downstream of the input plasma entrance nozzle at the reactor top. Inputs for additional gases for control of reactor atmosphere were at three positions in the upper part of the reactor. The gas produced in the reactor flowed through the connecting tube to the quenching chamber, which was created by a cylinder with the length of 2 m. At the upper entrance of the cylinder the gas was quenched by a spray of water from the nozzle, positioned at the top of the cylinder. The water flow rate in the spray was automatically controlled to keep the temperature of gas at the output of the quenching chamber at 300^oC. The gas then flows into the combustion chamber where it is combusted in the flow of the air. To prevent destruction of ceramic insulation wall the reactor was pre-heated prior to the experiments for 24 hours to temperature about 950^oC. Then the heating of the reactor walls to working temperature was made by plasma torch at arc power 110 kW.

The measuring system included monitoring of plasma torch operation parameters, temperatures in several positions inside the reactor and calorimetric measurements on cooling water loops. The temperature of inner wall of the reactor was measured in six positions by thermocouples. The flow rate of produced syngas was determined by two methods. Pitot flow meter was installed in the system downstream of the exit of quenching chamber and thus the total flow rate was measured of syngas and steam produced in quenching chamber with water spray. The flow rate was also determined from molar concentration of argon measured at the output of the reactor before quenching chamber in case when defined flow rate of argon was introduced into the reactor. Gas temperature was measured at the input and the output of the quenching chamber by thermocouples. The composition of produced gas was measured at the output of reactor before the gas enters the quenching chamber. The tube for collection of samples was cooled down by the water spray at the input of the quenching chamber.

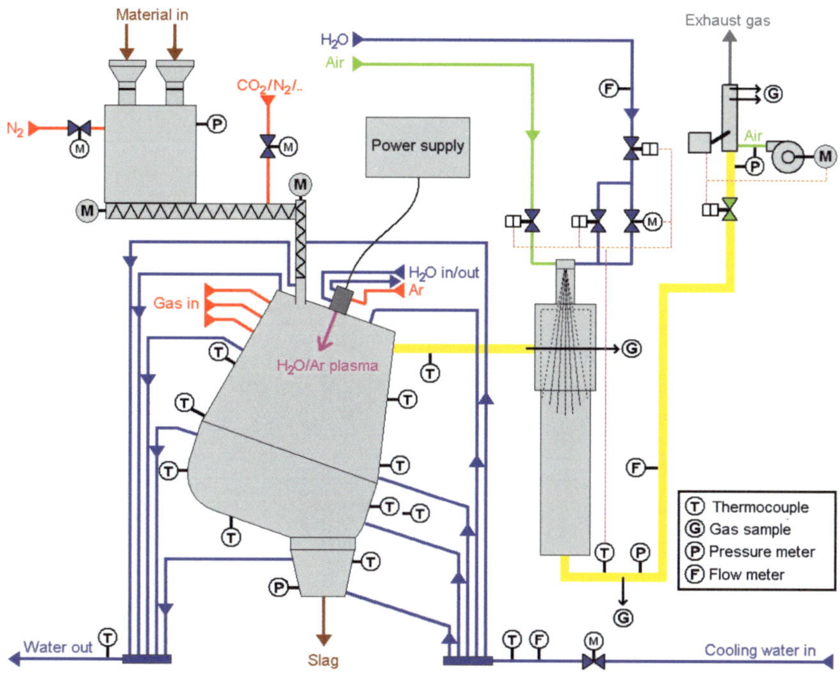

Fig. 11. Schematics of experimental reactor PLASGAS.

The main gas analysis was made by a quadruple mass spectrometer Balzers QMS 200. As the gas can contain some amount of steam which could after condensation block or damage the inputs of the mass spectrometer, the freezing unit was connected into the gas sample circuit.
Additional analyses of the composition of the produced syngas and the content of tar were made on samples of gas taken during the experiment by means of mass spectroscopy with cryofocusing, gas and liquid chromatography and FT infrared spectroscopy. Samples for tests of presence of tar in the gas were taken from the tube between the reactor and the quenching chamber. The samples were captured on the DSC-NH2 adsorbend or silica gel and analyzed by gas and liquid chromatography. The content of tar was below the sensitivity of the method, which was 1 mg/Nm^3.

4.2 Plasma generator with hybrid water/gas arc stabilization

Plasma was produced in the torch with a dc arc stabilized by combination of argon flow and water vortex. The torch generates an oxygen-hydrogen-argon plasma jet with extremely high plasma enthalpy and temperature. Typical arrangement of arc chamber with gas/water stabilization is shown in Fig. 12. The cathode part of the torch is arranged similarly like in gas torches. Gas is supplied along tungsten cathode tip, vortex component of gas flow that is injected tangentially, assures proper stabilization of arc in the cathode nozzle. Gas plasma flows through the nozzle to the second part of arc chamber, where arc column is surrounded by a water vortex. The chamber is divided into several sections, where water is injected tangentially. The inner diameter of the vortex is determined by the diameter of the holes in the segments between the sections. The sections with tangential water injection are separated by two exhaust gaps, where water is exhausted out of the arc chamber. Interaction of the arc column with the water vortex causes evaporation from the inner surface of the vortex. The steam mixes with the plasma flowing from the cathode section. An anode is created by a rotating copper disc with internal water cooling. Thus the arc column is composed of three sections. The cathode section is stabilized by a vortex gas flow. If gas with low enthalpy like argon is used, the voltage drop and power of this section is small. The most important section, which determines plasma properties, is the water-stabilized part, where the arc column interacts with the water vortex. The third part between the exit nozzle and the anode attachment is an arc column in a free jet formed from mixture of argon with steam.

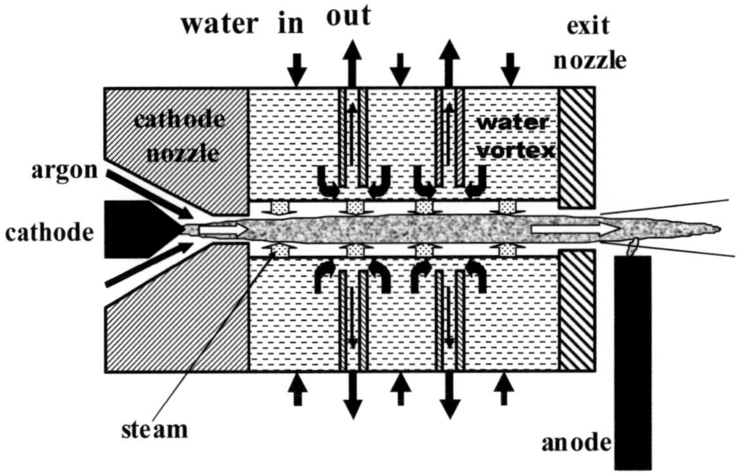

Fig. 12. Schematics of water/argon plasma torch.

As walls of stabilizing cylinder in the main arc chamber are created by water, arc can be operated at substantially higher power than in common gas stabilized torches. Figure 13 presents comparison of operation regimes of water stabilized torches and conventional gas stabilized torches, characterized by levels of arc power and plasma mass flow rate. Low mass flow rates of plasma for water torches follow from the energy balances of radial heat transfer. For gas torches mass flow rates can be controlled independently of arc power. However, lower limit of mass flow rate is given by a necessity to protect walls of arc chamber by gas flow. It can be seen that water plasma torches are characterized by very low mass flow rates. This fact results in high plasma enthalpies. Typical values of mean plasma enthalpies for dc arc torches are shown in Fig. 13. Figure 14 presents enthalpies of steam plasma compared with mixtures of nitrogen and argon with hydrogen, which are commonly used in gas plasma torches. High enthalpy of steam plasma represents capacity of plasma to carry energy. The other positive property of steam plasma for plasma processing is high heat conductivity. Thus, extreme properties of plasma jets generated in water stabilized and hybrid stabilized arc torches follows both from the properties of steam plasma and from the way of stabilization of arc by water vortex.

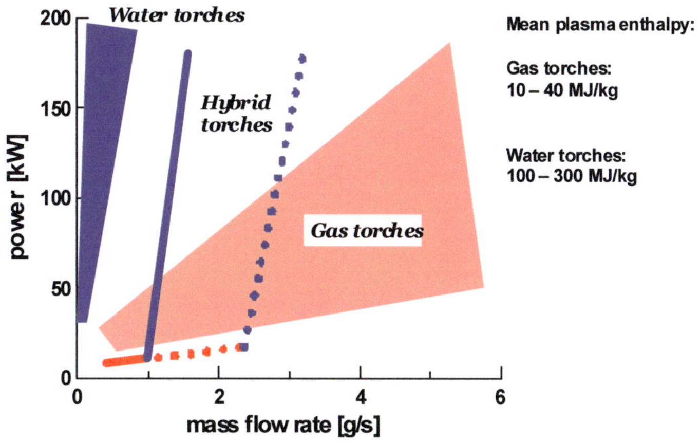

Fig. 13. Operation regimes of dc arc plasma torches.

The way how operation regime is established in a hybrid torch is illustrated in Fig. 13. In the cathode gas-stabilized section the power increases with gas flow rate slowly, if low enthalpy gas like argon is used (red part of characteristics in Fig. 13). Energy balance in the water

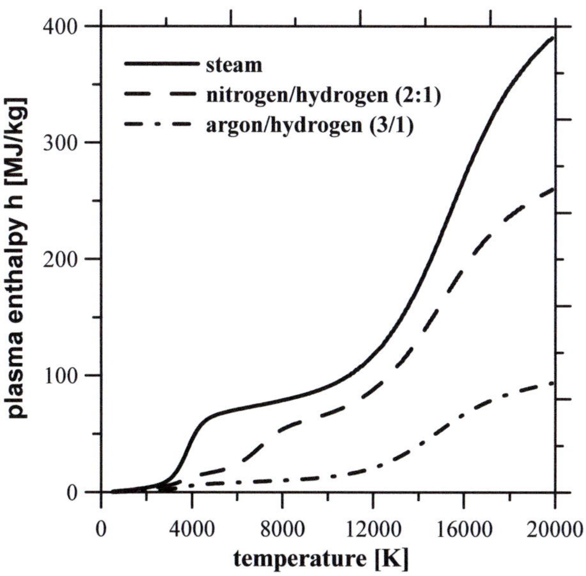

Fig. 14. Plasma enthalpy in dependence on temperature for steam and mixtures nitrogen/hydrogen (2:1 vol.) and argon hydrogen (3:1 vol.).

stabilized arc section is almost completely controlled by steam inflow and the arc in this section has electrical characteristics and power balances that are very close to the ones of water-stabilized torches. The power thus increases rapidly with mass flow rate as in the case of water torch (blue part of characteristics in Fig. 13).

High temperature plasma jet with high flow velocity is generated in the hybrid plasma torch. The centreline plasma flow velocity at the torch exit, which is increasing with both the arc current and the argon flow rate, ranges approximately from 1800 m/s to 7000 m/s. The centerline exit temperature is almost independent of argon flow rate and varies between 14 kK and 22 kK. In Fig. 15 measured profiles of plasma temperature for arc power 70 kW and 96 kW are presented. Temperature is increasing with arc current but does not depend much on argon flow rate, because thermal plasma parameters are determined by processes in water stabilized (Gerdien) arc part. Fig. 15 presents temperature profiles measured at position 2 mm downstream of torch nozzle. With increasing distance from the nozzle plasma jet temperature rapidly decreases due to mixing of plasma with ambient gas and due to intensive radial heat transfer to the jet surrounding.

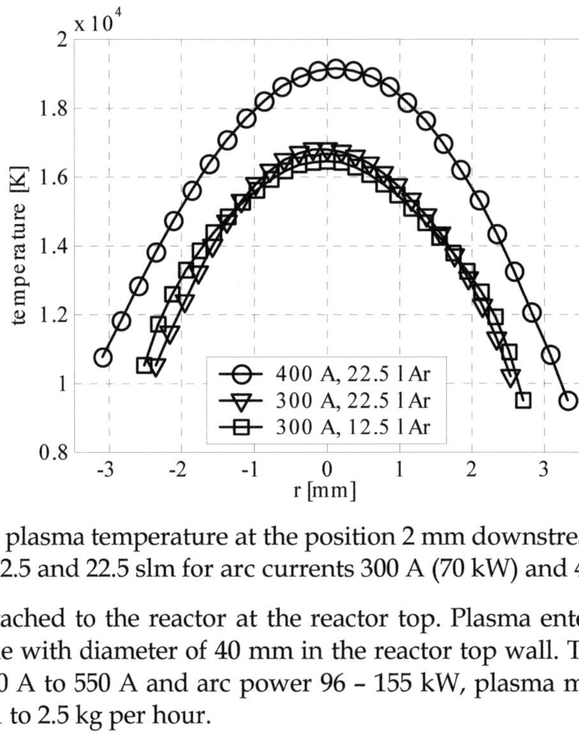

Fig. 15. Profiles of plasma temperature at the position 2 mm downstream of the torch exit for argon flow rates 12.5 and 22.5 slm for arc currents 300 A (70 kW) and 400 A (96 kW).

The torch was attached to the reactor at the reactor top. Plasma enters the reactor volume through the nozzle with diameter of 40 mm in the reactor top wall. The torch was operated at arc currents 350 A to 550 A and arc power 96 - 155 kW, plasma mass flow rates were in the range from 2.1 to 2.5 kg per hour.

4.3 Experimental results of plasma gasification of organic materials

Experiments with several materials at various conditions were performed with plasma reactor PLASGAS.

Table 1 presents examples of results obtained in experiments with gasification of wooden saw dust. The table gives values of basic operation parameters, i.e. plasma power, feed rate of wood, flow rates of gases added to the reactor (CO_2 and O_2) and averaged temperature T_r in the reactor. The temperature T_r given in the table is averaged temperature of the reactor

torch power	feed rate	CO_2	O_2	T_r	syngas	H_2	CO	CO2	O2	Ar	CH4	calorific value
[kW]	[kg/h]	[slm]	[slm]	[K]	[m^3/h]	%	%	%	%	%	%	[kW]
104	6.9	43	10	1360	7.13	27.7	60.8	5.4	0.7	4.9	0.5	21.11
104.3	6.9	20	10	1355	7.85	33.7	57.1	3.3	0.4	5.6	0.05	23.6
105.3	17	115	0	1345	30.42	31.5	59.5	4.9	0.1	2.3	1.6	92.2
106.1	17	115	30	1463	32.16	28.4	59.7	7.7	0.4	2.2	1.6	94.7
106.3	27.1	115	30	1417	34.41	22.3	68.3	2.4	4.8	1.4	0.8	105.4
152.5	27.1	115	30	1452		32.3	61.3	4.7	0.1	0.6	0.9	
95	28	16	0	1150	37.6	46.3	45.2	1.9	1.6	5.1	-	111.7
138	28	16	0	1200	32.6	42	44.3	3.4	2.5	7.8	0	101.6
107.7	47.2	115	30	1406	71.04	36	59.9	2.3	0.1	0.6	1.1	225.9
107.7	47.2	115	30	1364	76.36	37.3	60.1	1.8	0.1	0.2	0.4	246.3

Table 1. Basic operation parameters, composition, flow rate and calorific value of syngas produced by gasification of wood saw dust.

wall obtained as an average of inner wall temperatures measured at six positions in the reactor. The right hand side of the table presents flow rate of produced syngas, its composition and calorific value of syngas. The calorific value was calculated from measured flow rate of gas and its composition. It can be seen that for the highest feed rates the calorific value of produced syngas is almost 2.5 times higher then the torch power. The ratio of power available for material treatment (after all power losses were subtracted from the arc power) to total arc power increased with increasing arc power from 0.35 - 0.41 at arc power 95 - 100 kW to 0.41 - 0.46 for arc power higher then 130 kW for wall temperatures 1100 - 1200°C. The ratio was lower for higher wall temperatures. Most of the results in Table 1 were obtained at arc power 104 to 107 kW, some results for different power are also included. No effect of arc power on gas composition and flow rate was observed for tested feeding rates up to 47.2 kg/h. It can be concluded that maximum possible feeding rate at given power has not been reached.

The results of other test series of experimental gasification of wooden saw dust are presented in Table 2. The composition of produced syngas is compared with the composition determined by equilibrium computations which are presented in Fig. 2. In all test runs syngas with high concentrations of hydrogen and carbon monoxide was obtained. The concentration of CO_2 and CH_4 were small especially for higher feeding rates and higher flow rates of gases added for oxidation of surplus of carbon. The last column of Table 2 presents heating values of syngas calculated from the composition. It can be seen that the values of LHV and the composition are close to the results of equilibrium calculations.

	Test Parameters			Added gases		Syngas Composition						
	Feed [kg/h]	T_r [K]	Power [kW]	CO_2 [slm]	O_2 [slm]	H_2 %	CO %	CO_2 %	O_2 %	CH_4 %	Ar %	$LHV_{syn.}$ [MJ/m^3]
C	47	1350		115	30	42	56	0.3	0	0.4	1.0	11.72
E1	47.2	1364	108	115	30	37	60	1.8	0.1	0.4	0.2	11.76
E2	47.2	1420	108	115	30	36	59	2.9	0	1.5	0.6	11.84
E3	30	1280	110	15	0	43	44	7.2	0.1	1.3	3.3	10.81
E4	30	1360	110	15	0	42	49	4.7	0.1	1.7	2.5	11.33

Table 2. Measured (E) and computed (C) composition and LHV of syngas.

The differences between temperatures of inner wall measured at different positions within the reactor did not exceed 100°C. At all experiments the minimum measured wall

temperature was 1100°C. Under these conditions the change of wall temperature in the range of 1100 to 1450°C does not influence the flow rate and the composition of the produced gas, as can be seen in Tables 1 and 2.
The composition of produced gas was only slightly influenced by the material feeding rate and the power and was controlled by the ratio of mass of oxygen in supplied gases (O_2, CO_2), added for complete oxidation of carbon, to the feed rate of material. This is illustrated in Fig. 16 where molar fractions of gas components are plotted in dependence on ratio of oxygen mass flow rate to the material feed rate.

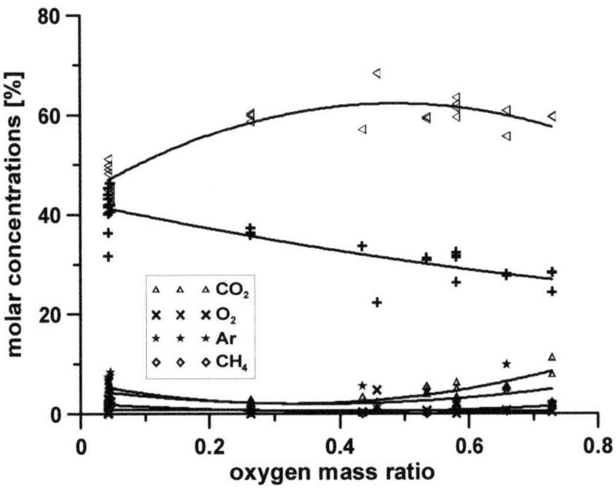

Fig. 16. Composition of syngas in dependence on mass ratio of oxygen in gases supplied into the reactor.

The degree of biomass gasification is characterized by the ratio of carbon content in syngas to the total amount of carbon supplied into the reactor in fed wood and in added gases. The ratio of carbon in gas phase to the supplied carbon is shown as carbon yield in Fig. 17. The ratios of carbon mass in syngas to the carbon mass in wood and to the total mass of supplied

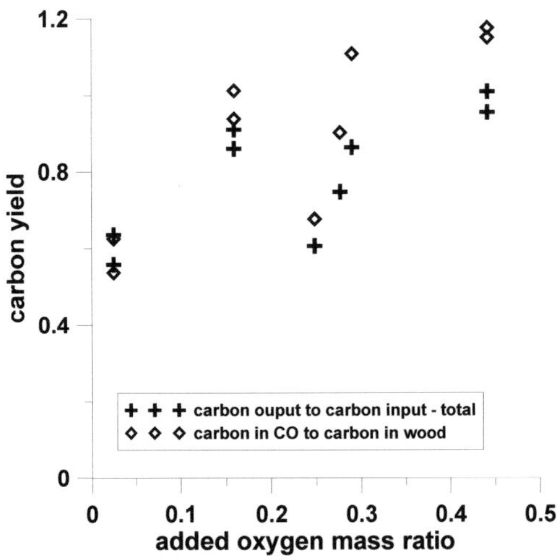

Fig. 17. Ratio of carbon in syngas to the supplied carbon in dependence on mass fraction of oxygen added into the reactor in O_2 and CO_2.

carbon including supplied gas species are plotted in dependence on ratio of mass of oxygen added into the reactor in the gas species (O_2 and CO_2) to the mass of wood. The carbon yield, defined on the basis of mass of wood, can be higher than 1 as carbon from supplied gas (CO_2) is added to syngas. It can be seen that for higher feeding rates almost all carbon was gasified. Lower values of carbon yield for lower material feeding rates are probably related to weak mixing of plasma with material and thus less intensive energy transfer to the material. The mixing is more intensive at higher feeding rates due to substantially higher amount of gas produced in the reactor volume at high feeding rates. The flow within the reactor is almost completely controlled by material gasification, especially for higher feeding rates, because the amount of gas produced by gasification is up to 120 Nm^3/h while the flow rate of plasma from the torch is 1.34 Nm^3/h.
The energy spent for the gasification of material at different feeding rates is shown in Fig. 18 in dependence on the feeding rate. Fig. 18 also gives the values of ratio of heating value of produced syngas (LHV), calculated from measured syngas composition and flow rate, to the energy spent for its production, corresponding to the torch power. It can be seen that for the highest values of the feeding rate this ratio, presented in Fig. 18 as energy gain, was 2.3.

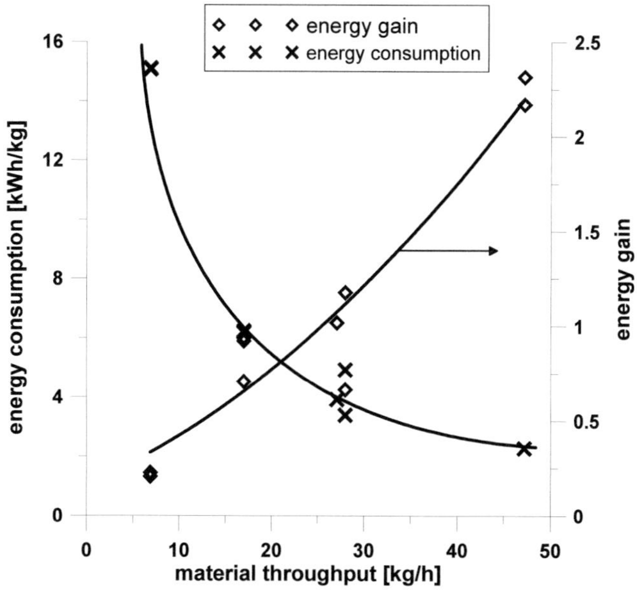

Fig. 18. Specific energy consumption for gasification and ratio of LHV of syngas to the torch power in dependence on feeeding rate.

The results of analysis of tar content in produced syngas are shown in Table 3. The overall content of tar was lower than 10 mg/Nm^3, which was under the detection limit of used TCD. This occurred even with toluene, and it is obvious that concentration of tar in produced gas is really low in comparison with other gasification technologies. Especially in the case of lower feeding rates of treated material the tar content was minimal. Low tar content is caused mainly by the high temperatures in the reactor and the fast quenching as well as by high level of uv radiation in the entrance of output gas tube, which was positioned close to the input for plasma jet.

Plasma torch power [kW]	107	107	107
CO_2 flow rate [slm]	5	10	60
Humidity of treated wood [w/w]	20.2	20.2	20.2
Wood flow rate [kg/hour]	10	20	50
Benzene [mg/Nm^3]	1,5	2,7	116,2
Toluene	< 1 mg/Nm^3		
Tar - SPE	< 10 mg/Nm^3		

Table 3. Content of benzene, toluene and tar in produced syngas.

Besides experiments with wood saw dust, gasification of several other organic materials was tested. Tables 4 and 5 show results of test runs of following four materials: wooden saw dust, wooden pellets 6 mm in diameter and 6 mm long, polyethylene balls of diameter 3 mm and waste polyethylene plastics composed of 80% high-density polyethylene and 20% low-density polyethylene. Gasification by reaction with CO_2, O_2 and mixture of the two gases was studied. Table 4 presents basic experimental parameters, feed rates of materials and flow rates of added gases. Arc current was 446 to 450 A and arc power between 130 and 140 kW. Small differences in arc current and power for various runs are caused by small fluctuations of arc voltage due to changes of temperature of water in the arc chamber.
Composition of syngas determined from the analysis by mass spectrometer is shown in Table 5. Amount of carbon transferred into gas phase was determined from syngas flow rate and gas composition. The gas yield of carbon represented by the ratio of amount of C in syngas to total amount of carbon in supplied material and gases is given in Table 5.

	I [A]	P[kW]	material	[kg/h]	CO2 [slm]	O2 [slm]	T_r [°C]
1	449	138	wood	41,1		64	1362
2	448	138	wood	41,1	125		1355
3	449	137	wood	25,2	125	43	1368
4	449	137	wood	25,2	125		1341
5	449	137	wood	25,2	86		1337
6	450	140	pellets	30		64	1493
7	450	140	pellets	30	248		1383
8	450	140	pellets	60	248		1286
9	446	140	PE	5,3	210	80	1539
10	446	140	PE	10,6	210	80	1559
11	448	131	plastics	11,2	300		1397

Table 4. Experimenal conditions and input parameters for several materials.

It can be seen that syngas with high concentrations of hydrogen and carbon monoxide was obtained in all runs. The CO_2 concentrations were small especially for wood saw dust and wood pelets (runs 4, 5, 7, 8), concentration of CH_4 was very low in all runs. Oxidation with CO_2 and O_2 led to the same composition (runs 1,2). Surplus of oxygen (run 3) resulted in increase of concentration of CO and reduction of H_2, probably due to formation of H_2O. Concentration of water in syngas could not be measured by mass spectrometer due to problems with condensation; water was removed in freezing unit. In the runs 5, 8 and 10 an

amount of supplied oxygen was close to stoichiometric values for oxidation of all carbon in material. Complete transformation of carbon into gas phase was found for wood saw dust and polyethylene. For wooden pellets and plastic waste the carbon yield was 0.7 – 0.8.
In all cases, like in case of wood saw dust, the content of tar and higher hydrocarbons in the produced gas was very low and substantially less than 10 mg/Nm³. This is lower than the tar content in most of non-plasma gasifiers, where the tar content for various types of reactors varies in the range from 10 mg/Nm³ to 100 g/Nm³.

	material	% H2	% CO	% CO2	%CH4	% O2	C_{out}/C_{in}
1	wood	44,8	39,2	15,0	0,9	0,1	1,0
2	wood	41,5	42,5	14,9	1,0	0,1	0,9
3	wood	34,6	51,4	12,6	0,4	1,0	1,0
4	wood	41,5	54,1	3,3	0,3	0,8	1,0
5	wood	43,6	52,0	3,3	0,3	0,8	1,0
6	pellets	48,1	40,0	11,0	0,1	0,8	0,7
7	pellets	36,5	59,1	3,4	0,1	1,0	0,8
8	pellets	41,5	52,7	4,8	0,2	0,8	0,8
9	PE	29,9	41,3	27,1	0,0	1,7	1,0
10	PE	35,3	41,5	21,7	0,1	1,4	1,0
11	plastics	41,6	49,7	7,4	0,0	1,3	0,7

Table 5. Composition of syngas and carbon yield for conditions in Table 4.

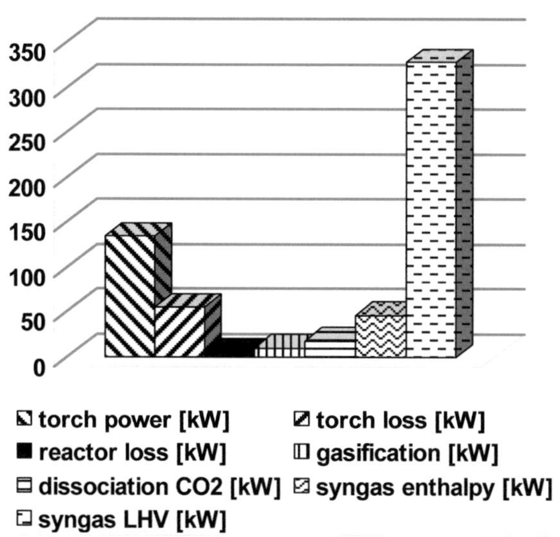

Fig. 19. Power balance of gasification of wooden pellets in the run 8.

Analysis of power balance for experimental run with the highest material feed rate (run 8) is shown in Fig. 19. Torch power, power loss in the torch, power loss to the reactor walls and total power spent for process of gasification were determined from current and voltage measurements and calorimetric measurements on cooling circuits of the system. Power spent for dissociation of CO_2 was calculated from flow rate of added CO_2, power corresponding to low heating value of syngas was calculated from measured composition and flow rate of syngas. Heating value of produced syngas is more than two times higher than power of the torch.

It can be seen that in case of gasification with CO_2 most of power needed for production of syngas was dissociation power of CO_2. Energy needed for dissociation of CO_2 is deposited in calorific value of produced syngas. The process thus can act as an energy storage – electrical energy is transferred to plasma energy and then stored in produced syngas. This can be used for storage of energy produced by new renewable sources of electrical energy that are often characterized by large fluctuations of energy production. Moreover, the process offers utilization and transformation of CO_2 produced by industrial technologies.

5. Conclusions

The research of plasma biomass gasification has been started as a response for a need of more efficient utilization of biomass for energy and fuel production. Classical ways of biomass gasification, based on partial combustion, do not produce synthesis gas with quality demanded by advanced technologies of fuel and energy production, mostly due to contamination of syngas by CO_2, methane, tars and other components. The necessity of production of clean syngas with controlled composition leads to technologies based on external energy supply for material gasification. Plasma is medium with the highest energy content and thus substantial lower plasma flow rates are needed to supply sufficient energy compared with other media used for this purpose. This result in minimum contamination of produced syngas by plasma gas and easy control of syngas composition. Especially high enthalpy steam plasma produced in water and water-gas torches offers excellent characteristics.

The experiments with gasification of wood, wooden pellets, polyethylene and plastic waste were performed on the reactor with hybrid gas-water plasma torch. The composition of produced syngas was close to the calculated equilibrium composition, determined for the case of complete gasification. The heating value of produced syngas was in good agreement with calculated equilibrium values. In all cases the content of tar and higher hydrocarbons in the produced gas was very low and usually less than 10 mg/Nm^3. This is substantially lower than the tar content in most of non-plasma gasifiers, where the tar content for various types of reactors varies in the range from 10 mg/Nm^3 to 100 g/Nm^3 [Hasler 1999, Jun Han 2008].

It has been experimentally verified that for small particles and higher feeding rates all supplied material was gasified. Heating value of produced syngas was for the highest material feed rates more than two times of power of plasma torch. In case of gasification with carbon dioxide as oxidizing medium, most of power needed for gasification process was power for dissociation of CO_2. The process can be used as an energy storage - electrical energy is transferred to plasma energy and then stored in produced syngas. This can be utilized for storage of energy produced by sources of electrical energy with large fluctuations of energy production. Moreover, the process offers utilization and transformation of CO_2 generated by industrial technologies.

If energy balances of plasma gasification are compared with the conventional autothermal reactors, where only very low power is supplied to ignite the process of partial combustion, the energy gain in plasma systems is smaller. However, the LHV of produced syngas for autothermal reactors is usually between 35% and 60% of its theoretical value, and moreover, quality of produced syngas is low especially due to the production of tars and other contaminants. Thus, plasma can offer advantages if high quality syngas with high heating value is needed. Moreover, possibility of electrical energy storage can be utilized in combination with new renewable power production technologies.

6. Acknowledgment

The author gratefully acknowledges the financial support of the Grant Agency of the Czech Republic under the project No. 205/11/2070.

7. References

Bird, R.B.; Stewart, W.E., Lightfoot, E.N. 2002. Transport Phenomena. J. Willey&Sons, Inc., New York/Chichester/Weinheim/Brisbane/Singapore/Toronto.

Boerrigter H.; van der Drift, B. 2005. "Biosyngas" key-intermediate for production of renewable transportation fuels, chemicals and electricity. *ECN report ECN-RX--05-181*, presented 14th European Biomass Conf.&Exhibition, Paris.

Boulos, M.I.; Fauchais, P., Pfender, E. 1994. Thermal Plasma Fundamentals and Applications. Plenum Press, New York-London.

Brothier, M., et al. 2007. Syngas production from the biomass gasification by plasma torch. *Proc. of 18th ISPC* (ed. K Tachibana et al), Kyoto, Book of Abstracts: 193, full paper on CD.

Coufal, O. 1994. Composition of the reacting mixture SF6 and Cu in the range from 298.15 to 3000 K and 0.1 to 2 Mpa. *High Temp. Chem. Processes*, 3: 117-139.

Coufal O.; Sezemsky P., Zivny O. 2005. Database system of thermodynamic properties of individual substances at high temperatures. *J. Phys. D: Appl. Phys.*, 38: 1265-1274.

Dietenberger M. 2002. Update for combustion properties of wood components. *Fire Mater.*, 26: 255 - 267.

Hasler P.; Nussbauer Th. 1999. Gas cleaning for IC engine applications from fixed bed biomass gasification. *Biomass and Bioenergy*, 16: 385-395.

Jun Han; Heejoon Kim 2008. *Renewable and Sustainable Energy Reviews*, 12: 397- 416.

Hrabovsky M.; Konrad M., Kopecky V., Hlina M. 2006. Pyrolysis of wood in arc plasma for syngas production. *J. of High Temperature Material Processes*, 10: 557-570.

Kezelis R.; Mecius V., Valinciute V., Valincius V. 2004. Waste and biomass treatment employing plasma technology. *J. of High Temp. Mat. Process.*, 8: 273-282.

Krenek P. 2008. Thermophysical properties of H2O-Ar plasmas at temperatures 400-50000K and pressure 0,1 MPa. *Plasma Chem. Pl. Process*, 28: 107-122.

Miller R.S.; Bellan J. 1997. A generalized biomass pyrolysis model based on superimposed cellulose, hemicellulose and lignin kinetics. *Comb. Sci. and Technol.*, 126: 97-137.

Rutberg P.G.; Bratsev A.N., Ufimtsev A.A. 2004. Plasmochemical technologies for processing of hydrocarbonic raw material with syngas production. *J. of High Temp. Mat. Process.*, 8: 433-446.

Tang L.; Huang H. 2005. Plasma pyrolysis of biomass for production of syngas and carbon adsorbent. *ENERGY & FUELS*, 19: 1174-1178.

Tang L.; Huang H. 2005. Biomass gasification using capacitively coupled RF plasma technology. *Fuel*, 84: 2055–2063.

Tu, Wen-Kai et al. 2008. Pyrolysis of rice straw using radio-frequency plasma. ENERGY&FUELS, 22: 24-30.

Xiu S.N.; Yi W.M., Li B.M. 2005. Flash pyrolysis of agricultural residues using a plasma heated laminar entrained flow reactor. *BIOMASS BIOENERG*, 29: 135-141.

Zasypkin I.M.; Nozdrenko G.V. 2001. Production of acetylene and synthesis gas from coal by plasma chemical methods. *Thermal Plasma Torches and Technologies*, Vol II., ed. O.P. Solonenko, Cambridge Interscience Publish.: 234-243.

Zhao Z.L.; Huang H.T., Wu C.Z., Li H.B., Chen Y. 2001. Biomass pyrolysis in an argon/hydrogen plasma reactor. *Chem. Engineering & Technology*, 24: 197-199.

Zhao Z.L. 2003. Plasma gasification of biomass in a downflow reactor. *Abstract of Papers of the American Chemical Society*, 226: U536-U536 048-FUEL Part 1.

4

Numerical Investigation of Hybrid-Stabilized Argon-Water Electric Arc Used for Biomass Gasification

J. Jeništa[1], H. Takana[2], H. Nishiyama[2], M. Bartlová[3], V. Aubrecht[3], P. Křenek[1], M. Hrabovský[1], T. Kavka[1], V. Sember[1] and A. Mašláni[1]
[1]*Institute of Plasma Physics, AS CR, v.v.i., Thermal Plasma Department, Praha*
[2]*Institute of Fluid Science, Tohoku University, Sendai, Miyagi,*
[3]*Brno University of Technology, Brno*
[1,3]*Czech Republic*
[2]*Japan*

1. Introduction

Plasma generators with arc discharge stabilization by water vortex exhibit special performance characteristics; such as high outlet plasma velocities (up to 7 000 $m \cdot s^{-1}$), temperatures (~ 30 000 K), plasma enthalpy and, namely, high powder throughput, compared to commonly used gas-stabilized (Ar, He) torches (Hrabovský et al., 1997). In a water-stabilized arc, the stabilizing wall is formed by the inner surface of water vortex which is created by tangential water injection under high pressure (~ 10 atm.) into the arc chamber. Evaporation of water is induced by the absorption of a fraction of Joule power dissipated within the conducting arc core. Further heating and ionization of the steam are the principal processes which produce water plasma. The continuous inflow and heating lead to an overpressure and plasma is accelerated towards the nozzle exit. The arc properties are thus controlled by the radial energy transport from the arc core to the walls and by the processes influencing evaporation of the liquid wall.

A combination of gas and vortex stabilization has been utilized in the so-called hybrid-stabilized electric arc, its principle is shown in Fig.1. In the hybrid H_2O-Ar plasma arc the discharge chamber is divided into the short cathode part where the arc is stabilized by tangential argon flow in the axial direction, and the longer part which is water-vortex-stabilized. This arrangement not only provides additional stabilization of the cathode region and protection of the cathode tip, but also offers the possibility of controlling plasma jet characteristics in wider range than that of pure gas or liquid-stabilized torches (Březina et al., 2001; Hrabovský et al., 2003). The arc is attached to the external water-cooled rotating disc anode a few mm downstream of the torch orifice. The characteristics of the hybrid-stabilized electric arc were measured and the effect of gas properties and flow rate on plasma properties and gas-dynamic flow characteristics of the plasma jet were studied. Experiments (Březina et al., 2001; Hrabovský et al., 2006) proved that plasma mass flow rate, velocity and momentum flux in the jet can be controlled by changing mass flow rate in the gas-stabilized section, whereas thermal characteristics are determined by processes in the water-stabilized section. The domain for numerical calculation is shown in Fig. 1 by a dashed line and includes the discharge area between the outlet nozzle for argon and the near-outlet region of the hybrid plasma torch.

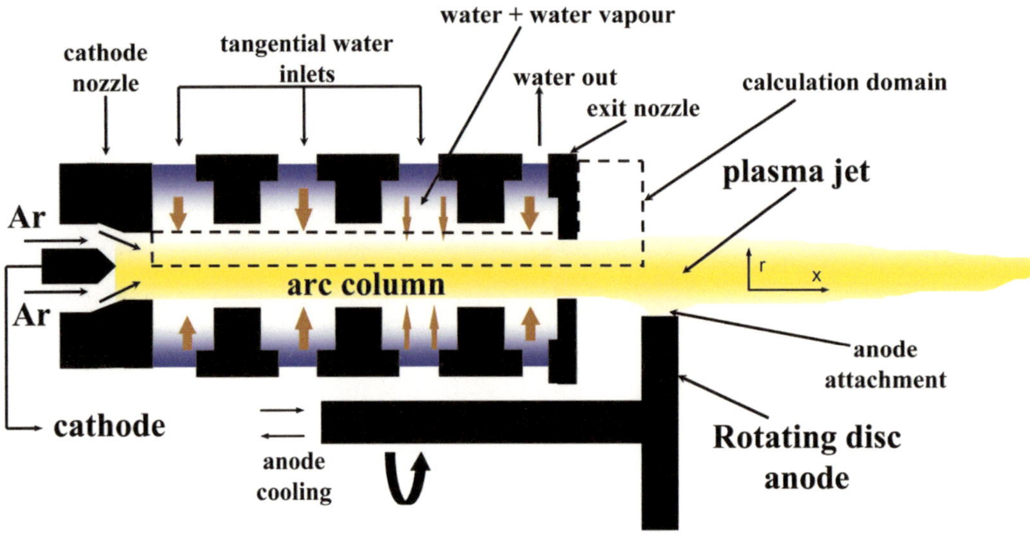

Fig. 1. Principle of hybrid plasma torch WSP®H with combined gas (Ar) and vortex (water) stabilizations. Water is injected tangentially and creates vortex in the chamber. The arc burns between the cathode, made of a small piece of zirconium pressed into a copper rod, and the water-cooled anode rotating disc. The calculation domain is shown by a dashed line.

The hybrid arc has been used at IPP AS CR, v.v.i., in the plasma spraying torch WSP®H (160 kW) for spraying metallic or ceramic powders injected into the plasma jet (Fig. 2). Recently, an experimental plasmachemical reactor PLASGAS (Fig. 3) equipped with the spraying torch WSP®H has been started for the innovative and environmentally friendly plasma treatment of waste streams with a view to their sustainable energetic and chemical valorization and to a reduction of the emission of greenhouse gases (Van Oost et al., 2006, 2008). Pyrolysis of biomass was experimentally studied in the reactor using crushed wood and sunflower seeds as model substances. Syngas with a high content of hydrogen and CO was produced.

This work aims to study properties and processes in the hybrid arc for high currents (300-600 A) and argon mass flow rates (22.5-40 standard liters per minute, slm). In contrast to our previous investigation (Jeništa, 2004; Jeništa et al., 2007), a special attention is devoted to the flow structure and temperature field in the discharge when the local Mach number is higher than one. Our former results indicated the possibility (Jeništa, 2004) and also proved the existence (Jeništa et al., 2008) of supersonic flow regime for currents higher or equal to 500 A. In addition, a detailed comparison of the calculated results with experiments is presented in this study.

Section 2 gives information about the model assumptions, plasma properties, boundary conditions and the numerical scheme. Section 3 reveals the most important findings such as thermal and fluid dynamic characteristics of plasma within the discharge and in the near-outlet regions, along with power losses from the arc and comparison of calculated results (temperature and velocity profiles near the nozzle exit) with experiments.

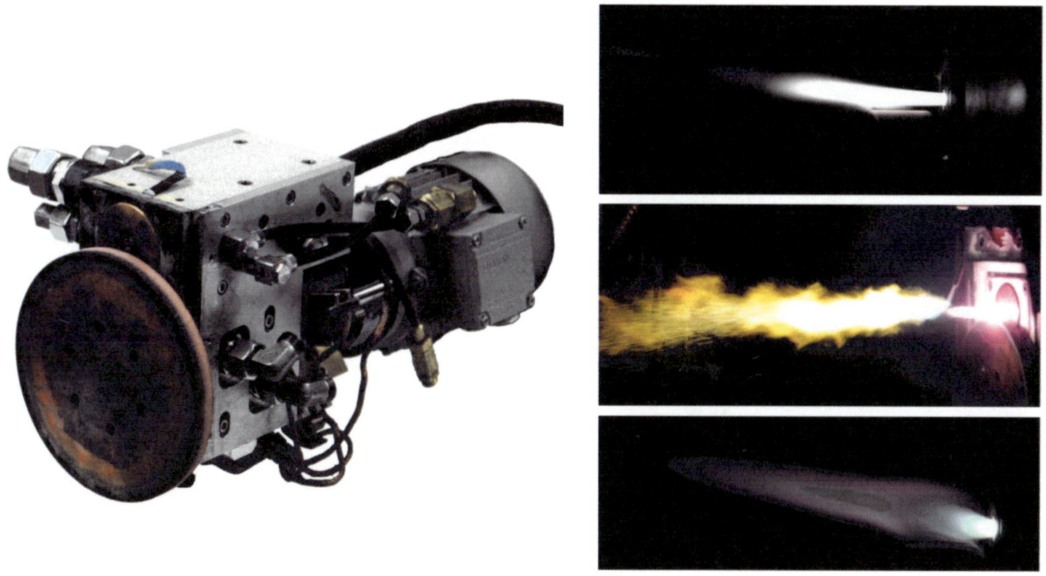

Fig. 2. The plasma spraying torch WSP®H with hybrid stabilization (left), i.e. the combined stabilization of arc by axial gas flow (Ar or N_2) and water vortex. The external rotating disk anode is made of copper. Images of plasma jets produced by WSP®H (right) from the mixture of steam and argon for different operational conditions: 300 A and 24 slm of argon (top), spraying of Cu particles at 500 A and 36 slm of argon (middle), supersonic jet at 300 A, 12 slm of argon at 10 kPa of surrounding atmosphere (bottom).

2. Physical model and numerical implementation

2.1 Assumptions and the set of equations

The following assumptions for the model are applied:

1. the numerical model is two-dimensional with the discharge axis as the axis of symmetry,
2. plasma flow is laminar/turbulent and compressible in the state of local thermodynamic equilibrium,
3. argon and water create a uniform mixture in the arc chamber,
4. only self-generated magnetic field by the arc itself is considered,
5. gravity effects are negligible,
6. the partial characteristics and the net emission coefficients methods (models) for radiation losses from the arc are employed,
7. all the transport, thermodynamic and radiation properties are dependent on temperature, pressure and argon molar content.

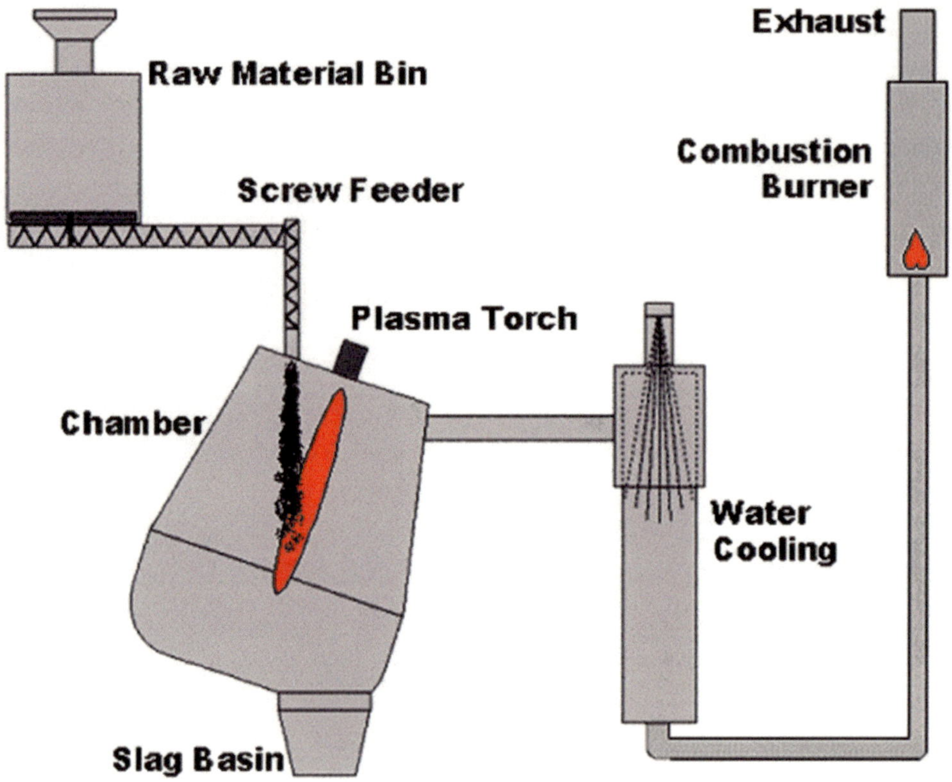

Fig. 3. Schematic diagram of the experimental reactor for plasma pyrolysis and gasification.

A few comments should be mentioned on these assumptions:

1. The cylindrical discharge chamber (Fig. 1) is divided into several sections by the baffles with central holes. Water is injected tangentially into the chamber by three sets of three inlet holes (totally 9 holes) placed equidistantly along the circumference at angles of 120°. The inner diameter of the water vortex is determined by the diameter of the holes in the baffles. Water is usually pumped under pressures of 0.39 MPa (0.6 MPa) with flow rates of 10 l min^{-1} (16 l min^{-1}). Higher pressures insure better hydrodynamic stability of the arc. Since water flows in a closed circuit, it is also exhausted at two positions along the arc chamber.

   In order to see the flow structure near the outlet, we included in our calculation domain also the near-outlet region which extends up to 20 mm from the nozzle exit. In experiment, the distance from the nozzle exit to the anode can be changed from 5 to 20 mm. It can be expected that regions close to the nozzle exit will remain undisturbed by the presence of the anode, while the more distant regions (15-20 mm) will be influenced by 3D effects (the anode jet and anode processes), provided the anode is placed somewhere 20 mm from the nozzle exit.

   It comes out from these considerations that the two-dimensional assumption is valid in major part of the domain due to a) cylindrical symmetry of the discharge chamber setup, b) tangential injection of water through the holes along the circumference, and c) the flexible distance between the nozzle exit and anode.

2. The assumption of laminar flow is based on experiments, showing the laminar structure of the plasma flowing out of the discharge chamber in the space between the nozzle exit and the anode. The laminar flow has been observed for currents up to 600 A. It comes out from our previous calculation that Reynolds number based on the outlet diameter 6 mm reaches in the axial region 13 000 at maximum and decreases to 300 in arc fringes. The type of flow inside the discharge chamber is questionable since no diagnostics is able to see inside the chamber and it is not clear if the laminar plasma stream is a result of laminarization of the plasma flow at the outlet. To check possible deviations from the laminar model, we have employed Large Eddy Simulation (LES) with the Smagorinsky sub-grid scale model. It was proved that simulations for laminar and turbulent regimes give nearly the same results, so that the plasma flow can be considered more or less laminar for the operating conditions and simplified discharge geometry in the present study. The maximum detected discrepancy between the turbulent and laminar models is 7 % for the relative temperature difference at the arc axis 2 mm downstream of the nozzle exit for 500 A and 40 slm of argon. For reasons of generality, all the results presented here were calculated using the LES turbulent model.
3. The assumption of a complete (uniform) mixing is a simplification of a reality since, based on experiments, argon and water species do not mix homogeneously in the hybrid torch, especially at lower currents. This assumption was discusssed in more detail in (Jeništa, 2004) and it was concluded that this assumption can underestimate temperature and velocity in the axial discharge region to some extent.

The complete set of conservation equations representing the mass, electric charge, momentum and energy transport of such plasma can be written in the vector notation as follows:
continuity equation:

$$\frac{\partial \rho}{\partial t}+\frac{1}{r}\frac{\partial}{\partial r}(\rho v r)+\frac{\partial}{\partial z}(\rho u)=0 \tag{1}$$

momentum equations:

$$\begin{gathered}\frac{\partial}{\partial t}(\rho u)+\frac{\partial}{\partial r}(\rho u v)+\frac{\partial}{\partial z}(\rho u u)=-\frac{\partial p}{\partial z}+j_r B_\theta-\frac{2}{3}\frac{\partial}{\partial z}\left[\mu\left(\frac{1}{r}\frac{\partial}{\partial r}(rv)+\frac{\partial u}{\partial z}\right)\right]+\\ \frac{\partial}{\partial z}\left(2\mu\frac{\partial u}{\partial z}\right)+\frac{1}{r}\frac{\partial}{\partial r}\left[r\mu\left(\frac{\partial u}{\partial r}+\frac{\partial v}{\partial z}\right)\right]\end{gathered} \tag{2}$$

$$\begin{gathered}\frac{\partial}{\partial t}(\rho v)+\frac{\partial}{\partial r}(\rho v v)+\frac{\partial}{\partial z}(\rho u v)=-\frac{\partial p}{\partial r}-j_x B_\theta-\frac{2}{3}\frac{\partial}{\partial r}\left[\mu\left(\frac{1}{r}\frac{\partial}{\partial r}(rv)+\frac{\partial u}{\partial z}\right)\right]+\frac{\rho w^2}{r}+\\ \frac{1}{r}\frac{\partial}{\partial r}\left(2\mu r\frac{\partial v}{\partial r}\right)-\frac{2\mu v}{r^2}+\frac{\partial}{\partial z}\left[\mu\left(\frac{\partial u}{\partial r}+\frac{\partial v}{\partial z}\right)\right]\end{gathered} \tag{3}$$

energy equation:

$$\begin{gathered}\frac{\partial e}{\partial t}+\frac{1}{r}\frac{\partial}{\partial r}\left[r\left\{(e+p)v-\lambda\frac{\partial T}{\partial r}\right\}\right]+\frac{\partial}{\partial z}\left[(e+p)u-\lambda\frac{\partial T}{\partial z}\right]=\\ \frac{1}{r}\frac{\partial}{\partial r}\left[r\left(\tau_{rr}v+\tau_{rz}u\right)\right]+\frac{\partial}{\partial z}\left(\tau_{rz}v+\tau_{zz}u\right)+j_r E_r+j_z E_z-R\end{gathered} \tag{4}$$

charge continuity equation:

$$\frac{1}{r}\frac{\partial}{\partial r}\left(r\sigma\frac{\partial\Phi}{\partial r}\right)+\frac{\partial}{\partial z}\left(\sigma\frac{\partial\Phi}{\partial z}\right)=0 \tag{5}$$

equation of state:

$$p=\rho R_g T\,. \tag{6}$$

Here z and r are the axial and radial coordinates, u, v and w are the axial, radial and tangential components of the velocity respectively, ρ is the mass density, μ is the viscosity (in the case of LES model, the turbulent contribution μ_{turb} is also added) p is the pressure, B_Θ is the magnetic field strength, e is the density of energy produced or dissipated in the unit volume (internal and kinetic), T is the temperature, τ is the viscous stress tensor, j_z and j_r are the axial and radial components of the current density, E_z and E_r are the axial and radial components of the electric field strength, Φ is the electric potential, R is the source term for the radiation losses and R_g is the molar gas constant. The magnetic field strength B_Θ is calculated from the Biot-Savart law, the current density $\vec{j}$ from the Ohm's law $\vec{j}=\sigma\cdot\vec{E}$.

2.2 Properties of argon-water plasma mixture

The water–argon mixture can be described by the formula $(H_2O)_{(1-q)}Ar_q$ where the argon molar amounts q were chosen from 0 to 1 with the step of 0.1. The total number of 35 chemical species was considered (Křenek, 2008). For the temperature range 400 – 20 000 K we supposed the following decomposition products: e (electrons), H, O, Ar, O^+, O^{2+}, O^{3+}, O^-, O_2, O_2^+, O_2^-, O_3, H^+, H^-, H_2, H_2^+, H_3^+, OH, OH^+, OH^-, HO_2, HO_2^-, H_2O, H_2O^+, H_3O^+, H_2O_2, Ar^+, Ar^{2+}, Ar^{3+}. For the temperature range 20 – 50 000 K the set of products is somewhat different, including also multiply charged ions: e (electrons), H, O, Ar, O^+, O^{2+}, O^{3+}, O^{4+}, O^{5+}, O^{6+}, H^+, Ar^+, Ar^{2+}, Ar^{3+}, Ar^{4+}, Ar^{5+}, Ar^{6+}. The calculations were performed using the modified Newton method for the solution of nonlinear equations system which is composed of equations of Saha and mass action law type expressing individual complex components by the help of basic ones (e,H,O,Ar). The system is completed by the usual particle and charge balance assuming quasineutrality and equilibrium.

The thermodynamic properties and the transport coefficients of this gas mixture were calculated according to the Chapman–Enskog method in the 4th approximation described e.g. in (Křenek & Něnička, 1984) for temperatures 400–50 000 K (Křenek, 2008) and pressures 0.1-0.3 MPa in the local thermodynamic equilibrium.

Two radiation models are implemented in the energy equation for energy losses from the argon-water plasma: 1) the net emission coefficients for the required arc radius of 3.3 mm, and 2) the partial characteristics method, both of them for different molar fractions of argon and water plasma species in dependence on temperature and pressure. Continuous radiation due to photorecombination and "bremsstrahlung" processes has been included in the calculation as well as discrete radiation consisting of thousands of spectral lines. Broadening mechanisms of atomic and ionic spectral lines due to Doppler, resonance and Stark effects have been considered. The numbers of oxygen and argon lines included in the calculation are O (93 lines), O^+ (296), O_2^+ (190), Ar (739), Ar^+ (2781), Ar^{2+} (403), Ar^{3+} (73). In addition, molecular bands of O_2 (Schuman-Runge system), H_2 (Lyman and Verner systems), OH (transition $A^2\Sigma^+ \rightarrow X^2\Pi_i$) and H_2O (several transitions) have been also implemented (Bartlová & Aubrecht, 2006). Absorption coefficient as a function of wavelength has been calculated from infrared to far ultraviolet regions and the tables of partial characteristics for 1 000 - 35 000 K. The net emission coefficients model used here is a special case of the partial characteristics model with zero partial sink, $\Delta Sim = 0$.

2.3 Boundary conditions and numerical scheme

The calculation region and the corresponding boundary conditions are presented in Fig. 4. The dimensions are 3.3 mm for the radius of the discharge region, 20 mm for the radius of the outlet region and 78.32 mm for the total length. These dimensions agree with the hybrid torch experimental setup.

a. *Inlet boundary (AB)* is represented by the nozzle exit for argon. Along this boundary we assume the zero radial velocity component, $v = 0$. Because of the lack of experimental data, the temperature profile $T(r, z = 0)$ and the electric field strength $E_z = -\partial\Phi/\partial z = const.$ for a given current are calculated at this boundary, before the start of the fluid-dynamic calculation itself, iteratively from the Elenbaas-Heller equation including the radiation losses from the arc (our previous numerical experiments proved a weak dependence of the form of the boundary temperature profile on the overall solution). The inlet velocity profile $u(r)$ for argon plasma for the obtained temperature profile $T(r, z = 0)$ is pre-calculated from the axial momentum equation under the assumption of fully developed flow.
b. *Axis of symmetry (BC).* The zero radial velocity and symmetry conditions for the temperature, axial velocity and electric potential are specified here, i.e. $\partial T/\partial r = \partial u/\partial r = \partial\Phi/\partial r = 0$, $v = 0$.
c. *Arc gas outlet plane (CD).* The zero electric potential $\Phi = 0$ (the reference value) and zero axial derivatives of the temperature and radial velocity are defined at CD, $\partial T/\partial z = \partial v/\partial z = 0$. Values of the axial velocity are interpolated from the inner grid points.
d. *Arc gas outlet plane (DE).* The zero radial velocity and zero radial derivatives of the temperature, axial velocity and electric potential are defined here, $\partial T/\partial r = \partial u/\partial r = \partial\Phi/\partial r = 0$, $v = 0$. Pressure is fixed at 1 atmosphere, p = 1 atm.
e. *Outlet wall and the nozzle (EF).* We specify no slip conditions for velocities, $u = v = 0$, constant values of E_r and E_z ($\partial\Phi/\partial z = \partial\Phi/\partial r = 0$) and $T(r, z) = 773\,K$ ($500°\,C$) for the temperature of the nozzle.
f. *Water vapor boundary (FA).* Along this line we specify the so-called "effective water vapor boundary", named in Fig. 4 as the "water vapor boundary" with a prescribed temperature of water vapor $T(R = 3.3mm, z) = 773$ K. This is a numerical simplification of a more complex physical reality assumed near the phase transition water-vapor in the discharge chamber. The shape of the phase transition between water and vapor in the discharge chamber is not experimentally known and it is unclear so far if the structure of the transition is simple or very complicated, for example, with a time-dependent form. Various irregularities in the transition such as

splitting of the phase transition or water drops in the vapor phase can increase complexity of the transition. In (Jeništa, 2003a) the iteration procedure for determination of the mass flow rate and the radius of the "water vapor boundary" for each current was proposed, based on comparison with available experimental temperature and velocity at the outlet and the electric potential drop in the chamber. It was concluded that the best fit between experiment and numerical simulation for all currents exists for a mean arc radius of 3.3 mm. The corresponding values of water mass flow rates are 0.228 g s^{-1} (300 A), 0.286 g s^{-1} (350 A), 0.315 g s^{-1} (400 A), 0.329 g s^{-1} (500 A), 0.363 g s^{-1} (600 A). The magnitude of the radial inflow velocity is calculated from the definition of mass flow rate

$$v(R)=\frac{\dot{m}}{2\pi R\sum_{\Delta z}\rho(R,z)\Delta z},$$

where $\rho(R,z)$ is a function of pressure and thus dependent on the axial position z, Δz is the distance between the neighboring grid points.

Because of practically zero current density in cold vapor region (no current goes outside of the lateral domain edges), the radial component of the electric field strength is put zero, i.e., $E_r = 0$. The axial velocity component is set to zero, $u = 0$. Since we do not solve here the equation for tangential velocity component w , distribution of w in the discharge for the presented results was taken from our previous calculations (Jeništa et al., 1999a) solved by the SIMPLER iteration procedure (Patankar, 1980) which enables calculation of w for axisymmetric case, i.e., w as a function of z and r coordinates. The w velocity acts here only through the centrifugal force $\rho w^2/r$ in the radial momentum equation (3).

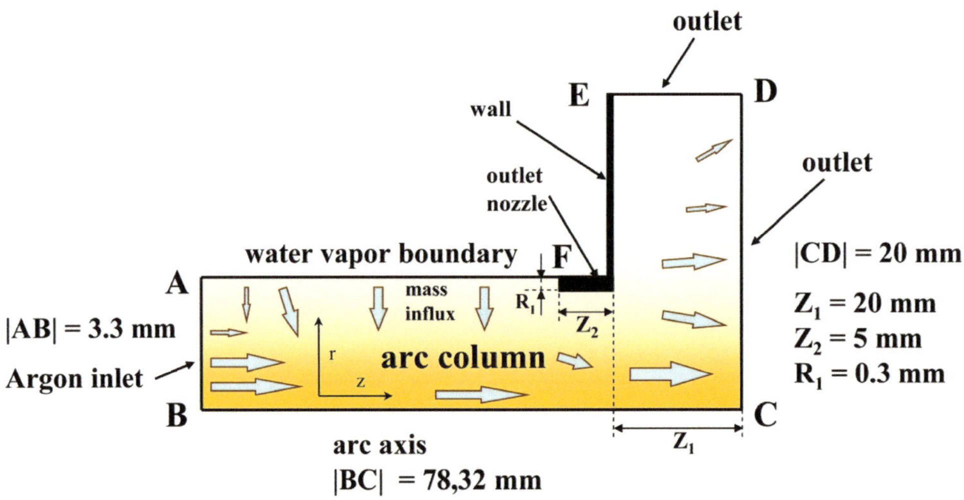

Fig. 4. Discharge area geometry. Inlet boundary (AB) is represented by the nozzle exit for argon. The length of the discharge region is 58.32 mm, the length of the outlet is 20 mm. Along the line FE we specify the outlet nozzle and the wall of the hybrid plasma torch equipment.

For time integration of (1)-(4), LU-SGS (Lower-Upper Symmetric Gauss-Seidel) algorithm (Jameson & Yoon, 1987; Yoon & Jameson, 1988), coupled with Newtonian iteration method are used for the integration of discretized equations in time and space. To resolve compressible phenomena accurately, the Roe flux differential method (Roe, 1981) coupled with the third-order MUSCL-type (Monotone Upstream-centered Schemes for Conservation Laws) TVD (Total Variation Diminishing) scheme (van Leer, 1979) are used for convective term. The electric potential from (5) is solved in a separate subroutine by the TDMA (Tri-Diagonal Matrix Algorithm) line-by-line method. From (1-4) we obtain ρ , $\rho\vec{u}$, e and Φ . Pressure is determined from the pressure dependence of the internal energy $U(p,T)=e(p,T)-0.5\rho|\vec{u}|^2$ and temperature is calculated from the equation of state (6) $p/\rho = R_g(p,T)\cdot T$, using the pre-calculated values of the product $R_g(p,T)\cdot T$ as a function of temperature, pressure and argon molar fraction in the mixture (Křenek, 2008).
The computer program is written in the FORTRAN language. The task has been solved on an oblique structured grid with nonequidistant spacing. The total number of grid points was 38 553, with 543 and 71 points in the axial and radial directions respectively.

3. Results of calculation

3.1 Thermal, fluid flow and electrical characteristics of the plasma

Calculations have been carried out for the currents 300, 400, 500 and 600 A. Mass flow rate for water-stabilized section of the discharge was taken for each current between 300 and 600 A from our previously published work (Jeništa, 2003a; Jeništa, 2003b), where it was determined iteratively as a minimum difference between numerical and experimental outlet quantities. The resulting values are 0.228 $g\cdot s^{-1}$ (300 A), 0.315 $g\cdot s^{-1}$ (400 A), 0.329 $g\cdot s^{-1}$ (500 A), 0.363 $g\cdot s^{-1}$ (600 A). Argon mass flow rate was varied in agreement with experiments in the interval from 22.5 slm to 40 slm, namely 22.5, 27.5, 32.5 and 40 slm. It was proved in experiments (Kavka et al., 2007) that part of argon is taken away before it reaches the torch exit because argon is mixed with vapor steam and removed to the water system of the torch. The amount of argon transferred in such a way from the discharge is at least 50 % for currents studied. Since the present model does not treat argon and water as separate gases and the mechanism of argon removal is not included in the model, we consider in the calculations that argon mass flow rate present in the discharge equals one-half of argon mass flow rate at the torch inlet. A relatively high values of argon mass flow rate, used also in experiment, were chosen here to demonstrate compressible phenomena.
Fig. 5 shows velocity, temperature, pressure and the Mach number in the outlet nozzle and near-outlet regions for 600 A, water mass flow rate of 0.363 $g\ s^{-1}$ and 40 slm of argon. The partial characteristics method for radiation losses is employed. The results shown here demonstrate the largest magnitude fluctuations of velocity, temperature, pressure and the Mach number just after the jet exhausts from the torch nozzle among all the studied currents and argon mass flow rates. Supersonic flow structure in the near-outlet region is obvious with clearly distinguished shock diamonds with the maximum Mach number about 1.6 with 10 500 $m\ s^{-1}$. The corresponding velocity and the Mach number maxima overlap with the temperature and pressure minima and vice versa. Since the pressure decreases at the torch exit to a nearly atmospheric pressure, the computed contours correspond to an under-expanded atmospheric-pressure plasma jet.

The corresponding axial profiles of the Mach number, pressure, temperature and velocity along the arc axis downstream from the nozzle orifice (the axial position 58.32 mm) for the same run are presented in Fig. 6. Several successive wave crests and troughs along the axis for each of the physical parameters is a typical feature of supersonic fluid flow. The fluctuation of presented quantities is between 1.1-1.7 for the Mach number, 0.7-1.4 atm. for the pressure, 7 200-10 000 m · s^{-1} for the velocity and 18 000-23 500 K for the temperature.

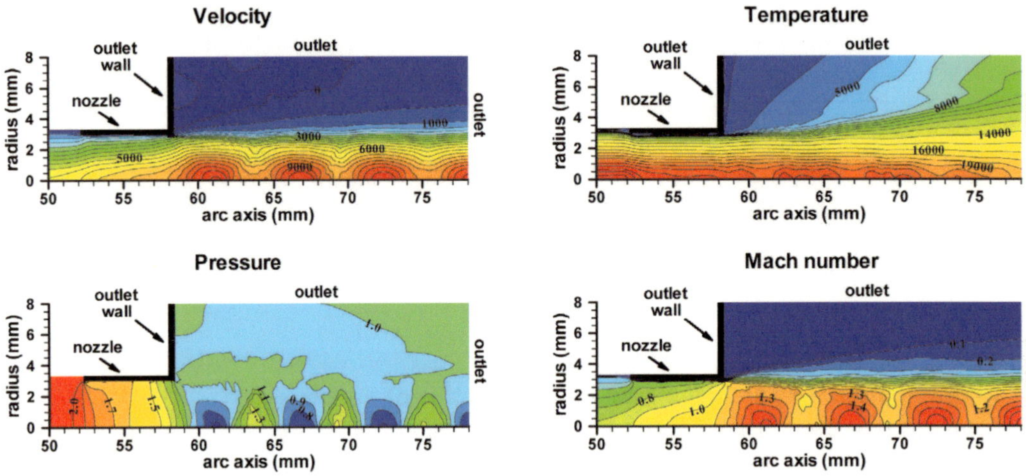

Fig. 5. Velocity, temperature, pressure and the Mach number contours in the outlet nozzle and near-discharge regions for the 600 A arc discharge. Water mass flow rate is 0.363 g s^{-1}, argon mass flow rate is 40 slm (standard liters per minute). Partial characteristics radiation model is employed. Supersonic flow structure is obvious with clearly distinguished shock diamonds. The maximum Mach number reaches 1.6. Contour increments are 500 m s^{-1} for velocity, 1 000 K for temperature, 0.1 atm for pressure and 0.1 for the Mach number.

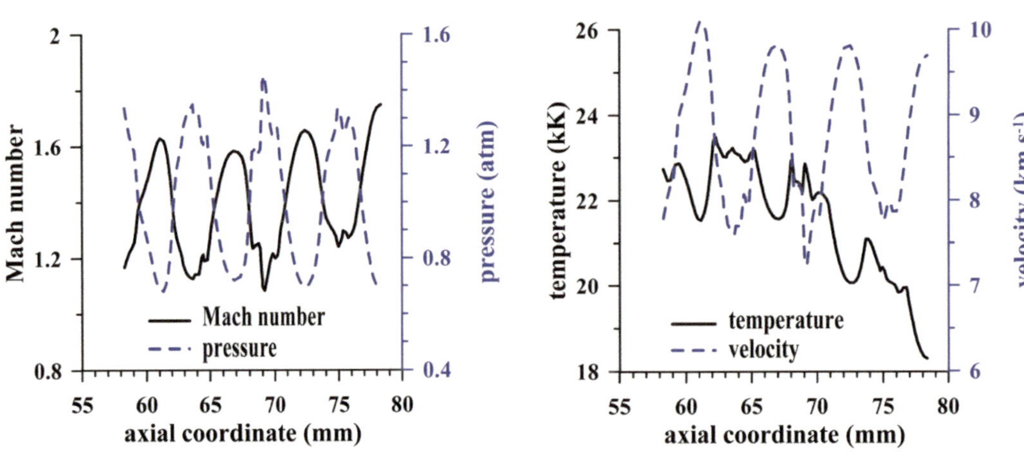

Fig. 6. Profiles of the Mach number, pressure, temperature and velocity along the arc axis from the nozzle orifice. Supersonic outlet with distinguished shock diamonds. 600 A, argon mass flow rate 40 slm, partial characteristics radiation model.

Fig. 7 displays temperature and velocity fields in the discharge and the near-outlet regions for a) 500 and b) 600 A with water mass flow rates of 0.329 $g \cdot s^{-1}$ (500 A), 0.363 $g \cdot s^{-1}$ (600 A) (Jeništa, 2003a) and argon mass flow rate of 0.554 $g \cdot s^{-1}$ (one half of 40 slm). The net emission coefficients radiation model is employed. Orientation of the calculation domain is the same as in Figs. 1, 4. Since the ratio of the axial to the radial dimensions of the calcualtion domain is ~ 24 the scaling of the radial and axial coordinates is not proportional to make the contours inside the discharge region clearly visible. Argon flows axially into the domain, whereas water evaporates in the radial direction from the "water vapor boundary". Both the results for 500 and 600 A exhibit supersonic under-expanded plasma flow regime but a progression from weak to highly-pronounced shock diamonds structure at 600 A is obvious. The maximum velocities are 7 200 $m \cdot s^{-1}$ (500 A) and 9 400 $m \cdot s^{-1}$ (600 A) near the axial position of 60 mm. Further downstream the velocity amplitudes decrease due to viscosity dissipation and due to the reduction of the difference between the jet static pressure and back pressure.

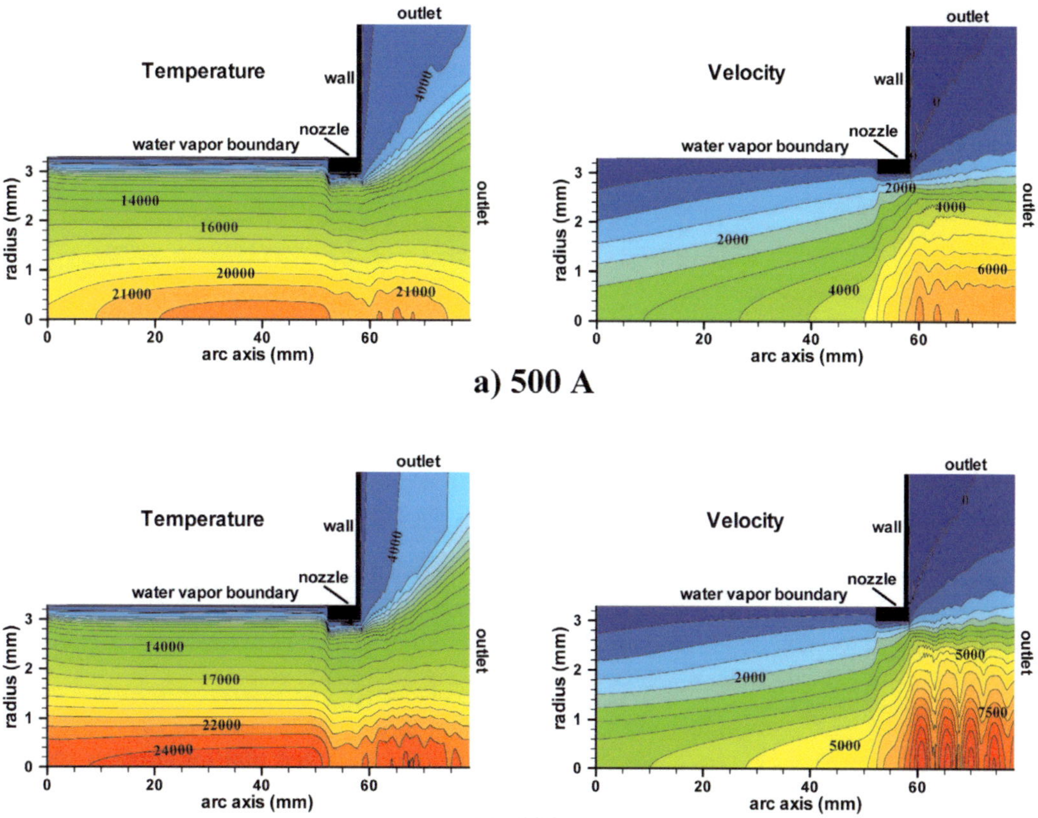

Fig. 7. Temperature and velocity contours for a) 500 A and b) 600 A arcs, net emission coefficients model. Water mass flow rates are 0.329 $g \cdot s^{-1}$ (500 A) and 0.363 $g \cdot s^{-1}$ (600 A); argon mass flow rate is 40 slm for both currents. Progression of a supersonic flow structure at the outlet is clearly visible. Contour increments are 1 000 K for temperature and 500 $m \cdot s^{-1}$ for velocity.

The impact of reabsorption of radiation on the distribution of temperature and velocity within the discharge and the near-outlet regions for 600 A is obvious in Fig. 8. The partial characteristics model gives lower temperatures and higher velocities at the outlet region. Similar results have been proved for all currents and argon mass flow rates.

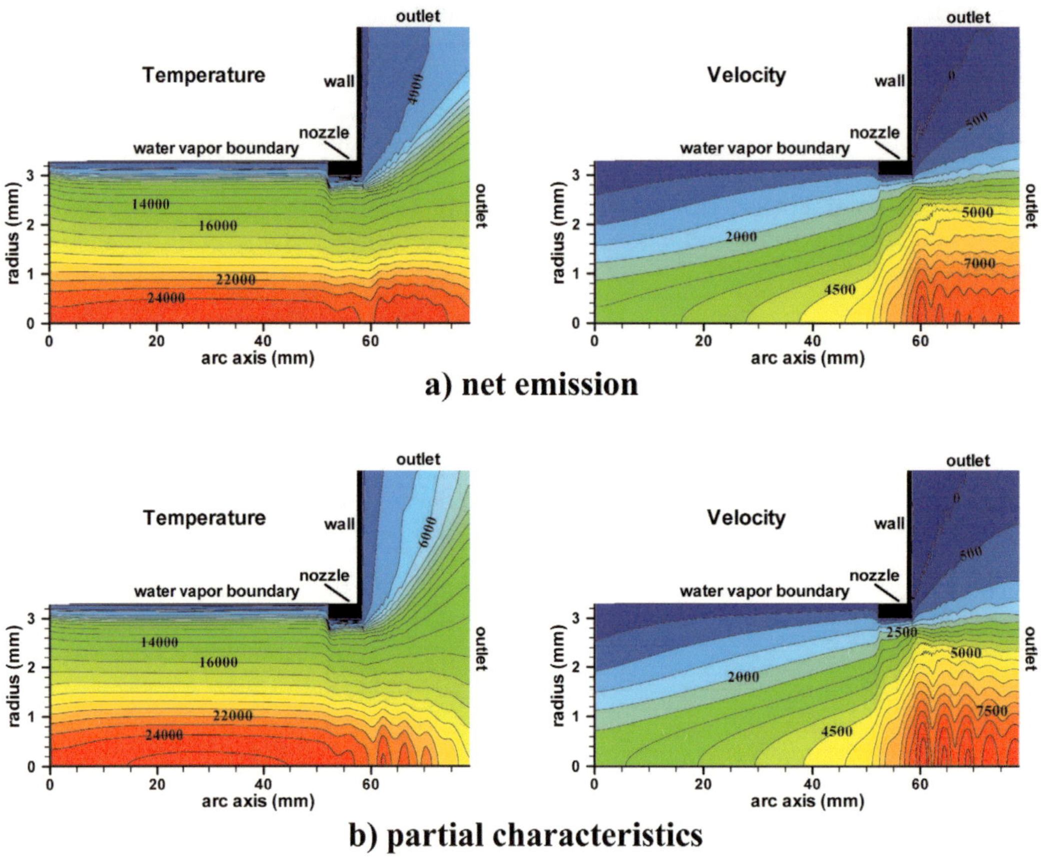

Fig. 8. Temperature and velocity contours for the net emission a) and partial characteristics b) models, 600 A, 27 slm of argon, water mass flow rate 0.363 $g \cdot s^{-1}$. The partial characteristics model gives lower temperatures and higher velocities at the outlet region. Contour increments are 1 000 K for temperature and 500 $m \cdot s^{-1}$ for velocity.

Fig. 9 presents the radial profiles of the Mach number 2 mm downstream of the nozzle exit with the argon mass flow rate of 40 slm. It is clearly demonstrated that for currents higher than 400 A, a supersonic rare plasma in the central parts of the discharge is surrounded by a subsonic, much denser but still hot plasma. Due to generally higher velocities and lower temperatures near the outlet the partial characteristics model gives higher values of the Mach number; the difference regarding the net emission coefficients model is below 0.1 at the arc axis.

Different flow structures for currents between 300 and 600 A and 32 slm of argon are visible in Fig. 10. Subsonic plasma flow at 300 A (Mach ~ 0.8) converts to transonic flow at 400 A (Mach ~ 1) at the outlet. The onset of supersonic flow structure formation is visible at 500 A

(Mach ~ 1.15) and the fully developed supersonic flow with shock diamonds is formed for 600 A (Mach ~ 1.4).

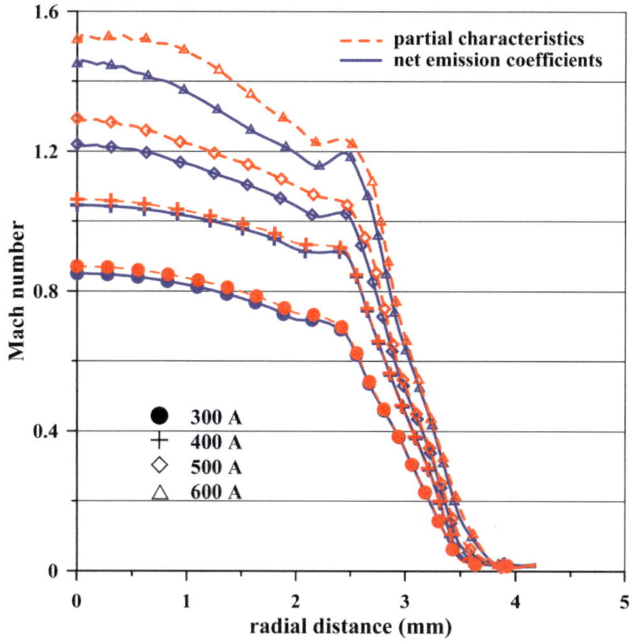

Fig. 9. Radial profiles of the Mach number 2 mm downstream of the nozzle exit, argon mass flow rate is 40 slm. The Mach numbers in the axial region are higher than 1 for currents equal to or higher than 400 A. Somewhat higher values of the Mach number provides the partial characteristics model.

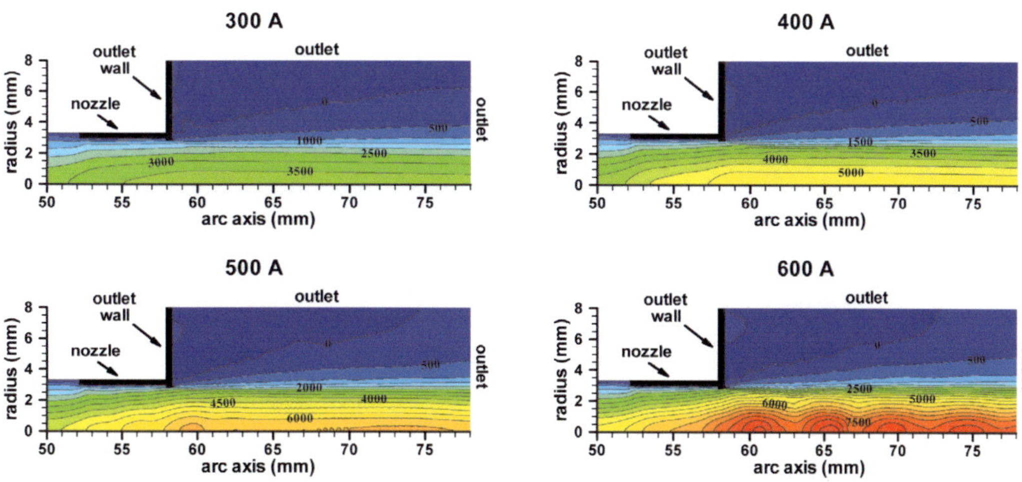

Fig. 10. Velocity contours in the outlet nozzle and near-discharge regions for 32 slm of argon. Partial characteristics radiation model is employed. Water mass flow rates are 0.228 $g \cdot s^{-1}$ (300 A), 0.315 $g \cdot s^{-1}$ (400 A), 0.329 $g \cdot s^{-1}$ (500 A), 0.363 g s^{-1} (600 A), contour increments are 500 m s^{-1}. Supersonic flow structure is obvious for 600 A.

Values of temperature, voltage drop, the Mach number and overpressure as functions of arc current (300-600 A) and argon mass flow rate (22.5-40 slm) for the partial characteristics model are displayed in Fig. 11. Temperature and the Mach number shown here are taken at the arc axis 2 mm downstream of the outlet nozzle, the voltage drop and overpressure refer to the length of the discharge region of the hybrid-stabilized electric arc (58.32 mm). The maximum

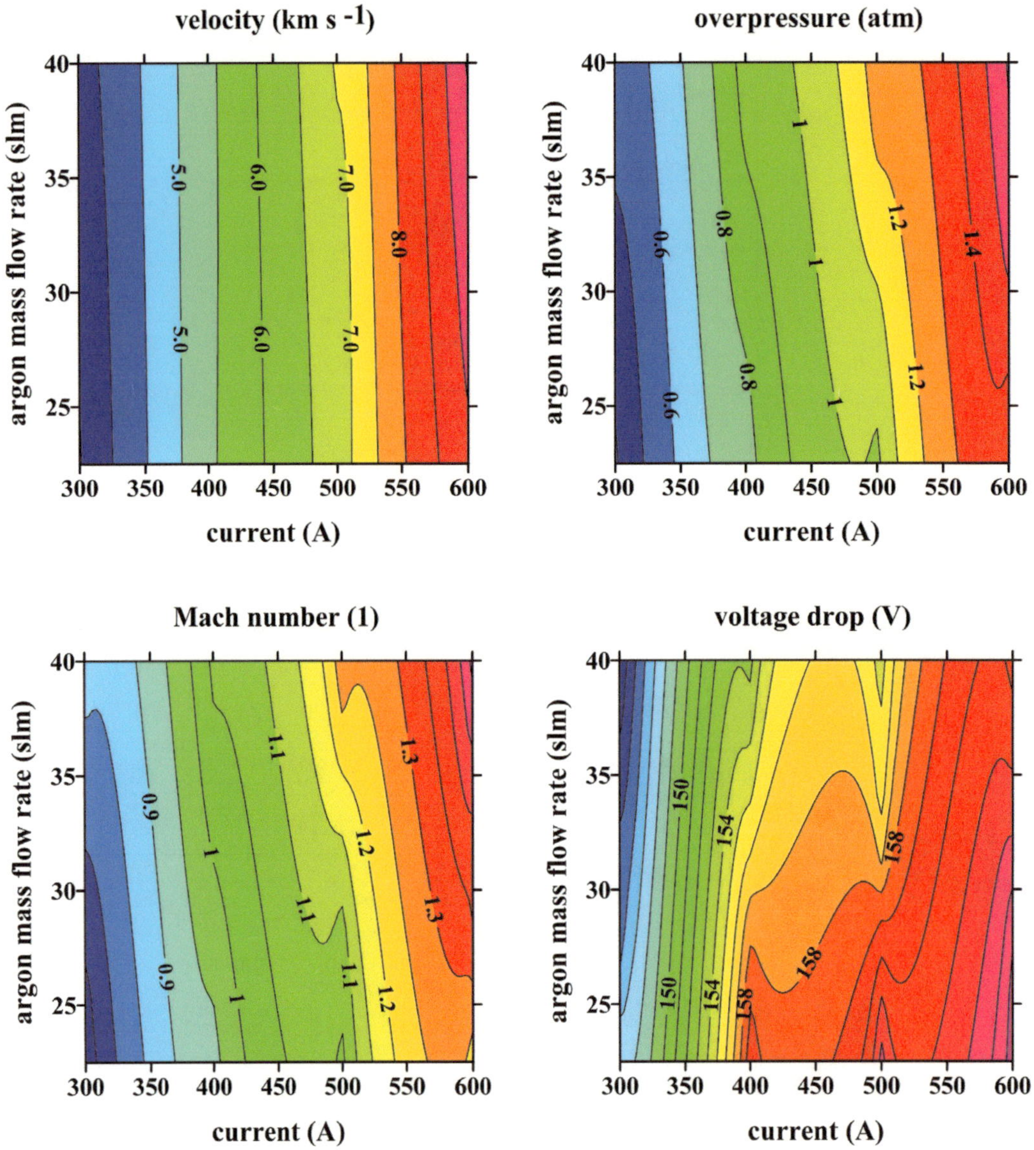

Fig. 11. Velocity (km s^{-1}), overpressure (atm), the Mach number and voltage drop (V) as functions of arc current (300-600 A) and argon mass flow rate (22.5-40 slm). Temperature and the Mach number are taken at the arc axis 2 mm downstream of the outlet nozzle, voltage drop and overpressure refer to the length of the discharge region of the hybrid-stabilized electric arc. Partial characteristics radiation model is employed.

values of the velocity (9 500 m s-1), overpressure (1.6 atm) and the Mach number (1.5) occur for 600 A and 40 slm of argon. It is evident from the slope of contours that overpressure and the Mach number increases with argon mass flow rate while the velocity increases only slightly. The transition from subsonic to supersonic flow occurs for currents around 400 A. The electric potential drop in the discharge chamber decreases with increasing argon mass flow rate, the maximum value reaches 164 V.

Temperature contours for the net emission and partial characteristics models are shown in Fig. 12. Temperature depicted is taken again at the arc axis 2 mm downstream of the outlet nozzle. Reabsorption of radiation increases temperature in arc fringes and decreases it near the arc axis, in the result the net emission model provides higher axial temperatures.

The slight increase of velocity with argon mass flow rate has thus an apparent explanation: on one hand, the increase of argon mass flow rate implies the corresponding increase of velocity. On the other hand, temperature decreases with argon mass flow rate, lowering thus the increase of plasma velocity.

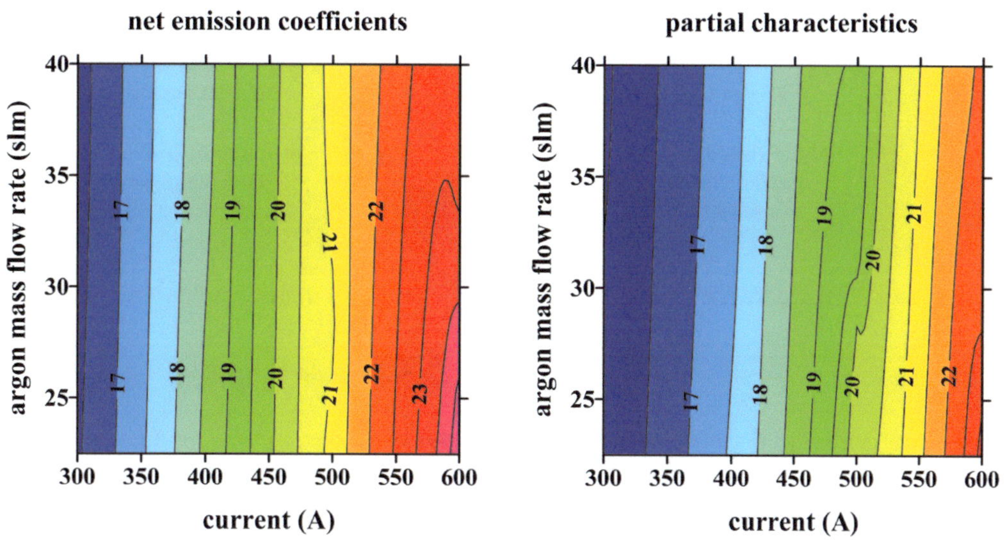

Fig. 12. Temperature contours (kK) in dependence of arc current (300-600 A) and argon mass flow rate (22.5-40 slm) for the net emission and partial characteristics models. Temperature depicted is taken at the arc axis 2 mm downstream of the outlet nozzle. Reabsorption of radiation slightly decreases axial temperatures.

3.2 Comparison of radial temperature and velocity profiles with experiments

A number of experiments have been carried out on the hybrid-stabilized electric arc in recent past for the currents 300-600 A with 22-40 slm of argon. Temperature is one of the fundamental plasma parameters, needed also for evaluation of the other quantities.

In experiment, the radial profiles of temperature at the nozzle exit were calculated from optical emission spectroscopy measurements. The procedure is based on the ratio of emission coefficients of hydrogen H_β line and four argon ionic lines using calculated LTE composition of the plasma for various argon mole fractions as a function of temperature (Křenek, 2008). From the calculated molar fractions of hydrogen and argon it is easy to obtain emission coefficients of H_β and argon lines. The temperature corresponding to an experimental ratio of emission coefficients is then found by cubic spline interpolation on the theoretical data.

Fig. 13 compares measured and calculated temperature profiles 2 mm downstream of the nozzle exit for 300-600 A and 22.5 slm of argon. Excellent agreement is demonstrated for 300 and 600 A where the measured profiles nearly coincide with the two profiles calculated using the net emission and partial characteristics radiation methods. For 400 and 500 A agreement is better for the profiles calculated by the net emission coefficients (black color).

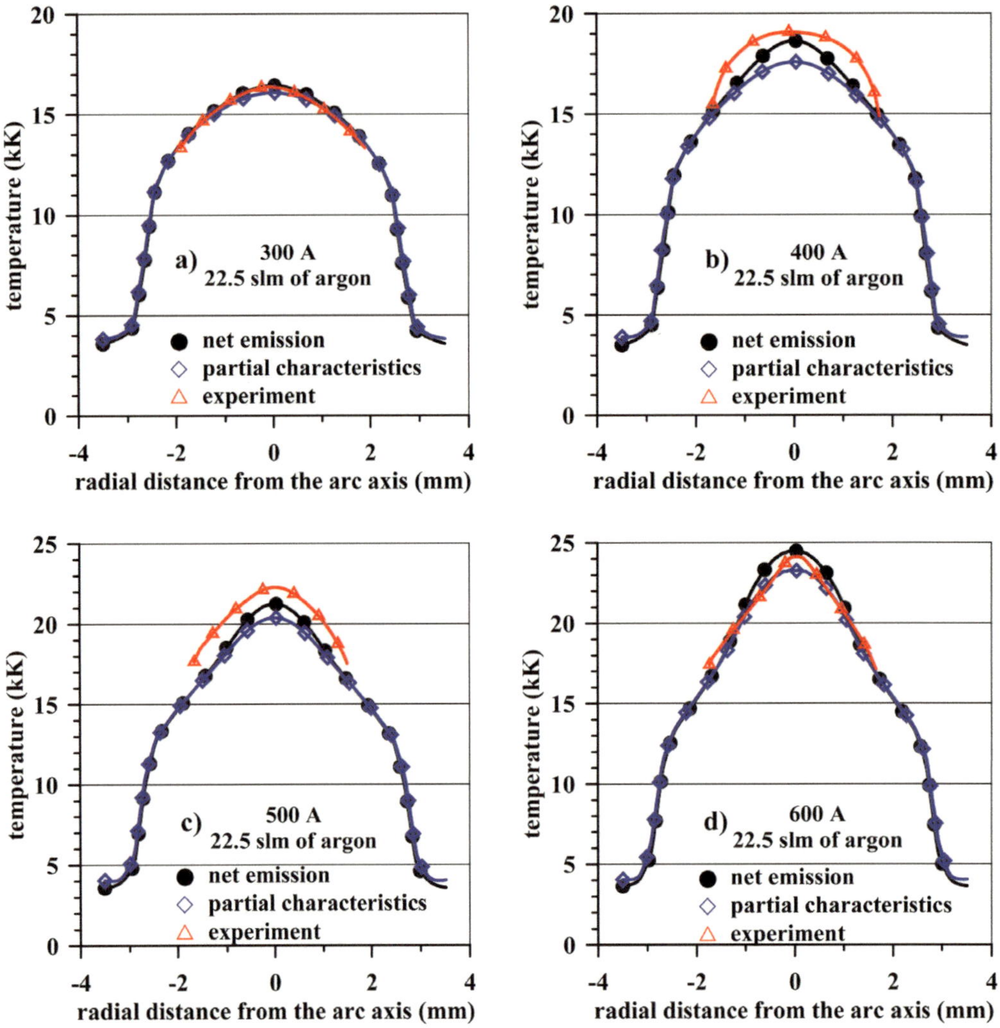

Fig. 13. Experimental and calculated radial temperature profiles 2 mm downstream of the nozzle exit for 300-600 A and 22.5 slm of argon.

Relative difference between the calculated and experimental values of temperature at the arc axis 2 mm downstream of the nozzle exit has been evaluated for broad range of currents and argon mass flow rates. The relative difference is defined here as

$$\Delta = 100 \cdot abs\left(T_{num} - T_{exp}\right) / T_{num} ,$$

where T_{num} $\left(T_{exp}\right)$ are the values of the calculated (experimental) temperature. It was proved that the maximum relative difference between the calculated and experimental temperature profiles is lower than 10% for the partial characteristics and 5% for the net emission radiation model used in the present calculation, i.e. the net emission radiation model gives better agreement with experiment as regards axial temperatures.

Comparison of the measured and calculated temperature profiles with our former calculations (Jeništa et al., 2010) is shown in Fig. 14 for 500 and 600 A. The set of profiles is calculated/measured again 2 mm downstream of the nozzle exit. The term "new model" introduced here refers to the present model with the assumptions described in Secs. 2.1, 2.2, while the "old model" means the former one with the following assumptions:

a. the transport and thermodynamic properties of the argon-water plasma mixture are calculated using linear mixing rules for non-reacting gases based either on mole or mass fractions of argon and water species (Jeništa et al., 2010),
b. all the transport and thermodynamic properties as well as the radiation losses are dependent on temperature, and argon molar content but NOT dependent on pressure,
c. radiation transitions of H_2O molecule are omitted.

In our present model 1) all the transport and thermodynamic properties are calculated according to the Chapman-Enskog method in the 4th approximation; 2) all the properties are dependent on pressure; 3) radiation transitions of H_2O molecule are considered. It is obvious that radial temperature profiles obtained by our "old model" give worse comparison with experiments - higher temperatures and flatter profiles compared to our present calculation. Similar results were obtained also for the net emission model. Improvements in the properties caused better convergence between the experiment and calculation.

More comprehensive view on the closeness of the calculated and experimental temperature profiles offers Fig. 15. The dots in the plot represent the so-called "average relative difference of temperature" defined as

$$\Delta_{av}^{T} = \frac{100}{N} \cdot \sum_{i=1}^{N} abs\left(T_{num}^{i} - T_{exp}^{i}\right) / T_{num}^{i} ,$$

estimating a sort of average relative difference along the temperature profile, N is the number of available coincident numerical T_{num}^{i} and experimental T_{exp}^{i} values of temperature along the radius. It is apparent that our present "new model" gives better comparison than the "old model" in all cases.

Besides temperature profiles, velocity profiles at the nozzle exit and mass and momentum fluxes through the torch nozzle are important indicators for characterization of the plasma torch performance. In experiment, velocity at the nozzle exit is being determined from the measured temperature profile and power balance assuming local thermodynamic euilibrium (Kavka et al., 2008). First, the Mach number M is obtained from the simplified energy equation integrated through the discharge volume (Jeništa, 1999b); second, the velocity profile is derived from the measured temperature profile using the definition of the Mach number

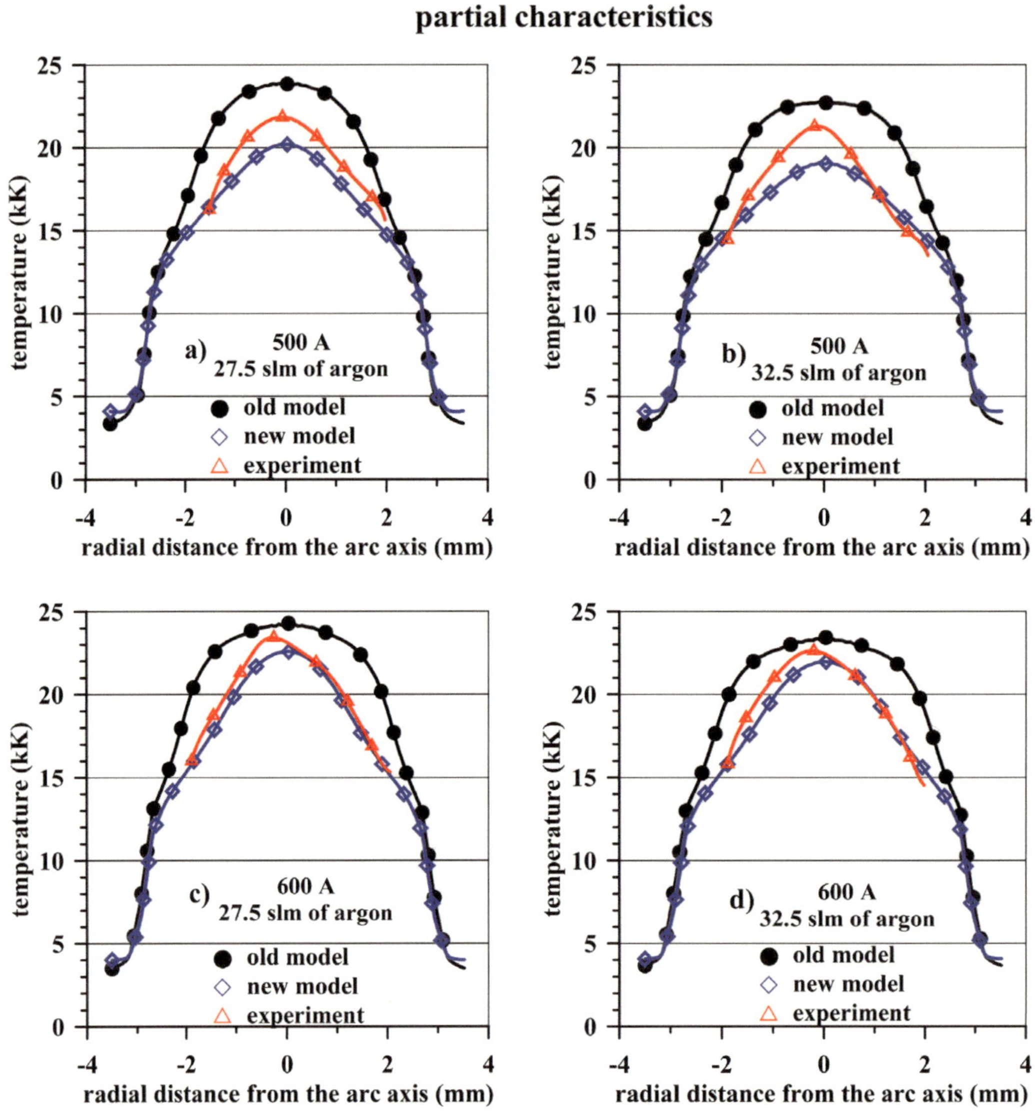

Fig. 14. Experimental and calculated radial temperature profiles 2 mm downstream of the nozzle exit for 500 and 600 A with 27 and 32 slm of argon, partial characteristics method. The so-called „new model" stands for the present model, the „old model" presents our previous model with simplified plasma properties (see the text).

$$u(r) = M \cdot c\{T(r)\},$$

where $c\{T(r)\}$ is the sonic velocity for the experimental temperature profile estimated from the T&TWinner code (Pateyron, 2009). The drawback of this method is the assumption of the constant Mach number over the nozzle radius. Nevertheless the existence of supersonic regime (i.e., the mean value of the Mach number over the nozzle exit higher than 1) using this method was proved for 500 A and 40 slm of argon, as well as for 600 A for argon mass flow rates higher than 27.5 slm. Similar results have been also reported in our previous work (Jeništa et al., 2008).

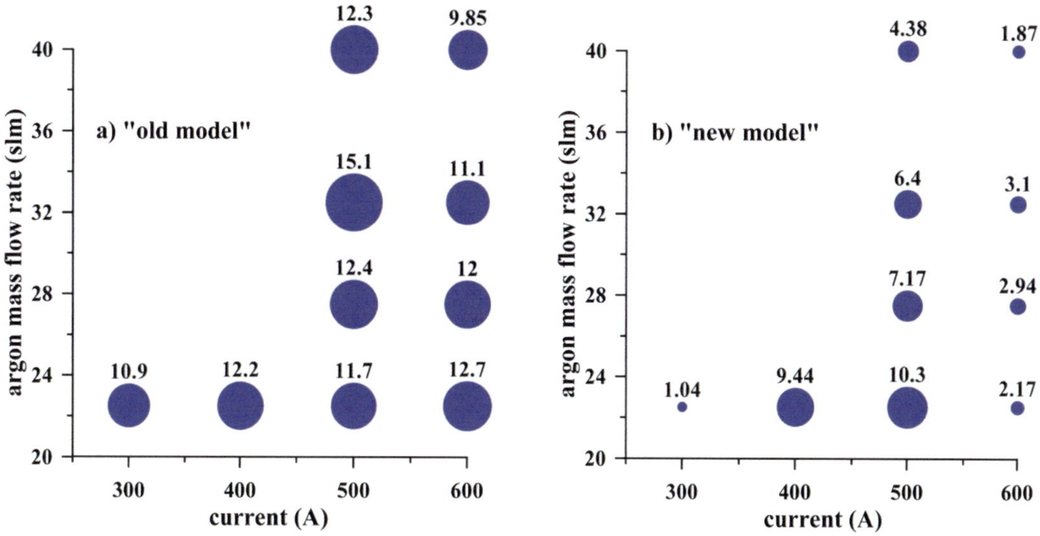

Fig. 15. Average relative difference (see the text) between the calculated and experimental radial temperature profiles, shown in %, at the axial position 2 mm downstream of the nozzle exit, partial characteristics. The so-called „new model" stands for the present model, exhibiting better agreement with experiments; the „old model" presents our previous model with simplified plasma properties (see the text).

For more exact evaluation of velocity profiles we employed the so-called "integrated approach", i.e., exploitation of both experimental and numerical results: velocity profiles are determined as a product of the Mach number profiles obtained from the present numerical simulation and the sonic velocity based on the experimental temperature profiles. The results for 300-600 A with 22.5 slm of argon for the partial characteristics method are displayed in Fig. 16. Each graph contains four curves - velocity profiles based on the "new" and "old" models (see above), the experimental velocity profile and the velocity profile obtained by the "integrated approach" (the blue curves), we will call it "re-calculated" velocity profile. It is clearly visible that agreement of such re-calculated experimental velocity profiles with the numerical ones is much better than between original experiments and calculation. High discrepancy between the "old" and "new" velocity profiles is also apparent, especially for lower currents.

Fig. 17 presents the same type of plot as is presented in Fig. 15 but with the analogous definition of the "average relative difference of velocity"

$$\Delta_{av}^{u} = \frac{100}{M} \cdot \sum_{i=1}^{M} abs\left(u_{re-exp}^{i} - u_{exp}^{i}\right) / u_{re-exp}^{i} ,$$

where u^i_{re-exp} is the re-calculated velocity and u^i_{exp} is the experimental velocity at the point i , M is the number of available points at which the difference is being evaluated. It is again evident that the present "new model" gives in most cases much lower relative difference than the "old model" for all studied cases.

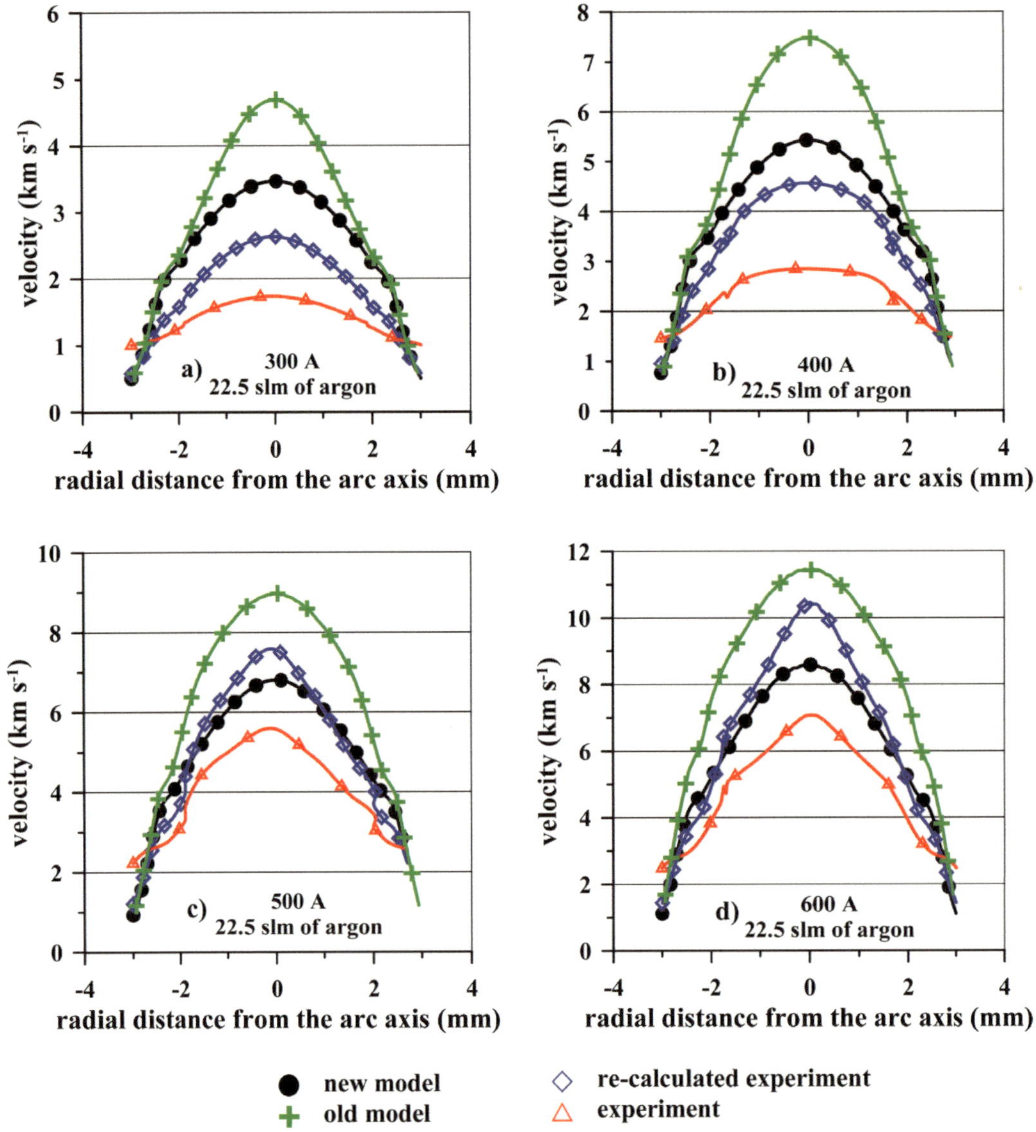

Fig. 16. Velocity profiles 2 mm downstream of the nozzle exit for 300 - 600 A with 22.5 slm of argon. Calculation – partial characteristics model, re-calculated experimental profile is based on the experimental temperature profile and calculated Mach number (see the text). The so-called „new model" stands for the present model, the „old model" presents our previous model with simplified plasma properties (see the text). The re-calculated velocity profiles show better agreement with the experiment.

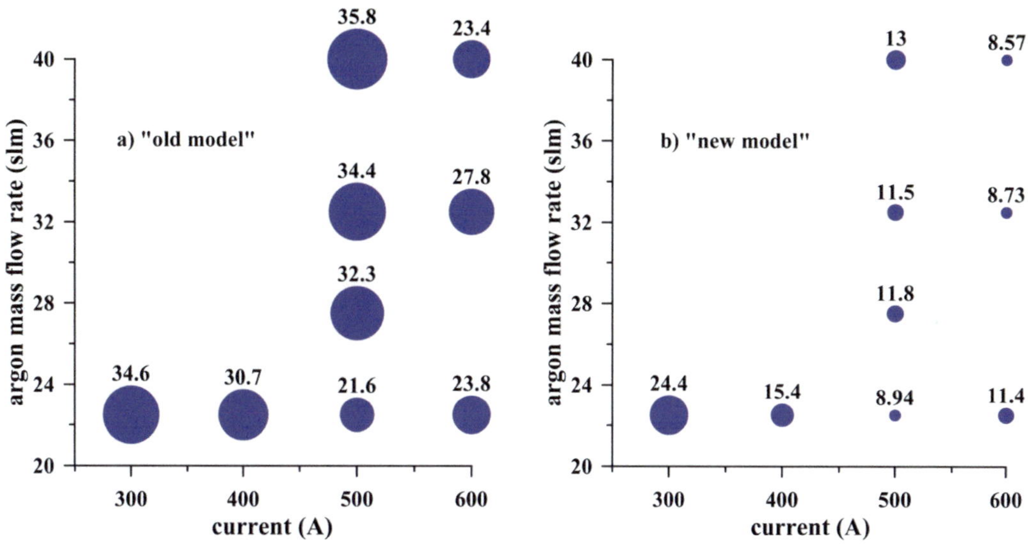

Fig. 17. Average relative difference (see the text) between the calculated and re-calculated (the experimental temperature profile and the calculated Mach number) radial velocity profiles, shown in % at the axial position 2 mm downstream of the nozzle exit, partial characteristics. The so-called „new model“ stands again for the present model and exhibits better agreement with experiments than the „old model“.

3.3 Power losses from the arc

Energy balance, responsible for performance of the hybrid-stabilized argon-water electric arc, is illustrated in the last set of figures. Fig. 18 (left) demonstrates the arc efficiency and the power losses from the arc discharge as a function of current for 40 slm of argon. The arc efficiency is defined here as $\eta = 1 - (\text{power losses})/(\Delta U \cdot I)$ with ΔU being the electric potential drop in the discharge chamber and I the current. The power losses from the arc stand for the conduction power lost from the arc in the radial direction and the radiation power leaving the discharge, which are considered to be the two principal processes responsible for the power losses. The ratio of the power losses to the input power in the discharge chamber $\Delta U \cdot I$ is indicated as the power losses in a per cent scale: the maximum difference of about 2-4 % between the net emission and partial characteristics methods is obviously caused by the amount of radiation reabsorbed in colder arc regions, the partial characteristics provides lower power losses. The arc efficiency is relatively high and ranges between 77-82 % for the net emission model and 80-84 % for the partial characteristics. The power losses slightly increases with increasing argon mass flow rate and with decreasing current, see Fig. 18 (right).

Fig. 19 (left) displays the typical radial profiles of temperature, divergence of radiation flux and radiation flux for 600 A and argon mass flow rate of 40 slm. Axial position is 4 cm from the argon inlet nozzle, i.e., inside the discharge chamber. Temperature reaches 24 700 K at the axis and declines to 773 K at the edge of the calculation domain. The radiation flux reaches $9.7 \cdot 10^6$ W $\cdot$ m^{-2} at the arc edge with the maximum magnitude $3.1 \cdot 10^7$ W $\cdot$ m^{-2} at the radial distance of 2.2 mm. The divergence of radiation flux becomes negative at the radial distance over 2.6 mm, i.e., the radiation is being reabsorbed in this region. Despite the negative values of the divergence of radiation flux in arc fringes are relatively small compared to the positive ones in the axial region, the amount of reabsorbed radiation is 32.4% (understand: ratio of the negative and positive contributions of the divergence of radiation flux, see below) because the plasma volume increases with the third power of radius.

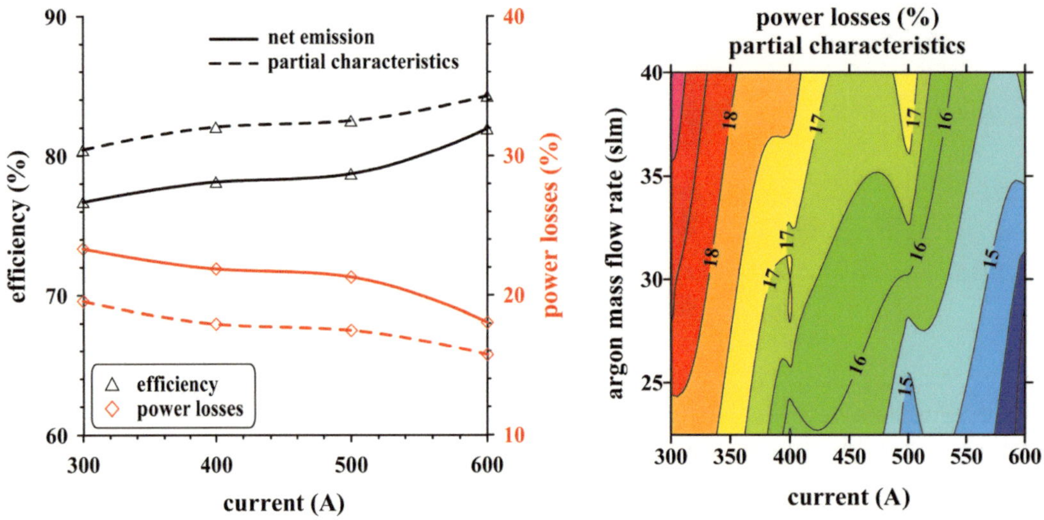

Fig. 18. Power losses and arc efficiency as functions of arc current for 40 slm of argon (left). The arc efficiency (%) is defined as $\eta = 1 - (\textit{power losses}) / (\Delta U \cdot I)$, where the power losses are due to radiation and radial conduction. Power losses in % is the ratio *power losses* / $(\Delta U \cdot I)$, shown also in dependence of current and argon mass flow rate (right).

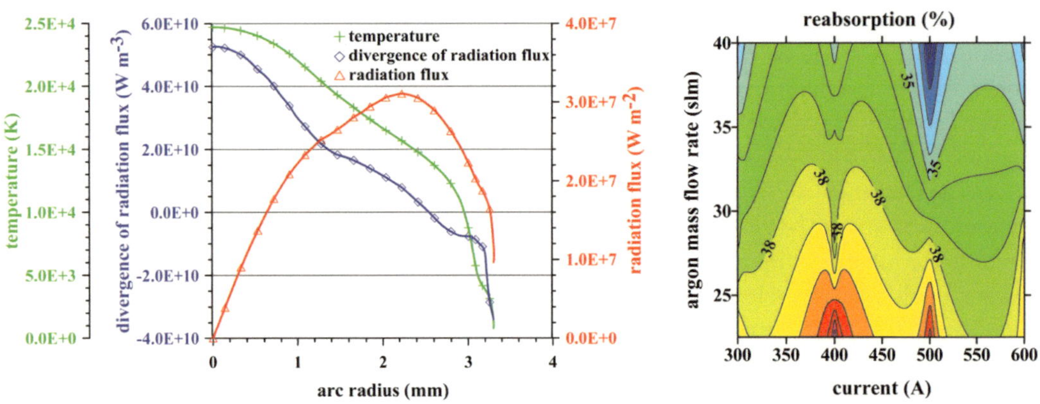

Fig. 19. Radial profiles of temperature, divergence of radiation flux and radiation flux for 600 A and argon mass flow rate of 40 slm inside the discharge chamber at the axial position of 4 cm (left); partial characteristics. Reabsorption of radiation occurs at ~ 2.6 mm from the axis. Reabsorption of radiation (right) for different currents and argon mass flow rates is defined as the ratio of the negative to the positive contributions of the divergence of radiation flux - it ranges between 30-45 % and slightly decreases for higher argon mass flow rates.

Fig. 19 (right) shows the amount of reabsorbed radiation (%) in argon-water mixture plasma within the arc discharge for the currents 300-600 A as a function of argon mass flow rate. The negative and positive parts of the divergence of radiation flux are integrated through the discharge volume. Reabsorption defined here is the ratio of the negative and positive contributions of the divergence of radiation flux - it ranges between 31-45 % and increases for lower contents of argon in the mixture.
Direct comparison of the amount of reabsorbed radiation with experiments is unavailable, however the indirect sign of validity of our results is a very good agreement between the experimental and calculated radial temperature profiles two millimeters downstream of the outlet nozzle presented above.

4. Conclusions

The numerical model for an electric arc in the plasma torch with the so-called hybrid stabilization, i.e., combined stabilization of arc by gas and water vortex, has been presented. To study possible compressible phenomena in the plasma jet, calculations have been carried out for the interval of currents 300-600 A and for relatively high argon mass flow rates between 22.5 slm and 40 slm. The partial characteristics as well as the net emission coefficients methods for radiation losses from the arc are employed. The results of the present calculation can be summarized as follows:

a. The numerical results proved that transition to supersonic regime starts around 400 A. The supersonic structure with shock diamonds occurs in the central parts of the discharge at the outlet region. The computed profiles of axial velocity, pressure and temperature correspond to an under-expanded atmospheric-pressure plasma jet.
b. The partial characteristics radiation model gives slightly lower temperatures but higher outlet velocities and the Mach numbers compared to the net emission model.
c. Reabsorption of radiation ranges between 31-45 %, it decreases with current and also slightly decreases with argon mass flow rate. The arc efficiency reaches up to 77-84%, the power losses from the arc due to radiation and radial conduction are between 16-24%.
d. It was proved that simulations for laminar and turbulent regimes give nearly the same results, so that the plasma flow can be considered to be laminar for the operating conditions and a simplified discharge geometry studied in this paper.
e. Comparison with available experimental data proved very good agreement for temperature - the maximum relative difference between the calculated and experimental temperature profiles is lower than 10% for the partial characteristics and 5% for the net emission radiation model used in the present calculation. Calculated radial velocity profiles 2 mm downstream of the nozzle exit show good agreement with the ones evaluated from the combination of calculation and experiment (integrated approach). Agreement between the calculated radial velocity profiles and the profiles analyzed purely from experimental data is worse. Evaluation of the Mach number from the experimental data for 500 and 600 A give values higher than one close to the exit nozzle, it thus proves the existence of the supersonic flow regime. The present numerical model provides also better agreement with experiments than our previous model based on the simplified transport, thermodynamic and radiation properties of argon-water plasma mixture.

The existing numerical model will be further extended to study the effect of mixing of plasma species within the hybrid arc discharge by the binary diffusion coefficients (Murphy, 1993, 2001) for three species - hydrogen, argon and oxygen.

5. Acknowledgments

J. Jeništa is grateful for financial support under the Fluid Science International COE Program from the Institute of Fluid Science, Tohoku University, Sendai, Japan, and their computer facilities. Financial support from the projects GA CR 205/11/2070 and M100430901 from the Academy of Sciences AS CR, v.v.i., is gratefully acknowledged. Our appreciation goes also to the Institute of Physics AS CR, v.v.i., for granting their computational resources (Luna/Apollo grids). The access to the METACentrum supercomputing facilities provided under the research intent MSM6383917201 is highly appreciated.

6. References

Bartlová, M. & Aubrecht, V. (2006). Photoabsorption of diatomic molecules. *Czechoslovak Journal of Physics*, Vol. 56, Suppl. B, (June 2006), pp. B632-B637, ISSN 0011-4626.

Březina, V.; Hrabovský, M.; Konrád M.; Kopecký, V. & Sember, V. (2001). New plasma spraying torch with combined gas-liquid stabilization of arc, *Proceedings of 15th International Symposium on Plasma Chemistry (ISPC 15)*, pp. 1021-1026, ISBN non-applicable, Orleans, France, July 9-13, 2001.

Hrabovský, M.; Konrád M.; Kopecký, V. & Sember, V. (1997). Processes and properties of electric arc stabilized by water vortex. *IEEE Transactions on Plasma Science*, Vol. 25, No.5, (October 1997), pp. 833-839, ISSN 0093-3813.

Hrabovský, M.; Kopecký, V. & Sember, V. (2003). Effect of Gas Properties on Characteristics of Hybrid Gas/Water Plasma Spraying Torch, *Proceedings of 16-th International Symposium on Plasma Chemistry (ISPC 16)*, on CD-ROM, ISBN non-applicable, Taormina, Italy, June 22-27, 2003.

Hrabovský, M.; Kopecký, V.; Sember, V.; Kavka, T.; Chumak, O. & Konrád, M. (2006). Properties of hybrid water/gas DC arc plasma torch, *IEEE Transactions on Plasma Science*, Vol. 34, No.4, (August 2006), pp. 1566-1575, ISSN 0093-3813.

Jameson, A. & Yoon, S. (1987). Lower-upper implicit schemes with multiple grids for the Euler equations. *AIAA Journal*, Vol. 25, No. 7, (July 1987), pp. 929-935, ISSN 0001-1452, E-ISSN 1533-385X.

Jeništa, J.; Kopecký, V. & Hrabovský, M. (1999a). Effect of vortex motion of stabilizing liquid wall on properties of arc in water plasma torch, In: *Heat and mass transfer under plasma conditions*, editors: Fauchais, P.; Mullen, van der J. & Heberlein, J., pp. 64-71, Annals of the New York Academy of Sciences, Vol. 891, ISBN 1-57331-234-7 (cloth), ISBN 1-57331-235-5 (paper), ISSN 0077-8923, New York.

Jeništa, J. (1999b). Water-vortex stabilized electric arc: I. Numerical model. *Journal of Physics D: Applied Physics*, Vol. 32, No. 21, (November 1999), pp. 2763-2776, ISSN 0022-3727 (print), ISSN 1361-6463 (online).

Jeništa, J. (2003a). Water-vortex stabilized electric arc: III. Radial energy transport, determination of water-vapour-boundary and arc performance. *Journal of Physics D: Applied Physics*, Vol. 36, No. 23, (December 2003), pp. 2995-3006, ISSN 0022-3727 (print), ISSN 1361-6463 (online).

Jeništa, J. (2003b). The effect of different regimes of operation on parameters of a water-vortex stabilized electric arc. *Journal of High Temperature Material Processes*, Vol. 7, No. 1, (March 2003), pp. 11-16, ISSN 1093-3611 (print), ISSN 1940-4360 (online).

Jeništa, J. (2004). Numerical modeling of hybrid stabilized electric arc with uniform mixing of gases. *IEEE Transactions on Plasma Science*, Vol. 32, No. 2, (April 2004), pp. 464-472, ISSN 0093-3813.

Jeništa, J.; Bartlová, M. & Aubrecht, V. (2007). The impact of molecular radiation processes in water plasma on performance of water-vortex and hybrid-stabilized electric arcs, *PPPS-2007 Proceedings of the 34th IEEE International Conference on Plasma Science and The 16th IEEE International Pulsed Power Conference*, pp. 1429-1432, ISBN 1-4244-0914-4, Albuquerque, New Mexico, USA, June 17-22, 2007.

Jeništa, J.; Takana, H.; Hrabovský, M. & Nishiyama, H. (2008). Numerical investigation of supersonic hybrid argon-water-stabilized arc for biomass gasification. *IEEE Transactions on Plasma Science*, Vol. 36, No. 4, (August 2008), pp.1060-1061, ISSN 0093-3813.

Jeništa, J.; Takana, H.; Nishiyama, H.; Bartlova, M.; Aubrecht, V. & Hrabovský, M. (2010). Parametric study of hybrid argon-water stabilized arc under subsonic and supersonic regimes. *Journal of High Temperature Material Processes*, Vol. 14, No. 1-2, (April 2010), pp. 63-76, ISSN 1093-3611 (print), ISSN 1940-4360 (online).

Kavka, T.; Chumak, O.; Sember, V. & Hrabovský, M. (2007). Processes in Gerdien arc generated by hybrid gas-water torch, *Proceedings of XXVIII International Conference on Phenomena in Ionized Gases (ICPIG 2007)*, pp. 1819-1822, ISBN 978-80-87026-01-4, Prague, Czech Republic, July 15-20, 2007.

Kavka, T.; Maslani, A.; Chumak, O. & Hrabovský, M. (2008). Character of plasma flow at the exit of DC arc gas-water torch, *Proceedings of 5-th International Conference on Flow Dynamics*, pp. OS8-11, ISBN non-applicable, Sendai, Japan, November 17-19, 2008.

Křenek, P. & Něnička, V. (1984). Calculation of transport coefficients of a gas mixture in the 4-th approximation of Enskog-Chapman theory. *Acta Technica CSAV*, Vol. 28, No. 4, (1984), pp. 420-433, ISSN 0001-7043.

Křenek, P. (2008). Thermophysical properties of H_2O–Ar plasmas at temperatures 400–50,000 K and pressure 0.1 MPa. *Plasma Chemistry and Plasma Processes*, Vol. 28, No. 1, (January 2008), pp. 107-122, ISSN 0272-4324.

Leer, van B. (1979). Towards the ultimate conservative difference scheme. V. A second-order sequel to Godunov's method. *Journal of Computational Physics*, Vol. 32, No. 1, (July 1979), pp. 101-136, ISSN 0021-9991.

Murphy, A. B. (1993). Diffusion in equilibrium mixtures of ionized gases, *Physical Review E*, Vol. 48, No. 5, (November 1993), pp. 3594-3603, ISSN 1539-3755 (print), ISSN 1550-2376 (online).

Murphy, A. B. (2001). Thermal plasmas in gas mixtures, *Journal of Physics D: Applied Physics*, Vol. 34, No. 20-21, (October 2001), pp. R151–R173, ISSN 0022-3727 (print), ISSN 1361-6463 (online).

Patankar, S. V. (1980). *Numerical Heat Transfer and Fluid Flow*, McGraw-Hill, ISBN 0-07-048740-5, New York.

Pateyron, B. & Katsonis, C. (2009). T&TWinner, available from: http://ttwinner.free.fr/.

Roe, P. L. (1981). Approximate Riemann solvers, parameter vectors, and difference schemes. *Journal of Computational Physics*, Vol. 43, No. 2, (October 1981), pp. 357-372, ISSN 0021-9991.

Van Oost, G.; Hrabovský, M.; Kopecký, V.; Konrád, M.; Hlína, M.; Kavka, T.; Chumak, O.; Beeckman, E. & Verstraeten, J. (2006). Pyrolysis of waste using a hybrid argon-water stabilized torch, *Vacuum*, Vol. 80, No. 11-12, (September 2006), pp. 1132-1137, ISSN 0042-207X.

Van Oost, G.; Hrabovský, M.; Kopecký, V.; Konrád, M.; Hlína, M.; Kavka, T. (2008). Pyrolysis/gasification of biomass for synthetic fuel production using a hybrid gas-water stabilized plasma torch. *Vacuum*, Vol. 83, No. 1, (September 2008), pp. 209-212, ISSN 0042-207X.

Yoon, S. & Jameson, A. (1988). Lower-upper symmetric-Gauss-Seidel method for the Euler and Navier-Stokes equations. *AIAA Journal*, Vol. 26, No. 9, (September 1988), pp. 1025-1026, ISSN 0001-1452, E-ISSN 1533-385X.

Part 2

Biomass Production

A Simple Analytical Model for Remote Assessment of the Dynamics of Biomass Accumulation

Janis Abolins and Janis Gravitis
University of Latvia,
Latvian State Institute of Wood Chemistry,
Latvia

1. Introduction

Efficient means for assessment of the dynamics and the state of the stocks of renewable assets such as wood biomass are important for sustainable supplies satisfying current needs. So far attention has been paid mainly to the economic aspects of forest management while ecological problems are rising with the expected transfer from fossil to renewable resources supplies of which from forest being essential for traditional consumers of wood and for emerging biorefineries. Production of biomass is more reliant on assets other than money the land (territory) available and suitable for the purpose being the first in the number. Studies of the ecological impacts (the "footprint") of sustainable use of biomass as the source of renewable energy encounter problems associated with the productivity of forest lands assigned to provide a certain annual yield of wood required by current demand for primary energy along with other needs.

Apart from a number of factors determining the productivity of forest stands, efficiency of land-use concomitant with growing forest depends on the time and way of harvesting (Thornley & Cannell, 2000). In the case of clear-cut felling the maximum yield of biomass per unit area is reached at the time of maximum of the mean annual increment (Brack & Wood, 1998; Mason, 2008). The current annual increment (rate of biomass accumulation by a forest stand or rate of growth) culminates before the mean annual increment reaches its peak value and there is a strong correlation between the maximums of the two measures. Knowing the time of growth-rate maximum (inflection point on a logistic growth curve) allows predicting the time of maximum yield (Brack & Wood, 1998). However, the growth-rate maximum is not available from field measurements directly. Despite the progress in development of sophisticated models simulating (Cournède, P. et al., 2009; Thürig, E. et al., 2005; Welham et al., 2001) and predicting (Waring et al., 2010; Landsberg & Sands, 2010) forest growth, there still remains, as mentioned by J. K. Vanclay, a strong demand for models to explore harvesting and management options based on a few available parameters without involving large amounts of data (Vanclay, 2010).

The self-consistent analytical model described here is an attempt to determine the best age for harvesting wood biomass by providing a simple analytical growth function on the basis of a few general assumptions linking the biomass accumulation with the canopy absorbing the radiation energy necessary to drive photosynthesis. A number of reports on employing remote sensing facilities (Baynes, 2004; Coops, et al., 1998; Lefsky et al., 2002; Richards & Brack, 2004; Tomppo E. et al., 2002 ; Waring et al.,2010) strongly support the optimism with regard to successful use of the techniques to detect the time of maximum yield of a stand well in advance by monitoring the expanding canopy.

According to the grouping of models suggested by K. Johnsen et al. in an overview of modeling approaches (Johnsen et al., 2001), the model described in this chapter belongs to simplistic traditional growth and yield models. It differs from other models of this kind by not incorporating mathematical representations of actual growth measurements over a period of time. Derived from a few essential basic assumptions the analytical representation rather provides the result that should be expected from measurements of growth under "traditional" (idealized) conditions. The chosen general approach of modeling the biomass production at the stand level allows obtaining compatible growth and yield equations (Vanclay, 1994) of a single variable - the age. Like with many other theoretical constructions the applicability of the model to reality is fairly accidental and restricted. However, since the derived equations are in good agreement with the universal growth curves obtained from measurements repeatedly confirmed and generally accepted as classic illustrations of biomass dynamics (Brack & Wood, 1998; Mason et al., 2008), it seems to offer a good approximation of the actual biomass accumulation by natural forest stands.

Equations representing the model are believed to reflect the simple assumptions made on the basis of common knowledge about photosynthesis and observations in nature: biomass is produced by biomass; the amount of produced biomass is proportional to the amount of absorbed active radiation; the absorbed radiation is proportional to effective light-absorbing area of the foliage (number and surface area of leaves) and limited by the ground area of the forest stand (the area determining the available energy flow). Projection of the canopy filling the ground area detectable by remote sensing is assumed to reflect dynamics and status (the stage) of forest growth. The height of the stand is another growth parameter accessible by remote sensing. Relationships of the latter with other measurable quantities determining the yield of accumulated biomass are well studied (Vanclay, 2009) and can be employed for remote assessment of the current annual increment and the state of forest stands (Lefsky et al., 2002; Ranson et al., 1997; Tomppo et al., 2002). The model presented hereafter has been developed to be aware of the current annual increment reaching the maximum merely from the data of remote observation of the dynamics of forest stand canopy while complemented by data of the average height would predict the yield.

2. General approach and basic equations

The analytical model offered to describe dynamics of the standing stock of wood biomass in natural forests is based on the obvious relationship between the rate of growth (rate of accumulation of biomass) y and the stock (amount of biomass) S stored in the forest stand (Garcia, 2005):

$$S(t) = \int y(t) \cdot dt \tag{1}$$

By turning to common knowledge that biomass is produced by biomass the rate of accumulation of new biomass in the first approximation can be assumed being proportional to the amount of biomass already accumulated:

$$y = \frac{dS}{dt} = aS \tag{2}$$

where a is a constant of the reciprocal time dimension and t is time. Rewriting the right-side equation of (2) in the form:

$$\frac{dS}{S} = adt\,, \tag{3}$$

and integrating it provides $\ln S = at$ and exponential growth of the stock of biomass:

$$S = const \cdot e^{at}\,, \tag{4}$$

which is unrealistic in the long run because of finite resources of nutrients and other limiting factors not taken into account in Eq. (2). The problem can be solved by setting an asymptotic limit to growth:

$$S(t) = S_{\infty} \cdot \left(1 - e^{-at}\right). \tag{5}$$

The rate of biomass accumulation y, Eq. (2), usually referred to as the current annual increment of stock measured by volume of wood mass per unit area (m^3/ha) (Brack & Wood, 1998) is not directly determined by the accumulated biomass stock. The uptake of CO_2 and photosynthesis of biomass rather depends on the total surface area of leaves determining the amount of absorbed radiation. The number of leaves and hence the light-absorbing area depend on the biomass accumulated by individual trees and the forest stand as a whole. The actual amount of the absorbed radiation that ultimately determines the rate of photosynthesis (and the annual increment) per unit area (a *hectare*) of a particular forest stand is limited regardless of the total surface area of leaves. So the concept of light-absorbing area should refer to the effective absorbing area limited by the particular area unit selected. It should be noticed here that further considerations are relevant to statistically significant numbers of individual trees and, consequently, to area units of stands comparable to hectare.

It seems to be reasonable to assume that accumulation of biomass in a forest stand occupying a large enough land area follows the same law as the rate at which the light-absorbing area (the canopy) of the growing stand expands with time. As noticed, the number and total surface area of leaves absorbing radiation is proportional to the accumulated biomass approaching some asymptotic limit L_{∞} of its own. However, the rate of expansion of the effective absorbing area also depends on the proportion of the free, unoccupied space available for expansion to intercept the radiation. Supposing the total light-absorbing area L as function of time being described by equation similar to Eq. (5):

$$L(t) = L_{\infty}\left(1 - e^{-at}\right), \tag{6}$$

the rate of expansion of the light-absorbing area expressed as:

$$\frac{dL}{dt} = const \cdot \left(L_{\infty} - L\right) \cdot L \tag{7}$$

can be written in the form:

$$\frac{dL}{dt} = const \cdot \left[L_\infty - L_\infty \cdot \left(1 - e^{-at}\right) \right] \cdot L_\infty \left(1 - e^{-at}\right) = const \cdot L_\infty^2 \cdot e^{-at} \cdot \left(1 - e^{-at}\right). \quad (8)$$

Dimension of the constant in Eq. (8) is the reciprocal of the product of area and time. Since area L_∞ also is constant it can be omitted for further convenience to focus attention on the time-dependent part of Eq. (8).

Assuming that the rate of biomass accumulation follows the rate of expansion of the light-absorbing area it can be described by equation similar to Eq. (8):

$$\frac{dS}{dt} = const \cdot \left(1 - e^{-at}\right) \cdot e^{-at}, \quad (9)$$

where the value of the constant factor (dimension of which here is the dimension of current increment) can be chosen to satisfy some selected normalizing condition, as will be done further.

The time-dependent part of Eq. (9) has a maximum at time t_m satisfying condition:

$$2e^{-at} - 1 = 0 \quad (10)$$

Wherefrom

$$at_m = \ln 2 \quad (11)$$

Exponent a determining the rate of growth in real time depends on the particular species and a number of other factors such as insolation and availability of water and nutrients at the site and has to be found from field measurements. However, existence of the maximum on the curve of the rate of growth (the curve of current annual increment often referred to as the growth curve) allows normalizing the time scale with respect to the time at which the maximum is reached. It is done by introducing dimensionless time variable

$$x = \frac{at}{\ln 2}, \quad (12)$$

or substituting at with $x \cdot ln2$ in Eq. (9), or just writing x instead of t and putting $a = ln2$. The current annual increment is normalized by choosing the constant factor to satisfy condition:

$$y_m = y(x=1) = const \cdot \left(1 - \frac{1}{2}\right) \cdot \frac{1}{2} = const \cdot \frac{1}{4} = 1. \quad (13)$$

The normalized rate of biomass accumulation expressed by current annual increment in time scale x normalized with respect to the time when it reaches its maximum now is presented by Eq. (9) where t is substituted by variable x:

$$y(x) \equiv \frac{dS}{dx} = 4 \cdot \left(1 - e^{-ax}\right) \cdot e^{-ax} \quad (14)$$

where $a = ln2$. Function $y(x)$ is shown in Fig. 1 (a).

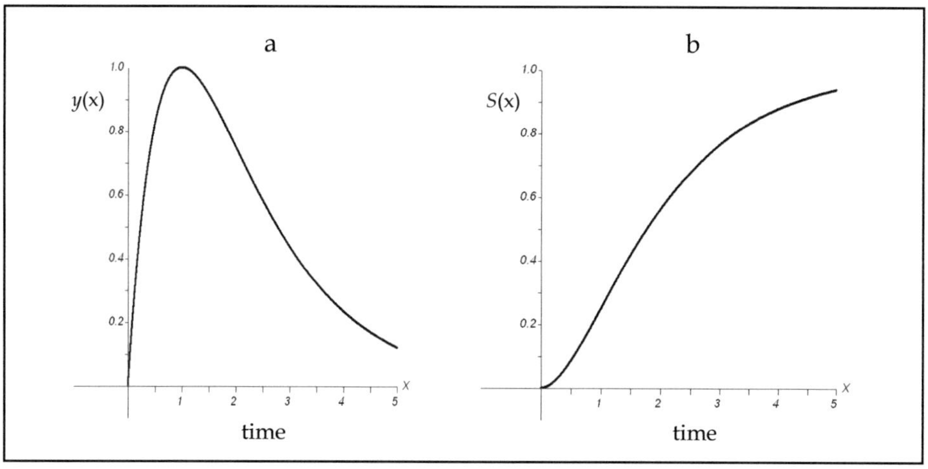

Fig. 1. a – rate of accumulation (current annual increment) of biomass $y(x)$ normalized with respect to its maximum value presented by Eq. (14) and b – stock normalized with respect to its asymptotic limit presented by Eq. (17) as functions of normalized time variable x.

Returning to Eq. (1) the biomass stored by time $x = x_c$ is expressed by definite integral:

$$S(x_c) = \int_0^{x_c} y(x)dx\,. \tag{15}$$

Substituting $y(x)$ from Eq. (14) into Eq. (15) and calculating the integral the stock S is presented as function of age explicitly:

$$S(x_c) = 4\int_0^{x_c}\left(1-e^{-ax}\right)\cdot e^{-ax}dx = 4\int_0^{x_c} e^{-ax}dx - 4\int_0^{x_c} e^{-2ax}dx = 4\left[-\frac{e^{-ax}}{a}+\frac{e^{-2ax}}{2a}\right]_0^{x_c} =$$

$$= \frac{2}{2a}\cdot\left[e^{-2ax}-2e^{-ax}\right]_0^{x_c} = \frac{2}{2a}\cdot\left(e^{-2ax_c}-2e^{-ax_c}+1\right) = \frac{2}{2a}\cdot\left(1-e^{-ax_c}\right)^2 \tag{16}$$

By normalizing the stock choosing its asymptotic limit as the normalized unit $S_\infty = 1$ the result of transformations in Eq. (16) can be summarized as

$$S(x_c) = \frac{2}{2a}\cdot\left(1-e^{-ax_c}\right)^2 = S_\infty\left(1-e^{-ax_c}\right)^2 \tag{17}$$

where, as previously in Eq. (14), $a = ln2$. Function (17) in the normalized time scale is presented in Fig. 1 (b).

3. Mean annual increment and productivity

The mean annual increment of a forest stand is an essential factor illustrating the overall productivity of the stand at a given age and is expressed by the ratio of stock to age of the stand (Brack & Wood, 1998). The stock being presented by Eq. (16) the mean annual increment Z is calculated in units of the current annual increment from

$$Z(x)=\frac{2}{a}\cdot\frac{\left(1-e^{-ax}\right)^2}{x} \tag{18}$$

where $a = ln2$. Function $Z(x)$ shown in Fig. 2 has a maximum at x satisfying condition:

$$\frac{e^{x\ln 2}}{x}=2\ln 2 \tag{19}$$

obtained from putting derivative of function (18) equal to zero. The value of $x \approx 1.81$ satisfying Eq.(19) is found from graphical solution of the equation (Fig. 3).

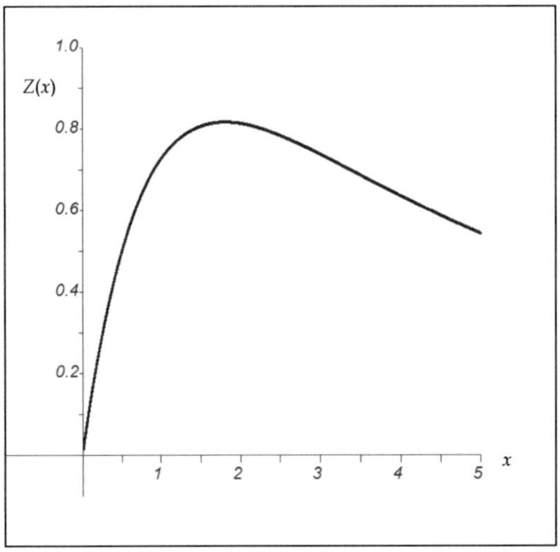

Fig. 2. Mean annual increment Eq. (18) as function of the normalized time variable x.

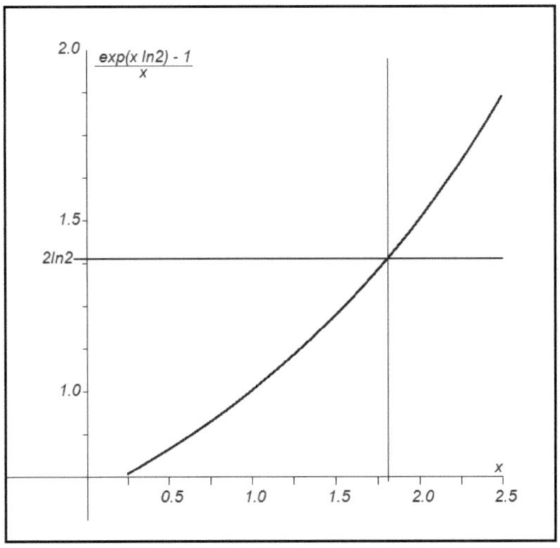

Fig. 3. Graphical solution of Eq. (19) determining position of the maximum of mean annual increment on the axis of the normalized time coordinate x.

In Fig. 4 the current annual increment (rate of biomass accumulation) and the mean annual increment are presented together wherefrom the mean annual increment is seen to reach the maximum value (equal to ≈ 0.8 of the peak value of current annual increment) at cross-point of the two curves.

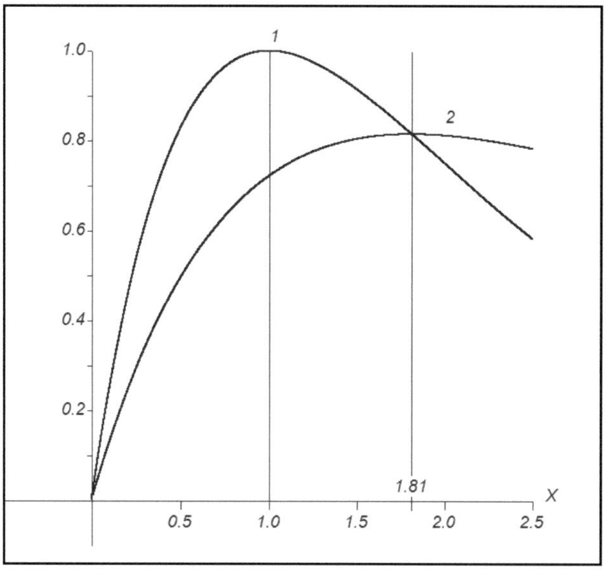

Fig. 4. Current (curve 1, Eq. 14) and mean (curve 2, Eq. 18) annual increments of biomass as functions of time x normalized with respect to the time of the growth-rate maximum chosen as the unit time interval. The mean annual increment (curve 2) is presented in the same scale as the current annual increment. The maximum of curve 2 is reached at the cross-point of the two curves at $x \approx 1.81$.

The reciprocal of the mean annual increment is a parameter characterizing the size of plantation for sustainable supply of biomass. The total area of a plantation for sustainable annual supply comprised of equal lots of stands of ages in sequence from one year to the cutting age is directly proportional to cutting age x_c and inversely proportional to the stock at cutting age $S(x_c)$:

$$A = const \cdot \frac{x_c}{S(x_c)} = const \cdot f(x_c)\,. \tag{20}$$

The constant is equal to the required annual yield of biomass; function $f(x_c)$ defined as

$$f(x_c) \equiv \frac{x_c}{S(x_c)} \tag{21}$$

is the reciprocal of the mean annual increment at cutting age.
At point $x \approx 1.81$ where the mean annual increment reaches maximum its reciprocal – function $f(x)$ has the minimum. If B_s is the demanded sustainable annual yield of biomass, $S(x_c)$ – the stock of biomass accumulated in the forest stand by the cutting age, and A_o – the area of the forest to be felled annually to satisfy the demand, then $B_s = S(x_c) \cdot A_o$ and the total

area of the plantation – $A = x_c \cdot A_o$. From here the yield per unit area of the whole plantation is found being proportional to the mean annual increment reaching the maximum at $x \approx 1.8$:

$$\frac{B_S}{A} = \frac{S(x_c) \cdot A_o}{x_c \cdot A_o} = \frac{S_c(x_c)}{x_c}. \tag{22}$$

As follows from Eq. (22), felling the forest at age corresponding to 1.8 units of the normalized time scale provides the maximum yield per unit area of a particular stand and hence of the whole plantation. In other words, the maximum productivity of land area under a forest is achieved when felling at the time of the mean annual increment peak.

4. Validation of the model

Neither the value of the current annual increment at maximum, nor the real time when a forest stand reaches the maximum is known *a priori*. Both parameters depend on the species and conditions represented by the quality class of the site and have to be determined by field measurements. However, the field measurements do not provide these quantities directly. They have to be found from periodic mean annual increments available from field measurements.

The growth-rate function given by Eq. (14) cannot be used directly to compare the model equation with experimental growth-rate data. For that purpose a different exponential equation can be employed containing variable parameters related to the quantities not measurable directly. The values of the variable parameters providing the best fit of the measured annual increments with the equation are chosen to evaluate the unknown quantities. A rather abundant database available for natural grey alder (*Alnus incana*) stands of up to 50 years old (Daugavietis, 2006) presents a good opportunity to test the model.

The 5-year mean annual increments available from field measurements (Daugavietis, 2006) are a good approximation for the current annual increment value at mid-time of the respective 5-year period (Fig. 5, a). By choosing a function of the type

$$y(t) = (c + kt) \cdot e^{-\frac{t}{a}} = k(b + t) \cdot e^{-\frac{t}{a}} \tag{23}$$

to describe the current annual increment it is possible to assign physical sense to variable parameters a and c and find the maximum value of the current annual increment and position of the maximum on the real-time axis by best fit of function (23) to the data from experimentally measured periodic mean increments. Under condition of taking coefficient k (of dimension y/t) equal to 1 function (23) has its maximum at time

$$t_m = a - \frac{c}{k} = a - b. \tag{24}$$

It should be noticed here that dimension of constant a in Eq. (23) is time, which is different from the constant a used in Eq. (2) with dimension of reciprocal time (frequency). The reason of choosing a different dimension of constant a in Eq. (23) is seen from Eq. (24).

By varying parameters a, b, and the maximum value of the current annual increment y_m (not available from any direct measurement) function (23) is varied for best fit to the set of experimental data normalized with respect to y_m.

The values of increments calculated from Eq. (23) coincide with the set of experimental data (Daugavietis, 2006) (Fig. 5) within standard deviation of 2.5 % of the maximum value, the correlation between the sets of calculated and experimental data being better than 0.99.
The normalized time scale is introduced by choosing variable x to satisfy condition

$$x = \frac{t}{t_m} = \frac{t}{a-b}. \tag{24}$$

By substituting the normalized time variable x for real time t in Eq. (23) the current annual increment is presented as

$$y(x) = \left[b + (a-b)\cdot x\right]\cdot e^{-\frac{a-b}{a}\cdot x}. \tag{25}$$

By defining new constant parameters $\alpha = \frac{a-b}{a}$ and $\beta = \frac{b}{a-b}$ Eq. (25) is rewritten as:

$$y(x) = (a-b)\cdot(\beta + x)\cdot e^{-\alpha x}. \tag{26}$$

Normalizing function (26) with respect to $y_m = (a - b)\cdot(\beta + 1)\cdot exp(-a)$ and taking into account that $\beta + 1 = \frac{b+a-b}{a-b} = \frac{a}{a-b}$ provide

$$y(x) = \alpha\cdot e^{\alpha}\cdot(\beta + x)\cdot e^{-\alpha x}. \tag{27}$$

By substituting $y(x)$ from Eq. (27) in Eq. (15) and calculating the integral the stock normalized to $S_\infty = \frac{a\cdot(a+b)}{(a-b)^2}$ as function of cutting age is expressed by:

$$S(x_c) = 1 - \left(1 + \frac{a-b}{a+b}\cdot x_c\right)\cdot e^{-\alpha x_c}. \tag{28}$$

The mean annual increment

$$\bar{y}(x) = \frac{S(x)}{x} = \frac{1}{x}\cdot\left[1 - \left(1 + \frac{a-b}{a+b}\cdot x\right)\cdot e^{-\alpha x}\right] \tag{29}$$

reaches maximum under condition

$$\exp(\alpha x) - \alpha x\cdot\left(1 + \frac{a-b}{a+b}x\right) = 1 \tag{30}$$

providing $x_m \approx 1.77$ corresponding to optimum cutting age of $x_c = 1.8$ or 18 years in case of grey alder.
After finding the age of the maximum of current annual increment, the set of experimental points (Fig. 5, a) can be put on the normalized time scale x and compared with function (14) as shown in Fig. 5, b. The variation of the value of growth-rate maximum at this point is still available for adjustment to improve the fit between

experimental data and Eq. (14). The curves presented by Eqs. (14) and (27) with best fit parameter values are practically identical within the normalized time interval $0.5 \leq x \leq 2.5$. Because of a nonzero initial growth-rate Eq. (27) provides higher values on the rise while lower at later time on the decline.

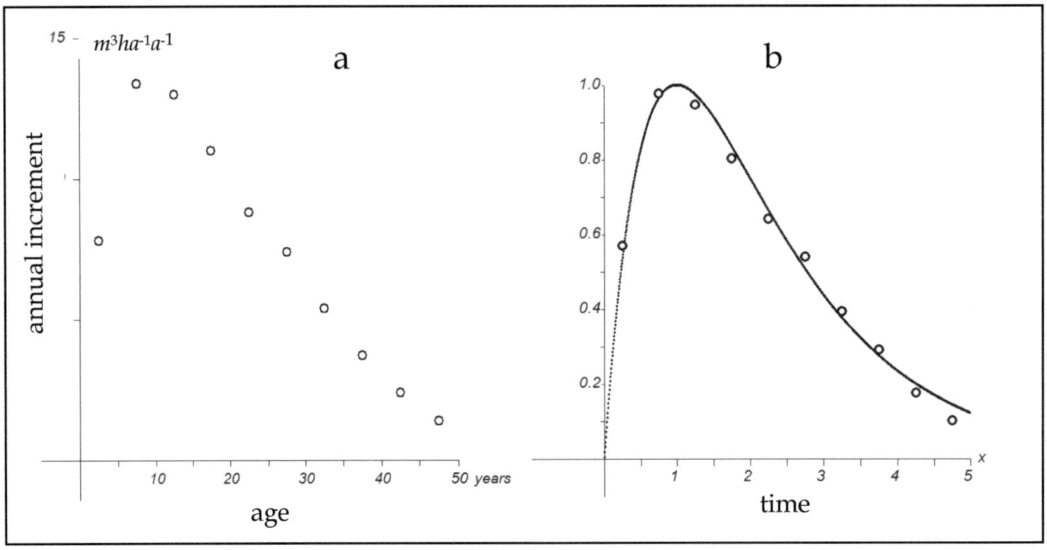

Fig. 5. a – current annual increments of grey alder stand calculated from measured 5-year periodic mean values with age (Daugavietis, 2006), in units of m^3 per *ha* per *annum*; b – best fit of Eq. (14) (solid curve) to experimental data (circles) normalized against the growth-rate maximum in the time scale of normalized age.

5. Rate of growth as function of light-absorbing area

Equation (9) describing the rate of biomass accumulation derived from Eq. (7) in section 1 is based on the assumption that dynamics of current annual increment follows dynamics of the expansion of light-absorbing area of the canopy. Returning to Eq. (7) it can be assumed to describe the relationship between the normalized rate of growth (y) and the normalized light-absorbing area (L):

$$y(L) = 4 \cdot L(1 - L) \tag{31}$$

shown in Fig. 6.

It has to be noticed that the pace at which the biomass is stored is not necessarily equal to the pace at which the light-absorbing area increases. The uptake of biomass (photosynthesis) depending on the effective light-absorbing area obviously should follow with some delay, which means that the normalized (intrinsic or specific) time scale of the equation derived from Eq. (8) to describe the rate of expansion of the light-absorbing area:

$$\frac{dL}{dx} = 4\left(1 - e^{-ax}\right) \cdot e^{-ax}, \; (a = ln2) \tag{32}$$

is different from that of Eq. (14) describing the rate of biomass accumulation.

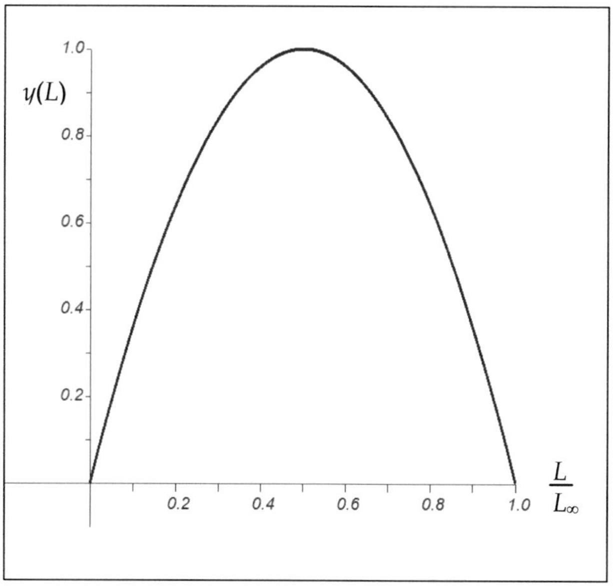

Fig. 6. Rate of biomass accumulation y as a function of the light-absorbing area L, Eq. (31).

Relationship between the units of the two normalized time variables – x_b describing the current annual increment (rate of biomass accumulation) and x_a describing the rate of expansion of the light-absorbing area can be concluded from knowing that maximum of the current annual increment is reached at $L/L_\infty = 0.5$ when $x_b = 1$. In units of time scale x_a the light-absorbing area L is expressed by integrating Eq. (32) the result of which is similar to Eq. (17):

$$L(x_a) = L_\infty \left(1 - e^{-ax_a}\right)^2 \tag{33}$$

where L is normalized in the same way as stock by taking the asymptotic limit L_∞ equal to 1. The "age" x_a at which the normalized light-absorbing area reaches the value 0.5, as follows from Eq. (33), satisfies equation:

$$1 - \frac{\sqrt{2}}{2} = e^{-ax_a} \tag{34}$$

wherefrom, remembering that $a = ln2$, the time in units of scale x_a corresponding to unit time of scale $x_b = 1$ is found being equal to

$$x_a = \frac{-\ln\left(1 - \frac{\sqrt{2}}{2}\right)}{\ln 2} \cong 1.77\,. \tag{35}$$

It means that a unit of the normalized time scale of the rate of expansion of the light-absorbing area is about 0.56 of the unit of the normalized time scale describing the rate of biomass accumulation. The units of the two normalized time scales presented in Fig. 7 are approximately equated by

$$x_b \cong 1.77 x_a \,. \quad (36)$$

As seen from Fig. 7, expansion of the light-absorbing area of the canopy (curve 1) proceeds ahead of the rate of biomass accumulation (curve 2) complying with the assumption that higher rates of the increase of the surface area (and the number) of leaves require a greater proportion of the gross product of photosynthesis lost after seasonal vegetation.

The size of the effective light-absorbing area expressed by the ratio to its asymptotic limit is presented in Fig. 7 on the lower time axis. The maximum rate of expansion dL/dx is reached at $x = x_b \approx 0.56$ ($x_a = 1$) when $L = 0.25L_\infty$ while the current annual increment reaches the maximum at $x = x_b = 1$ when $L = 0.5L_\infty$. By the time $x = x_b \approx 1.81$ when the mean annual increment reaches its maximum the effective light-absorbing area is equal to approximately $0.8\ L_\infty$. The current annual increment of biomass in the stand is maintained over 0.8 of the maximum value within the range of light-absorbing area between 0.28 and 0.8 of L_∞.

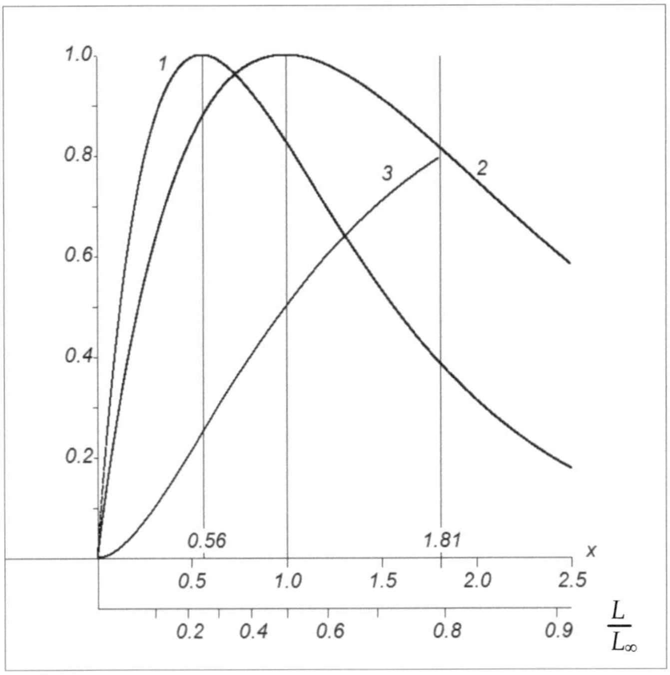

Fig. 7. Rate of expansion of the light-absorbing area (1), current annual increment (2), and the light-absorbing area (3) in time-scale $x = x_b$ normalized to the time of the current annual increment maximum. The lower axis shows the size of the light-absorbing area reached at the respective point on the time axis.

The basic components of the model - equations presenting current and mean annual increments, stock, and the rate of expansion of the light-absorbing area as functions of age expressed in the intrinsic time units are summarized in Fig. 8.

6. Conceptual remarks

The analytical expressions comprising the model are derived from rather general principles of biomass production by photosynthesis in living stands without taking into account factors affecting forest growth other than the effective light-absorbing area of the canopy. However, since dynamics of the latter is strongly dependent on availability of nutrients, water, and some other crucial factors, the model reflects the cumulative effect of all of them through the relationship between the rate of growth and the capacity to capture the active radiation. Therefore, monitoring the canopy dynamics can provide reliable information for conclusions about that capacity and the expected end product of photosynthesis.

Determining the best time for harvesting by observing expansion of the canopy from satellites is one of attractive practical applications of the model for management of even-age stands in concert with remote sensing. Even though the canopy projection measureable by remote sensing instruments is not quite equal either to the light-absorbing area or the leaf area index, the correlation between the three is strong enough to make corrections necessary for detecting the time (age) of growth-rate maximum from remote observations of the dynamics of canopy expansion.

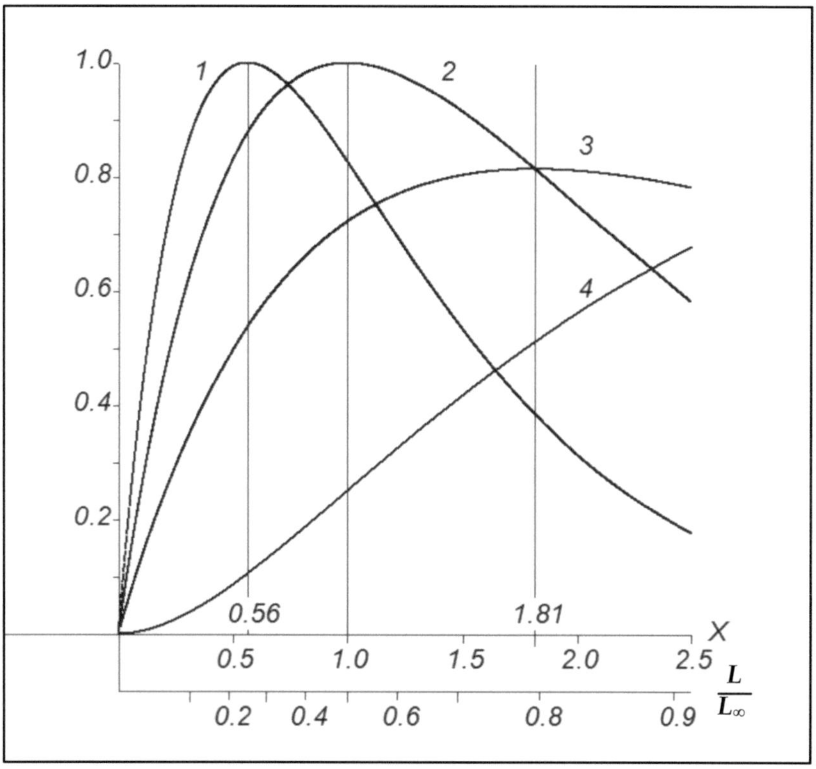

Fig. 8. Dynamics of the light-absorbing area (1), Eq. (7), the rate of production of above-ground biomass (2), Eq. (14), mean annual increment (3), Eq. (18), and the yield (4), Eq. (17), as functions of the intrinsic time provided by the rate of growth of a forest stand. The effective light-absorbing area as the ratio to its maximum value L/L_∞, Eq. (33), is presented by the lower abscissa. Note the inflection point of curve 4 being reached before 0.25 S_∞; at the time of maximum productivity $S \cong 0.5\ S_\infty$.

The obtained analytical expression, Eq. (17), for accumulated biomass of a stand as function of age is a particular case of the well-known Richards growth equation (Zeide, 2004):

$$y(t) = const \cdot \left(1 - e^{-bt}\right)^c \tag{37}$$

with parameter values b = ln2 and c = 2 describing sigmoid (logistic) growth.
A generalized differential form of sigmoid growth (the growth-rate function) has been considered by C. P. D. Birch (Birch, 1999) and a detailed formalistic analysis of the family of sigmoid growth equations is given by O. Garcia (Garcia, 2005). The sigmoid shape of the yield (stock) curve Eq. (17) in the present case is predetermined by the shape (the maximum) of the obtained growth-rate function Eq. (14).
The normalized time unit introduced to provide a dimensionless common measure to match the model with experimental data is the same intrinsic time unit suggested by B. Zeide as a unit provided by organisms themselves and clarifying the meaning of parameters of growth functions (Zeide, 2004). A number of other growth factors, such as biological potential of a particular species, the site quality, changing climate, etc. are reflected in the real-time equivalent of the intrinsic time unit. For instance, comparison of best fits to available measured data of grey alder stands at sites of different quality (Daugavietis, 2006) show the stands at sites of higher quality reaching the growth-rate maximum earlier (Kosmach, 2010). Since climate change is a factor affecting forest growth (Nakawatase & Peterson, 2006), the real-time equivalents of the intrinsic time unit obtained from monitoring the growth of stands of a given species hold information for potential assessment of the changing environment accessible by remote observations and retrospective studies of forest growth.
The Richards equation (37) predicts diminishing of the current increment to zero with the age of the stand while the effective light-absorbing area given by Eq. (6) approaches a constant maximum and, therefore, should be expected to provide a constant maximum increment of biomass. However, the real growth curves (at least of natural forest stands) rather comply with Richards equation even if the underlying models do not take into account factors, such as respiration or partition, diminishing the annual above-ground biomass production. In the present case they are somehow implied in the factor (L_∞ - L) restricting the rate of expansion of the effective light-absorbing area in Eq. (7), which ultimately determining the descent of the derived growth functions, Eqs. (14) and (17), can be attributed to shading. At large, the simplified models of this kind should not be expected to hold at the very short and far ranges of the time axis their application being limited by the range of the intrinsic time units between 0.5 and 3 – the interval of interest for commercial forest management. G. E. P. Box has likely hit the point with regard to the subject by writing in 1979: "All models are wrong, but some are useful" (cited in Vanclay, 2010).

7. Conclusions

The simple logistic analytical model of biomass accumulation by forest stands derived on the basis of general assumptions about photosynthesis comprises compatible equations of growth and yield as functions of time. The function describing dynamics of the rate of growth derived as function of the effective light-absorbing area of the canopy provides a growth function representing particular case of Richards equation and is in good agreement with data obtained from experimental measurements. The model contains two related parameters: the unit of the intrinsic (normalized with respect to peak current annual increment) time scale and the effective light-absorbing area of the canopy not equal but closely related to the leaf area index or to projection of the canopy. The latter accessible by remote sensing opens the use of remote sensing data for monitoring the growth of forest

stands to predict the culmination of current annual increment the age of the stand at which being known allows predicting the optimum age for harvesting.
The model has been developed for determining the land area and the optimum harvesting age of even-aged natural stands for sustainable supply of firewood and wood biomass to satisfy the needs of paper mills and biorefineries. It can be extended to consider solutions of the same problems with regard to timber products such as boards and other construction elements of buildings.
Some further studies are necessary to find out the relationship between remote observations of canopy dynamics and dynamics of the effective light-absorbing area to realize the benefits of using the model with the opportunities provided by remote sensing to forest management.

8. Acknowledgements

The presented model has been derived on the basis of studies supported by the National Research Programs of Renewable energy resources and rational management of deciduous tree forests.

9. References

Baynes, J. (2004). Assessing forest canopy density in a highly variable landscape using Landsat data and FCD Mapper software. *Australian Forestry*, Vol. 67, No. 4 (n. d.), pp. 247-253.

Birch, C. (1999). A New Generalized Logistic Sigmoid Growth Equation Compared with the Richards Growth Equation. *Annals of Botany*, Vol. 83, No. 6, (August 1887), pp. 713-723

Brack C. & Wood, G. (1998). Tree growth and increment. In: *http://online.anu.edu.au/Forestry/mensuration/S_GROWTH.HTM,* Accessed 2010.11.17, Available from: <http://fennerschool-associated.anu.edu.au/mensuration/BrackandWood1998/T_GROWTH.HTM>

Coops, N., Waring R., & Landsberg J. (1998). Assessing forest productivity in Australia and New Zealand using a physiologically-based model driven with averaged monthly weather data and satellite-derived estimates of canopy photosynthetic capacity. *Forest Ecology and Management,* Vol. 104, No. 1-3, (n. d. 1976), pp.113-127, ISSN: 0378-1127.

Cournède, P. et al. (2009). A Forest Growth Simulator Based on Functional-Structural Modelling of Individual Trees, In: *PMA '09: Proceedings of the 2009 Plant Growth Modeling, Simulation, Visualization, and Applications*. ISBN: 978-0-7695-3988-1, pp. 34-41, IEEE Computer Society, Washington, DC, USA

Daugavietis, M. (2006). Rate of grey alder growth, In: *Grey Alder in Latvia,* ed. K. Kalnina, pp. 90-96, Silava, ISBN 978-9984-39-131-1, Latvia [in Latvian].

Garcia, O. (2005) Unifying sigmoid univariate growth equations, In: *FBMIS,* Vol. 1, (2005), pp. 63–68. ISSN 1740-5955. Available from: http://www.fbmis.info/A/5 1 Garcia, O 1.pdf.

Johnsen, K. et al. (2001). Process Models as Tools in Forestry Research and Management. *Forest Science*, Vol. 47, No. 1, (n. d.), pp. 2-8.

Kosmach, J. (2010). , *Efficiency of energy from biomass of perennial species in case of grey alder (Alnus incana) growths,* BS Theses, University of Latvia, Riga, Latvia, [in Latvian].

Landsberg, J. & Sands, P. (2010). *Physiological Ecology of Forest Production,* Principles, Processes and Models, Elsevier, 2010, **ISBN 13:** 978-0-12-374460-9.

Lefsky, M. et al., (2002). Lidar remote sensing of above-ground biomass in three biomes. *Global Ecology & Biogeography,* Vol. 11, No. 5, (January 1998) , 393-399.

Mason, E., Woollons, R. & Manley, B. (2008). *Carbon Accounting: Forest Growth Rates and Changing Climates,* University of Canterbury, Christchurch, New Zeland.

Nakawatase, J. & Peterson, D. (2006). Spatial variability in forest growth - climate relationships in the Olympic Mountains, Washington. *Can. J. For. Res.* Vol. 36, No. 1, (March 1971), pp. 77–91.

Ranson, K. et al., (1997). Forest Biomass from Combined Ecosystem and Radar Backscatter Modeling. *Remote Sensing of Environment,* Vol. 59, No. 1, (March 1969) , pp. 118-133. ISSN: 0034-4257.

Richards, G. & Brack, C. (2004). A continental biomass stock and stock change estimation approach for Australia. *Australian Forestry,* Vol. 67, No. 4, (n. d.), pp. 284–288.

Thornley, J. & Cannell, M. (2000). Managing forests for wood yield and carbon storage: a theoretical study. *Tree Physiology,* Vol. 20, No. 7, (June 1986) 477-484.

Thürig, E. et al., (2005). Evaluation of the growth function of an empirical forest scenario model. *Forest Ecology and Management,* Vol. 204, No. 1, (n. d. 1976), 53–68. ISSN: 0378-1127.

Tomppo E. et al., (2002). Simultaneous use of Landsat-TM and IRS-1C WiFS data in estimating large area tree stem volume and aboveground biomass. *Remote Sensing of Environment,* Vol. 82, No. 1, (March 1969) , pp. 156-171. ISSN: 0034-4257.

Vanclay, J. (1994). *Modelling forest growth and yield,* CAB International, ISBN 0 85198 913 6, Wallingford, UK.

Vanclay, J. (2009). Tree diameter, height and stocking in even-aged forests. *Annals of Forest Science,* Vol. 66, No. 7, (n. d.), 702. ISSN: 1297-966X. Available online at: www.afs-journal.org.

Vanclay, J. (2010). Robust relationships for simple plantation growth models based on sparse data. *Forest Ecology and Management,* Vol. 259, No. 5, (n. d. 1976), 1050–1054. ISSN: 0378-1127.

Waring R., Coops N. & Landsberg J. (2010). Improving predictions of forest growth using the 3-PGS model with observations made by remote sensing, *Forest Ecology and Management,* Vol. 259, No. 9, (n. d. 1976), 1722–1729, ISSN: 0378-1127.

Welham, C., Seely, B., & Kimmins, H. (2001). Bringing biology into growth and yield: An overview of the ecologically based forest growth model. *FORECAST, SIFERP LINK newsletter,* Vol. 3, No. 2, pp. 6-7. Available at: <http://www.forestry.ubc.ca/ecomodels/moddev/forecast/summary/links_description.htm>

Zeide, B. (2004). Intrinsic units in growth modeling. *Ecological Modelling,* Vol. 175, No. 3, (May 1975), pp. 249–259.

6

Assessment of Forest Aboveground Biomass Stocks and Dynamics with Inventory Data, Remotely Sensed Imagery and Geostatistics

Helder Viana[1], Domingos Lopes[2] and José Aranha[2]
[1]*Polytechnic Institute of Viseu*
[2]*CITAB - University of Trás-os-Montes e Alto Douro*
Portugal

1. Introduction

Several issues, related with forest fires, forest disturbances (García-Martín et al., 2008), forest productivity (Chirici et al., 2007; Palmer et al., 2009), forest changes over time (Hu & Wang, 2008), or the role of forests in the global carbon balance cycle (Hese et al., 2005) are, nowadays, the focus of numerous studies and investigations. All these subjects demand the knowledge about aboveground biomass (AGB) stocks and/or its dynamics. Besides the availability of biomass, the information about the growth of forests is of increasing importance. This variable, which is related with the total biomass growth in a specific ecosystem, is called Net Primary Production (NPP). Annual NPP represents the net amount of carbon captured by plants through photosynthesis each year (Melillo et al., 1993; Cao & Woodward, 1998). In practice, NPP can be defined and measured in terms of either biomass or CO_2 exchange (Field et al., 1995). Waring et al. (1998) define NPP as the sum of live biomass periodic increment (ΔB) and dead biomass (losses, e.g. broken branches, fallen leaves) [$NPP = \Delta B + losses$]. NPP is an important ecological variable due to its relevance for accurate ecosystem management and for monitoring the impact of human activity on ecosystems vegetation at a range of spatial scales: local, regional and global (Melillo et al., 1993). It is one of the most complete and complex variables, since it reflects the growth of the entire ecosystem thus avoiding the analysis of only part of its components. NPP provides a complete view of the ecosystem including information, not only from the arboreal stratum, but also from the shrubs and all the litter produced from each stratum. Thus, the significance of NPP not only reflects the complexity of its measurement or estimation, but also its integrative ecological perspective ecosystems.

Mapping AGB stocks or NPP with the utmost accuracy and expedite methodologies is therefore a challenge. The need of continuous maps where the phenomenon under study can be individually analysed or used as auxiliary variable in a specific model requires that the spatial predictions are represented in the most accurate way. Over the years different spatial prediction methods have been explored in diverse data type (Isaaks & Srivastava, 1989; Goovaerts, 1997; Labrecque et al., 2006; Sales et al., 2007; Meng et al., 2009). Some approaches have a simple application methodology however others are sometimes complex in what concerns to their implementation, or the selection of the variables to be used.

Estimation of AGB has been made by a range of methods, from field measurements to remote sensing-based methods, as well GIS-based modelling approaches with auxiliary data (Lu, 2006). Traditionally, to predict the spatial distribution of AGB throughout the territory, the variables calculated based on the forest inventory dataset were usually assigned to the forest polygons, stratified by species, and mapped by aerial photos interpretation. Despite the field measurements being the most accurate methods for collecting biomass data, the level of precision of the resultant biomass map will depend of the land cover classification detail and of the sample intensity. In fact, the forest inventories data at regional or national scale are often not spatially exhaustive enough to generate continuous AGB estimates, thus limiting the use of this approach over large areas. An additional limitation is the long temporal resolution of these estimations, generally made in cycles of 10 or more years, which could not be compatible with the need of analysis and monitoring of the ecosystems' dynamics.
Remote sensing-based methods have been the most widely used approach to map AGB. The utility of the spectral information recorded by remote sensing for monitoring vegetation or gathering ecophysiological information over large areas is very well recognized, since satellite data became accessible for land cover dynamic studies. Different imagery data have been employed, such as coarse spatial-resolution data as SPOT-VEGETATION (Chirici et al., 2007; Jarlan et al., 2008), NOAA AVHRR (Häme et al., 1997; Atkinson et al., 2000), MODIS (Zheng et al., 2007, Muukkonen & Heiskanen, 2007); medium spatial-resolution data as ASTER (Muukkonen & Heiskanen, 2007), Landsat TM/ETM+ (Tomppo et al., 2002; Rahman et al., 2005; Meng et al., 2009); high spatial-resolution data as IRS P6 LISS-IV (Madugundu et al., 2008) and radar data (Hyde et al., 2007; Liao et al., 2009).
AGB can be estimated by means of Direct Radiometric Relationships (DRR), which consist in establishing regression relationships, such as ordinary least squares (OLS), between the satellite spectral data (e.g. individual spectral bands, band ratios, vegetation indices and other possible transformations) as independent variables, and the measured parameter at each corresponding inventory sample plot position in each forest cover strata. AGB can be directly predicted by multiple regression analysis between spectral data response and biomass amount (Labrecque et al., 2006; Muukkonen & Heiskanen, 2007); by nonparametric approaches including K nearest neighbour (KNN) (Tomppo, 1991; Meng et al., 2007), or by artificial neural network (ANN) (Liao et al., 2009); or indirectly predicted by using characteristic such as crown diameter or leaf area index (LAI). In this case, these variables are firstly derived from the imagery data and subsequently used in regression analysis to estimate AGB.
Spatial prediction models (algorithms) have been used for spatially predicting vegetation attributes. In general, these interpolation techniques are classified in deterministic and statistical (probabilistic) models (Isaaks & Srivastava, 1989; Goovaerts, 1997; Hengl, 2009). Attending that in the Earth sciences there is usually a lack of sufficient knowledge concerning how properties vary in space, a deterministic model may not be appropriate. Therefore, to make predictions at locations for which observations do not exist, with inherent uncertainty in predictions, the use of probabilistic models is necessary (Lloyd, 2007).
Spatial statistics and geostatistics were developed to describe and analyze the variation in both natural and man-made phenomena on above or below the land surface (Cressie, 1993). Largely developed by Matheron (1963) in the 1960s, to evaluate recoverable reserves for the mining industry, geostatistical models have been systematically applied in a wide range of fields (Cressie, 1993; Goovaerts, 1997). Today, geostatistics and the theory of regionalized variables (Matheron, 1971) are used to explore and describe the presence of spatial variation that occur in most natural resource variables. Introduced to remote sensing by Woodcock et al. (1988) and by Curran (1988), geostatistical models have been used to design optimum sampling

schemes for image data and ground data; to increase the accuracy in which remotely sensed data can be used to classify land cover; or to estimate continuous variables. Geostatistical models are reported in numerous textbooks (e.g. Isaaks & Srivastava, 1989; Cressie 1993; Goovaerts, 1997; Deutsch & Journel, 1998; Webster & Oliver, 2007; Hengl, 2009; Sen, 2009) such as Kriging (plain geostatistics); environmental correlation (e.g. regression-based); Bayesian-based models (e.g. Bayesian Maximum Entropy) and hybrid models (e.g. regression-kriging).
Despite Regression-kriging (RK) is being implemented in several fields, as soil science, few studies explored this approach to spatially predict AGB with remotely sensed data as auxiliary predictor. Hence, this research makes use of RK and remote sensing data to analyse if spatial AGB predictions could be improved.
This research presents two case studies in order to explore the techniques of remote sensing and geostatistics for mapping the AGB and NPP. The first, aims to compare three approaches to estimate *Pinus pinaster* AGB, by means of remotely sensed imagery, field inventory data and geostatistical modeling. The second aims to analyse if NPP of *Eucalyptus globulus* and *Pinus pinaster* species can easily and accurately be estimated using remotely sensed data.

2. Case study I – Aboveground biomass prediction by means of remotely sensed imagery, field inventory data and geostatistical modeling

2.1 Study area

This study was carry out in Portugal (Continental), extending from the latitudes of 36° 57′ 23″ and 42° 09′ 15″N and the longitudes of 09° 30′ 40″ and 06° 10′ 45″ W (Figure 1). This area

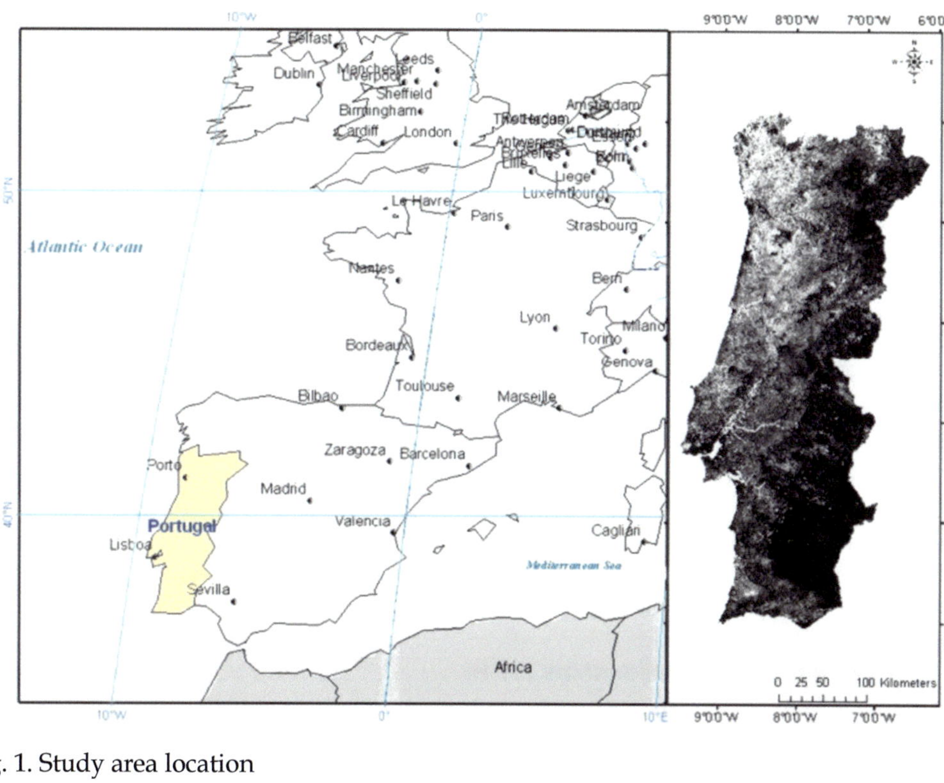

Fig. 1. Study area location

includes two distinctive bioclimatic regions: a Mediterranean bioclimate in everywhere except a small area in the North with a temperate bioclimate. With four distinct weather seasons, the average annual temperatures range from about 7 °C in the highlands of the interior north and center and about 18 ° C in the south coast. Average annual precipitation is more than 3000 mm at the north and less than 600 mm at the south.

Due to a 20 years of severe wild fires during summer time, and intense people movement from rural areas to sea side cities or county capital, forestry landscape changed from large trees' stands interspersed by agricultural lands, to a fragmented landscape. The land cover is fragmented with small amount of suitable soils for agriculture and the main areas occupied by forest spaces. Forest activity is a direct source of income for a vast forest products industry, which employs a significant part of the population.

2.2 Methods and data

2.2.1 GIS and field data

In a first stage a GIS project (ArcGis 9.x), was created in order to identify *Pinus pinaster* pure stands, over a Portuguese Corine Land Cover Map (CLC06, IGP, 2010). In a second stage, GIS project database was updated with the dendrometric data collected during Portuguese National Forestry Inventory (AFN, 2006), in order to derive AGB allometric equations, with Vegetation Indices values as independent variable. A total of 328 field plots of pure pine stands were used. The inventory dataset was further used in spatial prediction analysis, to create continuous AGB maps for the study area.

2.2.2 Biomass estimation from the forest inventory dataset

In order to calculate the biomass exclusively from the forest inventory, the biomass values measured in each field plot were spatially assigned to the pine stands land cover map polygons. In the cases where multiple plots were coincident with the same polygon, weighted averages were calculated proportionally to the area of occupation in that polygon.

2.2.3 Remote sensing imagery

In this research we used the Global MODIS vegetation indices dataset (h17v04 and h17v05) from the Moderate Resolution Imaging Spectroradiometer (MODIS) from 29 August 2006: (MOD13Q1.A2006241.h17v04.005.2008105184154.hdf; and MOD13Q1.A2006241.h17v05.005.2008105154543.hdf), freely available from the US Geological Survey (USGS) Earth Resources Observation and Science (EROS) Center. The Global MOD13Q1 data includes the MODIS Normalized Difference Vegetation Index (NDVI) and a new Enhanced Vegetation Index (EVI) provided every 16 days at 250-meter spatial resolution as a gridded level-3 product in the Sinusoidal projection.
(https://lpdaac.usgs.gov/lpdaac/products/modis_products_table/vegetation_indices/16_day_l3_global_250m/mod13q1).

MODIS data was projected to the same Portuguese coordinate system (Hayford-Gauss, Datum of Lisbon with false origin) used in the GIS project.

2.2.4 Direct Radiometric Relationships (DRR)

Using GIS tools, field inventory dataset was updated with information from MODIS images. The spectral information extracted (NDVI and EVI) was then used as independent variables for developing regression models. Linear, logarithmic, exponential, power, and second-order polynomial functions were tested on data relationship analysis. The best model achieved was then applied to the imagery data, and the predicted aboveground biomass map was produced. In some pixels where Vegetation index values were very low, the biomass values predicted by the regression equations were negative, so these pixels were removed, because in reality negative biomass values are not possible.

2.2.5 Geostatistical modeling

Regression-kriging (RK) (Odeh et al., 1994, 1995) is a hybrid method that involves either a simple or multiple-linear regression model (or a variant of the generalized linear model and regression trees) between the target variable and ancillary variables, calculating residuals of the regression, and combining them with kriging. Different types or variant of this process, but with similar procedures, can be found in literature (Ahmed & De Marsily, 1987; Knotters et al.; 1995; Goovaerts; 1999; Hengl et al.; 2004, 2007), which can cause confusion in the computational process.
In the process of RK the predictions $\left(\hat{z}_{rk(S_0)}\right)$ are combined from two parts; one is the estimate $\hat{m}(s_0)$ obtained by regressing the primary variable on the k auxiliary variables $q_k(s_0)$ and $q_0(s_0)=1$; the second part is the residual estimated from kriging $\left(\hat{e}_{(S_0)}\right)$. RK is estimated as follows (Eqs. 1 and 2):

$$\hat{z}_{rk}(s_0)=\hat{m}(s_0)+\hat{e}(s_0) \tag{1}$$

$$\hat{z}_{rk}(s_0)=\sum_{k=0}^{v}\hat{\beta}_k\cdot q_k(s_0)+\sum_{i=1}^{n}w_i(s_0)\cdot e(s_i) \tag{2}$$

where $\hat{\beta}_k$ are estimated drift model coefficients ($\hat{\beta}_0$ is the estimated intercept), optimally estimated from the sample by some fitting method, e.g. ordinary least squares (OLS) or, optimally, using generalized least squares (GLS), to take the spatial correlation between individual observations into account (Cressie, 1993); w_i are kriging weights determined by the spatial dependence structure of the residual and $e(s_i)$ are the regression residuals at location s_i.
RK was performed using the GSTAT package in IDRISI software (Eastman, 2006) both to automatically fit the variograms of residuals and to produce final predictions (Pebesma, 2001 and 2004). The first stage of geostatistical modeling consists in computing the experimental variograms, or semivariogram, using the classical formula (Eq. 3):

$$\hat{\gamma}(h)=\frac{1}{2N(h)}\sum_{i=1}^{N(h)}\left[z(x_i)-z(x_i+h)\right]^2 \tag{3}$$

where $\hat{\gamma}(h)$ is the semivariance for distance h, $N(h)$ the number of pairs for a certain distance and direction of h units, while z(xi) and $Z(x_i + h)$ are measurements at locations x_i and $x_i + h$, respectively.
Semivariogram gives a measure of spatial correlation of the attribute in analysis. The semivariogram is a discrete function of variogram values at all considered lags (e.g. Curran 1988; Isaaks & Srivastava 1989). Typically, the semivariance values exhibit an ascending

behaviour near the origin of the variogram and they usually level off at larger distances (the sill of the variogram). The semivariance value at distances close to zero is called the nugget effect. The distance at which the semivariance levels off is the range of the variogram and represents the separation distance at which two samples can be considered to be spatially independent.
For fitting the experimental variograms we tested the exponential, the gaussian and the spherical models, using iterative reweighted least squares estimation (WLS, Cressie, 1993). Finally, RK was carried out according to the methodology described in http://spatial-analyst.net. The EVI image was used as predictor (auxiliary map) in RK. GSTAT produces the predictions and variance map, which is the estimate of the uncertainty of the prediction model, i.e. precision of prediction.

2.2.6 Validation of the predicted maps

The validation and comparison of the predicted AGB maps were made by examining the discrepancies between the known data and the predicted data. The dataset was, prior to estimates, divided randomly into two sets: the prediction set (276 plots) and the validation set (52 plots). According to Webster & Oliver (1992), to estimate a variogram 225 observations are usually reliable. The prediction approaches were evaluated by comparing the basic statistics of predicted AGB maps (e.g., mean and standard deviation) and the difference between the known data and the predicted data were examined using the mean error, or bias mean error (ME), the mean absolute error (MAE), standard deviation (SD) and the root mean squared error (RMSE), which measures the accuracy of predictions, as described in Eqs. (4-7).

$$SD = \sqrt{\frac{1}{N-1}\sum_{i=1}^{N}(e_i - \overline{e})^2} \tag{4}$$

$$ME = \frac{1}{N}\sum_{i=1}^{N}(\hat{e}_i - e_i) \tag{5}$$

$$MAE = \frac{1}{N}\sum_{i=1}^{N}\left|\hat{e}_i - e_i\right| \tag{6}$$

$$RMSE = \sqrt{\frac{1}{N}\sum_{i=1}^{N}(\hat{e}_i - e_i)^2} \tag{7}$$

where: N is the number of values in the dataset, $\hat{e}_i$ is the estimated biomass, e_i is the biomass values measured on the validation plots and $\overline{e}$ is the mean of biomass values of the sample.

2.3 Results and discussion

2.3.1 *Pinus pinaster* stands characteristics

The descriptive statistics of pine stands data are presented in Table 1, where: N is the number of trees; t is the forestry stand age; h_{dom} is the dominant height; dbh_{dom} is the dominant diameter at breast height; SI is the site index; BA is the basal area; V is the stand volume and AGB is the biomass in the sample plot.

The pine stands are highly heterogeneous with ages ranging from 8 to 110 years old and the biomass per hectare ranging from 0.9 to 136.1 ton ha^{-1}. The values of Biomass present a normal distribution with mean m = 52.12 ton ha^{-1} and standard deviation σ = 32.32 ton ha^{-1} (Figure 2).

	Pine stands plots							
	N (trees ha^{-1})	t (year)	h_{dom} (m)	dbh_{dom} (cm)	SI (m)	BA (m^2 ha^{-1})	V (m^3 ha^{-1})	AGB (ton ha^{-1})
Mean	566	31	13.4	25.3	11.8	14.39	99.46	52.12
Min	20	8	4.6	8.9	0.0	0.41	1.37	0.85
Max	2219	110	36.5	59.0	69.0	38.34	259.03	136.09
SD	405.2	15.9	4.0	8.0	11.5	7.64	61.86	32.32

Table 1. Descriptive statistics of data measured in the forest inventory dataset

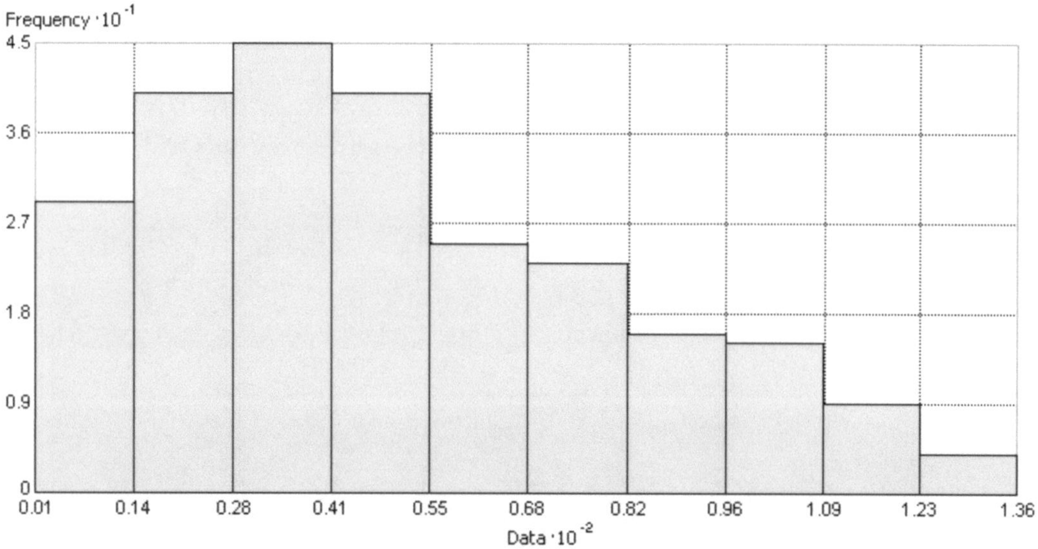

Fig. 2. Histogram of the distribution of the AGB (ton ha^{-1}) in the forest inventory dataset

2.3.2 Aboveground biomass estimation from the inventory dataset

The estimates based in the inventory dataset were achieved by assigning the 328 field plot biomass values (weighted by each polygon area) into all the polygons of the pine cover class. After the global calculation, the dataset used for training (276 plots) was used to make a first validation of this approach. Hence, a regression was established between the biomass values, measured in the field plots, and the forest inventory polygon data. In Figure 3 it is presented the positive relationship between the measured and the predicted data with a coefficient of determination (R^2) of 0.71.

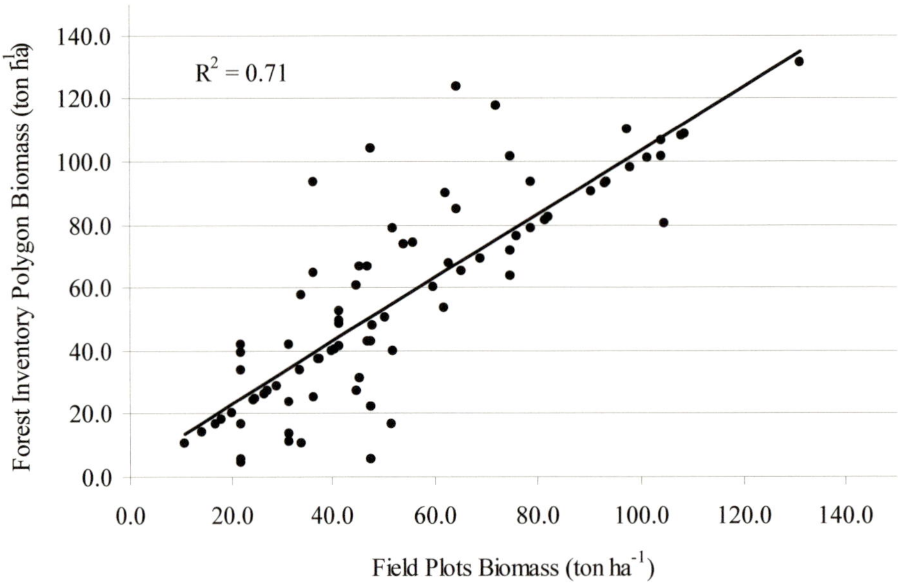

Fig. 3. Relationship between the biomass data measured in field plots and the predicted data extracted in the polygons of land cover map

2.3.3 Aboveground biomass estimation from DRR

After performing correlation analyses, between AGB and Vegetation indices, several regression models were developed using stand-wise forest inventory data and the MODIS vegetation indices (NDVI and EVI) as predictors.

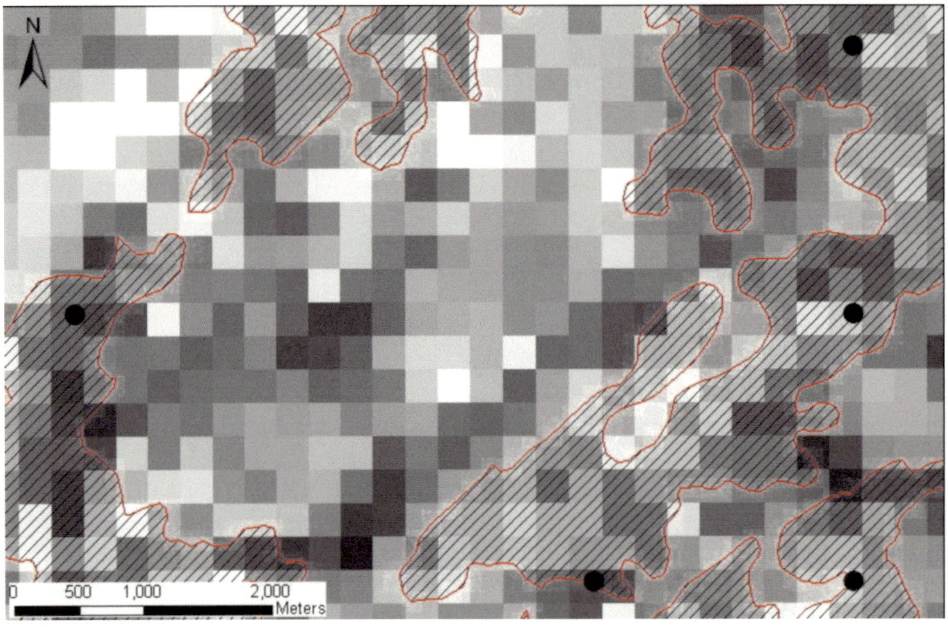

Fig. 4. MODIS image showing the effect of pixels (250m) in the edge of polygons

The best correlation was obtained with EVI as independent variable as (Eq. 8):

$$AGB = 322.4(EVI) - 39.933 \ (R^2 = 0.32) \quad (8)$$

The AGB was then estimated for the entire study area. The low correlation achieved is explained, in part, by the heterogeneity of pine stands and the high effect of mixed pixels (Burcsu et al., 2001) in coarse resolution MODIS data (250 m).

As it can be seen in Figure 4, the reflectance value recorded in the boundary pixels of the polygons limits is not pure, they record both pine stands, and the neighbouring land cover classes reflectance values.

2.3.4 Aboveground biomass estimation from geostatistical methods

To spatially estimate the AGB by geostatistical approach, the first step consisted in the modeling and analysis of the experimental semivariograms (Eq. 3). The directional semivariograms of the residuals showed anisotropy at 38.6º, so at this direction were fitted Exponential, Gaussian and Spherical models. Based on experimentation, the exponential variogram model was fitted better (nugget of 703.75 and a partial sill of 390.17 reaching its limiting value at the range of 43,9Km) to the calculated biomass pine stands data (Figure 5). The present data showed a low spatial autocorrelation. The high nugget effect, visible in the figure, which under ideal circumstances should be zero, suggests that there is a significant amount of measurement error present in the data, possibly due to the short scale variation.

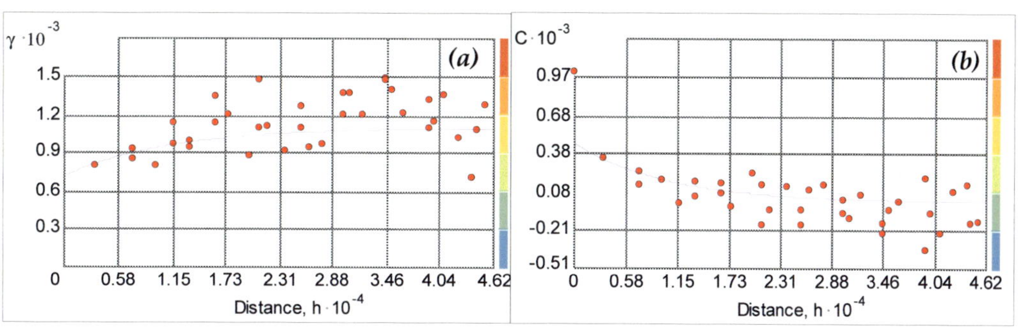

Fig. 5. Directional experimental semivariogram (38.6º) with the exponential model fitted (a) and covariance (b)

2.3.5 Validation and comparison of the aboveground biomass estimation approaches

The validation of the AGB estimation approaches was made by comparing the calculated basic statistics (Table 2) in the 52 validation random samples. Training and validation sets were compared, by means of a Student's *t* test (t = 0.882 ns), in order to check if they provided unbiased sub-sets of the original data.

As expected, the Inventory Polygons method produced the best statists. The mean error (ME), which should ideally be zero if the prediction is unbiased, shows a bias in the three approaches, being lower in the Inventory polygons method, and higher in the DRR method. The analysis of the root mean squared errors (RMSE), shows that Inventory Polygons present the lower discrepancies in the estimations (RMSE=33.53%), and RK achieve estimations under lower errors (RMSE=51.95%) than the DRR approach (RMSE=61.62%). Despite this, the errors from the two prediction approaches are very high, which can be explained by the low correlation found between the vegetation indices data, as explained above. This limitation can be overcome by using remote sensing data with higher spatial resolution. Moreover, the work area must also be sectioned into smaller areas, to minimize the heterogeneity that is observed in very large landscapes.

Method	Estimated AGB (average - ton ha^{-1})	ME (ton ha^{-1})	MAE (ton ha^{-1})	RMSE (ton ha^{-1})	SD (ton ha^{-1})	RMSE %
Inventory Polygons	53.94	-3.11	11.26	18.09	27.70	33.53
DRR	50.23	-6.83	25.84	30.95	22.03	61.62
RK	52.01	-5.05	22.70	27.02	19.67	51.95

Table 2. Statistics of validation plots for the AGB prediction methods

In order to determine the significance of the differences between interpolation methods, analysis of variance (ANOVA) was performed (Table 3). The results show that, at alpha level 0.05, do not exist significant differences between the biomass values, predicted by the different methods.

Source	DF	SS	MS	F	P
Between	2	122.86	61.432	0.123	0.884
Within	243	113453.67	497.604		
Total	245	113576.54			

Table 3. Results from ANOVA to compare the differences between the means of the different prediction methods

A quantitative comparison of the complete AGB maps, estimated by the three approaches, was additionally made. The estimates (ton ha^{-1}) are shown in the Table 4. In order to better preserve the land cover areas, the maps were brought to the resolution of 50x50m, and then clipped by the pine land cover mask.

Method	Pixels	Area (ha)	AGB (average – ton ha^{-1})	Std (ton ha^{-1})	B (tonnes)
Inventory Polygons		300446	53.8	30.8	15564351
DRR	1191597	297899	53.8	20.0	16020055
RK	1189213	297303	52.8	21.3	15711245

Table 4. Summary statistics of predicted pine AGB maps

The three AGB maps originates very similar average values (ton ha^{-1}), and the differences between the maximum and minimum values of total biomass (tonnes) estimated by the different methods varies less than 1.6%.

Although there has been a low discrepancy between the total biomass values, estimated by three maps, the analysis of the correlation coefficient of regressions, carried out between the three maps, show low to moderate correlation between *Inventory Polygons x DRR* and *Inventory Polygons x RK* methods (R = 0.27 and 0.40, respectively). Only DRR x RK methods present high correlation values (R = 0.95) indicating a very similar biomass estimation at individual pixels (Figure 6).

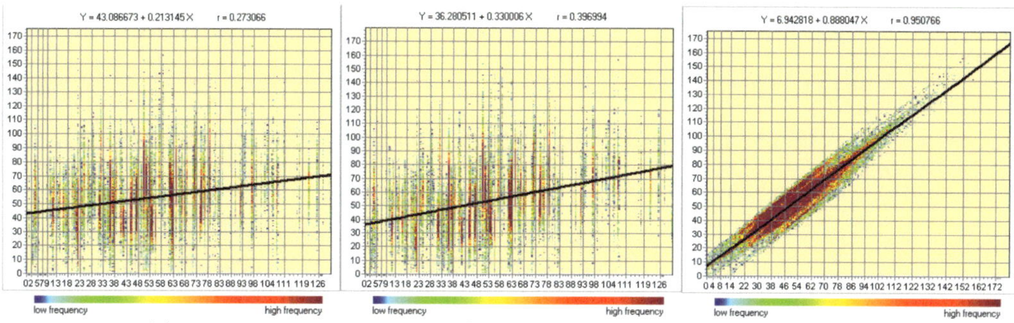

(a) (b) (c)

Fig. 6. Regression performed between AGB maps (a) *Inventory Polygons x DRR; (b) Inventory Polygons x RK*; (c) *DRR x RK*

Based in the calculated statistics of the validation dataset and in the global biomass estimations for entire area, we can consider that the Regression-kriging geostatistical prediction approach, with remotely sensed imagery as auxiliary variable, increases the classifications accuracy when compared with estimates based merely in the Direct Radiometric Relationships (DRR). Furthermore, the accuracy of these estimations could increase by using imagery data with higher spatial resolution, and if the work region is more homogeneous.

The biomass maps derived by the three methods (Inventory Polygons, Direct Radiometric Relationships and Regression-Kriging) for the whole study area are presented in Figure 7.

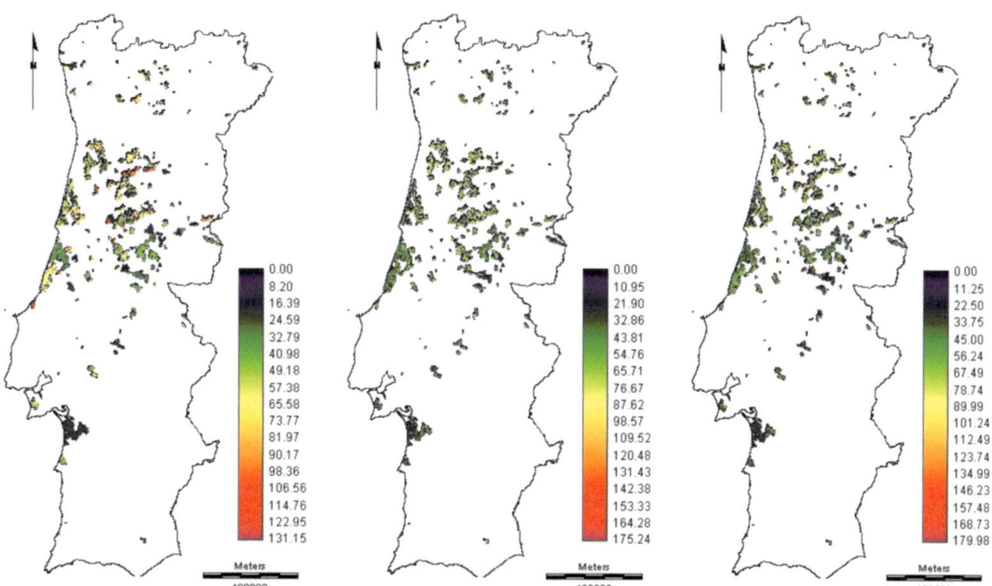

(a) (b) (c)

Fig. 7. Aboveground biomass maps (a) Inventory Polygons (b) DRR and (c) RK

3. Case study II – Biomass growth (NPP) of *Pinus pinaster* and *Eucalyptus globulus* stands, in the north of Portugal. Estimations by means of LANDSAT ETM+ images

3.1 Study area

This research took place within an area in the northern part of Portugal where *Pinus pinaster* Ait. and *Eucalyptus globulus* Labill constitute the two most important forest species in terms of forested area (Figure 8).

The *P. pinaster* study area is a 60 km^2 rectangle (10 km × 6 km) with extensive stands of this species located at the north of Vila Real (41°39'N, 7°35'W) and the *E. globulus* study area is a 24km^2 rectangle (4 km × 6 km) of extensive stands of this species located at west of Vila Real (41°2'N, 7°43'W).

Both species are ecologically well adapted, despite *E. globulus* being an exotic tree, and the case study areas are representative of these ecosystems in Portugal. The *P. pinaster* forest is very heterogeneous in canopy density, has experienced only limited human intervention, and covers a wide range of structures, varying widely in terms of number of trees per hectare, average dimensions, and age groups. The *E. globulus* forest is much more homogeneous and has been more extensively investigated to enable greater timber production, which is very valuable for pulp production.

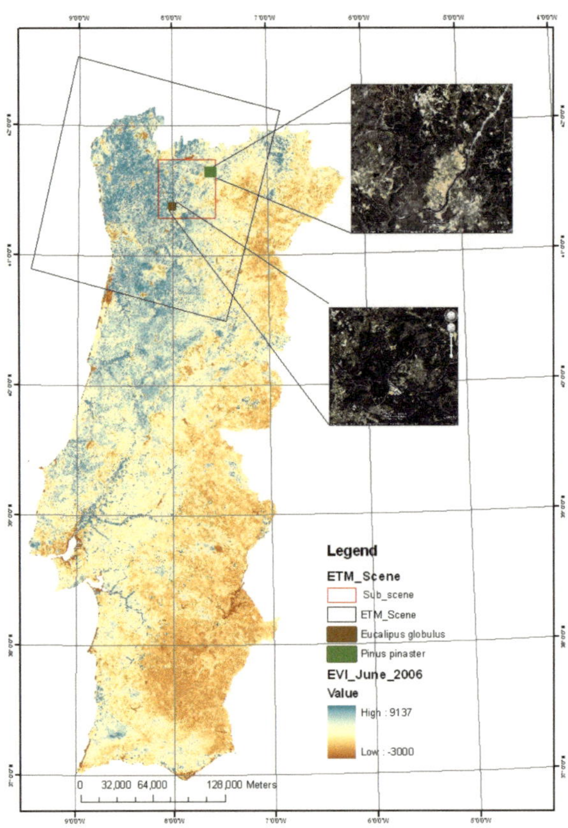

Fig. 8. Study area.

3.2 Methods and data

3.2.1 Methodology used in geometric and radiometric corrections

The available LANDSAT-7 ETM+ Image was acquired on the 15th of September 2001 at 10:02:13 (UTC). The image was geometrically and radiometrically corrected using MiraMon ("WorldWatcher"). This software allows displaying, consulting and editing raster and vector maps and was developed by the Autonomous University of Barcelona (UAB) remote sensing team. The software allows for the geometric correction of raster (e.g., IMG and JPG: satellite images, aerial photos, scan maps) or vector maps (e.g., VEC, PNT, ARC and POL and NOD), based on ground control points coordinates.

In the present research the ground control points were collected from Portuguese topographic maps on a 1/25000-scale, using the original ETM+ Scene. Twenty-five control points were collected (Toutin, 2004) to allow image correction and eleven control points were used for its validation. A first-degree polynomial correction was chosen for the geometric correction, using the nearest neighbour option for the resampling process.

Two Digital Elevation Models (DEMs) were constructed for each study area (*Pinus pinaster* and *Eucaliptus globulus* - see Figure 8), based on 10 m contour lines. The first DEM had a spatial resolution of 15 m and was used to correct the panchromatic band, mainly to allow identification of the ground control points due to its better spatial resolution. The second DEM had a spatial resolution of 20 m and was used for the correction of the LANDSAT ETM+ bands 1, 2, 3, 4, 5, and 7. Those 20 m DEMs were merged with a altitude model for Europe, with a pixel size of 1 Km. The radiometric correction was based on the lowest radiometric value for each band which is well known as the *kl*, and should be collected from the histogram analysis (Pons & Solé-Sugrañes, 1994 and Pons, 2002).

3.2.2 Methodology used to calculate vegetation indices

Within the study area, 31 sampling plots for the *Eucalyptus globulus* and 34 for the *Pinus pinaster* were surveyed and the coordinates of the centre of each plot recorded by Global Positioning System (GPS). The plots' location could then be identified on the geo-corrected images and reflectance data extracted for each ETM+ band. These data were then used to calculate a series of vegetation indices (Table 5), which were further used to analyse potential relationships with the forest variables.

In table 5, G represents the reflectance on the green wavelength; R is the reflectance in the red wavelength; NIR is the reflectance in the near infrared wavelength; and MIR1 and MIR2 are the reflectance in the two middle infrared bands from LANDSAT ETM+ image.

3.2.3 Model adjustment and selection

The available data (31 sampling plots for the *Eucalyptus globulus* and 34 for the *Pinus pinaster*) were divided in two groups, one for the adjustment of mathematical models and the other for the validation. An overall analysis of the correlation matrix allowed to identify the variables strongest related to NPP, which were then selected to establish regression models to Estimate NPP. The best NPP prediction models were selected based in the following statistics: the coefficient of determination (R^2); the adjusted coefficient of determination (R^2adj.); the root mean square error (RMSE); and the percentage root mean square error (RMSE%).

	Designation	Mathematical expression	Source
1	NDI(MIR1)	$\frac{(NIR-MIR1)}{(NIR+MIR1)}$	Lucas (1995)
2	NDI(MIR2)	$\frac{(NIR-MIR2)}{(NIR+MIR2)}$	Lucas (1995)
3	NDVI	$\frac{(NIR-R)}{(NIR+R)}$	Rouse *et al.* (1974); Bouman (1992); Malthus *et al.* (1993); Xia (1994); Nemani *et al.* (1993); Baret *et al.* (1995); Hamar *et al.* (1996); Fassnacht *et al.* (1997); Purevdorj *et al.* (1998); Todd *et al.* (1998); and Singh *et al.* (2003)
4	MVI1	$\frac{MIR1}{MIR2}$	Fassnacht *et al.* (1997)
5	MVI2	$\frac{NIR}{MIR2}$	Fassnacht *et al.* (1997)
6	RVI1	$\frac{NIR}{R}$	Tucker (1979); Xia (1994); Baret *et al.* (1995); Hamar *et al.* (1996); Fassnacht *et al.* (1997); and Xu *et al.* (2003).
7	TVI1	$\sqrt{\frac{NIR}{R}}$	Tucker (1979)
8	TVI2	$\frac{(NIR+R)}{(NIR-R)}$	Tucker (1979)
9	TVI9	$\sqrt{\frac{(G-R)}{(G+R)}+0{,}5}$	Tucker (1979)

Table 5. Vegetation indices used in the research

3.2.4 Comparison of the NPP images

NPP images obtained from different methodologies were compared by the *Kappa* index of agreement. Kappa was adopted by the remote sensing community as a useful measure of classification accuracy Rossiter (2004). The *Kappa* coefficient (*K*) measures pairwise agreement among a set of coders making category judgments, thus correcting values for expected chance of agreement (Carletta, 1996).

The overall *kappa* statistic, defining the overall proportion of area correctly classified, or in agreement, is calculated by the mathematical expression defined by Eq. 9 (Stehman, 1997; Rossiter, 2004):

$$\hat{k} = \frac{\sum_{i=1}^{k} p_{ii} - \sum_{i=1}^{k} P_{i+}.P_{+i}}{1 - \sum_{i=1}^{k} P_{i+}.P_{+i}} \quad (9)$$

where:

k = number of land-cover categories

$\sum_{i=1}^{k} p_{ii}$ represents the overall proportion of area correctly classified

$\sum_{i=1}^{k} P_{i+}.P_{+i}$ is the expected overall accuracy if there were chance agreement between reference and mapped data

According to Green (1997) when there is complete agreement between two maps K=1, and a kappa value of zero, the two maps are said to be unrelated.

Moss (2004) considers that when Kappa is less than 20 the strength of agreement between both images is poor; between 0.21 and 0.40 fair; between 0.41 and 0.60 moderate; between 0.61 and 0.80 good; higher than 0.81 very good. However, according to Green (1997), kappa lower than 0.40 indicates a low degree of agreement; between 0.40 and 0.75 a fair to good degree of agreement; and higher than 0.75 a high degree of agreement.

3.3 Results and discussion

3.3.1 Identification of the best prediction variables

In order to identify whether if it was possible to directly or indirectly estimate NPP from the remote sensing data, the Vegetation Index better correlated with NPP was identified from the general correlation matrix and analysed. The most relevant results are summarised in Table 6.

	Pinus NPP	*Eucalyptus* NPP
DN_B	-0.179	-0.739
DN_G	-0.268	-0.692
DN_R	-0.194	-0.688
DN_NIR	0.344	-0.280
DN_MIR1	-0.078	-0.605
DN_MIR2	-0.174	-0.614
TVI2	-0.142	-0.535
TVI9	0.030	0.288
MVI1	0.486	0.427
MVI2	0.435	0.318
NDVI	0.280	0.519
NDI(MIR1)	0.181	0.386
NDI(MIR2)	0.232	0.466

Table 6. Correlation between NPP and the reflectance from each individual band and some vegetation indices

As presented in Table 6, *Pinus* NPP shows the higher correlation (positive) with the near infrared wavelength band, while *Eucalyptus* NPP is better correlated (negatively) whit the middle infrared wavelength band.

The NDVI and TVI2 are the best correlated indices for the *Eucalyptus* and the MVI1 and MVI2 for the *Pinus*. These results reflect the initial observation when only reflectance from each individual band was analysed.
The best correlated vegetation indices were selected as independent variables for adjusting regression models to estimate NPP.

3.3.2 Models for the NPP *Eucalyptus globulus* estimation

The best mathematical models to estimate the NPP for the *Eucalyptus* stands and the basic statistics (ME and MAE) calculated from the validation dataset are presented in Table 7.

Mathematical models	NPP adjusted models statistics				Validation dataset statistics	
	R^2	$R^2_{adj.}$	s_{yx}	s_{yx} (%)	ME	MAE
$NPP=27.644-0.243B-0.0007GR^2-0.00014R^2$	0.613	0.558	2.988	22.5	-1.631	2.758
$NPP_{arboreal}=89.260NDVI^2-117.195NDVI^3$ $NPP=-13.114+12.271NPP_{arboreal}-1.818(NPP_{arboreal})^2+0.091(NPP_{arboreal})^3$	0.936 0.694	0.933 0.695	1.654 2.656	35.4	0.116 -1.198	1.238 3.098
$NPP=3.593+167.750NDVI^2-233.667NDVI^3$	0.493	0.447	3.342	25.2	-0.340	2.959
$NPP_{litter}=56.584NDVI^2-69.233NDVI^3$ $NPP=7.893(NPP_{litter})^{0.412}$	0.812 0.678	0.805 0.666	2.088 2.484	53.0 18.7	-0.150 -0.589	1.309 2.834
$NPP=17.672-0.611TVI2^2+0.048TVI2^3$	0.422	0.370	3.567	26.9	-0.347	2.903
$G=13.431-155.484NDVI+648.846NDVI^2-635.713NDVI^3$ $NPP=-5.787+4.652G-0.339G^2+0.008G^3$	0.657 0.634	0.608 0.581	4.170 2.908	33.1 21.6	1.121 -0.779	2.687 3.347
$G=38.150-0.300GR-0.174MIR1$ $NPP=-5.787+4.652G-0.339G^2+0.008G^3$	0.793 0.634	0.774 0.581	3.168 2.908	33.7 21.6	-1.754 -2.199	2.754 3.662

Table 7. Selected models to estimate *Eucalyptus* NPP, and validation dataset statistics

The observed standard error of the estimates are lower in the model using as independent variable the blue, the green and the red reflectances, and in the model using the NDVI, respectively. However, the model with NDVI as independent variable reveals a lower ME. Additionally, this model has a superior applicability since the individual bands reflectance have a great variation along the year, thus varying from image to image.
Based in the field measurements and in the estimated NPP, by the model using only the NDVI directly as independent variable (R^2=0.493), two images were created for the entire study area (Figures 9a and 9b).
After the classification into four classes (1 - NPP < 5 ton $ha^{-1}year^{-1}$; 2- 5≤ NPP <10 ton $ha^{-1}year^{-1}$; 3 - 10 ≤ NPP < 15 ton $ha^{-1}year^{-1}$; and 4 - NPP > 15 ton $ha^{-1}year^{-1}$) the cross tabulation was carried out and the matrix error table analysed.
Kappa statistic showed a slight agreement around 37%. However, for a first approach these results are a good indicator for further studies. From the analyses of the *Eucalyptus* NPP map, obtained from fieldwork, it can be observed that there are no areas with an NPP lower than 5 ton $ha^{-1}year^{-1}$, and almost the whole *Eucalyptus* stand presents NPP figures between 10 and 15 ton $ha^{-1}year^{-1}$.

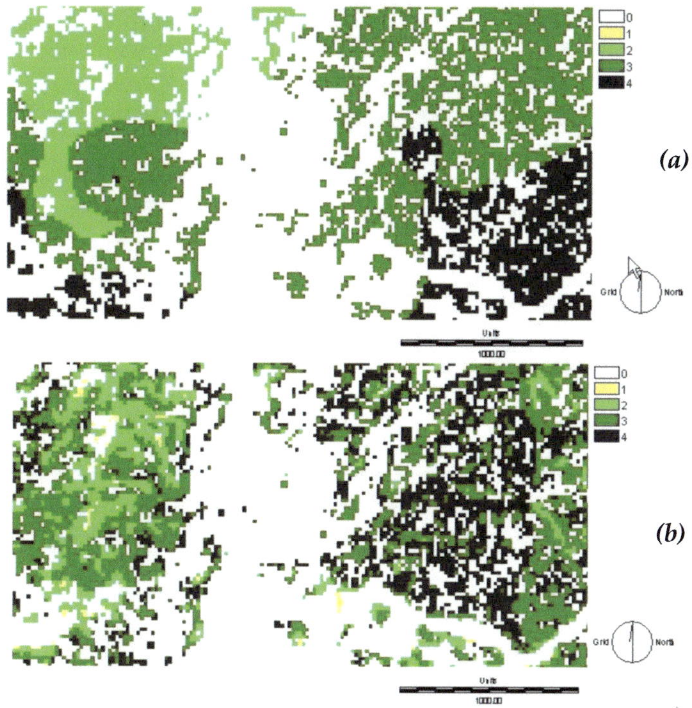

Fig. 9. *Eucalyptus* NPP estimations from field measurements (a) and NDVI model (b).

A significant result to estimate *Eucalyptus* NPP was obtained with the basal area (G) as independent variable (R^2=0.634). In this case, the basal area can be estimated with acceptable confidence, using the NDVI or MIR1 as independent variables (R^2=0.657 and 0.793, respectively). In alternative, *Eucalyptus* NPP can also be estimated indirectly, with acceptable accuracies, by the litter present in the *Eucalyptus* stands (R^2=0.678). A strong relationship was found between NPP from litter and NDVI (R^2=0.812). The same methodology can be used by estimating, in a previous stage, the NPP arboreal with the NDVI as independent variable (R^2=0.936) and subsequently, indirectly estimate the *Eucalyptus* NPP (R^2=0.694).

3.3.3 Models for the NPP *Pinus pinaster* estimation

The best mathematical models to estimate the NPP for the *Pinus* stands and the basic statistics (ME and MAE) calculated from the validation dataset are presented in Table 8. The observed standard error of the estimates, as well the ME achieved from the validation dataset shows that the best model is obtained in the model using as independent variable the MVI1 for estimate the NPP of shrubs. The NPP of pine is subsequently estimated indirectly using this variable.

As in the *Eucalyptus* predictions the same methodology was implemented to compare the final maps achieved for the Pinus stands. The *Pine* NPP model using only the MVI1 as independent variable was used (R^2=0.417). The two created maps for the entire study area (Figures 10a and 10b), were classified into four classes (1 - NPP < 5 ton $ha^{-1}year^{-1}$; 2- 5≤ NPP <10 ton $ha^{-1}year^{-1}$; 3 - 10 ≤ NPP < 15 ton $ha^{-1}year^{-1}$; and 4 - NPP > 15 ton $ha^{-1}year^{-1}$), a cross tabulation was carried out and the matrix error table analysed. Kappa statistic showed an agreement around 48%, slightly better than in *Eucalyptus* estimations. However, it was observed that the achieved model was not able to identify and locate the extreme values of NPP (e.g. neither the most productive areas nor the least productive ones).

Mathematical models	NPP adjusted models statistics				Validation dataset statistics	
	R^2	$R^2_{adj.}$	s_{yx}	s_{yx}(%)	ME	MAE
$NPP=51.288-32.080MVI1+6.787MVI1^2$	0.417	0.369	4.617	31.7	-0.902	1.974
$NPP_{shrubs}=-0.516MVI1^2+0.414MVI1^3$	0.816	0.809	2.614	71.3	-0.279	2.146
$NPP=10.629+1.071NPP_{shrubs}$	0.649	0.635	3.508	27.5	-0.317	1.677
$NPP_{shrubs}=1.146+0.142MVI2^2$	0.486	0.466	3.196	83.8	-0.490	2.268
$NPP=10.629+1.071NPP_{shrubs}$	0.649	0.635	3.508	27.6	-0.842	2.276

Table 8. Selected models to estimate *Pinus* NPP and validation dataset statistics

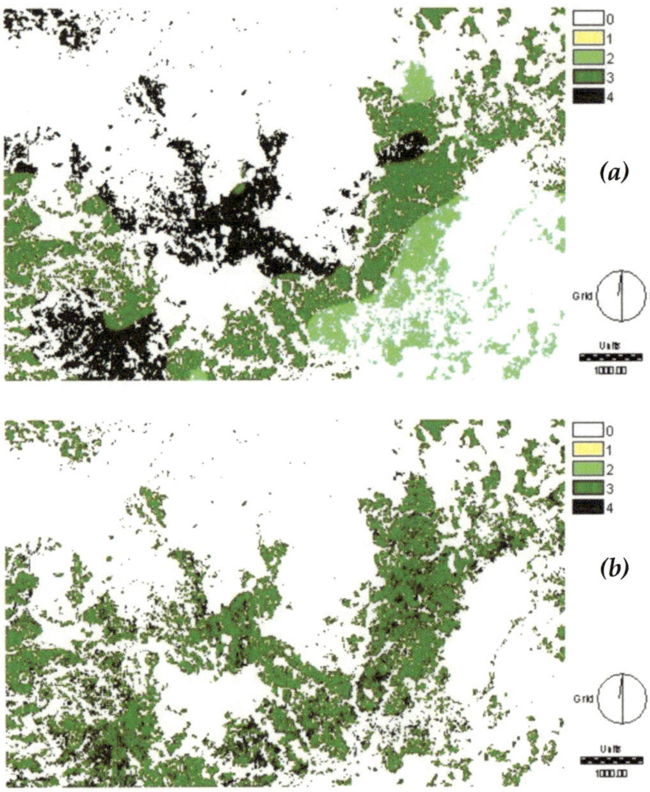

Fig. 10. Pinus NPP estimations from field measurements (a), and the MVI1 model (b).

For the *Pinus* stands, it was possible to estimate the total NPP (R^2=0.816) knowing only the NPP from shrubs. In this case, the NPP from shrubs was predicted using the MVI as auxiliary variable (R^2=0.645).

4. Conclusions

In this research, AGB and NPP estimates were carried out by means of forest inventory data remote sensing imagery and geostatistical modeling. The general conclusions are:

In the case study I, tree Aboveground biomass (AGB) mapping approaches were compared: Inventory Polygons; Direct Radiometric Relationships (DRR) and Regression-kriging (RK). Pure pine stands were mapped and AGB estimates were achieved using data collected in the National Forest inventory dataset. The Inventory polygons method was used since the field plots of forest inventory dataset fall within all the polygons of the forest cover map. At the same time, this approach was used to compare and validate DRR and RK methods.

The results showed that DRR and RK, using Vegetation Indices transformed from MODIS remotely sensed data, can be used for biomass mapping purposes. However, it should be pointed out that, in the present research, the coarse resolution of MODIS (250m) data associated with small polygons of the pine landcover class did not allow to extract the pure spectral response of this vegetation type. Hence, the correlation between AGB and NDVI as independent variable is not as high as desired.

This limitation can be overcome by using images with higher spatial resolution. Moreover, these methodologies can be applied with greater accuracy in areas where land cover polygons are large enough to minimize, as much as possible, the effect of edging.

The analysis of statistical parameters of validation dataset such as the mean error (ME), the mean absolute error (MAE), standard deviation (SD) and the root mean squared error (RMSE) show that RK, making use of geostatistical modeling techniques, combined with remote sensing data as auxiliary variable improves the predictions when compared to DRR. Furthermore, RK has the advantage of generating estimates for the spatial distribution of AGB and its uncertainty for the study area. The uncertainty maps allow the evaluation of the reliability of estimates by identifying the sites with major uncertainties which can be useful to select different estimation methods for those areas.

In the case study II, some simplified methodologies were proposed to estimate NPP. For the *Eucalyptus* ecosystem using the basal area or the NPP from litter, and for the *Pinus* ecosystem using the NPP from shrubs.

Despite the direct NPP estimation from remote sensing data did not provide very promising results, it was possible to establish indirect relationships between some vegetations indices calculated from Landsat ETM+ imagery data and the litter NPP, shrubs NPP and from basal area of the studied forest stands.

Those simplifications can be extremely important when time and economic resources are limited. The importance of those methodologies could become more relevant as NPP is a variable very difficult to obtain, consuming time and demanding hard fieldwork.

The loss in accuracy is certainly compensated by decrease of fieldwork. The balance between both should only be taken in each particular case, considering the general context of each situation (e.g., time and funds available, human resources available, objectives of the research).

5. Acknowledgements

Authors would like to express their acknowledgement to the Portuguese Science and Technology Foundation (FCT), programmes SFRH/PROTEC/49626/2009 and FCT FCOMP-01-0124-FEDER-007010 (PTDC/AGR-CFL/68186/2006).

6. References

AFN (2006). Dados do Inventário Florestal Nacional (05/06) Information retrieval tool. Autoridade Florestal Nacional. Ministério da Agricultura do Desenvolvimento Rural e das Pescas, Lisboa.

Ahmed, S., & de Marsily, G (1987). Comparison of geostatistical methods for estimating transmissivity using data on transmissivity and specific capacity. *Water Resources Research* 23(9): 1717-1737.

Atkinson, P. M.; Foody, G. M.; Curran, P. J., & Boyd D. S. (2000). Assessing the ground data requirements for regional scale remote sensing of tropical forest biophysical properties. *International Journal of Remote Sensing* 21(13-14): 2571-2587.

Baret, F., Clevers, J.G. and Steven, M.D. (1995). The robustness of canopy gap fraction estimates from red and near-infrared reflectances: A comparison of approaches. *Remote Sensing of Environment* 54: 141-151.

Bouman, B.M. (1992). Accuracy of estimating the leaf area index from vegetation indices derived from crop reflectance characteristics: a simulation study. *International Journal of Remote Sensing* 13(16): 3069-3084.

Burcsu, T. K., Robeson, S. M., & Meretsky, V.J. (2001). Identifying the Distance of Vegetative Edge Effects Using Landsat TM Data and Geostatistical Methods. *Geocarto International* 16(4): 61-70.

Cao, M. and Woodward, F.I. (1998). Net primary and ecosystem production and carbon stocks of terrestrial ecosystems and their responses to climate change. *Global Change Biology* 4: 185-198.

Carletta, J. (1996). Assessing agreement on classification tasks: the kappa statistic. *Computational linguistics* 22(2): 1-6.

Chirici, Gherardo; Barbati, Anna, & Maselli, Fabio (2007). Modelling of Italian forest net primary productivity by the integration of remotely sensed and GIS data. *Forest Ecology and Management* 246: 285-295.

Cressie, N. A. C. (1993). *Statistics for Spatial Data.* New York, USA: John Wiley & Sons. pp. 416.

Curran, P.J. (1988). The semivariogram in remote sensing: An introduction. *Remote Sensing of Environment* 24: 493-507.

Deutsch, C. V., & Journel A. G. (1998). *Geostatistical Software Library and User's Guide.* (2nd ed.). New York, USA: Oxford University Press, pp. 369.

Eastman, J. R. (2006). *Idrisi Andes. Guide to GIS and Image Processing.* USA: Clark Labs. Clark University, pp. 328.

Fassnacht, K.S., Gower, S.T., MacKenzie, M.D., Nordheim, E.V. and Lillesand, T.M. (1997). Estimating the leaf area index of north central Wisconsin forests using the Landsat Thematic Mapper. *Remote Sensing of Environment* 61: 229-245.

Field, C.B., Randerson, J.T. and Malmstrom, C.M. (1995). Global net primary production: combining ecology and remote sensing. *Remote Sensing of Environment* 51: 74-88.

García-Martín, A., Pérez-Cabello, F., de la Riva, J. R., & Montorio, R. (2008). Estimation of crown biomass of Pinus spp. from Landsat TM and its effect on burn severity in a Spanish fire scar. *IEEE Journal of Selected Topics in Applied Earth Observations and Remote Sensing* 1(4): 254-265.

Goovaerts, P. (1997). *Geostatistics for Natural Resources Evaluation.* New York, USA: Oxford University Press, Inc., pp. 496.

Goovaerts, P. (1999). Using elevation to aid the geostatistical mapping of rainfall erosivity. *Catena 34*(3-4): 227-242.

Green, A.M. (1997). Kappa statistics for multiple raters using categorical classifications, *Proceedings of the twenty-second annual SAS @Users Group International Conference*, Cary, NC: SAS Institute, Inc: 1110-1115.

Hamar, D., Ferencz, C., Lichtenberger, J., Tarcsa, G. and Ferencz-Árkos, I. (1996). Yield estimation for corn and wheat in the hungarian great plain using landsat MSS data. *International Journal of Remote Sensing* 17(9): 1689-1699.

Häme, T., Salli, A., Andersson, K., & Lohi, A. (1997). A new methodology for estimation of biomass of conifer-dominated boreal forest using NOAA AVHRR data. *International Journal of Remote Sensing* 18: 3211-3243.

Hengl T., Heuvelink G. M. B., & Stein A. (2004). A generic framework for spatial prediction of soil variables based on regression-kriging. *Geoderma 122*(1-2): 75-93.

Hengl T., Heuvelink, G. B. M. & Rossiter, D. G. (2007). About regression-kriging: from equations to case studies. *Computers and Geosciences 33*(10): 1301-1315.

Hengl, T. (2009). *A Practical Guide to Geostatistical Mapping*. Amsterdam, ISBN 978-90-9024981-0.

Hese, S., Lucht, W., Schmullius, C., Barnsley, M., Dubayah, R., Knorr, D., Neumann, K., Riedel, T., & Schröter, K. (2005). Global biomass mapping for an improved understanding of the CO2 balance-the Earth observation mission Carbon-3D. *Remote Sensing of Environment* 94(1): 94-104.

Hu, Huifeng, & Wang, G. G. (2008). Changes in forest biomass carbon storage in the South Carolina Piedmont between 1936 and 2005. *Forest Ecology and Management* 255(5-6): 1400-1408.

Hyde, P., Nelson, R., Kimes, D., & Levine, E. (2007). Exploring LiDAR–RaDAR synergy-predicting aboveground biomass in a southwestern ponderosa pine forest using LiDAR, SAR and InSAR. *Remote Sensing of Environment* 106(1): 28-38.

IGP (2010). *CLC2006 – Cartografia CORINE Land Cover 2006 para Portugal Continental; 2009.* <http://www.igeo.pt/gdr/projectos/clc2006/> (verified 01.01.2010).

Isaaks, E. H., & Srivastava, R. M. (1989). *An Introduction to Applied Geostatistics*. New York, USA: Oxford University Press, pp. 542.

Jarlan, L., Mangiarotti, S., Mougin E., Mazzega, P., Hiernaux, P., & Le Dantec, V. (2008). Assimilation of SPOT/VEGETATION NDVI data into a sahelian vegetation dynamics model. *Remote Sensing of Environment* 112(4): 1381-1394.

Knotters, M., Brus, D. J., & Voshaar, J. H. O. (1995). A comparison of kriging, co-kriging and kriging combined with regression for spatial interpolation of horizon depth with censored observations. *Geoderma* 67(3-4): 227–246.

Labrecque, S., Fournier, R. A., Luther, J. E., & Piercey, D. (2006). A comparison of four methods to map biomass from Landsat-TM and inventory data in western Newfoundland. *Forest Ecology and Management* 226, 129-144.

Liao, J., Dong, L., & Shen, G. (2009). Neural network algorithm and backscattering model for biomass estimation of wetland vegetation in Poyang Lake area using Envisat ASAR data. *Geoscience and Remote Sensing Symposium*, IEEE International, IGARSS 2009.

Lloyd, C. D. (2007). *Local Models for Spatial Analysis*. Boca Raton: CRC Press, Taylor & Francis Group.

Lu, D. (2006). The potential and challenge of remote sensing-based biomass estimation. *International Journal of Remote Sensing* 27: 1297-1328.

Lucas, N.S. (1995). *Coupling remotely sensed data to a forest ecosystem simulation model.* Thesis for the Degree of Doctor of Philosophy, University of Wales, Swansea, England, pp. 375.

Madugundu, R.; Nizalapur, V. & Jha, C. S. (2008). Estimation of LAI and above-ground biomass in deciduous forests: Western Ghats of Karnataka, India. *International Journal of Applied Earth Observation and Geoinformation* 10(2): 211-219.

Malthus, T.J., Andrieu, B., Danson, F.M., Jaggard, K.W. and Steven, M.D. (1993). Candidate high spectral resolution infrared indices for crop cover. *Remote Sensing of Environment* 46: 204-212.

Matheron G. (1963). Principles of geostatistics. *Economic Geology* 58: 1246–1266.

Matheron G. (1971). *The theory of regionalised variables and its applications. Les Cahiers du Centre de Morphologie Mathematique de Fontainblue.* Paris: Ecole Nationale Superiore de Paris, pp. 212.

Melillo, J.M, McGuire, A.D., Kicklighter, D.W., Moore III, B., Vorosmarty, C.J. and Schloss, A.L. (1993). Global climate change and terrestrial net primary production. *Nature* 363: 234-240.

Meng, Q., Cieszewski, C. J., Madden, M., & Borders, B. E. (2007). K Nearest Neighbor method for forest inventory using remote sensing data. *GIScience & Remote Sensing* 44(2): 149-165.

Meng, Q.; Cieszewski, C. J. & Madden, M. (2009). Large area forest inventory using Landsat ETM+: A geostatistical approach. *ISPRS Journal of Photogrammetry and Remote Sensing* 64: 27-36.

Moss, S. (2004). Kappa statistics. The Institute of Cancer Research, http://www.icr.ac.uk/cseu/bc_eqa_kappa.html.

Muukkonen, P., & Heiskanen, J. (2007). Biomass estimation over a large area based on standwise forest inventory data and ASTER and MODIS satellite data: A possibility to verify carbon inventories. *Remote Sensing of Environment* 107: 617–624.

Nemani, R., Pierce, L., Running, S. and Band, L. (1993). Forest ecosystem processes at the watershed scale: sensitivity to remotely-sensed leaf area index estimates. *International Journal of Remote Sensing* 14(13): 2519-2534.

Odeh, I. O. A., McBratney, A. B., & Chittleborough, D. J. (1994). Spatial prediction of soil properties from landform attributes derived from a digital elevation model. *Geoderma,* 63: 197-214.

Odeh, I.O. A., McBratney, A. B., & Chittleborough, D. J. (1995). Further results on prediction of soil properties from terrain attributes: heterotopic cokriging and regression-kriging. *Geoderma* 67(3-4): 215–226.

Palmer, D. J.; Höck, B. K.; Kimberley, M. O.; Watt, M. S.; Lowe, D. J. & Payn, T. W. (2009). Comparison of spatial prediction techniques for developing Pinus radiate productivity surfaces across New Zealand. *Forest Ecology and Management* 258: 2046-2055.

Pebesma, E. (2001). *Gstat user's manual.* University of Utrecht, pp. 108.

Pebesma, E. (2004). Multivariable geostatistics in S: the gstat package. Computers & *Geosciences* 30(7): 683-691.

Pons, X and Solé-Sugrañes, L. (1994). A simple radiometric correction model to improve automatic mapping of vegetation from multispectral satellite data. *Remote Sensing of Environment* 48: 191-204.

Pons, X. (2002). *MiraMon. Geographic information system and remote sensing software* Centre de Recerca Ecològica i Aplicacions Forestals, CREAF. Bellaterra. ISBN: 84-931323-5-7.

Purevdorj, T., Tateishi, R., Ishiyamas, T., and Honda, Y. (1998). Relationships between percent cover and vegetation indices. *International Journal of Remote Sensing* 19(18): 3519-3535.

Rahman, M. M.; Csaplovics, E., & Koch, B. (2005). An efficient regression strategy for extracting forest biomass information from satellite sensor data. *International Journal of Remote Sensing* 26(7): 1511-1519.

Rossiter, D.G. (2004). *Statistical method for accuracy assessment of classified thematic map.* International Institute for Geo-information Science and Earth, Department of Earth Systems Analysis, Enschede, NL, pp. 46.

Rouse, J.W., Haas, R.H., Schell, J.A., Deering, D.W. and Harlan, J.C. (1974). *Monitoring the vernal advancement retrogradiation of natural vegetation.* Final Report Type III, NASA/GSFC, Greenbelt, MD., USA, pp. 371.

Sales, M. H., Souza Jr., C. M., Kyriakidis, P. C.; Roberts, D. A., & Vidal, E. (2007). Improving spatial distribution estimation of forest biomass with geostatistics: A case study for Rondônia, Brazil. *Ecological Modelling* 205: 221-230.

Sen, Zekai (2009). *Spatial Modeling Principles in Earth Sciences.* London, New York: Springer Dordrecht Heidelberg, pp. 351.

Singh, R.P., Roy, S., and Koogan, F. (2003). Vegetation and temperature condition indices from NOAA AVHRR data for drought monitoring over India. *International Journal of Remote Sensing* 24 (22): 4393-4402.

Stehman, S.V. (1997). Selecting and interpreting measures of thematic classification accuracy. *Remote Sensing of Environment* 62: 77-89.

Todd, S.W., Hoffer, R.M., and Milchunas, D.G. (1998). Biomass estimation on grazed and ungrazed rangelands using spectral indices. *International Journal of Remote Sensing* 19 (3): 427-438.

Tomppo, E. (1991). Satellite imagery-based national inventory of Finland. *International Archives of Photogrammetry and Remote Sensing 28*(7-1): 419-424.

Tomppo, E., Nilsson, M., Rosengren, M., Aalto, P., Kennedy, P. (2002). Simultaneous use of Landsat-TM and IRS-1c WiFS data in estimating large area tree stem volume and aboveground biomass. *Remote Sensing of Environment* 82: 156–171.

Toutin, T. (2004). Review article: Geometric processing of remote sensing images: models, algorithms and methods. *International Journal of Remote Sensing* 25(10): 1893-1924.

Tucker, C.J. (1979). Red and photographic infrared linear combinations for monitoring vegetation. *Remote Sensing of Environment* 8: 127-150.

Waring, R.H., Landsberg, J.J. and Williams, M. (1998). Net primary production of forests: a constant fraction of gross primary production? *Tree Physiology* 18: 129-134.

Webster, R. & Oliver, M. A. (1992). Sample adequately to estimate variograms of soil properties. *Journal of Soil Science* 43: 177-192.

Webster, R. & Oliver, M. A. (2007). *Geostatistics for Environmental Scientists.* (2nd ed.), England: John Wiley & Sons Ltd, pp.332.

Woodcock, C. E, Strahler, A. H., & Jupp, D. L. B. (1988). The use of variograms in remote sensing: II. Real digital images. *Remote Sensing of Environment* 25: 349-379.

Xia, L., 1994. A two-axis adjusted vegetation index (TWVI). *International Journal of Remote Sensing* 15(7): 1447-1458.

Xu, B., Gong, P., and Pu, R. (2003). Crown closure estimation of oak savannah in a dry season with Landsat TM imagery: comparison of various indices through correlation analysis. *International Journal of Remote Sensing* 24 (9): 1811-1822.

Zheng, D., Heath, L. S., & Ducey, M. J. (2007). Forest biomass estimated from MODIS and FIA data in the. Lake States: MN, WI and MI, USA. *Forestry* 80(3): 265-278.

Part 3

Metal Biosorption and Reduction

Hexavalent Chromium Removal by a *Paecilomyces sp* Fungal

Juan F. Cárdenas-González and Ismael Acosta-Rodríguez
Universidad Autónoma de San Luis Potosí, Facultad de Ciencias Químicas, Centro de Investigación y de Estudios de Posgrado, Laboratorio de Micología Experimental S.L.P. México

1. Introduction

The strong impact of hexavalent chromium on the environment and on the human health demand suitable technologies to neutralize the hazard of chromium. The traditional technologies used for the remediation of environment contaminated with Cr (VI) are based on physical and chemical approaches, which require large amounts of chemical substances and energy. Such methodologies have proved complete expensive on a large-scale application at contaminated sites, and also they have generated hazardous by-products (Cervantes *et. al.*, 2001). Bioremediation, a strategy that uses living microorganisms, is essentially proposed to clean up the environment from organic pollutants. However, since there is an evidence that several microorganisms possess the capability to reduce Cr (VI) to relatively toxic Cr (III), bioremediation gives immense opportunities for the development of technologies for the detoxification of soil contaminated with Cr (VI) as an alternative to existing physical-chemical remediation technologies (Cervantes *et al.*, 2001).

Chromium is an essential micro-nutrient in the diet of animals and humans, as it is indispensable for the normal sugar, lipid and protein metabolism of mammals. Its deficiency in the diet causes alteration in lipid and glucose metabolism in animals and humans. Chromium is included in the complex named glucose tolerance factor (GFC) (Armienta-Hernández and Rodríguez-Castillo, 1995). On the other hand, no positive effects of chromium are known in plants and microorganisms. However, elevated levels of chromium are always toxic, although the toxicity level is related to the chromium oxidation state. Cr (VI) not only is highly toxic to all forms of living organisms. It is mutagenic for bacteria, mutagenic and carcinogenic for humans and animals, but also, it is involved in causing birth defects and the decrease of reproductive health (Marsh and Mclnerney, 2001). This metal may cause death in animals and humans, if ingested in large doses. The LD_{50} for oral toxicity in rats is from 50 to 100 mg/kg for Cr (VI) and 1900-3000 mg/kg for Cr (III). Cr (VI) toxicity is related to its easy diffusion across the cell membrane in prokaryotic and eukaryotic organisms and subsequent Cr (VI) reduction in cells, which gives free radicals that, may directly cause DNA alterations as well as toxic effects. Cr (III) has been estimated to be from 10 to 100 times less toxic than Cr (VI), because cellular membranes appear to be quite impermeable to most Cr (III) complexes. Nevertheless, intracellular Cr (III), which is the terminal product of the Cr (VI)-reduction, forms amino acid nucleotide complexes *in vivo*, whose mutagenic potentiality is not fully known (Gutiérrez Corona, *et al.*, 2010).

It is well known that prokaryotes are more resistant to Cr (VI) than eukaryotes. Toxic chromium effects on bacteria, algae and plants have been reviewed by Wong and Trevors (1988). On the contrary, scant information is available about the impact of the chromium on the structure and diversity of soil microbial communities. In many studies, it has been difficult to assess the toxicity of chromium to soil microorganisms, because the environments examined were often polluted at the same time with organic pollutants and/or different heavy metals. In a soil chronically polluted with chromium (about 5000 mg/kg of soil) by leather tannery activity, the oxygenic phototrophic microorganisms and heterotrophic bacterial communities were both affected by chromium. Nitrogen-fixing cyanobacteria were not detected in contaminated soil with Cr using the MPN test, and data obtained from enriched cultures for nitrogen-fixing cyanobacteria showed that this, belonging to the Nostoc group was present, but they had a low number of heterocyst. The size of the cultivable heterotrophic bacterial community was not affected by chromium pollution, but there was a relationship between the percentage of chromate-tolerant bacteria and the level of chromium in the soil (Anjana *et al.*, 2007). The ability of some microorganisms for interact with different Cr forms makes them attractive in the context of environmental biotechnology. In this sense, the use of microbial biomass for the removal of Cr from industrial wastewater and polluted water has already been recognized. The properties of some microorganisms for both: tolerate and reduce Cr (VI) enable their application in biotechnological process focusing on detoxification of Cr (VI). Cr resistance has been described in bacteria and fungi isolated from Cr-polluted environments. Yeast strains isolated include *Candida* and *Rhodosporidium* genera, but in these, the general mechanism of chromate resistance is related to limited ion uptake, rather than to chemical reduction of the toxic species (Baldi, *e. al.*, 1990; Pepi and Baldi, 1992). However, other yeasts such as *Candida utilis* (Muter, *et al.*, 2001) and *Candida maltose* (Ramírez-Ramírez, *et al.*, 2004), showed partial ability to reduce Cr (VI) and also the capability to accumulate Cr in the biomass. Recent reports have also examined Cr (III) and Cr (VI) uptake and accumulation by different filamentous fungi (Acevedo-Aguilar, *et al.*, 2008; Fukuda, *et al.*, 2008; Srivastava and Thakur, 2007; Morales-Barrera and Cristiani-Urbina, 2008). The present study report the isolation and identification of a *Paecilomyces* sp fungal strain that exhibits high resistance level, resistance, biosorption and reduction potential to Cr (VI).

2. Materials and methods

2.1 Microorganism and chromate resistance test

A chromate-resistant filamentous fungus was isolated from polluted air with industrial vapors, in Petri dishes containing modified Lee's minimal medium (LMM, Lee, *et al.*, 1975) [with 0.25% KH_2PO_4, 0.20% $MgSO_4$, 0.50% $(NH_4)_2SO_4$, 0.50% NaCl, 0.25% glucose] supplemented with 500 mg/L $K_2Cr_2O_7$; the pH of the medium was adjusted and maintained at 5.3 with 100 mMol/L citrate-phosphate buffer. The cultures were incubated at 28°C for 7 days. The strain was identified based on their morphological structures such color, diameter of the mycelia, and microscopic observation of formation of spores (Kirk, *et al.*, 2001). Chromate-resistant tests of the isolated strain, filamentous fungus *Paecilomyces* sp, were performed on liquid LMM containing the appropriate nutritional requirements and different concentrations of Cr (VI) (as potassium dichromate), and determining the dry weight. The isolation was carried out near of Chemical Science Faculty, located in the city of San Luis Potosí, Mexico.

2.2 Preparation of biosorbent

The biomass was obtained by growth the fungus in thioglycollate broth (8 g/L) at 28°C with constant shaking (100 rpm). After of 4 days of incubation, the fungal biomass was obtained by filtration on Whatman filter paper No. 1. Later, the fungal biomass was centrifuged (3000 rpm, 5 min), washed 3 times with trideionized water, dried (80°C, 12 h) in bacteriological stove, ground in mortar and stored in amber glass bottles at room environment until use.

2.3 Preparation of stock solution

An aqueous stock solution (1000 mg/L) of Cr (VI) ions was prepared using $K_2Cr_2O_7$ salt. pH of the solution was adjusted using 0.1 N HCl or NaOH. Fresh dilutions were used for each study.

2.4 Biosorption studies

The biosorption capacity of fungal biomass was determined by contacting various concentrations (100 - 1000 mg/L) of 100 mL Cr (VI) solution in 250 ml Erlenmeyer glass flasks, with 1 g of biomass. The mixture was shaken in a rotary shaker at 120 rpm followed by filtration using Whatman filter paper No. 1. The filtrate containing the residual concentration of Cr (VI) was determined spectrophotometrically at 540 nm after complexation with 1, 5 Diphenylcarbazide (Eaton *et. al.*,1995), Cr (III) with Chromazurol S (Pantaler and Pulyaeva,1985) and Cr total by total by Electrothermal Atomic Absorption Spectroscopy (Eaton *et. al.*,1995). For the determination of rate of metal biosorption by biomass from 100 mL (at 200, 400, 600, 800 and 1000, mg/L), the supernatant was analyzed for residual Cr (VI) after the contact period of 1-12 hours. The effect of pH and temperature on Cr (VI) sorption by fungal biomass, was determined at pH values of 1, 2, 3, and 4, 28°C, 40°C, 50°C and 60°C, respectively. The effect of different doses of biomass ranging from 1 to 5 g/L, with 100 mg/L of Cr (VI) concentrations was determined.

2.5 Culture conditions in liquid media

Cultures in 100 mL of sterile LMM [amended with 50 mg/L Cr (VI)] inoculated with 5×10^5 spores/mL were incubated at 28°C for 48 h. Then, cells were aseptically separated by centrifugation at 2,000 rpm (4°C) for 10 min, and washed twice with sterile trideionized water to eliminate culture medium components and cell debris. The cell pellet was resuspended in 3 mL of sterile trideionized water by shaking in a vortex mixer for 30s, and was then transferred to 100 mL of fresh LMM amended with 50 mg/L Cr (VI). At various times during the course of incubation, 1 mL aliquots were removed and centrifuged at 5,000 rpm for 10 min to sediment the cells; the supernatant fluid was used to determine the concentration of hexavalent, trivalent or total Cr.

2.6 Determination of hexavalent, trivalent, and total Cr

Hexavalent Cr and trivalent Cr were quantified by a spectrophotometric method employing diphenylcarbazide and chromazurol S, respectively (Eaton *et al.*, 1995; Pantaler and Pulyaeva, 1985), total Cr was determined by electrothermal atomic absorption spectroscopy (Eaton *et al.*, 1995).

The values shown in the results section are the mean from three experiments carried out by triplicate.

2.7 Bioremediation assay

Two 250 ml Erlenmeyer glass flasks, with 5 g of fungal biomass, were add with 20 g of contaminated earth with 50 mg Cr (VI)/g earth of tannery (Celaya , Guanajuato, México), and the volume was complete to 100 mL with trideionized water. The mixture was shaken in a rotary shaker at 120 rpm followed by filtration using Whatman filter paper N°1. The filtrate containing the residual concentration of Cr (VI) was determined with 1, 5 diphenylcarbazide (Eaton *et al.*, 1995).

3. Results and discussion

3.1 Isolation and identification of a fungal strain capable of removing Cr (VI)

The fungal strain isolated was able to growth on LMM supplemented with 2000 mg/L of Cr (VI) (Figure 1). This indicates that this fungus developed the Cr (VI) resistance and probably the Cr (VI) is being reduced in the polluted air. A variety of microorganisms with the Cr (VI) resistance and Cr (VI) reducing ability have been isolated from effluents of tanneries (Seng and Wang, 1994; Dark, *et al.*, 2004; Fukuda, *et.al.*, 2008). Colonies of the isolated fungal strain grew rapidly and mature within 3 days. *Paecilomyces* sp are thermopile and can grow well at temperatures as high as 50° and 60°C. The colonies are flat, powdery or velvety in texture. The initial color is white, and becomes yellow, yellow-green, pink, or violet. The reverse is dirty white or buff. A sweet aromatic color may be associated with older cultures. Septate hyaline hyphae, conidiophores, phialides, conidia, and chlamidospores are observed. Conidiophores (3-4 μm wide and 400-600 μm long) are often branched and carry the phialides at their tips. The phialides are swollen at their bases and taper towards their apices. They are usually grouped in pair or brush-like clusters. Conidia are unicellular, hyaline to darkly colored, smooth or rough, oval to fusoid, and form long chains. Chlamidospores are occasionally present. With different concentrations of Cr (VI) include changes in morphologies, showing slower growth and least conidiation (Figure 2) (Kirk, *et al.*, 2001).

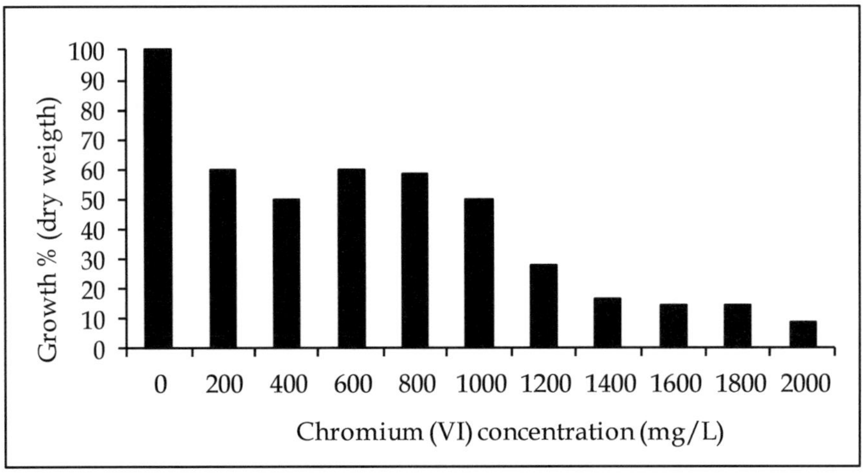

Fig. 1. Growth in dry weight of *Paecilomyces* sp with different concentrations of Cr (VI). $1x10^5$ spores/mL, 28°C, 7 days of incubation, 100 rpm.

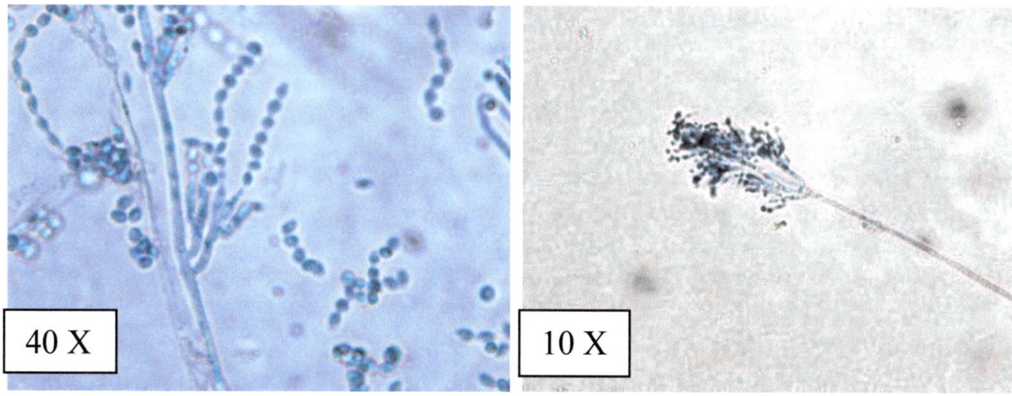

Fig. 2. Microscopic morphology of the fungus *Paecilomyces* sp. In absence and presence of 500 mg/L of Cr (VI), respectively.

3.2 Studies with fungal biomass

3.2.1 Effect of pH and incubation time

Figure 3 shows the adsorption of Cr (VI) by 1.0 g/100 mL of fungal biomass as a function of time at pH of 1.0, 2.0, 3.0 and 4.0, for initial Cr (VI) concentration of 100 mg/L. The metal removal was found to be 100% at 9 hours and 79.2% at 10 hours, with pH 1.0 and 2.0, respectively. Aqueous phase pH governs the speciation of metals and also the dissociation of active functional sites on the sorbent. Hence, metal sorption is critically linked with pH.

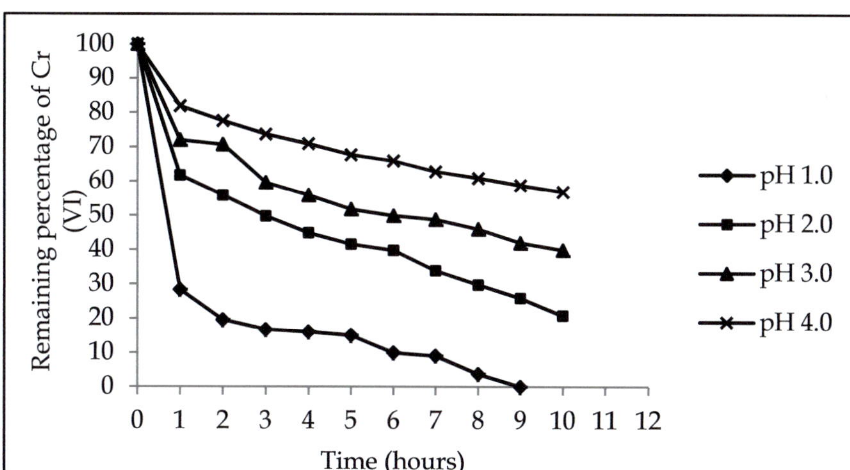

Fig. 3. Effect of pH and incubation time on the removal of 100 mg/L Cr (VI). 28°C. 1 g of fungal biomass. 100 rpm.

Not only different metals show different pH optima for their sorption but may also vary from one kind of biomass to the other (Tewari *et al.*, 1995; Ucun *et al.*, 2002). It can be observed from the figure that the uptake of Cr (VI) decreases with increase in pH. In general, the Cr (VI) adsorption by different biosorbents have shown similar trend and the optimum pH 1.0 has been reported (Nourbakhsh *et.al.*, 1994). The literature has reported an optimal pH for the removal of Cr (VI) by the fungi *Rhizopus arrhizus* and *Saccharomyces cerevisiae* in a range of 1.5-2.5, at 4 h (Nourbakhsh *et al.*, 1994), although most show a pH optimum of removal in the range of 2.0 to 3.0 (Tewari *et al.*, 2005; with *Mucor hiemalis*; Sag and Aktay, 2002, for *Rhizopus arrhizus*, both at 24 h, Bai and Abraham, 2001; with *Rhizopus nigricans*, at 8 h). The highest sorption capacity of mandarin shell for Chromium (VI) was at pH 1.0 and the decrease in sorption capacity with increase in pH may be attributed to the changes in metal speciation and the dissociation of functional groups on the sorbent. Ucun *et al.*, (2002) have reported that the pH dependence of metal uptake could be largely related to the various functional groups on the adsorbent surface along with metal solution chemistry.

3.2.2 Effect of temperature

Temperature dependence of the adsorption process is associated with several thermodynamic parameters. Figure 4 shows an increasing trend of Cr (VI) removal with the rise in temperature from 28 to 60°C. Results that is consistent with those of Park *et al.*, (2004), who observed that at 45°C and 24 h, adsorption occurs for the same metal with *Aspergillus niger*, and Leyva-Ramos *et al.*, (2005) for the removal of cadmium (II) with corn cob (40 °C and 5 days), but differ from 35°C and 24 h reported by Sag and Aktay (2002) for *Rhizopus arrhizus*, and with those reported for mandarin flax husk (Zubair, *et al.*, 2008). The increase in Cr (VI) uptake may be due to creation of some new sorption sites on the sorbent surface or the increased rate of intraparticle diffusion of sorbate ions into the pores of adsorbent at higher temperature, as diffusion is an endothermic process (Das, *et al.*, 2000).

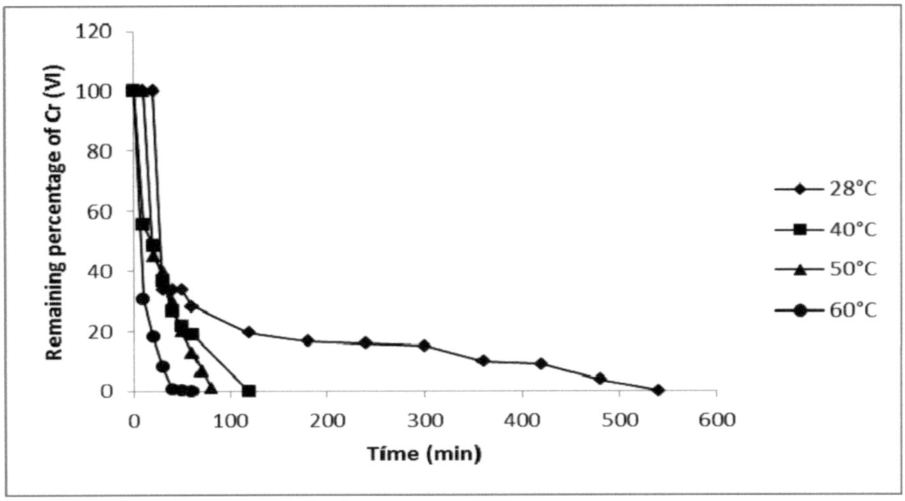

Fig. 4. Effect of temperature on the removal of 100 mg/L Cr (VI). 1 g of fungal biomass. 100 rpm

3.2.3 Effect of Cr (VI) concentration

The time taken to remove 200 mg/L chromium solution was 70 min. But as the chromium concentration increased, the percentage of chromium biosorption progressively decreased from 100% in 100 mg/L to 80% in 1000 mg/L solution, to 60°C (Figure 5a), and to 28°C, 200 and 1000 mg/L of the metal was remove in 9 and 12 hours, respectively (Figure 5b). This appears to be due to the increase in the number of ions competing for the available binding sites in the biomass and also due to the lack of binding sites for the complexation of Cr ions at higher concentration levels. At lower concentrations, all metal ions present in the solution would interact with the binding sites and thus facilitated 100% adsorption. At higher concentrations, more Cr ions are left unabsorbed in solution due to the saturation of binding sites (Ahalya *et al.* 2005).

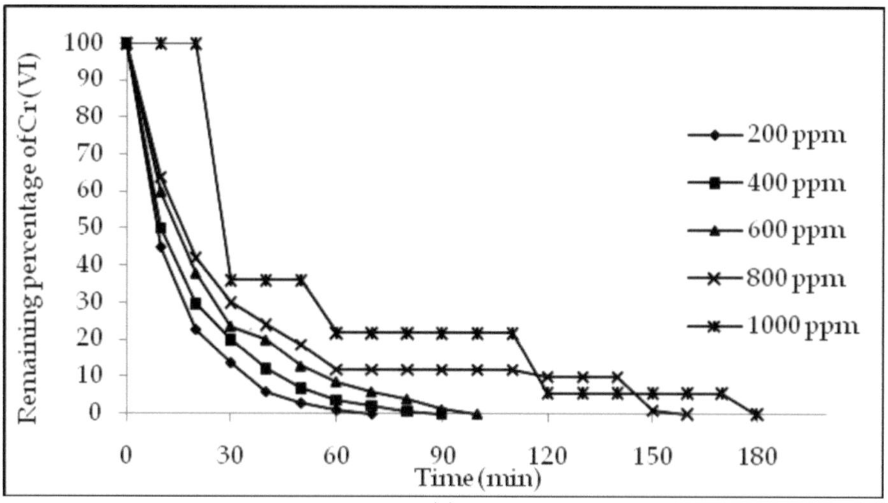

(a)

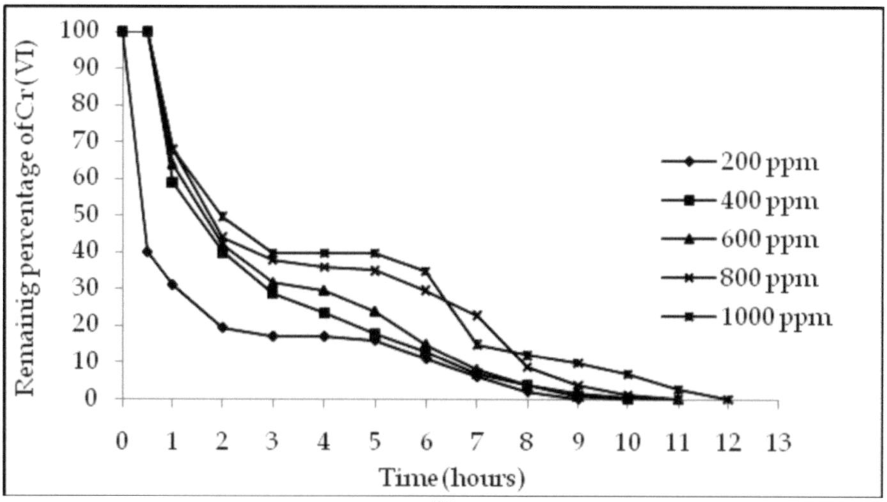

(b)

Fig. 5. Effect of Cr (VI) concentration on the removal of the metal. 1 g of fungal biomass. 100 rpm. a. - 60°C. b. - 28°C.

3.2.4 Effect of biomass concentration

We studied the removal of 1000 mg/L of Cr (VI) with various concentrations of fungal biomass at 60°C, finding that to higher concentration of biomass, is better the biosorption of Cr (VI), because the metal is removed at 70 minutes using 5.0 g of biomass (Figure 6). If we increasing the amount of biomass, also increases the removal of Cr (VI) in solution, since there are more metal biosorption sites, because the amount of added biosorbent determines the number of binding sites available for metal biosorption (Cervantes *et al.*, 2001). Similar results have been reported for biomass *Mucor hiemalis* and *Rhizopus nigricans*, although the latter with 10 g of biomass (Tewari *et al.*, 2005, Bai and Abraham, 2001), but are different from those reported by Zubair *et al.*, (2008), for mandarin flax husk biomass, who report an optimal concentration of biomass of 100 mg/L.

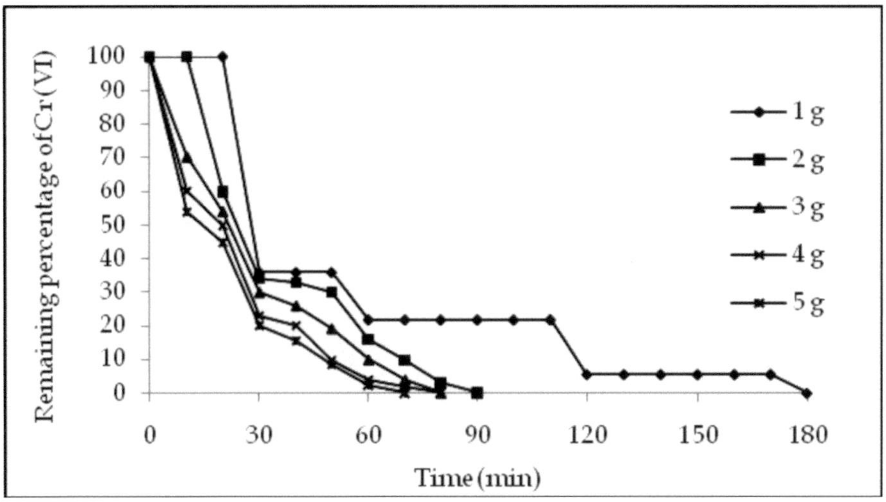

Fig. 6. Effect of biomass concentration on the removal of 1.0 g/L of Cr (VI). 100 rpm. 60°C.

Finally, Table 1 shows the adsorption efficiency of Cr (VI) by different biomass of microorganisms which shows that the biomass of *Paecilomyces* sp reported in this study is the most efficient in the removal of metal.

3.3 Studies with fungal alive

3.3.1 Effect of pH

Figure 7 shows the effect of varying pH (4.0, 5.3, and 7.0, maintained with 100 mMol/L citrate-phosphate buffer.) on the rate of Cr (VI) removal. The rate of chromium uptake and the extent of that capture were enhanced as the pH falls from 7.0 to 4.0. The maximum uptake was observed at pH 4.0 (96% at 7 days), 96%, Liu *et. al.*, (2007) and Bai and Abraham, (2001) reported maximum removal at 100 mg/L Cr (VI) solution using *Mucor racemosus* and *Rhizopus nigricans* with pH optimum of 0.5-1.0, and 2.0 respectively, Sandana Mala *et.al.*, (2006) at pH 5.0 for Cr (VI) with *Aspergillus niger* MTCC 2594, Rodríguez *et. al.*, (2008) at pH 3.0-5.0 for Pb^{+2}, Cd^{+2} and Cr^{+3} with the yeast *Saccharomyces cerevisiae*, Park *et. al.*, (2004) at pH 1-5 for Cr (VI) with brown seaweed *Ecklonia*, Higuera *et. al.*, (2005) at pH 5.0 for Cr (VI) with the brown algae *Sargassum* sp, and Fukuda *et. al.*, (2008) at pH 3.0 for Cr (VI) with *Penicillium* sp. In contrast to our observations, Prasenjit and Sumathi (2005), reported maximum uptake of Cr (VI) at pH 7.0 with *Aspergillus foetidus*, Puranik and Paknikar (1985) reported an enhanced uptake of lead, cadmium, and zinc, with a shift in pH from 2.0 to 7.0 using a *Citrobacter* strain, and a decrease at higher pH values. Al-Asheh and Duvnjak (1995) also demonstrated a positive effect of increasing pH in the range 4.0-7.0 on Cr (III) uptake using *Aspergillus carbonarius*. At low pH, the negligible removal of chromium may be due to the competition between hydrogen (H+), and metal ions Srivasta and Thakur (2007). At higher pH (7.0), the increased metal removal may be due to the ionization of functional groups and the increase in the negative charge density on the cell surface. At alkaline pH values (8.0 or higher), a reduction in the solubility of metals may contribute to lower uptake rates.

Biosorbent	Capacity of adsorption (mg/g)	References
Aspergillus foetidus	2	Prasenjit and Sumathi (2005)
Aspergillus niger	117.33	Khambhaty et al. (2009)
Aspergillus sydowi	1.76	Kumar et al. (2008)
Rhizopus nigricans	47	Bai and Abraham (2001)
Rhizopus oligosporus	126	Ariff et al. (1999)
Rhizopus arrhizus	11	Bai and Abraham (1998)
Rhizopus arrhizus	78	Aksu and Balibek (2007)
Rhizopus sp.	4.33	Zafar et al. (2007)
Mucor hiemalis	53.5	Tewari et al. (2005)
Paecilomyces sp	1000	(Present study)
Bacillus coagulans	39.9	Srinath et al. (2002)
Bacillus megaterium	30.7	Srinath et al. (2002)
Zoogloea ramigera	2	Nourbakhsh et al. (1994)
Streptomyces noursei	1.2	Mattuschka and Straube (1993)
Chlorella vulgaris	3.5	Nourbakhsh et al. (1994)
Cladophora crispate	3	Nourbakhsh et al. (1994)
Dunaliella sp.	58.3	Donmez and Aksu (2002)
Pachymeniopis sp.	225	Lee et al. (2000)

Table 1. Capacity of biosorption of different microbial biomass for removal Cr (VI) in aqueous solution.

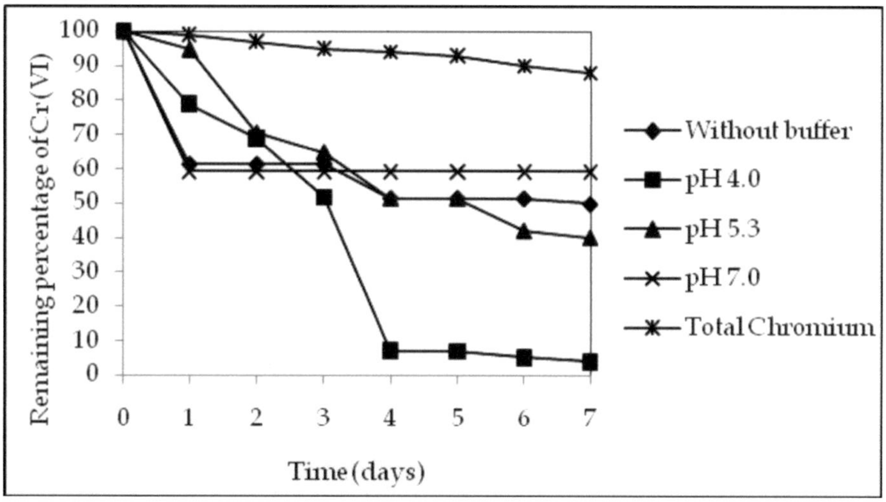

Fig. 7. The effect of pH on Chromium (VI) removal by *Paecilomyces* sp. 50 mg/L Cr (VI), 100 rpm, 28°C.

3.3.2 Effect of cell concentration

The influence biomass in the removal capacity of Cr (VI) was depicted in Figure 8. From the analyzed (38, 76, and 114 mg of dry weight) the removal capacity was in the order of 99.17%, 97.95%, and 97.25%, respectively. In contrast to our observations, the most of the reports in the literature observe at higher biomass dose resulted in an increase in the percentage removal [1, 3, 7, 8, 19, and 22]. To higher biomass concentration, there are more binding sites for complex of Cr (VI) (e.g. $HCrO^{-4}$ and Cr_2O7^{-2} ions) (Seng and Wang, 1994; Cervantes *et. al.*, 2001). However it did not show in our observations.

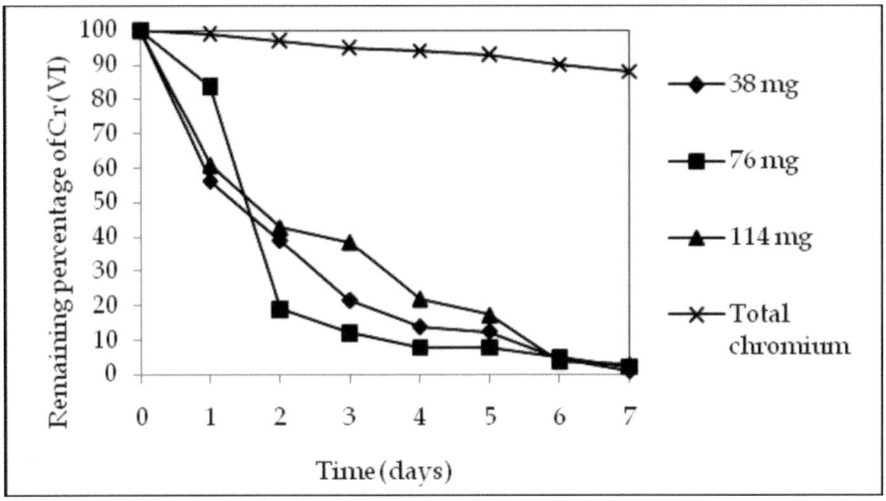

Fig. 8. The effect of cell concentration on the removal of 50 mg/L Cr (VI), 100 rpm, 28°C, pH 1.0.

3.3.3 Effect of initial Cr (VI) concentration

As seen in Figure 9, when the initial Cr (VI) ions concentration increased from 50 mg/L to 200 mg/L, the percentage removal of metal ions decreased. This was due to the increase in the number of ions competing for the available functions groups on the surface of biomass. Our observations are like to the most of the reports in the literature (Bai and Abraham, 2001; Seng and Wang, 1994; Beszedits, 1988; Park *et. al.*, 2004; Sahin and A. Öztürk, 2005; Liu, *et. al.*, 2007; Rodríguez, *et. al.*, 2008; Park *et. al.*, 2004; Higuera Cobos *et. al.*, 2005).

3.3.4 Effect of carbon source

Figures 10a and 10b, shows that the decrease of Cr (VI) level in culture medium of *Paecilomyces* sp occurred exclusively in the presence of a carbon source, either fermentable (glucose, sucrose, fructose, citrate) or oxidable (glycerol). In the presence of glucose, other inexpensive commercial carbon sources like unrefined sugar and brown sugar or glycerol, the decrease in Cr (VI) levels occurred at a similar rate, at 7 days of incubation are of 99.17%, 100%, 94.28%, 81.5, and 99%, respectively, and the other carbon sugar were fewer effectives. On the other hand, incubation of the biomass in the absence of a carbon source did not produce any noticeable change in the initial Cr (VI) concentration in the growth medium. These observations indicated that in culture of the fungus a carbon source is required to provide the reducing power needed to decrease Cr (VI) in the growth medium. Our observations are like to the report of Acevedo-Aguilar, *et. al.*, (2008) and Prasenjit and Sumathi (2005), with glucose like carbon source, and are different to the observations of Srivasta and Thakur (2007) with *Aspergillus* sp and *Acinetobater* sp, who observed how the main carbon source the sodium acetate.

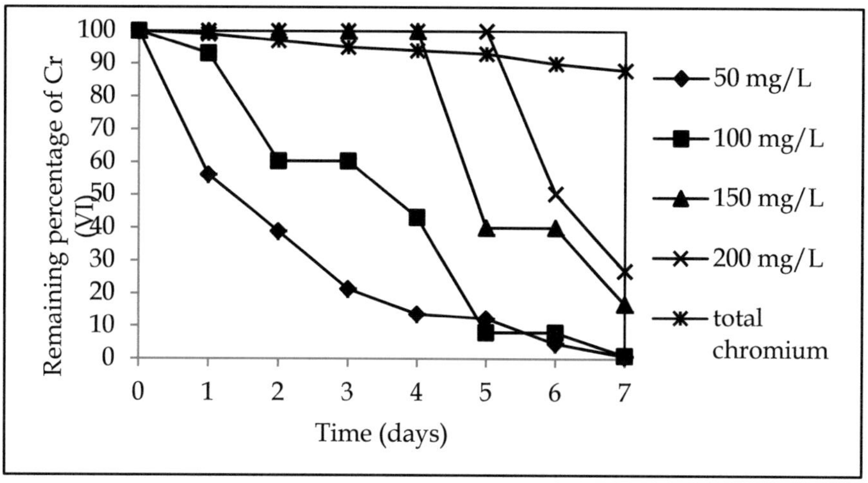

Fig. 9. The effect of the concentration of Cr (VI) in solution on the removal, 100 rpm. 28°C, pH 4.0.

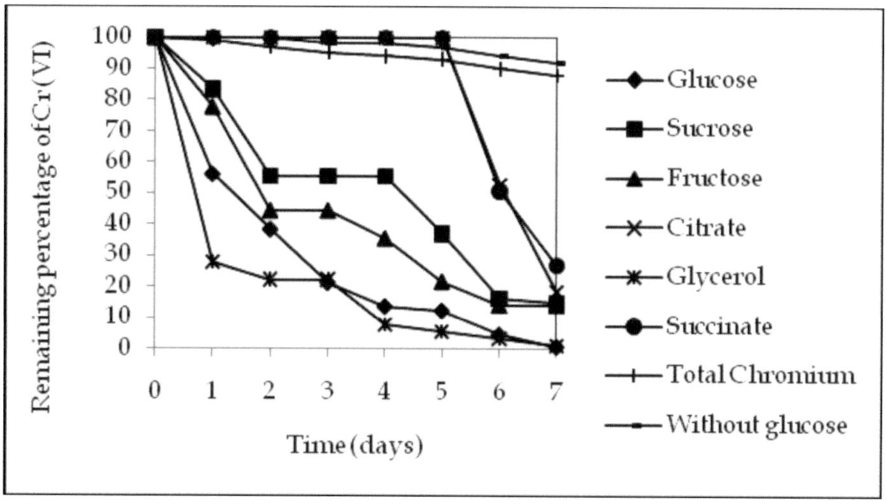

Fig. 10. (a) Influence of carbon source on the capability of *Paecilomyces* sp to decrease Cr (VI) levels in the growth medium. 100 rpm, 28°C, pH 4.0

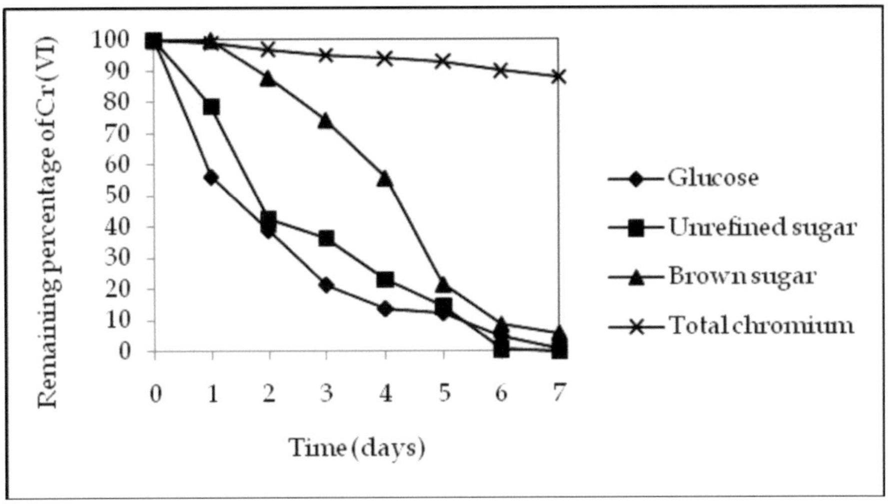

Fig. 10. (b) Influence of commercial carbon sources and salt on the capability of *Paecilomyces* sp to decrease Cr (VI) levels in the growth medium. 100 rpm, 28°C, pH 4.0

3.3.5 Time course of Cr (VI) decrease and Cr (III) production

The ability of the isolated strain to lower the initial Cr (VI) of 50 mg/L, and Cr (III) production in culture medium was analyzed. Figure 11A show that *Paecilomyces* sp exhibited a remarkable efficiency to diminish Cr (VI) level with the concomitant production of Cr (III) in the growth medium (indicated by the formation of a blue-green color and a white precipitate, and its determination by Cromazurol S, (Figure No. 11 B) (Pantaler and Pulyaeva, 1985). Thus, after 7 days of incubation, the fungus strain caused a drop in Cr (VI) from its initial concentration of 50 mg/L to almost undetectable levels. As expected, total Cr concentration remained constant over time, in medium without inoculum. These observations indicate that *Paecilomyces* sp strain is able to reduce Cr (VI) to Cr (III) in growth medium amended with chromate. There are two mechanisms by which chromate could be reduced to a lower toxic oxidation state by an enzymatic reaction. Currently, we do not know whether the fungal strain used in this study express and Cr (VI) reducing enzyme(s). Further studies are necessary to extend our understanding of the effects of coexisting ions on the Cr (VI) reducing activity of the strain reported in this study. Cr (VI) reducing capability has been described in some reports in the literature (Smith et. al., 2002; Sahin and A. Öztürk, 2005; Muter *et. al.*, 2001; Ramírez-Ramírez *et. al.*, 2004; Acevedo-Aguilar, *et. al.*, 2008; Fukuda et. al., 2008). Biosorption is the second mechanism by which the chromate concentration could be reduced, and 1 g of fungal biomass of *Paecilomyces* sp is able to remove 1000 mg/L of Cr (VI) at 60°C, at 3 hours of incubation (Figure 4), because the fungal cell wall can be regarded as a mosaic of different groups that could form coordination complexes with metals, and our observations are like to the most of the reports in the literature (Bai and Abraham, 2001; Seng and Wang, 1994; Ramírez-Ramírez *et. al.*, 2004; Acevedo-Aguilar, *et. al.*, 2008; Fukuda *et. al.*, 2008; Prasenjit and Sumathi, 2005).

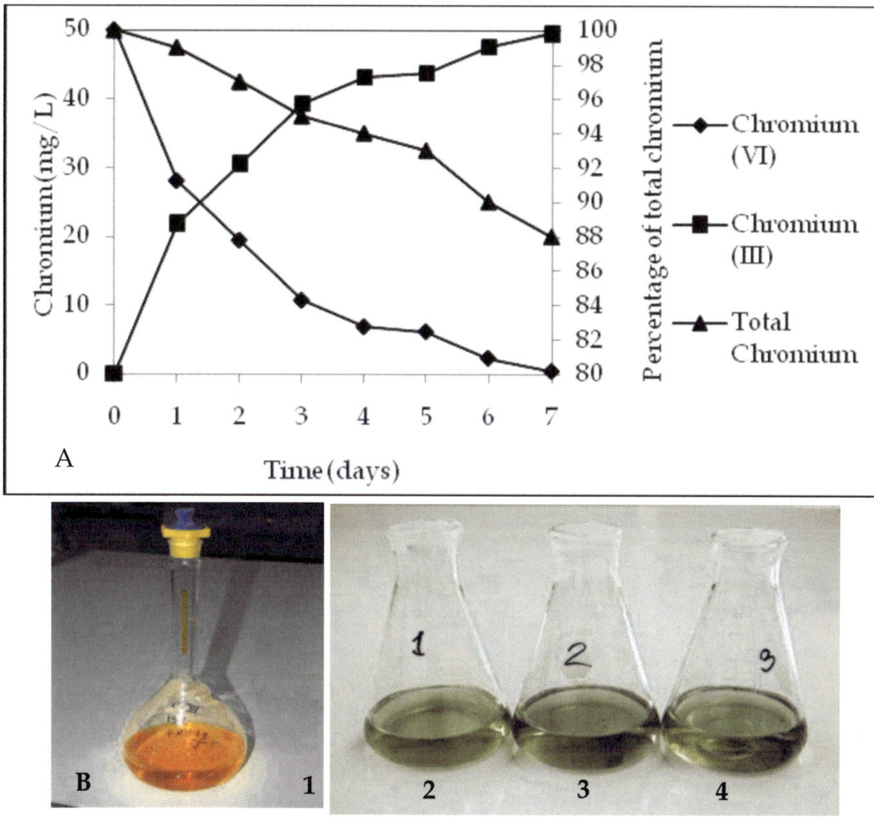

Fig. 11. Time-course of Cr (VI) decrease and Cr (III) production in the spent medium of culture initiated in Lee´s minimal medium, amended with 50 mg/L Cr (VI), 100 rpm, 28°C, pH 4.0 (**A**). B. - Appearance of the solutions. Total Cr coupled with the biomass, after different incubation times in the presence of Cr (VI). 1. - Standard solutions of Cr (VI) (1.0 g/L, pH= 1.0). 2.-25 mg/L 3.-50 mg/L 4.-100 mg/L

3.3.6 Removal of Cr (VI) in industrial wastes with fungal biomass

We adapted a water-phase bioremediation assay to explore possible usefulness of strain of *Paecilomyces* sp, for eliminating Cr (VI) from industrial wastes, the mycelium biomass was incubated with non sterilized contaminated soil containing 50 mg Cr (VI)/g, suspended in LMM, pH 4.0. It was observed that after eight days of incubation with the *Paecilomyces* sp biomass, the Cr (VI) concentration of soil sample decrease fully (Figure 12), and the decrease level occurred without change significant in total Cr content, during the experiments. In the experiment carried out in the absence of the fungal strain, the Cr (VI) concentration of the soil samples decreased by about of 18% (date not shown); this might be caused by indigenous microflora and (or) reducing components present in the soil. The chromium removal abilities of *Paecilomyces* sp are equal or better than those of other reported strains, for example *Candida maltose* RR1 (Ramírez-Ramírez *et. al.*, 2004). In particular, this strain was superior to the other strains because it has the capacity for efficient chromium reduction under acidic conditions. Most other Cr (VI) reduction studies were carried out at neutral pH (Fukuda *et. al.*, 2008; Greenberg *et. al.*, 1992). *Aspergillus niger* also has the ability to reduce and adsorb Cr (VI) (Fukuda *et. al.*, 2008). When the initial concentration of Cr (VI) was 500 ppm, *A. niger* mycelium removed 8.9 mg of chromium/g dry weight of mycelium in 7 days. In the present study, *Paecilomyces* sp, remove 50 mg/g, (pH, 4.0 and 8 days).

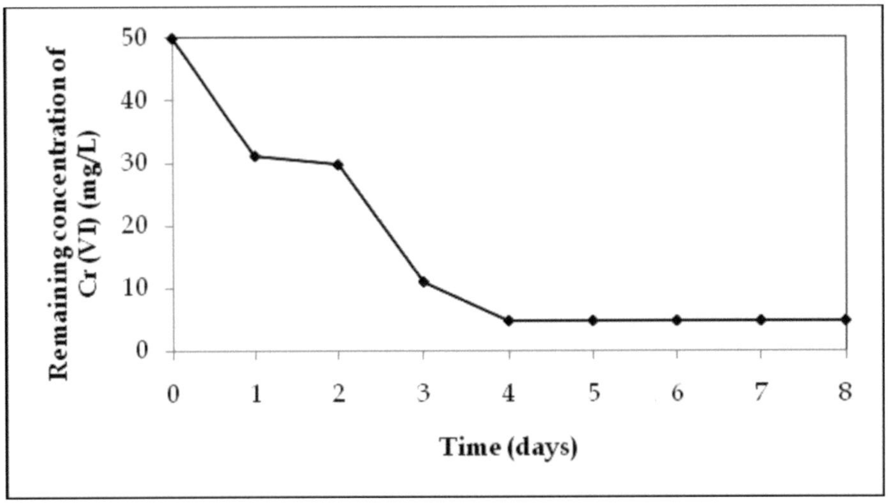

Fig. 12. Removal of Chromium (VI) in industrial wastes incubated with the fungal biomass. 100 rpm, 28°C, pH 4.0, 50 g of contaminated soil (50 mg Cr (VI)/g soil).

Reports on applications of microorganisms for studies of bioremediation of soils contaminated with chromates are rare. One such study involved the use of unidentified bacteria native to the contaminated site, which are used in bioreactors to treat soil contaminated with Cr (VI). It was found that the maximum reduction of Cr (VI) occurred with the use of 15 mg of bacterial biomass per g of soil (wet weight), 50 mg per g of soil molasses as carbon source, the bioreactor operated under these conditions, completely reduced 5.6 mg/Cr (VI) per g of soil at 20 days (Jeyasingh and Philip, 2004). In another study using unidentified native bacteria-reducing Cr (VI) of a contaminated site, combined with *Ganoderma lucidum,* the latter used to remove by biosorption Cr (III) formed. The results showed that the reduction of 50 mg/L of Cr (VI) by bacteria was about 80%, with 10 g / L of peptone as a source of electrons and a hydraulic retention time of 8 h. The Cr (III) produced was removed using a column with the fungus *G. lucidum* as absorber. Under these conditions, the specific capacity of adsorption of Cr (III) of *G. Lucidum* in the column was 576 mg/g (Krishna and Philip, 2005). In other studies, has been tested the addition of carbon sources in contaminated soil analyzed in column, in one of these studies was found that the addition of tryptone soy to floor to add to with 1000 mg/L of Cr (VI) increase reduction ion, due to the action of microorganisms presents in the soil, although such action is not observed in soil with higher concentrations (10.000 mg/L) of Cr (VI) (Tokunaga *et al.*, 2003). Another study showed that the addition of nitrate and molasses accelerates the reduction of Cr (VI) to Cr (III) by a native microbial community in microcosms studied, in batch or in columns of unsaturated flow, under conditions similar to those of the contaminated zone. In the case of batch microcosms, the presence of such nutrients caused reduction of 87% (67 mg/L of initial concentration) of Cr (VI) present at the start of the experiment, the same nutrients, added to a column of unsaturated flow of 15 cm, added with 65 mg/L of Cr (VI) caused the reduction and immobilization of the10% of metal, in a period of 45 days (Oliver *et al.*, 2003).

4. Conclusion

A fungal strain resistant to Cr (VI) and capable of removing the oxyanion from the medium was isolated from the environment near Chemical Science Faculty, located in the city of San Luis Potosí, Mexico. The strain was identified as *Paecilomyces* sp, by macro and microscopic characteristics. It was concluded that application of this biomass on the removal of Cr (VI) in aqueous solutions can be used since 1 g of fungal biomass remove 100 and 1000 mg/100 mL of this metal after one and three hours of incubation, and remove 297 mg Cr (VI) of waste soil contaminated, and this strain showed the capacity at complete concentrations reduction of 50 mg/L Cr (VI) in the growth medium after 7 days of incubation, at 28°C, pH 4.0, 100 rpm and a inoculum of 38 mg of dry weight. These results suggest the potential applicability of *Paecilomyces* sp for the remediation of Cr (VI) from polluted soils in the Fields.

5. References

Acevedo-Aguilar, F.A., Wrobel, K. Lokits, K., Caruso, J.A., Coreño Alonso, A., Gutiérrez-Corona, J.F. & Wrobel, K. 2008. Analytical speciation of chromium in in-vitro cultures of chromate-resistant filamentous fungi. *Analytical Bioanalytical Chemistry*, Vol: 392, No. 1-2, (September, 2008), 269-276, ISSN 1618-2642.

Ahalya, N., Kanamadi, R.D. & Ramachandra, T.V. 2005. Biosorption of chromium (VI) from aqueous solutions by the husk of Bengal gram (*Cicer arientinum*). *Electronic Journal of Biotechnology*, Vol: 8, No. 3, (December, 2005), 1-7, ISSN 0717-3458.

Al-Asheh S. & Duvnjak, Z. 1995. Adsorption of copper and chromium by *Aspergillus carbonarius*. *Biotechnology Progress*, Vol: 11, No. 6, (November-December, 1995), 638-642, ISSN 1520-6033.

Anjana, K., Kaushik, A., Kiran, B. & Nisha, R. 2007. Biosorption of Cr (VI) by immobilized biomass of two indigenous strains of cyanobacteria isolated from metal contaminated soil. *Journal of Hazardous Materials*, Vol: 148, No.1-2, (September, 2007), 383-386, ISSN 0304-3894.

Ariff, A.B., Mel, M., Hasan, M.A., Karim, M.I.A. 1999. The kinetics and mechanism of lead (II) biosorption by powderized *Rhizopus olgisporus*, *World Journal of Microbiology Biotechnology*, Vol: 15, No. 2, (April, 1999), 291-298, ISSN 0959-3993.

Aksu, Z. & Balibek, E. 2007. Chromium (VI) biosorption by dried *Rhizopus arrhizus*: effect of salt (NaCl) concentration on equilibrium and kinetic parameters. *Journal of Hazardous Materials*, Vol: 145, No. 2, (January, 2007), 210-220, ISSN 0304-3894.

Armienta-Hernández, M. & Rodríguez-Castillo, R. 1995. Environmental exposure to Chromium compounds in the valley of León, México. *Environmental Health Perspectives*, Vol: 103, No. 12, (December, 1995), 47- 51, ISSN 10222227.

Bai, R.S. & Abraham, T.E. 2001. Biosorption of chromium (VI) from aqueous solution by *Rhizopus nigricans*. *Bioresource Technology*, Vol: 79, No. 1, (September, 2001), 73-81, ISSN 09608524.

Baldi, F., Vaughan, A.M. & Olson, G.J., 1990. Chromium(VI)-resistant yeast isolated from a sewage treatment plant receiving tannery wastes. *Applied and Environmental Microbiology*, Vol: 56, (February, 1990), 913–918, ISSN 1098-5336.

Beszedits, S. 1988. Chromium removal from industrial wastewaters. In: *Chromium in the natural and human environments*. J.O. Nriagu and E. Nieboer (Eds.). 232-263, 1988. New York: John Wiley.

Cervantes, C., Campos-García, J., Devars, S., Gutiérrez-Corona, F., Loza-Tavera, H., Torres-Gúzman, J.C. & Moreno-Sánchez, R. 2001. Interactions of chromium with microorganisms and plants. *FEMS Microbiology Review*, Vol: 25, No. 3, (July, 2001), 335-347. ISSN 1574-6976.

Das, D.D., Mahapatra, R., Pradhan, J., Das, S.N. & Thakur, R.S. 2000. Removal of Cr (VI) from aqueous solution using activated cow dung carbon. *Journal of Colloids and Interface Science*, Vol: 232, No. 2, (December, 2000), 235–240, ISSN 0021-9797.

Donmez, G.C. & Aksu, Z. 2002. Removal of chromium (VI) from saline wastewaters by *Dunaliella* species. *Process Biochemistry*, Vol: 38, No. 4, (December, 2002), 751-762, ISSN 1359-5113.

Eaton, A.D., Clesceri, L.S. & Greenberg, A.E. 1995. *Standard Methods for the Examination of Water and Wastewater,* 19th ed. American Public Health Association Washington, DC, 1325, 3.58, 3.60, ISSN 0875530478.

Fukuda, Y., Ishino, A., Ogawa, K., Tsutsumi, X. & Morita, H. 2008. Cr(VI) reduction from contaminated soils by *Aspergillus* sp. N2 and *Penicillium* sp. N3 isolated from chromium deposits. *Journal of General and Applied Microbiology*, Vol: 54, No. 5, (September, 2008), 295-303, ISSN 1349-8037.

Gadd, G.M. 1989. Accumulation of metals by microorganisms and algae. In: *Biotechnology: a comprehensive treatise.* Rhem H.J., Reed, G. (eds). VCH, Weinheim, Germany, Vol: 6B, 401-433.

Greenberg, A.E., Clesceri, L.S. & Eaton, A.D. 1992. *Standard methods for the examination of water and wastewater,* 18a ed. 58-3.60, 187-190, (1992). American Public Health Association, Washington, D.C. ISSN 0875530478.

Gutiérrez Corona, J.F., Espino Saldaña, A.E., Coreño Alonso, A., Acevedo Aguilar, F.J., Reyna López, G.E., Fernández, F.J., Tomasini, A., Wrobel, K. & Wrobel, K. 2010. Mecanismos de interacción con cromo y aplicaciones biotecnológicas en hongos. *Revista Latinoamericana de Biotecnología Ambiental y Algal*, Vol: 1, No. 1, (Mayo, 2010), 47-63 ISSN. En trámite.

Higuera Cobos, O.F., Escalante Hernández, H. & Laverde, D. 2005. Reducción del cromo contenido en efluentes líquidos de la industria del cuero, mediante un proceso adsorción-desorción con algas marinas. *Scientia et Técnica,* Año XII, No. 29, (Abril, 2005), 115-120, ISSN 0122-1701.

Jeyasingh, J., & Philip, L. 2005. Bioremediation of chromium contaminated soil: optimization of operating parameters under laboratory conditions. *Journal of Hazardous Materials*, Vol: 118 No. 1-3, (January, 2005), 113-120, ISSN 0304-3894.

Khambhaty, Y., Mody, K., Basha, S. & Jha, B. 2009. Kinetics equilibrium and thermodynamic studies on biosorption of hexavalent chromium by dead fungal biomass of marine *Aspergillus niger. Chemical Engineering Journal*, Vol: 145, No. 1, (January, 2009), 489-495, ISSN 1385-8947.

Kirk, M.P., Cannon, F.P., David, C.J. & Stalpers, A.J. 2001. *Dictionary of the fungi*, 51-52, 385-387, (2001). CABI Publishing, UK.

Krishna, K.R., & Philip, L. 2005. Bioremediation of Cr (VI) in contaminated soils. *Journal of Hazardous Materials,* Vol: 121, No. 1-3, (January, 2005), 109-117, ISSN 0304-3894.

Kumar, R., Bishnoi, N.R., Garima, A. & Bishnoia, B. 2008. Biosorption of chromium (VI) from aqueous solution and electroplating wastewater using fungal biomass. *Chemical Engineering Journal*, Vol: 135, No. 3 (February, 2008), 202-208, ISSN 1385-8947.

Lee, K., Buckley, L. & Campbell, C.C. 1975. An aminoacid liquid synthetic medium for the development of mycelial and yeast forms of *Candida albicans. Journal of Medicine Veterinary and Mycology*, Vol: 13, (February, 1975), 148-153, ISSN 0268-1218.

Lee, D.C., Park, C.J., Yung, J.E. & Jeong, Y.H. 2000. Screening of hexavalent chromium biosorbent from marine algae. *Applied Microbiology and Biotechnology*, Vol: 54, No. 3, (September, 2000), 445-448, ISSN 1432-0614.

Leyva-Ramos, R., Bernal-Jacome, L.A. & Acosta-Rodriguez, I. 2005. Adsorption of cadmium (II) from aqueous solution on natural and oxidized corncob. *Separation and Purification Technology*, Vol: 45, No. 1 (September, 2005), 41-49, ISSN 1383-5866.

Liu, T., Li, H., Li, Z., Xiao, X., Chen, L. & Deng, L. 2007. Removal of hexavalent chromium by fungal biomass of *Mucor racemosus*: influencing factors and removal mechanism. *World Journal of Microbiology and Biotechnology*, Vol: 23, No. 12, (December, 2007), 1685-1693, ISSN 0959-3993.

Lofroth, G. & Ames, B.N. 1978. Mutagenicity of inorganic compounds in *Salmonella typhimurium*: Arsenic, chromium and selenium. *Mutation Research*, Vol: 53, No. 2, (September, 1978), 65-66, ISSN 1383-5742.

Marsh, T.L. & McInerney, M.J. 2001. Relationship of hydrogen bioavailability to chromate reduction in aquifer sediments. *Applied and Environmental Microbiology*, Vol: 67, No. 4, (April, 2001), 1517-1521, ISSN 1098-5336.

Mattuschka, B. & Straube, G. 1993. Biosorption of metals by waste biomass. *Journal of Chemical Technology Biotechnology*, Vol: 58, No. 1 (January, 1993), 57-63, ISSN 1097-4660.

Morales-Barrera, L. & Cristiani-Urbina, E. 2008. Hexavalent Chromium Removal by a *Trichoderma inhatum* Fungal Strain Isolated from Tannery Effluent. *Water Air Soil Pollution*, Vol: 187, (April, 2001), 327-336, ISSN 1573-2932.

Muter O., Patmalnieks A. & Rapoport A. 2001. Interrelations of the yeast *Candida utilis*: metal reduction and its distribution in the cell and medium. *Process Biochemistry*, Vol: 36, (April, 2001), 963-970, ISSN 0032-9592.

Nourbakhsh, M., Sag, Y., Ozer, D., Aksu, Z., Kutsal T.A. & Caglar, A. 1994. A comparative study of various biosorbents for removal of chromium (VI) ions from industrial wastewater. *Process Biochemistry*, Vol: 29, No. 1, (January, 1994), 1-5, ISSN 0032-9592.

Pantaler, R.P. & Pulyaeva, I.V. 1985. A spectrophotometric study of complexation between chromium and chromazurol S. *Journal of Analytical Chemistry* (Moscow), Vol: 40, (May, 1985), 1634-1639, ISSN 1061-9348.

Oliver, D.S., Brockman, F.J., Bowman, R.S., Thomas L. & Kieft, T.L. 2003. Microbial reduction of hexavalent chromium under vadose zone conditions. *Journal of Environmental Quality*, Vol: 32, No. 1, (January, September, 2003), 317-324, ISSN 0047-2425.

Park, D., Yung, Y.S. & Park, J.M. 2004. Reduction of Hexavalent chromium with the brown seaweed *Ecklonia* Biomass. *Environmental Science Technology*, Vol: 38, No. 18, (September, 2004), 4860-4864, ISSN 1735-2630.

Park, D., Yun, Y.S., Cho, H.Y. & Park, J.M. 2004. Chromium biosorption by thermally treated biomass of the brown seaweed, *Ecklonia* sp. *Industrial and Engineering Chemistry Research*, Vol: 43, No. 26, (December, 2004), 8226-8232, ISSN 0888- 5885.

Pas, M., Milacic, R., Drasar, K., Pollack, N. & Raspor, P. 2004. Uptake of chromium (III) and chromium (VI) compounds in the yeast cell. *Biometals*, Vol: 17, No. 1, (January, 2004), 25-33, ISSN 0966-0844.

Pepi M. & Baldi F. 1992. Modulation of chromium (VI) toxicity by organic and inorganic sulfur species in yeasts from industrial Wastes. *BioMetals*, Vol: 5, No. 2, (April, 2004), 179–185, ISSN 0966-0844.

Prasenjit, B. & Sumathi, S. 2005. Uptake of chromium by *Aspergillus foetidus*. *Journal of Material Cycles and waste Management*, Vol: 7, No. 2, (August, 2005), 88-92, ISSN 1438-4957.

Puranik, P.R. & Paknikar, K.M. 1999. Biosorption of lead, cadmium, and zinc by *Citrobacter* strain MCM B-181: characterization studies. *Biotechnology Progress*, Vol: 15, No. 2, (march-april 1999), 228-237, ISSN 1520-6033.

Ramírez-Ramírez, R., Calvo-Méndez, C., Avila-Rodríguez, M., Lappe, P., Ulloa, M., Vázquez-Juárez, R. & Gutiérrez-Corona, J.F. 2004. Cr (VI) reduction in a Chromate-resistant strain of *Candida maltose* isolated from the leather industry. *Antonie van Leeuwenhoek*, Vol: 85, No. 1, (January, 2004), 63-68, ISSN 0000020151.

Rodríguez, M.E., Miranda, R.C., Olivas, R. & Sosa, C.A. 2008. Efectos de las condiciones de operación sobre la biosorción de Pb^{++}, Cd^{++} y Cr^{+++} en solución por *Saccharomyces cerevisiae* residual. *Información Tecnológica*, Vol: 19, No. 6, (Noviembre-Diciembre, 2008), 47-55, ISSN 0716-87.

Sag, Y. & Aktay, Y. 2002. Kinetic studies on sorption of Cr (VI) and Cu (II) ions by chitin, chitosan and *Rhizopus arrhizus*. *Biochemical Engineering Journal*, Vol: 12, No. 2, (March, 2002), 143-153, ISSN 1369-703X.

Sahin, Y. & Öztürk, A. 2005. Biosorption of chromium (VI) ions from aqueous solution by the bacterium *Bacillus thuriengensis*. *Process Biochemistry*, Vol: 40, No. 5, (April, 2001), 1895-1901, ISSN 1359-5113.

Sandana Mala, J.G., Unni Nair, B. & Puvanakrishnan, R. 2006. Bioaccumulation and biosorption of chromium by *Aspergillus niger* MTCC 2594, *Journal of General and Applied Microbiology*, Vol: 52, No. 3, (September, 2006), 179-186, ISSN 1349-8037.

Seng, H. & Wang, Y.T. 1994. Biological reduction of chromium by *E. coli*. *Journal of Environmental Engineering*, Vol: 120, No. 4, (July, 1994), 560-572, ISSN 0733-9372.

Smith, W.A., Apel, W.A., Petersen, J.N. & Peyton, B.M. 2002. Effect of carbon and energy source on bacterial chromate reduction. *Bioremediation Journal*, Vol: 6, No. 1, (March, 2002), 205-215, ISSN 1547-6529.

Srinath, T., Verma, T., Ramteke, P.W., & Garg, S.K. 2002. Chromium (VI) biosorption and bioaccumulation by chromate resistant bacteria, *Chemosphere*, Vol: 48, No. 4, (July, 2002), 427-435, ISSN 0045-6535.

Srivastava, S. & Thakur, I.S. 2007. Evaluation of biosorption potency of *Acinetobacter* sp. for removal of hexavalent chromium from tannery effluent. *Biodegradation*, Vol: 18, No. 5, (October, 2007), 637-646, ISSN 0923-9820.

Tewari, N., Vasudevan, P. & Guha, B. 2005. Study on biosorption of Cr (VI) by *Mucor hiemalis*. *Biochemical Engineering Journal*, Vol: 23, No. 2, (April, 2005), 185-192, ISSN 1369-703X.

Tokunaga, T.K, Wan, J., Firestone, M.K., Hazen, T.C., Olson, K.R., Donald, J. Herman, D.J. Sutton, S.R. & Lanzirotti, A. 2003. Bioremediation and Biodegradation. In situ reduction of chromium(VI) in heavily contaminated soils through organic carbon amendment. *Journal of Environmental Quality*, Vol: 32, No. 5, (September, 2003), 1641-1649, ISSN 0047-2425.

Ucun, H., Bayhan, Y.K., Kaya, Y., Cakici, A. & Algur, O.F. 2002. Biosorption of chromium (VI) from aqueous solution by cone biomass of *Pinus sylvestris*. *Bioresource Technology*, Vol: 85, No. 2, (November, 2002), 155–158, ISSN 09608524.

Wong, P.T. & Trevors, J.T. 1988. Chromium toxicity to algae and bacteria. In: *Chromium in the Natural and Human Environments*. Nriagu, J.O., Nieboer, E. Eds Wiley, New York, 305-315.

Zafar, S., Aqil, F. & Ahmad, I. 2007. Metal tolerance and biosorption potential of filamentous fungi isolated from metal contaminated agricultural soil. *Bioresource Technology*, Vol: 98, No. 13, (September, 2007), 2557-2561, ISSN 09608524.

8

Biosorption of Metals: State of the Art, General Features, and Potential Applications for Environmental and Technological Processes

Robson C. Oliveira, Mauricio C. Palmieri and Oswaldo Garcia Jr.
Instituto de Química, Universidade Estadual Paulista (UNESP), Araraquara, Brazil

1. Introduction

The interactions among cells and metals are present since the life origin, and they occur successfully in the nature. These interactions are performed on cellular envelope (walls and membranes) and in cellular interior. They are based on the adsorption and absorption of metals by cells for the production of biomolecules and in vital metabolic processes (Palmieri, 2001). Some metals such as calcium, cobalt, copper, iron, magnesium, manganese, nickel, potassium, sodium, and zinc are required as essential nutrients to life existence. The principal functions of metals are: the catalysis of biochemical reactions, the stabilization of protein structures, and the maintenance of osmotic balance. The transition metals as iron, copper, and nickel are involved in redox processes. Other metals as manganese and zinc stabilize several enzymes and DNA strands by electrostatic interactions. Iron, manganese, nickel, and cobalt are components of complex molecules with a diversity of functions. Sodium and potassium are required for the regulation of intracellular osmotic pressure (Bruins et al., 2000).

The interactions among metals and biomasses are performed through different mechanisms. For instance, on cellular envelope, the metal uptake occurs via adsorption, coordination, and precipitation due to the interaction among the surface chemical groups and metals in aqueous solution. Similar mechanisms are related in the exopolymeric substances (EPS). On the other hand, specific enzymes in some biomasses can change the oxidation state of the noxious metals followed by formation of volatile compounds, which removes the metal from aqueous solution. Finally, the life maintenance depends on the metal absorption by active transport according with the nutritional requirements of the biomass (Gadd, 2009; Palmieri, 2001; Sen & Sharadindra, 2009).

The removal of metallic ions of an aqueous solution from cellular systems is carried out by passive and/or active forms (Aksu, 2001; Modak & Natarajan, 1995). As such live cells as dead cells do interact with metallic species. The bioaccumulation term describes an active process that requires the metabolic activity of the organisms to capture ionic species. In the active process the organisms usually tend to present tolerance and/or resistance to metals when they are in high concentrations and/or they are not part of the nutrition (Godlewska-Zylkiewicz, 2006; Zouboulis et al., 2004).

Group	Occurrence	pKa
Carboxylate	Uronic acid	3-4.4
Sulfate	Cisteyc acid	1.3
Fosfate	Polysaccharides	0.9-2.1
Imidazol	Hystidine	6-7
Hydroxyl	Tyrosine-phenolic	9.5-10.5
Amino	Cytidine	4.1
Imino	Peptides	13

Table 1. Some chemical groups involved in the metal-biomass interactions and their pK_as. Source: Eccles, 1999.

Biosorption is a term that describes the metal removal by its passive linkage in live and dead biomasses from aqueous solutions in a mechanism that is not controlled by metabolic steps. The metal linkage is based on the chemical properties of the cellular envelope without to require biologic activity (Gadd, 2009; Godlewska-Zylkiewicz, 2006; Palmieri et al., 2000; Valdman et al., 2001; Volesky, 2001). The process occurs through interaction among the metals and some active sites (e.g. carboxylate, amine, sulfate, etc.) on cellular envelope. Some of these chemical groups and their respective pK_as are described in the Table 1.

2. Biosorption of metals: general features

Usually, metallic species are not biodegradable and they are removed physically or chemically from contaminated effluents (Ahluwalia & Goyal, 2007; Hashim & Chu, 2004; Tien, 2002). The biosorption is a bioremediation emerging tool for wastewater treatment that has gained attention in the scientific community in the last years (Chu, 2004). It is a promising biotechnological alternative to physicochemical classical techniques applied such as: chemical precipitation, electrochemical separation, membrane separation, reverse osmosis, ion-exchange or adsorption resins (Ahluwalia & Goyal, 2007; Deng & Bai, 2004; Vegliò et al., 2002; Vegliò et al., 2003; Zouboulis et al., 2004). The conventional methods (Table 2) involve or capital and operational high costs, or they are inefficient at low metal concentration (1-100 ppm), or they can be associated to production of secondary residues that present treatment problems (Aksu, 2001; Ahluwalia & Goyal, 2007).

The initial incentives of biosorption development for industrial process are: (a) low cost of biosorbents, (b) great efficiency for metal removal at low concentration, (c) potential for biosorbent regeneration and metal valorization, (d) high velocity of sorption and desorption, (e) limited generation of secondary residues, and (f) more environmental friendly life cycle of the material (easy to eliminate compared to conventional resins, for example) (Crini, 2005; Kratochvil & Volesky, 2000; Volesky & Naja, 2005). Therefore the use of dead biomasses is generally preferred since it limits the toxicity effects of heavy metals (which may accumulate at the surface of cell walls and/or in the cytoplasm) and the necessity to provide nutrients (Modak & Natarajan, 1995; Sheng et al., 2004; Volesky, 2006).

Methodology	Disadvantages	Advantages
Chemical precipitation	a. Hard separation; b. Generation of secondary residues; c. Commonly inefficient in low metal concentration	a. Simple procedures; b. Generally presents low costs
Electrochemical treatment	a. Possibility of application in high metal concentration; b. Technique is sensible under determined conditions, as the presence of interfering agents	a. Successful metal recuperation
Reverse osmosis	a. Application of high pressures; b. Membranes that can foul or peel; c. High costs	a. Effluent purification that become available to recycle
Ion-exchange	a. It is sensible to the presence of particulate materials; b. Resins with high costs	a. Effective; b. Possibility of metal recuperation
Adsorption	No efficiency for some metals	Conventional adsorbents (e.g. activated carbon and zeolites)

Table 2. Conventional methods of metal removal from aqueous systems. Source: Zouboulis et al., 2004.

The mechanisms involved in metal accumulation on biosorption sites are numerous and their interpretation is made difficult because the complexity of the biologic systems (presence of various reactive groups, interactions between the compounds, etc.) (Gadd, 2009; Godlewska-Zylkiewicz, 2006; Palmieri, 2001). However, in most cases, metal binding proceeds through electrostatic interaction, surface complexation, ion-exchange, and precipitation, which can occur individually or combined (Yu et al., 2007a; Zouboulis et al., 2004). The uptake of metallic ions starts with the ion diffusion to surface of the evaluated biomasses. Once the ion is diffused to cellular surface, it bonds to sites that display some affinity with the metallic species (Aksu, 2001).

In general, literature describes that the biosorption process takes in consideration: (a) the temperature does not influence the biosorption between 20 and 35 °C; (b) the pH is a very important variable on process, once it affects the metal chemical speciation, the activity of biomass functional groups (active sites), and the ion metallic competition by active sites; (c) in diluted solutions, the biomass concentration influences on biosorption capacity: in lower concentrations, there is an increase on biosorption capacity; and (d) in solutions with different metallic species there is the competition of distinct metals by active sites (Vegliò & Beolchini, 1997).

The biosorption performance is influenced by physicochemical parameters as: (a) the biomass nature: the physical structure (porosity, superficial area, particle size) and the chemical nature of functional groups (diversity and density); (b) the chemical and the availability of the adsorbate; and (c) the solution conditions, such as: ionic force, pH, temperature and adsorbate concentration (Gadd, 2009; Godlewska-Zylkiewicz, 2006; Crini, 2005).

3. Environmental and technological demands

Environmental demands have received a great focus in public policies of different world's nations in the last decades. This is resulted of the external pressures of distinct areas as such the media vehicles, the scientific researches, and the greater conscious of the civil society about the environmental topics (Karnitz Jr., 2007; Volesky, 2001). These pressures have intensified the creation of regulatory laws as the water control and handling from anthropogenic activities. The mining and metallurgy wastewaters are considered the big resources of heavy metals contamination (cadmium, chromium, mercury, lead, zinc, copper, etc.) that are noxious in low concentrations (Sen & Sharadindra, 2009). The heavy metal recuperation from industrial effluents is extremely important due the society current requirements by the metal recycling and conservation (Hashim & Chu, 2004). The need for economic and effective methods for heavy metals removal from aqueous systems has resulted in the development of new technologies of concentration and separation (Hashim & Chu, 2004; Karnitz Jr., 2007; Sen & Sharadindra, 2009).

The biosorption of metals is established as research area since the 80s. The literature is mainly associated to the bioremediation of industrial wastewaters with low metal concentration. These works have been focused in the uptake of heavy metals because the metal ions in the environment bioaccumulate and are biomagnified along the food chain (Ahluwalia & Goyal, 2007; Vegliò et al., 2003; Volesky, 2001).

Besides the studies on environmental field of biosorption processes, others applications were investigated in the last few years led to develop the recovery of high demand and/or aggregated value metals such as gold, silver, uranium, thorium, and recently rare earth metals (RE) (Palmieri, 2001). The selection of interest metals in order to apply biosorption processes for recovery have to consider: (a) the environmental risk based on the technologic uses and the market value; and (b) the depletion rate of the metal resources, which is used as an indicator of variations on metal prices (Zouboulis et al., 2004). The price variations of interesting metals are essentially related to the market demands, environmental legislation, and energetic costs (Diniz & Volesky, 2005).

4. Biosorbents

There is a great variety of biomasses used to achieve the biosorption of metals as such micro and macroalgae, yeasts, bacteria, crustacean, etc. The use of adsorbents from dead organisms has an attractive economic cost because they are originated in less expansive materials in comparison to the conventional technologies. Other economic advantage is the possibility of biosorbent reuse from agro-industrial and domestic wastes (e.g., fermentation processes in breweries and pharmaceutics, activated sludge, sugarcane bagasse, etc.) (Godlewska-Zylkiewicz, 2006; Karnitz Jr., 2007; Pagnanelli et al. 2004; Palmieri et al., 2002).

Commonly, the biosorption studies describe applications with native biomasses and with products obtained from biomasses, which are generally biopolymers (polysaccharides and glycoproteins).

The use of biosorbents in native forms from microbial biomasses (e.g. yeasts, microalgae, bacteria, etc.) present a series of problems: the difficulty on separation of cells after the biosorption, the mass loss during the separation, and the low mechanic resistance of the cells (Arica et al., 2004; Sheng et al., 2008; Vegliò & Beolchini, 1997; Vullo et al., 2008). The biomass immobilization makes possible a material with more appropriated size, greater mechanic resistance, and desirable porosity to use in fixed-bed columns (Sheng et al., 2008; Zhou et al., 2005). Besides the immobilization provides the metal recuperation and the column reuse (Sheng et al., 2008; Zhou et al., 2005).

The most common immobilization procedures are: (a) the adsorption on inert supports by preparation of biofilms; (b) the encapsulation in polymeric matrices as calcium alginate, polyacrylamide, polysulfone, and polyhydroxyetilmetacrilate; (c) the covalent linkage on supports by chemical agents; and (d) the cross-linking by chemical agents that form stable cellular aggregated. The most common chemical agents used are formaldehyde, glutaraldheyde, divinylsulfone, and formaldehyde-urea mixture (Vegliò & Beolchini, 1997).

An important area that has been developed is the surface modification of biomasses by the insertion of additional chemical groups to increase the biosorption uptake process (Yang & Chen, 2008; Yu et al., 2007a; Yu et al., 2007b). This procedure is used for biomasses with low uptake capacities and in numerous cases the chemical modification also provides the cellular immobilization.

Since the 80s several biosorption processes have been developed in commercial scale. Some commercial applications are described by Wang & Chen (2009):

a. B. V. SORBEX Inc.: several biosorbents of different biomaterials from biomass as such *Sargassum natans, Acophylum nodosum, Halimeda opuntia, Palmira pamata, Chondrus crispus,* and *Chlorella vulgaris,* which can adsorb a broad range of metals and can be regenerated easily;
b. Advance Mineral Technologies Inc.: biosorbents based in *Bacillus* sp., but that finished their operations in 1988;
c. AlgaSORB (Bio-recovery Systems Inc.): biomass *Chlorella vulgaris* immobilized in silica and polyacrylamide gels that adsorb metals of diluted solution with concentrations between 1-100 mg/L and can undergo more than 100 biosorption-desorption cycles;
d. AMT-BIOCLAIM™ (Visa Tech Ltd.): biosorbent from *Bacillus subtilis* immobilized in polyethyleneimine and glutaraldheyde beads, which removes efficiently metals as gold, cadmium, and zinc from cyanide solutions. The biosorbent is not selective, but it presents high metal recuperation (99%) and can be regenerated by sodium hydroxide or sulfuric acid solutions;
e. BIO-FIX (U. S. Bureau of Mines): biosorbent based in several biomasses, including *Sphagnum* peat moss, yeast, bacteria, and/or aquatic flora immobilized in high density polysulfone. The biosorbent is selective for heavy metals and it is applied in acid mine drainages. The metals can be eluted more than 120 recycles with solutions of hydrochloric acid and nitric acid.

Additionally the Table 3 presents some biosorbents and their applications in biosorption purposes.

Metal	Biosorbent	Reference
Gd	Several microorganisms (fungal and bacteria) from sand	Andrès et al., 2000
Hg, Cd, and Zn	Ca-alginate and immobilized wood-rotting fungus *Funalia trogii*	Arica et al., 2004
Sm and Pr	*Sargassum* sp.	Oliveira et al., 2011
Cu	*Sargassum* sp. immobilized in poly(vinyl alcohol) cryogel beads	Sheng et al., 2008
Co and Ni	*Ulva reticulate, Turbinaria ornata, Sargassum ilicifolium, Sargassum wightii, Gracilaria edulis*, and *Geledium* sp.	Vijayaraghavan et al., 2005
Cd, Zn, and Pb	*Laminaria hyperborea, Bifurcaria bifurcata, Sargassum muticum*, and *Fucus spiralis*	Freitas et al., 2008
Cu and Pb	Activated sludge	Xuejiang et al., 2006
La, Nd, Eu, and Gd	*Sargassum* sp.	Oliveira & Garcia Jr., 2009
Pb and Zn	*Phanerochaete chrysosporium* immobilized in Ca-alginate	Arica et al., 2003
Pb	*Streptomyces rimosus*	Selatnia et al., 2004
Pb	Cellulose/chitin beads	Zhou et al., 2005
Ni	*Sargassum wightii*	Vijayaraghavan et al., 2006
Cr	*Sargassum* sp.: raw and chemically modified (treated with NaOH, HCl, $CaCl_2$, formaldehyde, or glutaraldehyde)	Yang & Chen, 2008
Cu	Sugarcane bagasse: raw and chemically modified (treated with NaOH and/or citric acid)	Dos Santos et al., 2011
Cu, Mo, and Cr	Chitosan: flakes, beads, and modified beads (treated with glutaraldehyde)	Dambies et al., 2000
Ag	*Lactobacillus* sp.	Lin et al., 2005
Cd, Cu, and Ni	Aerobic granules	Xu & Liu, 2008
Cr and V	Waste crab shells	Niu & Volesky, 2006
Cd and Pb	Modified baker's yeast (treated with glutaraldehyde and cystine)	Yu et al., 2007a
Eu	Alfafa	Parsons et al.,2002
Pb, Zn, Cd, Fe, La, and Ce	Cross-linked *Laminaria japonica* (treated with propanol and HCl)	Ghimire et al., 2008
U, La, Ce, Pr, Nd, Sm, Eu, Gd, Tb, Dy, Ho, Er, Tm, Yb, and Lu	*Dictyota dichotoma, Ecklonia stolonifera, Undaria pinnatifida, Sargassum honeri*, and *Sargassum hemiphyllum*	Sakamoto et al., 2008

Table 3. Biosorbents used in some biosorption purposes.

5. Biosorption in batch systems

The quantitative information in the biosorption purposes can be obtained from equilibrium analysis on batch experiments (Volesky, 2003). In these experiments are assayed the optimal conditions to perform a more effective biosorption and they may be used in the research of physicochemical models that describe the metal-biomass interactions. Despite of the continuous operation in columns to be the preferential mode for amplifying the biosorption process to a pilot scale (Volesky, 2003), the batch systems serve as pre-stage for an initial evaluation of adsorption phenomena and operational conditions before the application of the process on continuous systems (Gadd, 2009). The main difference between the operational modes refers to transport phenomena involved: in batch systems the diffusive and convective resistances for the adsorption are pronouncedly diminished in relation to column systems, which exhibit smaller mass transfer rates due to dependence of the combination of several parameters.
The physicochemical modeling is based on the analysis of the metal uptake capacity (according with Eq. (1)) as function of the assay time (biosorption kinetics) or the equilibrium concentration of adsorbed metal (biosorption isotherms).

$$q = (C_0 - C_{EQ})V/M \tag{1}$$

where q is the metal uptake that represents the amount of accumulated metal by mass unity or matter moiety of biomass; V is the solution volume; C_0 e C_{EQ} are the initial and equilibrium concentrations (in the liquid phase), respectively; and M is the biomass mass.
Physicochemical models differ with regard to the number of adsorbed layers, the type of interactions among the active sites and metals, and the possibility to use the equilibrium constants among the solid and liquid phases. The criteria for choosing an isotherm or a kinetic equation for biosorption data is mainly based on the best adjustment of curve fitting which is often evaluated by statistical analysis. The model chosen should be the one reflecting the best the biosorption mechanisms (Liu & Liu, 2008; Vegliò et al., 2003). The next items exemplify the use of batch systems as much in the optimization of operational parameters as in the physicochemical modeling for the biosorption of metals.

5.1 Biosorption isotherms

The study of the phase equilibrium is a part of the thermodynamic that relate the equilibrium composition of two phases and it is represented by graphics of concentration in the stationary phase (expressed in biosorption purposes in terms of metal uptake, q) versus the concentration in the mobile phase, both at equilibrium (Godlewska-Zylkiewicz, 2006). Usually the mechanisms of adsorption and ion-exchange are the most used because their concepts are easily extended to other mechanisms of metal retention. The adsorption models in liquid-solid equilibrium are derived of models for gas-solid equilibrium from the Gibbs isotherm and assuming an equation of state for the adsorbed phase. The Table 4 displays some adsorption models used in biosorption studies and the advantages and disadvantage in their utilization.
These models (Table 4) differ in the amount of adsorbed layers, the interaction between the binding sites and the metal (adsorbent-adsorbate, adsorbate-adsorbate, and adsorbent-adsorbent), and the possibility to apply equilibrium constants equations between the liquid and solid phases. Obviously, these considerations for biosorption systems do not explain the

mechanisms of metal uptake due to the complexity of the biologic systems, but it supplies parameters that are utilized to evaluate the biosorption performance, such as the maximum metal uptake and the affinity of the active sites by metallic ions (Kratochvil & Volesky, 2000; Palmieri, 2001).

Biosorption of metals in the mostly cases of equilibrium isotherms is modeled according to non-linear functions that are described by Brunauer-Emmet-Teller (BET) type-I isotherms with hyperbolic shape (Guiochon et al., 2006). The general form of the curve $q = f(C_{EQ})$ is showed on Fig. 1.

Adsorption Model	Equation	Advantages	Disadvantages
Langmuir	$q = (q_{MAX}bC_{EQ})/(1+bC_{EQ})$	Interpretable parameters	Not structured; Monolayer Adsorption
Freundlich	$q = K_F C_{EQ}^{1/n}$	Simple expression	Not structured; No leveling off
Combination Langmuir-Freundlich	$q = (q_{MAX}bC_{EQ}^{1/n})/(1+bC_{EQ})$	Combination of above	Unnecessarily complicated
Radke-Prausnitz	$1/q = 1/(aC_{EQ})+1/(bC_{EQ}^{\beta})$	Simple expression	Empirical, uses 3 parameters
Brunaer-Emmet-Teller	$q = (BCQ^0)/\{[C_s-C][1+(B-1)C/C_S]\}$	Multilayer adsorption; Inflection point	No total capacity equivalent

Table 4. Examples of physicochemical models of adsorption. Source: Volesky, 2003.

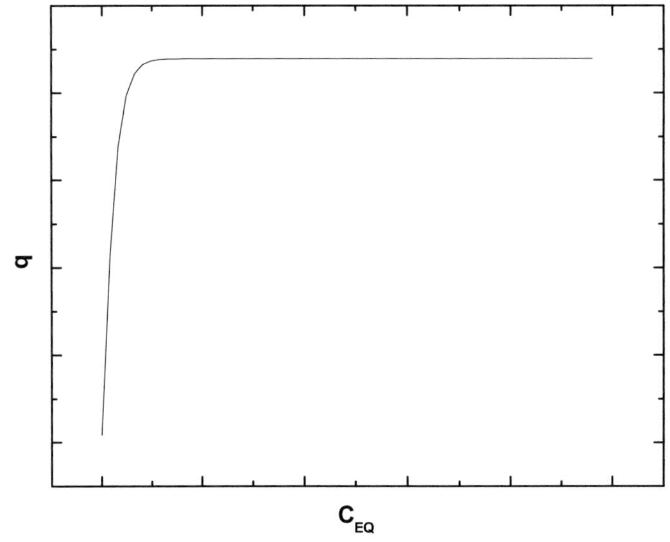

Fig. 1. Typical curve of an adsorption isotherm. Source: Oliveira, 2011.

These isotherms generally are associated mainly to Langmuir and Freundlich besides other models derived of these firsts. The Freundlich model suggests adsorbed monolayers, where the interactions among adjacent molecules that are adsorbed: the energy distribution is heterogeneous due to the diversity of the binding sites and the nature of the adsorbed metallic ions. The Langmuir model considers an adsorbed monolayer with homogeneous distribution of binding sites and adsorption energy, without interaction among the adsorbed molecules (Selatnia et al., 2004).

For instance, on biosorption of Sm(III) and Pr(III) by *Sargassum* sp. biomass described by Oliveira et al. (2011), the Langmuir adsorption model has been founded very accurate, that is approximated for liquid-solid equilibrium by the Eq. (2) and it can be observed in the Fig. 2.

$$q = (q_{MAX} b C_{EQ})/(1+bC_{EQ}) \quad (2)$$

where q is the metal uptake; q_{MAX} is the maximum biosorption uptake that is reached when biomass active sites are saturated by the metals; b is a constant that can be related to the affinity between the metal and the biomass; and C_{EQ} is the metal concentration in the liquid phase after achieving the equilibrium.

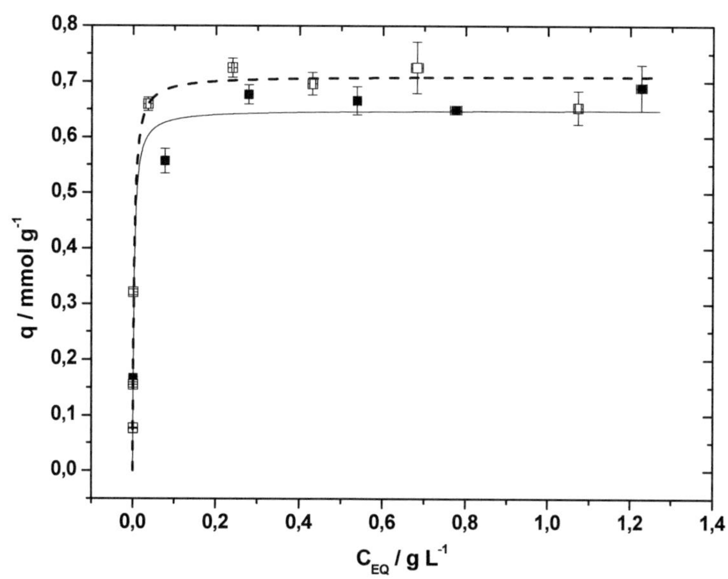

Fig. 2. Biosorption isotherms for Sm(III) and Pr(III) solutions by *Sargassum* sp. described by the Langmuir adsorption model. Symbols: (-■-) Sm(III) and (--□--) Pr(III).
Source: Oliveira et al., 2011.

Additionally, it is noteworthy that the shape of the biosorption isotherms (Fig. 2) approaches the profile of irreversible isotherms: the initial slope is very steep and the equilibrium plateau is reached at low residual concentration. This can be correlated to the great affinity of Sm(III) and Pr(III) for the biosorbent (Oliveira et al., 2011).

The models presented on Table 4 are applied for mono-component systems. For systems with more than one metallic species, the mathematical modeling must be modified to take into account the competition of metal by the binding sites (Aksu & Açikel, 2000). Some approaches are listed on Table 5.

Adsorption Model	Equation	Advantages	Disadvantages
Langmuir	$q_i=(q_{MAX,i}b_iC_{EQ,i})/\ (1+\sum_{i=1}^{n} bC_{EQ,i})$	Constants have physical meaning; Isotherms levels off at maximum saturation	Not structured; Does not reflect the mechanism well
Combination Langmuir-Freundlich	$q_i = (a_iC_{EQ,i}{}^{1/n_i})/(1+\sum_{i=1}^{n} b_iC_{EQ,i}{}^{1/n_i})$	Combination of above	Unnecessarily complicated
Surface complexation model	$q \sim f(C_{EQ})$, could follow e.g. Langmuir	Model more structured: intrinsic equilibrium constant could be used	Equilibrium constants have to be established for different types of binding

Table 5. Examples of physicochemical multi-component models of adsorption. Source: Volesky, 2003.

5.2 Biosorption kinetics

Biosorption processes tend to occur rapidly, taking from few minutes to a couple of hours and it takes account transfer mass processes and adsorption processes. The biosorption kinetics is controlled mainly by convective and diffusive processes. In a first stage occurs the metal transference from solution to adsorbent surface neighborhood; then in the next step, the metal transference from adsorbent surface to intraparticle active sites; and finally, the metallic ion removal by the active sites via complexation, adsorption, or intraparticle precipitation. The first and second steps represent the resistance to convective and diffusive mass transferences and the last one is quick and non-limiting for the overall biosorption velocity (Selatnia et al., 2004).

Analogously to the biosorption isotherms, the biosorption kinetics in general present hyperbolic shape (as the Fig. 1) and they are described by various models. The models more used in biosorption studies are presented on Table 6.

Adsorption model	Differential equation	Integral equation	Initial adsorption velocity
Pseudo-first-order	$dq_t/dt = k_1(q_{EQ} - q_t)$	$\ln(q_{EQ} - q_t) = \ln q_{EQ} - k_1t$	$v_1 = k_1q_{EQ}$
Pseudo-second-order	$dq_t/dt = k_2(q_{EQ} - q_t)^2$	$q_t = t/[1/(k_2q_{EQ}{}^2)+t/q_{EQ}]$	$v_2 = k_2q_{EQ}{}^2$

Table 6. Examples of kinetics models used in biosorption studies. Source: Wang & Chen, 2009.

The pseudo-second-order model is preferred for biosorption of RE (Oliveira & Garcia Jr., 2009; Oliveira et al., 2011) and is represented by the integral Eq. (3).

$$q_t = t/[1/(k_2 q_{EQ}^2)+t/q_{EQ}] \quad (3)$$

where q_t is the biosorption uptake in the t time of assay; q_{EQ} is the equilibrium metal uptake; and k_2 is a constant that represent the metal access rate to biomass in the pseudo-second-order kinetic model. Fig. 3 displays the modeling of samarium and praseodymium biosorption kinetics in *Sargassum* sp. by the pseudo-second-order kinetics model.

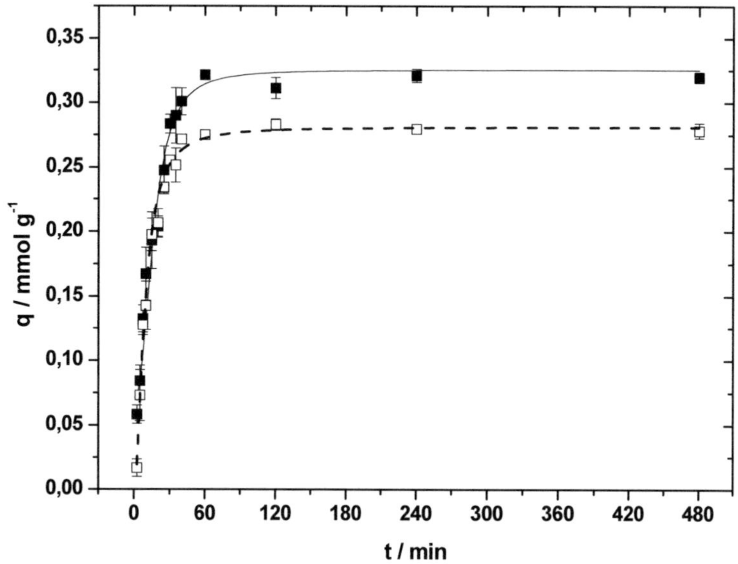

Fig. 3. Biosorption kinetics for Sm(III) and Pr(III) solutions by *Sargassum* sp. described by the pseudo-second-order kinetics model. Symbols: (-■-) Sm(III) and (--□--) Pr(III). Source: Oliveira et al., 2011.

5.3 Chemical speciation and pH

Generally the biosorption carried out in low pH values (smaller than 2.0) has a non-effective metal uptake (for the cases that metallic cationic species are involved) because the high hydronium concentration makes the competition among these protons more favorable than the metals in solution by the biomass active sites. Moreover the acidic groups in low pH should be protonated according with their pKa values as can be seen on Table 1.

The metal uptake is increased when the acidic groups tend to be deprotonated from their pKa values (Table 1) and the metallic ion presents a chemical speciation that provides greater adsorption performance. In the case of RE biosorption for *Sargassum* sp. biomass, Palmieri et al. (2002) and Diniz & Volesky (2005) founded that the biosorption of La(III), Eu(III), and Yb(III) is more effective in crescent pH values (2.00 to 5.00) because the quantity of negative ligands is increased, and consequently the increase of the attraction among the ligands and the metallic cations. The optimal pH for *Sargassum* founded about 5.0. In this pH the carboxyl pK_as of mannuronic and guluronic acid residues (3.38 and 3.65, respectively) in the alginate biopolymer (main component of brown algae cellular envelope) are suppressed; so all carboxyl sites should be more available for the adsorption. Towards the RE speciation in distinct pH ranges: (a) in pH < 6.0 prevail the presence of RE^{3+}; (b) between about 6.0 < pH < 9.5 there is the generation of $RE(OH)^{2+}$ and $RE(OH)^{2+}$ that remain solubilized or suspended in solution; and (c) from pH ~ 8.5 occurs the precipitation of RE hydroxide. Biosorption of anionic species are very less common and occurs when a metallic complex is formed with a negative global charge, e.g. the AMT-BIOCLAIM™ is able to adsorb gold, zinc, and cadmium from cyanide solution (i.e. cyanide complexes with the metals) in metal-finishing operations (Atkinson et al., 1998).

5.4 Temperature

In general, the literature describes that the biosorption process is not influenced between 20 and 35ºC (Vegliò & Beolchini, 1997). However some biosorbent present considerable differences on biosorption performance as function of the temperature. For instance, Ruiz-Manríquez et al. (1998) studied the biosorption of copper on *Thiobacillus ferrooxidans* [sic] considering temperatures of 25 and 37 °C: the results indicate that there was a positive effect in the biosorption uptake when the temperature was increased, where the increase in the biosorption was of 68%.

Besides the evaluation of the optimal temperature to be used in biosorption purposes, the batch procedures commonly can be utilized to find thermodynamic parameters as enthalpy (ΔH), entropy (ΔS), and Gibbs free-energy (ΔG) through the Eqs. (4) and (5).

$$\Delta G = -RT\ln K_{EQ} \tag{4}$$

$$\Delta G = \Delta H - T\Delta S \tag{5}$$

where R is the gas constant (8.314 J/(K mol)), T is the temperature, and K_{EQ} is equilibrium constant in determined temperature that corresponds the ratio between the equilibrium metal concentration in the liquid (C_{EQ}) and solid phases (q_{EQ}). In this context, Dos Santos et al. (2011) verified that the chemical modification of the sugarcane bagasse by different treatments lead a more energetically favorable adsorption of copper in comparison with raw material, because the negative increase of the Gibbs free-energy.

5.5 Presence of counter-ions

The binding of metallic ions biomasses is influenced by other ionic species, such as cations and anions present in solution. Benaissa & Benguella (2004) describe the influence of the presence of cations (Na^+, Mg^+, and Ca^{2+}) and anions (Cl^-, SO_4^{2-}, and CO_3^{2-}) on cadmium biosorption for chitin. The presence of these ions in solution inhibits the uptake of cadmium by chitin to different degrees: sodium and chloride ions have no significant. For magnesium, calcium, sulfate, and carbonate ions the effects ranged from a large inhibition of cadmium by calcium and carbonate to a weak inhibition by magnesium and sulfate. These interferences in cadmium biosorption are resulted of the competition among the interesting metal and the counter-ion by the binding sites.

Additionally, Palmieri et al. (2002) studied the lanthanum biosorption by *Sargassum fluitans* in solution with chloride and sulfate ions: at same pH it was observed higher maximum metal uptake values for the biosorption on presence of chloride, as such can be seen on Fig. 4. In the case of lanthanides, the formation of complexes with chloride or sulfate affects the coordination sphere of metal, leading to an influence on the net charge of the cation. Chloride ions are reported to have an outer sphere character with a less disturbance in the hydration sphere. On the other hand, sulfate and carboxylate anions present inner sphere character more pronounced in the complex formation with lanthanum. The biosorption uptake of lanthanum presents higher value for chloride-based solutions than sulfate-based solutions could suggest that the fewer disturbances on the inner coordination sphere caused by chloride anion facilitate the interaction with carboxylate groups present in the biomass.

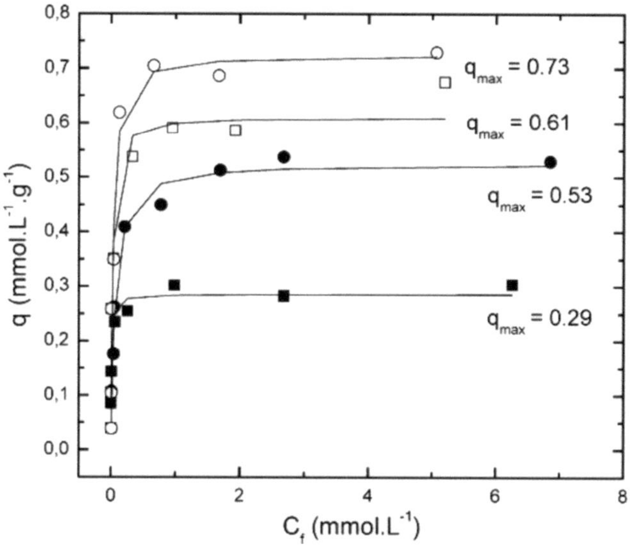

Fig. 4. Bisorption isotherms for La(III) on *Sargassum fluitans* from chloride or sulfate-based solutions at different pHs. Symbols: chloride-based solutions at (□) pH 4 and (○) pH 5; and sulfate-based solutions at (■) pH 4 and (●) pH 5. Source: Palmieri et al., 2002.

5.6 Desorption

After the metal removal from aqueous solutions by the biomass, it is important the metal recuperation from biomass. In this point, it is achieved the metal desorption process, whose aim is the weakening the metal-biomass linkage (Modak & Natarajan, 1995). Generally it can be applied diluted mineral acids and complexing agents as desorbents. Biosorption and desorption isotherms present close behavior characteristic of Langmuir modeling, which has at equilibrium equivalent kinetic rates (Palmieri et al., 2002).

Diniz & Volesky (2005) evaluate the reversibility of the adsorption reaction for the biosorption of lanthanum, europium, and ytterbium by *Sargassum polycystum* using the desorbent agents: nitric and hydrochloric acids, calcium nitrate and chloride salts, EDTA, oxalic and diglycolic acids. This work as such other studies founded the hydrochloric acid as the best agent for brown algae, with percentage of recovery between 95-100%.

5.7 Biomass characterization from analytic and spectroscopic methodologies

Beyond the perspectives of application of the biosorption in order to optimize the process, the understanding of the mechanisms involved in the biosorption is justifiable for better comprehension of the process and of itself scale-up. This is carried out from qualitative and/or quantitative characterizations by potentiometric titrations, and spectroscopic and microscopic techniques as such FTIR (Fourier transform infrared spectroscopy), SEM (scanning electron microscope), EDX (energy-dispersive X-ray spectroscopy), XPS (X-ray photoelectron spectroscopy), etc. The main objective of the biosorbent characterization has

been to indentify the chemical groups involved in the biosorption and the way that these groups perform the metal binding.
The most common technique used is the potentiometric titration, which evaluate the existence of stoichiometric relationships among the metals and the binding sites, and to determine the pK_as values of the chemical groups on biomass cellular envelope. The Table 1 summarizes the characteristics of the protonated *Sargassum* sp. biomass before and after samarium and praseodymium biosorption.

Material	Strong acid groups (mmol/g)	Total amount of acid groups (mmol/g)	Weak acid groups (mmol/g)	Occupancy of binding sites (%)
Protonated biomass	0.15	1.77	1.62	-
Sm(III) - loaded biomass	0.07	1.26	1.19	29
Pr(III) - loaded biomass	0.07	1.18	1.11	33

Table 7. Acid-base properties of protonated *Sargassum* sp. before and after Sm(III) and Pr(III) biosorption. Source: Oliveira et al., 2011.

The strong acid groups counted for only 0.15 mmol/g on protonated biomass, and decreased to 0.07 mmol/g after the biosorption of either Sm(III) or Pr(III). These groups of lowest pK_a have been identified as the ester sulfate groups of the fucoidan, which are present on the cell wall of brown seaweeds. Weak acid groups are mainly constituted by carboxylate groups from alginate compounds, which represent more than 90 % of total acid groups, i.e., 1.62 mmol/g. After metal biosorption the titration identified 1.19 and 1.11 mmol/g of weak acid groups for Sm(III) and Pr(III), respectively. Thereby only around 30 % of the acid groups were involved in metal binding (Oliveira et al., 2011).
Another example of the biomass characterization can be observed on Fig. 5, which displays the analysis of SEM-EDX of *Sargassum* sp. biomass after lanthanum biosorption. The lanthanum presence in the X-ray spectra confirms the adsorption of the metal on the biosorbent surface. In the SEM micrography also is evident the surface colonization by diatoms as well as the assignments of chemical elements from the marine environment (calcium, aluminum, silicon).

6. Biosorption in fixed-bed columns

Despite of the biosorption in batch systems to available parameters to understand the metal-biomass interaction and to select the best operational condition, the procedures in columns are generally the preferential mode for the biosorption application in the industrial scale-up, once that the process can be performed continuously (Vieira et al., 2008; Volesky, 2003). This operational mode is more appropriate for large-scale applications in industry than other types of reactors as such agitated tanks, fluidized-bed columns, etc. The fixed-bed columns have a series of advantages: they have simple operation, they achieve large yields, and they have ease scale-up from procedures in laboratorial scale (Borba et al., 2006; Borba et al., 2008; Valdman et al., 2001; Vijayaraghavan et al., 2005; Vijayaraghavan & Prabu, 2006). The use of fixed-bed columns allow to avoid separation difficulties between the biosorbent and the effluent (Kentish & Stevens, 2001). This experimental procedure has as limiting step the mass transfer of metal from solution to the biosorbent, since the adsorption reactions do not limit the process due to the fast kinetics (Aksu, 2001; Crini, 2005; Volesky, 2001).

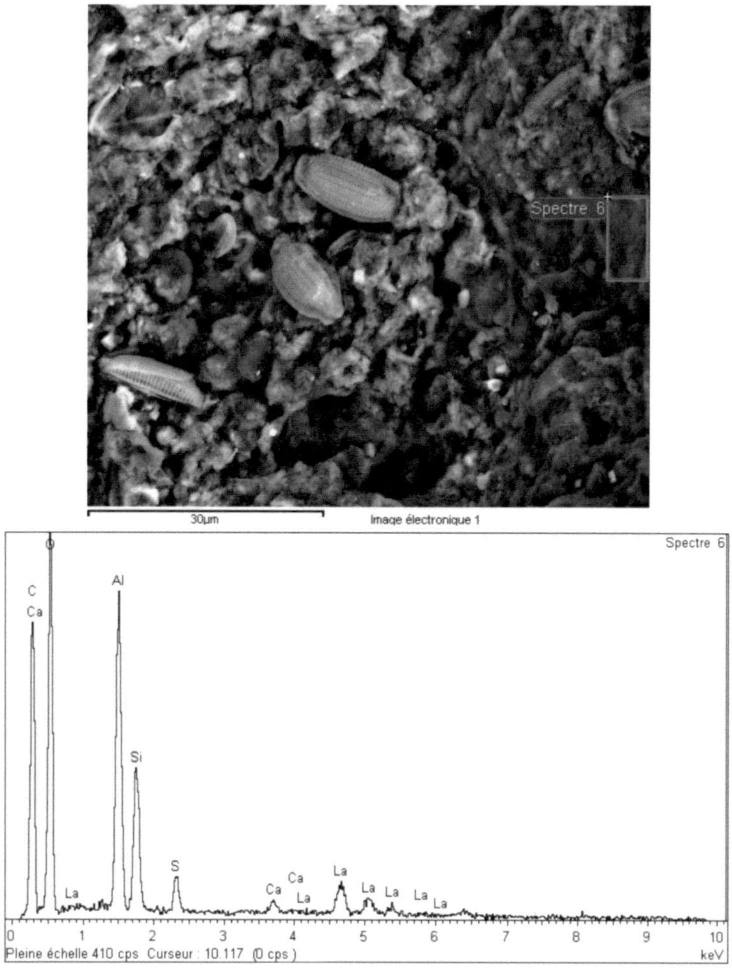

Fig. 5. Scanning electronic micrography of *Sargassum* sp. biomass after lanthanum biosorption and related X-ray spectra. Source: Oliveira, 2011.

6.1 Biosorption: frontal analysis and breakthrough curves

The main methodologies for the concentration, separation, and purification of metals involve a great number of equilibriums and phase transferences, such as the methodologies listed in Table 8.

Methodology	Concentration applied (g/L)
Solvent extraction	0.5–500
Microporous membranes	10^{-2}–10
Emulsified or supported liquid membranes	10^{-4}–10
Ion-exchange	10^{-6} - 1
Biosorption	10^{-6} - 0.1

Table 8. Separation technologies and concentration ranges applied. Source: Kentish & Stevens, 2001.

The biosorbents should have several mechanisms of metal uptake, but for column biosorption perspectives these mechanisms are approximated to mainly ion-exchange or adsorption. Generally the chromatographic separations by fixed-bed columns occur by two ways: the frontal analysis and the displacement elution.

On frontal analysis is carried out the metal adsorption for a percolated volume of solution in the column, which produces a mixed zone of metallic ions that spreads to a distance across the column according to the individual and competitive interactions among the metals and the adsorbent. In this process, the mixed zone is composed by several equilibriums among the displaced ions and the retained ions and it moves across the column without to alter your volume. After the mixed zone is displaced to sufficient distance across the column, it is reached an equilibrium which the components are resolved in differentiate heights, i.e. in distinct or enriched zones for each one of the components (Fritz, 2004). Thus the greater interaction among the metals and the biosorbent represents a greater retention of these metals across the column. Therefore, a greater number of distinct affinities of the percolated metals by the adsorbent mean a better possibility of the system to resolve the metals in differentiate heights.

Commonly the frontal analysis performance is mathematically quantified and modeled from the application of approximations and boundary conditions on non-linear material balance equations based mainly for biosorption columns on equilibrium dispersive model (Guiochon et al., 2006). The model assumes that all conditions are due to a non-equilibrium, which is treated into a term of apparent axial dispersion, where it is considered that the dispersion coefficients of the components remain constants. The column is considered unidimensional and radially homogenous, i.e. the properties are constants in a same cross section. When a fixed-bed column is occupied by fluid with a constant linear velocity, the differential mass balance involved is given by the Eq. (6).

$$\partial q(t,z)/\partial t + v[\partial C(t,z)]/\partial z] + [(1-\varepsilon)/\varepsilon][\partial q(t,z)/\partial t] - D_L[\partial^2 C(t,z)]/\partial z^2] = 0 \quad (6)$$

where t is the time; z is the axial coordinate with origin on column entrance; q is the metal uptake in the stationary phase; C is the concentration in the mobile phase; v is the linear velocity; $(1-\varepsilon)/\varepsilon$ is the phase ratio (mobile phase volume/stationary phase volume) and ε is the adsorbent porosity; and D_L is a parameter that includes the contributions of the axial dispersion (due to molecular diffusion), the non-homogeneity of the flux (eddy diffusion), and the bed tortuosity.

The terms on Eq. (6) represent respectively: (a) the accumulation in the stationary phase; (b) the convective phenomena; (c) the accumulation in the mobile phase, and (d) the diffusive phenomena. Some approximations should be achieved as such: (a) the column should be considered radially homogenous only in isothermic or isobaric operations; (b) the compressibility of the mobile phase is neglected between 0 and 200 bar in the mostly cases if the volume is altered between 0.5 and 2%; (c) the viscosity in the mobile phase is constant; (d) since the pump provides constant flow rate, the velocity is also constant; (e) the parameter D_L is constant; (f) the partial molar volume of the sample components is constant in both phases; (g) the solvent is not adsorbed; (h) constant operational conditions: temperature, pressure, flow rate, physicochemical parameters, porosities, etc. (Guiochon et al., 2006).

There are several parameters that govern the adsorption, which may be modified to find a more effective adsorption and/or a separation with better resolution of the components as such: (a) the column geometry that considers the height and the cross section area of the bed; (b) the homogeneity or the heterogeneity of the adsorbent; (c) the particle diameter and their implications on porosity, packing, and tortuosity of the bed; (d) the number of theoretical plates; (e) the concentration and composition of the solute on mobile and stationary phases; (f) the presence of additives on feeding, e.g. complexing agents, buffers, etc.; (g) the column flow rate; etc. (Guiochon et al., 2006).

In biosorption isotherms, the concentration profiles in the liquid and solid phases change in space and time. Thereby the development and performance of adsorption columns are difficult to reach without an approximated quantitative modeling of the Eq. (6). From perspective of design and optimization of the column processes, the behavior in fixed-bed is described by the effluent concentration profile (C/C_0, where C and C_0 are the concentration of eluate and eluent, respectively) in function of the time or percolated volume (Nadaffi et al., 2007), i.e. by breakthrough curve, which is showed on Fig. 6. The curve shape is given by a sigmoid function and it is determined by the shape of the equilibrium isotherm, i.e. it is influenced by the transport processes and the adsorbent nature (Chu, 2004).

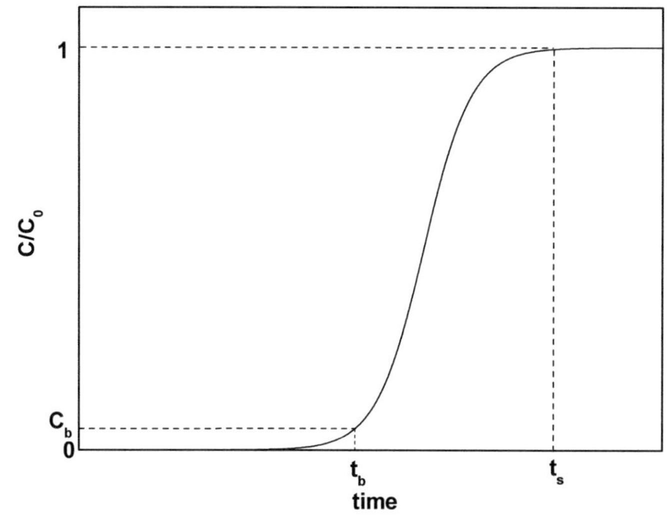

Fig. 6. Schematic representation of the breakthrough curve. Source: Oliveira, 2011.

In the breakthrough curves (Fig. 6) are determined the breakthrough and saturation times (t_b and t_s, respectively). The breakthrough time indicates the instant in which the metallic ion is effectively discharged on eluate, and the saturation time corresponds to the instant of metal mass saturation on biomass. The breakthrough time is arbitrarily inferred for C/C_0 at 0.05; while the saturation time is defined ideally when C/C_0 values reach 1.0 (generally at 0.90-0.95). All optimized system in columns is based on accurate prediction of the breakthrough time under selected operational conditions. When the eluate concentration reaches a predefined level, the column operation is finalized; in this point the regeneration process may be achieved to activate the column for a next operation cycle (Kentish & Stevens, 2001). In order to investigate the alternatives for the separation of metallic species, the breakthrough time is crucial because it represents the interaction between the metal and the biomass; so if the breakthrough time is great, this indicates that the interaction between the metal and the biomass is greater.

The variation between the breakthrough and saturation times depends on the capacity of the column toward the quantity of applied metal (Aksu, 2001). A more efficient adsorption performance will be obtained as greater is the curve slope, i.e. as smaller is the gap between the breakthrough and saturation times (Fig. 6) (Chu, 2004). This gap corresponds to the extension of the mass transfer zone (MTZ) on bed (Nadaffi et al., 2007), which is the bed active region where the adsorption occurs as can be seen on Fig. 7. So the column efficiency will be better in smaller values of height of mass transfer zone which indicate a behavior near to ideality; in that case a step function where the curve inclination between the breakthrough and the saturation tends to zero.

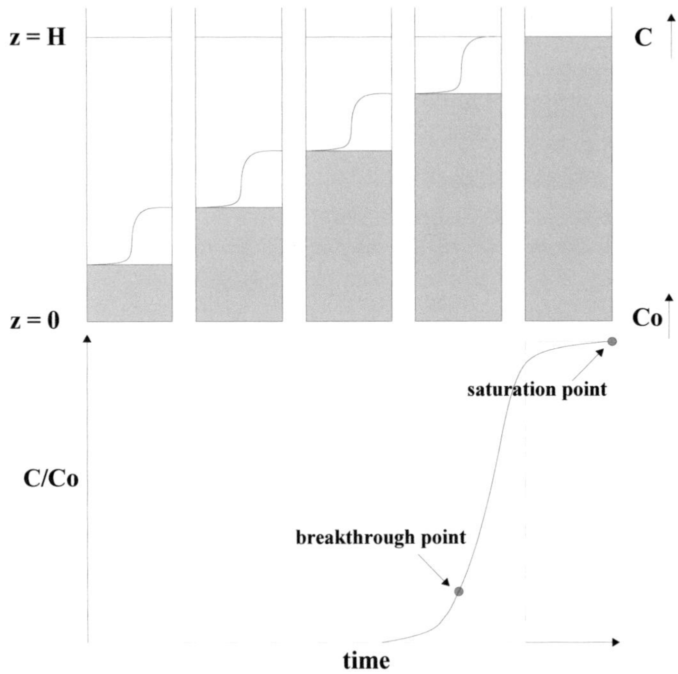

Fig. 7. Schematic representation of the movement of the mass transfer zone in fixed-bed column. Symbols: (—) ideal and (--) real cases. Source: Oliveira, 2011.

Several derivations may be used from the material balance in the Eq. (6) to perform the breakthrough curves such as the models of Thomas, Bohart-Adams, Yoon-Nelson, etc. Some models are described in function of operational and kinetic parameters (e.g. Thomas and Bohart-Adams); in other hand, there are models related to adjustment purely mathematic according with the sigmoid function (e.g. Yoon-Nelson model). For instance the Thomas model is expressed Eq. (7).

$$C/C_0 = 1/\{1+\exp[(k_{Th}/Q)(q_{MAX}M-C_0V)]\qquad (7)$$

where k_{Th} is the Thomas constant; Q is the flow rate; q_{MAX} is maximum biosorption uptake; M is the dry mass of biomass; and V is the volume percolated. The Fig. 8 presents the experimental data for column biosorption of lanthanum by *Sargassum* sp. adjusted by the Thomas model.

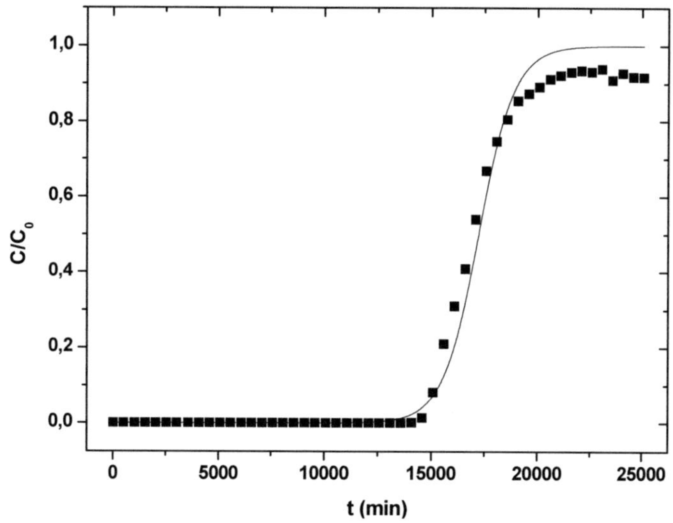

Fig. 8. Modeling of breakthrough curve in the column biosorption of La(III) for *Sargassum* sp. biomass by the Thomas model. Symbols: (■) data of metal concentration on eluate and (--) curve fit for Thomas model. Source: Oliveira, 2011.

6.2 Dependence of the operational parameters

There is broad literature that describes the effects of operational parameters to augment and to improve the biosorption in fixed-bed columns (Chu, 2004; Hashim & Chu, 2004; Kratochvil & Volesky, 2000; Naddafi et al., 2007; Oliveira, 2007; Oliveira, 2001; Valdman et al., 2001; Vieira et al., 2008; Vijayaraghavan et al., 2005; Vijayaraghavan et al., 2008; Vijayaraghavan & Prabu, 2006; Volesky et al., 2003). These parameters modified mainly related are: flow rate, feeding concentration, height of packed-bed column, porosity, mass of biomass, etc. Vijayaraghavan & Prabu (2006) evaluate some variables as the bed height (15 to 25 cm), flow rate (5 to 20 mL/min), and copper concentration (50 to 100 mg/L) in *Sargassum wightii* biomass from breakthrough curves: each variable evaluated was changed and the others were fixed. Continuous experiments revealed that the increasing of the bed height and inlet solute concentration resulted in better column performance, while the lowest flow rate favored the biosorption (Vijayaraghavan & Prabu, 2006)

Naddafi et al. (2007) studied the biosorption of binary solution of lead and cadmium in *Sargassum glaucescens* biomass from the breakthrough curves modeled according with the Thomas model (eq. (7)). Under selected flow rate condition (1.5 L/h) the experiments reached a selective biosorption. The elution of the metals in distinct breakthrough times with biosorption uptake in these times at 0.97 and 0.15 mmol/g for lead and cadmium, respectively.

6.3 Desorption: chromatographic elution and biomass reuse

Column desorption is used for the metal recovery, but this procedure under selected conditions may be operated to carry out chromatographic elution by the displacement of the adsorbed components in enriched fractions containing each metal (Diniz & Volesky, 2006). This is resulted of the simple drag of the previous separation on frontal analysis. Nevertheless the eluent may present differential affinity by the adsorbed solutes, so there is

the possibility to use the procedure to promote a more effective separation of the components. The chromatographic elution is dependent of the parameters referred to frontal analysis and of the composition and concentration of the displacement solution. Desorption profiles are given as bands or peaks whose modeling are associated directly to mathematic approximations by Gaussian functions that may be modified or not exponentially (Guiochon et al., 2006).
A typical column desorption with hydrochloric acid from *Sargassum* sp. previously submitted to biosorption of lanthanum is showed on Fig. 9, which is represented by lanthanum concentration in eluate as function of the volume.

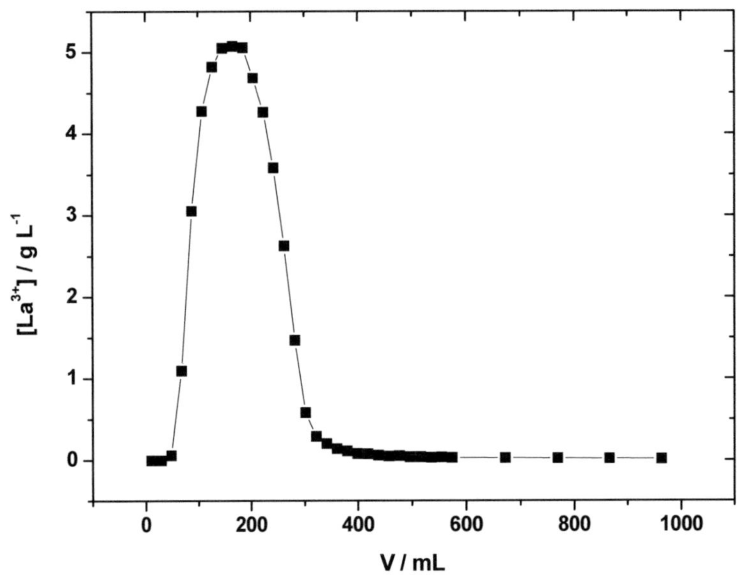

Fig. 9. Column desorption of La(III) from *Sargassum* sp. biomass with HCl 0.10 mol/L. Symbols: (-■-) metal concentration on eluate. Source: Oliveira, 2011.

On Fig. 9 can be seen that after the start of the acid percolation occurs a quick increase of concentration until the maximum to 5.08 g/L for lanthanum. Parameters as the recovery percentage (p) and concentration factor (f) are obtained from biosorption and desorption curves. The recovery percentage is resulted of the ratio between the values of metal recovery on desorption and maximum metal uptake on biosorption, while the concentration factor refers to the ratio between the saturation volume on biosorption and the effective recovery volume on desorption. Both measure the efficiency of the desorbing agents in the metal recovery. For instance, these parameters obtained from Fig. 9 were 93.3% and 60.4 times of recovery percentage and concentration factor, respectively; which are expressive and satisfactory for the column biosorption purposes (Oliveira, 2011).
For biosorption and desorption processes, other important aspect is the biosorbent reuse for recycles biosorption-desorption according the cost benefit between the biosorption capacity loss during desorption steps and the metal recuperation operational yield (Diniz & Volesky, 2006; Gadd, 2009; Godlewska-Zylkiewicz, 2006; Gupta & Rastogi, 2008; Volesky et al., 2003). Oliveira (2007) performed the neodymium column biosorption by *Sargassum* sp. and the subsequent desorption in three recycles. In these experiments was observed that occurs a

decrease in mass metal accumulation through the cycles. Accumulation decrease from first to third cycle in 22%, which is due to the partial destruction of binding sites on desorption procedures, and the binding sites blocking by neodymium ions strongly adsorbed. The result showed that the biomass may be used for recycle finalities.
The loss in performance of the adsorption during the recycles can has numerous origins. Generally they are associated to the modifications on chemistry and structure of the biosorbent (Gupta & Rastogi, 2008), and the changes of access conditions of the desorbent to the metal and mass transfer. Low-grade contaminants in the solutions used in these procedures may accumulate and to block the binding sites or to affect the stability of these molecules (Volesky et al., 2003).

7. References

Ahluwalia, S. S. & Goyal, D. (2007). Microbial and plant derived biomass for removal of heavy metals from wastewater. *Bioresource Technology*, Vol.98, No.12, (September 2007), pp. 2243-2257, ISSN 0980-8524.

Aksu, Z. (2001). Equilibrium and kinetic modeling of cadmium (II) biosorption by *C. vulgaris* in a batch system: effect of temperature. *Separation and Purification Technology*, Vol.21, No.3, (January 2001), pp. 285-294, ISSN 1383-5866.

Aksu, Z. & Açikel, Ü. (2000). Modeling of a single-staged bioseparation process for simultaneous removal for iron(III) and chromium(VI) by using *Chlorella vulgaris*. *Biochemical Engineering Journal*, Vol.4, No.3, (February 2000), pp. 229-238, ISSN 1369-703X.

Andrès, Y.; Thouand, G.; Boualam, M. & Mergeay, M. (2000). Factors influencing the biosorption of gadolinium by microorganisms and its mobilization from sand. *Applied Microbiology and Biotechnology*, Vol.54, No.2, (August 2000), pp. 262-267, ISSN 0175-7598.

Arica, M.; Bayramoglu, G.; Yilmaz, M.; Bektas, M. & Genç, O. (2004). Biosorption of Hg^{2+}, Cd^{2+}, and Zn^{2+} by Ca-alginate and immobilized wood-rotting fungus *Funalia trogii*. *Journal of Hazardous Materials*, Vol.109, No.1-3, (June 2004), pp. 191-199, ISSN 0304-3894.

Atkinson, B. W.; Bux, F. & Kasan, H. C. (1998). Considerations for application of biosorption technology to remediate metal-contaminated industrial effluents. *Water S.A.*, Vol.24, No.2, (April 1998), pp. 129-135, ISSN 0378-4738.

Benaissa, H & Benguella, B. (2004). Effect of anions and cations on cadmium sorption kinetics from aqueous solutions by chitin: experimental studies and modeling. *Environmental Pollution*, Vol.130, No.2 , (July 2004), pp. 157-163, ISSN 0269-7491.

Bruins, M. R.; Kapil, S. & Oehme, F. W. (2000). Microbial resistance to metals in the environment. *Ecotoxicology and Environmental Safety*, Vol.45, No.3, (March 2000), pp. 198-207, ISSN 0147-6513.

Chu, K. H. (2004). Improved fixed bed models for metal biosorption. *Chemical Engineering Journal*, Vol.97, No.2-3, (February 2003), pp. 233-239, ISSN 1385-8947.

Crini, G. (2005). Recent developments in polysaccharide-based materials used as adsorbents in wastewater treatment. *Progress in Polymer Science*, Vol.30, No.1, (January 2005), pp. 38-70, ISSN 0079-6700.

Dambies, L.; Guimon, C.; Yiacoumi, S. & Guibal, E. (2000). Characterization of metal ion interactions with chitosan by X-ray photoelectron spectroscopy. *Colloids and Surfaces A: Physicochemical and Engineering Aspects*, Vol.177, No.2-3, (February 2000), pp. 203-214, ISSN 0927-7757.

Deng, S. & Bai, R. (2004). Removal of trivalent and hexavalent chromium with aminated polyacrylonitrile fibers: performance and mechanisms. *Water Research*, Vol.38, No.9, (May 2004), pp. 2424-2432, ISSN 0043-1354.

Diniz, V. & Volesky, B. (2005). Biosorption of La, Eu and Yb using *Sargassum* biomass. *Water Research*, Vol.39, No.1, (January 2005), pp. 239-247, ISSN 0043-1354.

Diniz, V. & Volesky, B. (2006). Desorption of lanthanum, europium and ytterbium from *Sargassum*. *Separation and Purification Technology*, Vol.50, No.1, (June 2006), pp. 71-76, ISSN 1383-5866.

Dos Santos, V. C. G., De Souza, J. V. T .M., Tarley, C. R. T., Caetano, J. & Dragunsky, D. C. (2011). Copper ions adsorption from aqueous medium using the biosorbent sugarcane bagasse in natura and chemically modified. *Water, Air & Soil Pollution*, Vol.216, No.1-4, (March 2011), pp. 351-359, ISSN 0049-6979.

Eccles, H. (1999). Treatment of metal-contamined wastes: why select a biological process? *Trends in Biotechnology*, Vol.17, No. 12, (December 1999),pp. 462-465, ISSN 0167-7799.

Freitas, O. M. M.; Martins, R. J. E.; Delerue-Matos, C. M. & Boaventura, R. A. R. (2008). Removal of Cd(II), Zn(II) and Pb(II) from aqueous solutions by brown marine macro algae: Kinetic modeling. *Journal of Hazardous Materials*, Vol.153, No.1-2, (May 2008), pp. 493–501, ISSN 0304-3894.

Gadd, G. M. (2009). Biosorption: critical review of scientific rationale, environmental importance and significance for pollution treatment. *Journal of Chemical Technology & Biotechnology*, Vol.84, No.1, (January 2009), pp. 13-28, ISSN 1097-4660.

Ghimire, K. N.; Inoue, K.; Ohto, K. & Hayashida, T. (2008). Adsorption study of metal ions onto crosslinked seaweed *Laminaria japonica*. *Bioresource Technology*, Vol.99, No.1, (January 2008), pp. 32-37, ISSN 0980-8524.

Godlewska-Zylkiewicz, B. (2006). Microorganisms in inorganic chemical analysis. *Analytical and Bioanalytical Chemistry*, Vol.384, No.1, (January 2006), pp. 114-123, ISSN 1618-2642.

Guiochon, G.; Felinger, A.; Shirazi, D. G. & Katti, A. M. (2006). *Fundamentals of preparative and nonlinear chromatography* (2nd ed.), Academic Press, ISBN 978-0-12-370537-2, Boston, USA.

Gupta, V. K. & Rastogi, A. (2008). Equilibrium and kinetic modeling of cadmium(II) biosorption by nonliving algal biomass *Oedogonium* sp. from aqueous phase. *Journal of Hazardous Materials*, Vol.153, No.1-2, (May 2008),pp. 759-766, ISSN 0304-3894.

Hashim, M. A. & Chu, K. H. (2004). Biosorption of cadmium by brown, green, and red seaweeds. *Chemical Engineering Journal*, Vol. 97, No.2-3, (February 2004), pp. 249-255, ISSN 1385-8947.

Karnitz Jr., O. (2007). *Modificação química do bagaço de cana e celulose usando anidrido do EDTA. Uso destes materiais na adsorção de metais pesados em solução aquosa*, MSc Thesis,

Instituto de Ciências Exatas e Biológicas, Universidade Federal de Ouro Preto, Ouro Preto, Brazil.

Kentish, S. E. & Stevens, G. W. (2001). Innovations in separations technology for the recycling and re-use of liquid waste streams. *Chemical Engineering Journal*, Vol.84, No.2, (October 2001), pp. 149-159, ISSN 1385-8947.

Kratochvil, D. & Volesky, B. (2000). Multicomponent biosorption in fixed beds. *Water Research*, Vol.34, No.12, (August 2000), pp. 3186-3196, ISSN 0043-1354.

Lin, Z.; Zhou, C.; Wu, J.; Zhou, J. & Wang, L. (2005). A further insight into the mechanism of Ag^+ biosorption by *Lactobacillus* sp. strain A09. *Spectrochimica Acta Part A: Molecular and Biomolecular Spectroscopy*, Vol. 61, No.6, (April 2005), pp. 1195-1200, ISSN 1386-1425.

Liu, Y. & Liu, Y-. J. (2008). Biosorption isotherms, kinetics and thermodynamics. *Separation and Purification Technology*, Vol.61, No.3, (July 2008), pp. 229-242, ISSN 1383-5866.

Modak, J. M. & Natarajan, K. A. (1995). Biosorption of metals using nonliving biomass: a review. *Mineral and Metallurgical Processing*, Vol.12, No.4, (September 1995), pp. 189-196, ISSN 0747-9182.

Naddafi, K.; Nabizadeh, R.; Saeedi, R.; Mahvi, A. H.; Vaezi, F.; Yahgmaeian, K.; Ghasri, A. & Nazmara, S. (2007). Biosorption of lead(II) and cadmium(II) by protonated *Sargassum glaucescens* biomass in a continuous packed bed column. *Journal of Hazardous Materials*, Vol.147, No.3, (August 2007), pp. 785-791, ISSN 0304-3894.

Niu, H. & Volesky, B. (2006). Biosorption of chromate and vanadate species with waste crab shells. *Hydrometallurgy*, Vol.84, No.1-2, (October 2006), pp. 28-36, ISSN 0304-386X.

Oliveira, R. C. (2007). *Estudo da concentração e recuperação de íons lantânio e neodímio por biossorção em coluna com a biomassa Sargassum sp.* MsC Thesis, Instituto de Química, Universidade Estadual Paulista, Araraquara, Brazil.

Oliveira, R. C. (2011). *Biossorção de terras-raras por Sargassum sp.: estudos preliminares sobre as interações metal-biomassa e a potencial aplicação do processo para a concentração, recuperação e separação de metais de alto valor agregado em colunas empacotadas.* PhD Thesis, Instituto de Química, Universidade Estadual Paulista, Araraquara, Brazil.

Oliveira, R. C. & Garcia Jr., O. (2009). Study of biosorption of rare earth metals (La, Nd, Eu, Gd) by *Sargassum* sp. biomass in batch systems: physicochemical evaluation of kinetics and adsorption models. *Advanced Materials Research*, Vol.71-73, (May 2009), pp. 605-608, ISSN 1022-6680.

Oliveira, R. C.; Jouannin, C.; Guibal, E. & Garcia Jr., O. (2011). Samarium(III) and praseodymium(III) biosorption on *Sargassum* sp.: Batch study. *Process Biochemistry*, Vol.46, No.3, (March 2011), pp. 736-744, ISSN 1359-5113.

Pagnanelli, F.; Vegliò, F. & Toro, L. (2004). Modelling of the acid-base properties of natural and synthetic adsorbent materials used to heavy metal removal from aqueous solutions. *Chemosphere*, Vol.54, No.7, (February 2004), pp. 905-915, ISSN 0045-6535.

Palmieri, M. C.; Garcia Jr., O & Melnikov, P. (2000). Neodymium biosorption from acidic solutions in batch system. *Process Biochemistry*, Vol.36, No.5, (December 2000),pp. 441-444, ISSN 1359-5113.

Palmieri, M. C.; Volesky, B. & Garcia Jr., O. (2002). Biosorption of lanthanum using *Sargassum fluitans* in batch system. *Hydrometallurgy*, Vol.67, No.1, (December 2002), p. 31-36, ISSN 0304-386X.

Palmieri, M. C. (2001). *Estudo da utilização de biomassas para biossorção de terras-raras*. PhD Thesis, Instituto de Química, Universidade Estadual Paulista, Araraquara, Brazil.

Parsons, J. G.; Gardea-Torresdey, J. L.; Tiemann, K. J.; Gonzalez, J. H.; Peralta-Videa, J. R.; Gomez, E. & Herrera, I. (2002). Absorption and emission spectroscopic investigation of the phytoextraction of europium(III) nitrate from aqueous solutions by alfafa biomass. *Microchemical Journal*, Vol.71, No.2-3, (April 2002), pp. 175-183, ISSN 0026-265X.

Ruiz-Manríquez, A; Magaña, P. I.; López, V. & Guzmán, R. (1998). Biosorption of Cu by *Thiobacillus ferrooxidans*. *Bioprocess and Biosystems Engineering*, Vol.18, No.2, (February 1998), pp. 113-118, ISSN 1615-7591.

Sakamoto, N.; Kano, N. & Imaizumi, H. (2008) Biosorption of uranium and rare earth elements using biomass of algae. *Bioinorganic Chemistry and Applications*, Vol.2008, (December 2008), pp. 1-8, ISSN 1565-3633.

Selatnia, A.; Boukazoula, A.; Kechid, N.; Bakhti, M. Z.; Chergui, A. & Kerchich, Y. (2004). Biosorption of lead (II) from aqueous solution by a bacterial dead *Streptomyces rimosus* biomass. *Biochemical Engineering Journal*, Vol.19, No.2, (July 2004), pp. 127-135, ISSN 1369-703X.

Sen, R. & Sharandindra, C. (2009). Biotechnology - applications to environmental remediation in resource exploitation. *Current Science*, Vol.97, No.6, (September 2009), pp. 768-775, ISSN 0011-3891.

Sheng, P. X.; Ting, Y. P.; Chen, J. P. & Hong, L. (2004). Sorption of lead, copper, cadmium, zinc, and nickel by marine algal biomass: characterization of biosorptive capacity and investigation of mechanisms. *Journal of Colloid and Interface Science*, Vol.275, No.1, (July 2004), pp. 131-141, ISSN 0021-9797.

Tien, C. J. (2002). Biosorption of metal ions by freshwater algae with different surface characteristics. *Process Biochemistry*, Vol.38, No.4, (December 2002), pp. 605-613, ISSN 1359-5113.

Valdman, E.; Erijman, L.; Pessoa, F. L. P. & Leite, S. G. F. (2001). Continuous biosorption of Cu an Zn by immobilized waste biomass *Sargassum* sp. *Process Biochemistry*, Vol.36, No.8-9, (March 2001), pp. 869-873, ISSN 1359-5113.

Vegliò, F. & Beolchini, F. (1997). Removal of metal by biosorption: a review. *Hydrometallurgy*, Vol.44, No.3, (March 1997), pp. 301-316, ISSN 0304-386X.

Vegliò, F.; Esposito, A. & Reverberi, A. P. (2002). Copper adsorption on calcium alginate beads: equilibrium pH-related models. *Hydrometallurgy*, Vol.69, No.1, (July 2002), pp. 43-57, ISSN 0304-386X.

Vegliò, F.; Esposito, A. & Reverberi, A. P. (2003) Standardization of heavy metal biosorption tests: equilibrium and modeling study. *Process Biochemistry*, Vol.38, No.6, (January 2003), pp. 953-961, ISSN 1359-5113.

Vieira, M. G. A.; Oisiovici, R. M.; Gimenes, M. L. & Silva, M. G. C. (2008). Biosorption of chromium(VI) using a *Sargassum* sp. packed-bed column. *Bioresource Technology*, Vol.99, No.8, (May 2008), pp. 3094-3099, ISSN 0980-8524.

Vijayaraghavan, K.; Jegan, J.; Palanivelu, K. & Velan, M. (2005). Biosorption of cobalt(II) and nickel(II) by seaweeds: batch and column studies. *Separation and Purification Technology*, Vol.44, No.1, (July 2005), pp.53-59, ISSN 1383-5866.

Vijayaraghavan, K.; Padmesh, T.V.N.; Palanivelu, K. & Velan, M. (2006). Biosorption of nickel(II) ions onto *Sargassum wightii*: Application of two-parameter and three-parameter isotherm models. *Journal of Hazardous Materials*, Vol.133, No.1-3, (May 2006), pp. 304–308, ISSN 0304-3894.

Vijayaraghavan, K. & Prabu, D. (2006). Potential of *Sargassum wightii* biomass for copper(II) removal from aqueous solutions: application of different mathematical models to batch and continuous biosorption data. *Journal of Hazardous Materials*, Vol.137, No.1, (September 2006), pp. 558-564, ISSN 0304-3894.

Volesky, B. (2001). Detoxification of metal-bearing effluents: biosorption for the next century. *Hydrometallurgy*, Vol.59, No.2, (February 2001), pp. 203-216, ISSN 0304-386X.

Volesky, B. (2003). Biosorption process simulation tools. *Hydrometallurgy*, Vol.71, No.1-2, (October 2003), pp. 179-190, ISSN 0304-386X.

Volesky, B. & Naja, G. (2005). Biosorption: application strategies, In: *IBS-2005, South Africa*, 20.09.2006, Available from <http://biosorption.mcgill.ca/publication/index.htm>.

Volesky, B.; Weber, J. & Park, J. M. (2003). Continuous-flow metal biosorption in a regenerable *Sargassum* column. *Water Research*, Vol.37, No.2, (January 2003), pp. 297-306, ISSN 0043-1354.

Vullo, D. L.; Ceretti, H. M.; Daniel, M. A.; Ramirez, S. A. M. & Zalts, A. (2008). Cadmium, zinc and copper biosorption mediated by *Pseudonomas veronii* 2E. *Bioresource Technology*, Vol.99, No.13, (September 2008), pp. 5574-5581, ISSN 0980-8524.

Xu, H. & Liu, Y. (2008) Mechanisms of Cd^{2+}, Cu^{2+} and Ni^{2+} biosorption by aerobic granules. *Separation and Purification Technology*, Vol.58, No.3, (January 2008), pp. 400-411, ISSN 1383-5866.

Wang, J. & Chen, C. (2009). Biosorbents for heavy metals removal and their future. *Biotechnology Advances*, Vol.27, No.2 , (March-April 2009), pp. 195–226, ISSN 0734-9750.

Wang, X.; Chen, L; Siqing, X.; Zhao, J.; Chovelon, J. –M. & Renault, N. J. (2006). Biosorption of Cu(II) and Pb(II) from aqueous solutions by dried activated sludge. *Minerals Engineering*, Vol.19, No.9 , (July 2006), pp. 968–971, ISSN 0892-6875.

Yang, L. & Chen, J. P. (2008). Biosorption of hexavalent chromium onto raw and chemically modified *Sargassum* sp. *Bioresouce Tecnology*, Vol.99, No.2, (January 2008), pp. 297-307, ISSN 0980-8524.

Yu, J.; Tong, M.; Sun, S. & Li, B. (2007a). Cystine-modified biomass for Cd (II) and Pb (II) biosorption. *Journal of Hazardous Materials*, Vol.143, No.1-2, (May 2007), pp. 277-284, ISSN 0304-3894.

Yu, J.; Tong, M.; Xiaomei, S. & Li, B. (2007b). Biomass grafted with polyamic acid for enhancement of cadmium(II) and lead(II) biosorption. *Reactive & Functional Polymers*, Vol.67, No.6, (June 2007), pp. 564-572, ISSN 1381-5148.

Zhou, D.; Zhang, L. & Guo, S. (2005). Mechanisms of lead biosorption on cellulose/chitin beads. *Water Research*, Vol.39, (October 2005), No.16, pp. 3755-3762, ISSN 0043-1354.

Zouboulis, A. I.; Loukidou, M. X. & Matis, K. A. (2004). Biosorption of toxic metals from aqueous solutions by bacterial strain isolated from metal-polluted soils. *Process Biochemistry*, Vol.39, No.8, (April 2004), pp. 909-916, ISSN 1359-5113.

Part 4

Waste Water Treatment

9

Investigation of Different Control Strategies for the Waste Water Treatment Plant

Hicham EL Bahja[1], Othman Bakka[2] and Pastora Vega Cruz[1]
[1]*Faculty of Sciences, Dept. Automatica y Informatica, Universidad de Salamanca*
[2]*University Cady Ayyad, Faculty of science semlalia Marrakech*
[1]*Spain*
[2]*Morocco*

1. Introduction

Wastewater treatment is just one component in the urban water cycle; however, it is an important component since it ensures that the environmental impact of human usage of water is significantly reduced. It consists of several processes: biological, chemical and physical processes. Wastewater treatment aims to reduce: nitrogen, phosphorous, organic matter and suspended solids. To reduce the amount of these substances, wastewater treatment plants (WWTP) consisting of (in general) four treatment steps, have been designed. The steps are: a primarily mechanical pre-treatment step, a biological treatment step, a chemical treatment step and a sludge treatment step. See Figure 1.

The quality of water is proportional to the quality of life and therefore in modern world the sustainable development concept is to save water. The goal of a wastewater treatment plant is to eliminate pollutant agents from the wastewater by means of physical and (bio) chemical processes. Modern wastewater treatment plants use biological nitrogen removal, which relies on nitrifying and denitrifying bacteria in order to remove the nitrogen from the wastewater.

Biological wastewater treatment plants are considered complex nonlinear systems due to large variations in their flow rates and feed concentrations. In addition, the microorganisms that are involved in the process and their adaptive behaviour coupled with nonlinear dynamics of the system make the WWTP to be really challenging from the modelling and control point of view [Clarke D.W], [Dutka.A& Ordys], [Grimblea & M. J], [H.Elbahja & P.Vega],[H.Elbahja & O.Bakka] and [O.Bakka & H.Elbahja].

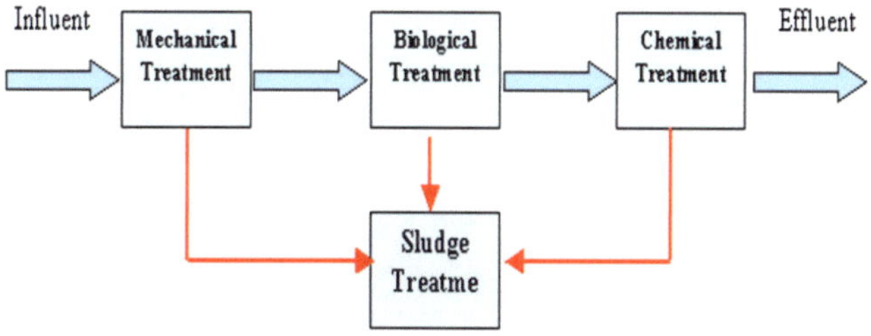

Fig. 1. Layout of a typical wastewater treatment plant

The paper is organized as follows. The modelling of the continuous wastewater treatment is detailed in Section 2. Section 3 is dedicated to the non linear predictive control technique. Observer based Regulator Problem for a WWTP with Constraints on the Control in Section 4. In Section 5 the efficiency of the two controls schemes are illustrated via simulation studies. Finally Section 6 ends the paper.

2. Process modelling

A typical, conventional activated sludge plant for the removal of carbonaceous and nitrogen materials consists of an anoxic basin followed by an aerated one, and a settler (figure 2). In the presence of dissolved oxygen, wastewater that is mixed with the returned activated sludge is biodegraded in the reactor. Treated effluent is separated from the sludge is wasted while a large fraction is returned to anoxic reactor to maintain the appropriate substrate-to-biomass ratio. In this study we consider six basic components present in the wastewater: autotrophic bacteria X_A, heterotrophic bacteria X_H, readily biodegradable carbonaceous substrates S_S, nitrogen substrates S_{Nh}, S_{NO} and dissolved oxygen S_O.
In the formulation of the model the following assumptions are considered: the physical properties of fluid are constant; there is no concentration gradient across the vessel; substrates and dissolved oxygen are considered as a rate-limiting with a bi-substrate Monod-type Kinetic; no bio-reaction takes place in the settler and the settler is perfect. Based on the above description and assumptions, we can formulate the full set of ordinary differential equations (mass balance equations), making up the IAWQ AS Model NO.1 [Henze].

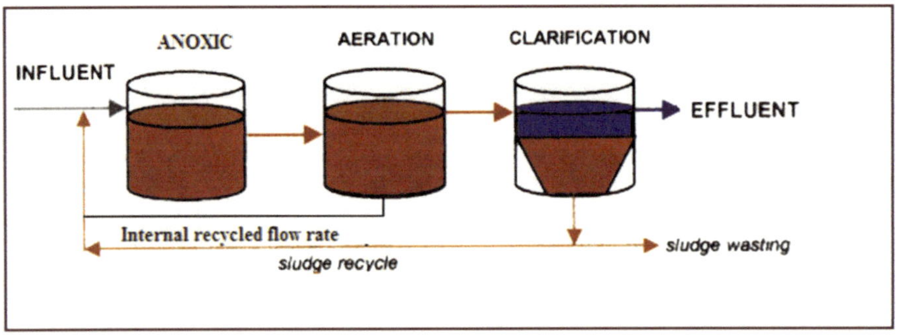

Fig. 2. Pre-denitrification plant design

2.1 Modeling of the aerated basin

$$\dot{X}_{A,nit}(t) = (1 + r_1 + r_2) \cdot D_{nit}(X_{A,denit} - X_{A,denit}) + (\mu_{A,nit} - b_A)X_{A,nit} \quad (1)$$

$$\dot{X}_{H,nit}(t) = (1 + r_1 + r_2) \cdot D_{nit}(X_{H,denit} - X_{H,nit}) + (\mu_{H,nit} - b_H)X_{H,nit} \quad (2)$$

$$\dot{S}_{S,nit}(t) = (1 + r_1 + r_2) \cdot D_{nit}(S_{S,denit} - S_{S,nit}) + (\mu_{H,nit} - \mu_{Ha,nit})\frac{X_{H,nit}}{Y_H} \quad (3)$$

$$\dot{S}_{NH,nit}(t) = (1 + r_1 + r_2)D_{denit}(S_{NH,denit} - S_{NH,nit})(i_{xb}\ 1/Y_A)\mu_{A,nit}X_{A,nit}$$
$$-(\mu_{H,nit} - \mu_{Ha,nit})i_{xb}X_{H,nit} \quad (4)$$

$$\dot{S}_{NO,nit}(t) = (1 + r_1 + r_2)D_{nit}(S_{NO,denit}S_{NO,nit}) + \frac{\mu_{A,nit}X_{A,nit}}{Y_A} - \frac{1-Y_H}{2.86Y_H}\mu_{Ha,nit}X_{H,nit} \quad (5)$$

$$\dot{S}_{O,nit}(t) = (1 + r_1 + r_2)D_{nit}(S_{O,denit}S_{O,nit}) + a_0 Q_{air}(C_S - S_{O,nit})$$

$$-\frac{(4.57-Y_A)\mu_{A,nit}X_{A,nit}}{Y_A} - \frac{1-Y_H}{Y_H}\mu_{Ha,nit}X_{H,nit} \quad (6)$$

Where:

$$\mu_{A,nit} = \mu_{max,A} \cdot \frac{S_{NH,nit}}{(K_{NH,A}+S_{NH,nit})} \cdot \frac{S_{O,nit}}{(K_{O,A}+S_{O,nit})}$$

$$\mu_{H,nit} = \mu_{max,H} \cdot \frac{S_{S,nit}}{(K_S+S_{S,nit})} \cdot \frac{S_{NH,nit}}{(K_{NH,H}+S_{NH,nit})} \cdot \frac{S_{O,nit}}{(K_{O,H}+S_{O,nit})}$$

$$\mu_{Ha,nit} = \mu_{max,H} \cdot \frac{S_{S,nit}}{(K_S+S_{S,nit})} \cdot \frac{S_{NH,nit}}{(K_{NH,H}+S_{NH,nit})} \cdot \frac{K_{O,H}}{(K_{O,H}+S_{O,nit})} \cdot \frac{S_{NO,nit}}{(K_{NO}+S_{NO,nit})} \cdot \eta_{NO}$$

$\mu_{A,nit}$ and $\mu_{H,nit}$ are the growth rates of autotrophy and heterotrophy in aerobic conditions and $\mu_{Ha,nit}$ is the growth rate of heterotrophy in anoxic conditions.

2.2 Modeling of the anoxic basin

$$\dot{X}_{A,denit}(t) = D_{denit}(X_{A,in} + r_1 X_{A,nit}) - (1 + r_1 + r_2) \cdot D_{denit}X_{A,denit}$$

$$+\alpha \cdot r_2 D_{denit}X_{rec} + (\mu_{A,denit} - b_A)X_{A,denit} \quad (7)$$

$$\dot{X}_{H,denit}(t) = D_{denit}(X_{H,in} + r_1 X_{H,nit}) - (1 + r_1 + r_2) \cdot D_{denit}X_{H,denit}$$

$$+(1-\alpha)r_2 D_{denit}X_{rec} + (\mu_{H,denit} - b_H)X_{H,denit} \quad (8)$$

$$\dot{S}_{S,denit}(t) = -(\mu_{H,denit} - \mu_{Ha,denit})\frac{X_{H,denit}}{Y_H} - (1 + r_1 + r_2) \cdot D_{denit}S_{S,denit}$$

$$+D_{denit}(S_{S,in} - r_1 S_{S,nit}) \quad (9)$$

$$\dot{S}_{NH,denit} = D_{denit}(S_{NH,in} - r_1 S_{NH,nit}) - (1 + r_1 + r_2) \cdot D_{denit}S_{NH,denit}$$

$$-(i_{xb} + 1/Y_A)\mu_{A,denit}X_{A,denit} - (\mu_{H,denit} - \mu_{Ha,denit})i_{xb}X_{H,denit} \quad (10)$$

$$\dot{S}_{NO,denit}(t) = D_{denit}(S_{NO,in} - r_1 S_{NO,nit}) - (1 + r_1 + r_2) \cdot D_{denit}S_{NO,denit}$$

$$+\frac{\mu_{A,denit}X_{A,denit}}{Y_A} - \frac{1-Y_H}{2.86Y_H} \quad (11)$$

Where:

$$\mu_{A,denit} = \mu_{max,A} \cdot \frac{S_{NH,denit}}{(K_{NH,A}+S_{NH,denit})}$$

$$\mu_{H,denit} = \mu_{max,H} \cdot \frac{S_{S,denit}}{(K_S + S_{S,denit})} \cdot \frac{S_{NH,denit}}{(K_{NH,H} + S_{NH,denit})}$$

$$\mu_{Ha,denit} = \mu_{max,H} \cdot \frac{S_{S,denit}}{(K_S + S_{S,denit})} \cdot \frac{S_{NH,denit}}{(K_{NH,H} + S_{NH,denit})} \cdot \frac{S_{NO,denit}}{(K_{NO} + S_{NO,denit})} \cdot \eta_{NO}$$

2.3 Modeling of the setller

$$\dot{X}_{rec} = (1 + r_2) D_{dec} (X_{A,nit} + X_{H,nit}) - (r_2 + \omega) D_{dec} X_{rec} \tag{12}$$

r_1, r_2 and ω represent respectively, the ratio of the internal recycled flow Q_{r1} to influent flow Q_{in} ,the ratio of the recycled flow Q_{r2} to the influent flow, C_S is the maximum dissolved oxygen concentration. D_{nit} , D_{denit} and D_{dec} are the dilution rates in respectively, nitrification, denitrification basins and settler tank; X_{rec} is the concentration of the recycled biomass. The other variables and parameters of the system equations (1)-(13) are also defined.

3. Control of global nitrogen and dissolved oxygen concentrations

The implementation of efficient modern control strategies in bioprocesses [Hajji, S., Farza, Hammouri, H., & Farza, Shim, H.], highly depends on the availability of on-line information about the key biological process components like biomass and substrate. But due to lack or prohibitive cost, in many instances, of on-line sensors for these components and due to expense and duration (several days or hours) of laboratory analyses, there is a need to develop and implement algorithms which are capable of reconstructing the time evolution of the unmeasured state variables on the base of the available on-line data. However, because of the nonlinear feature of the biological processes dynamics and the usually large uncertainty of some process parameters, mainly the process kinetics, the implementation of extended versions of classical observers proves to be difficult in practical applications, and the design of new methods is undoubtedly an important research matter nowadays. In that context, Extended Kalman Filter (EKF) is presented in this work.

3.1 Method presentation of the Extended Kalman Filter

The aim of the estimation procedure is to compute estimated values of the unavailable state variables of the process [$X_{A,nit}(t)$, $X_{A,denit}(t)$, $X_{H,nit}(t)$, $X_{H,denit}(t)$, $S_{S,nit}(t)$, $S_{S,denit}(t)$, $X_{rec}(t)$] and the specific growth rate $\mu(t)$ using the concentrations [$S_{NH,nit}(t)$, $S_{NH,denit}(t)$, $S_{NO,nit}(t)$, $S_{NO,denit}(t)$, $S_{O,nit}(t)$] as measurable variables. The EKF estimator uses a non-linear mathematical model of the process and a number of measures for estimating the states and parameters not measurable. The estimation is realised in three stages: prediction, observation and registration.

The EKF estimator uses a non-linear mathematical model of the process and a number of measures for estimating the states and parameters not measurable. The estimation is realised in three stages: prediction, observation and registration.

Let a dynamic non-linear system be characterised by a model in the state space form as following:

$$\frac{dX(t)}{dt} = f(X(t), u(t), t) + v(t) \tag{13}$$

Where:
$X(t)$: Represents the state vector of dimension n.
$f(.)$: Non-linear function of $X(t)$ and $u(t)$.
$u(t)$: Represents the input vector of dimension m.
$v(t)$: Vector of noise on the state equation of dimension n, assumed Gaussian white noise, medium null and covariance matrix known $q(t) = Var(v(t))$.
The state of the system is observed by m discrete measures related to the state X (t) by the following equation of observation:

$$Z(t_k) = h(X(t_k), t_k) + \omega(t_k) \tag{14}$$

Where:
$Z(t_k)$: Represents the observation vector of dimension n.
$h(.)$: Observation matrix of dimension mxn.
t_k: Observation instant.
$\omega(t_k)$: Vector of noise on the measure, of dimension m, independent of, $v(t)$ assumed Gaussian white noise, medium null and covariance matrix known $r(t) = V(\omega(t))$.

- The EKF algorithm corresponding to the continuous process in discreet observation, where the measurements are acquired at regular intervals, is given by [17]:
- Initialisation filter $t = t_0$:

$$X(t_0) = E(X(t_0)) \tag{15}$$

$$L(t_0) = Var(X(t_0)) \tag{16}$$

- Between two instant of observation:
- The estimated state $\hat{X}(t)$ and its associated covariance matrix $L(t)$ are integrated by the equations:

$$\frac{d\hat{X}(t)}{dt} = f(\hat{X}(t), u(t), t) \tag{17}$$

$$\frac{dL(t)}{dt} = F(t)L(t) + L(t)F^T(t) + q(t) \tag{18}$$

$$F(t) = \frac{\partial f(\hat{X}(t), u(t), t)}{\partial \hat{X}} \tag{19}$$

Then we have, before the observation at $t = t_{k-}$, an estimated of $\hat{X}(t_{k-})$ and its covariance matrix $L(t_{k-})$.

- Updating the gain

$$K(t_k) = L(t_{k-})H^T(\hat{X}(t_{k-}), t_k)\left[H(\hat{X}(t_{k-}), t_k)L(t_{k-})H^T(\hat{X}(t_{k-}), t_k) + r(t_k)\right]^{-1} \tag{20}$$

- Update of the estimated state

$$\hat{X}(t_k) = \hat{X}(t_{k-}) + K(t_k)[Z(t_k) - h(\hat{X}(t_{k-}), t_k)] \tag{21}$$

- Update of the covariance matrix

$$L(t_{k+}) = L(t_{k-}) - K(t_k)H(\hat{X}(t_{k-}),t_k)L(t_{k-}) \tag{22}$$

$$H(\hat{X}(t_{k-})) = \frac{\partial h(\hat{X}(t_{k-}),t_k)}{\partial \hat{X}(t_{k-})} \tag{23}$$

The estimator EKF is an iterative algorithm. The final results of each step of calculation are used as initial conditions for the next step.

3.2 The non linear GPC

The control objective is to make the effluent organics concentration below certain regulatory limits. A multivariable non linear generalized predictive control strategy based on NH_4, NO_3 and O_2 measurements is developed, enabling the control of the nitrogen and the dissolved oxygen concentrations, by acting on the internal flow and aeration flow rates, Q_{r1} and Q_{air}, at desired levels. The dynamics of the WWTP are represented by the equations below. The system is discretized using Euler integration method and re-arranged into the state dependent coefficient form the state-space model [10, 11].
State and control dependent matrices in general may be formulated in an infinite number of ways. Finally we can write the discrete model in the following matrix form:

$$x_{n+1} = \widetilde{A}(x_n)x_n + \widetilde{B}(x_n)u_n \tag{24}$$

$$y_n = \widetilde{C}(x_n)x_n \tag{25}$$

The state dependant form of the model, in state space format is substituted to the traditional GPC format, allowing for inherent integral action within the model, including the control increment as system input to the state space model.
Thus, an extra system state is included.

$$\chi_{n+1} = A(\chi_n)\chi_n + B(\chi_n)\Delta u_n \tag{26}$$

$$y_n = C(\chi_n)\chi_n \tag{27}$$

Where:

$$A(\chi_n) = \begin{bmatrix} A(x_n) & B(x_n) \\ 0 & I \end{bmatrix}, B(\chi_n) = \begin{bmatrix} B(x_n) \\ I \end{bmatrix},$$

$$C(\chi_n) = [C(x_n) \;\; 0],\ \chi_n = \begin{bmatrix} x_n \\ u_{n-1} \end{bmatrix},\ \ \Delta u_n = u_n - u_{n-1}$$

To derive the non-linear predictive control algorithm the assumption on the future trajectory of the system must be made. For a moment assume, that the future trajectory for the state of the system is known. State-space model (26), (27) matrices may be re-calculated for the future using the future trajectory. The resulting state-space model may be seen as a time-varying linear model and for this model the controller is designed. Therefore the following notation for state dependent matrices $A_n = A(\chi_n)$ $B_n = B(\chi_n)$ $C_n = C(\chi_n)$ is used in the remaining part of the paper. Now, the future trajectory for the system has to be

determined. In the classic predictive control strategy the vector of current and future controls is calculated. For the receding horizon control technique only the first control is used for the plant inputs manipulation, remaining part is not.
But this part may be employed in the next iteration of the algorithm to predict the future trajectory.
The cost function of the GPC controller here is defined as:

$$J_n = \sum_{i=1}^{Ne}\{(sp_{n+i} - y_{n+i})^T \Lambda_E^i (sp_{n+i} - y_{n+i})\} + \sum_{i=1}^{Nu}\{\Delta u_{n+i-1}^T \Lambda_u^i \Delta u_{n+i-1}\} \tag{28}$$

Where sp_n is a vector of size n_y of set point at time n, $\Lambda_E^i, i = 1 \dots Ne$ and $\Lambda_E^j, j = 1 \dots Nu$ are weighting matrices (symmetric) and Ne Nu are positive integer numbers greater or equal one.
Next the following vectors containing current and future values of the control u_n, and future values of statecx_n, and output y_n are introduced:

$$\begin{aligned} X_{n+1,Ne} &= [\chi_{n+1}^T, \dots, \chi_{n+Ne}^T]^T, \\ \Delta U_{n,Nu} &= [\Delta u_n^T, \dots, \Delta u_{n+Nu-1}^T]^T \\ Y_{n+1,Ne} &= [y_{n+1}^T, \dots, y_{n+Ne}^T]^T, \\ R_{n+1,Ne} &= [sp_{n+1}^T, \dots, sp_{n+Ne}^T]^T \end{aligned} \tag{29}$$

The cost function (28) with notation (29) may be written in the vector form:

$$J_n = (R_{n+1,Ne} - Y_{n+1,Ne})^T \Lambda_E (R_{n+1,Ne} - Y_{n+1,Ne}) + \Delta U_{n,Nu}^T \Lambda_u \Delta U_{n,Nu} \tag{30}$$

With:

$$\Lambda_E = diag(\Lambda_E^1, \Lambda_E^2, \dots, \Lambda_E^{Ne}), \Lambda_u = diag(\Lambda_u^1, \Lambda_u^2, \dots, \Lambda_u^{Nu})$$

It is possible now to determine the future state prediction. For j = 1, ...,Ne the future state predictions may be obtained from:

$$\begin{aligned} \chi_{n+j} = {} & [A_{n+j-1}A_{n+j-2} \dots A_n]\chi_n + [A_{n+j-1}A_{n+j-2} \dots A_{n+1}] \cdot \\ & B_n \Delta u_n + [A_{n+j-1}A_{n+j-2} \dots A_{n+2}] B_{n+1} \Delta u_{n+1} + \cdots \\ & + [A_{n+j-1}A_{n+j-2} \dots A_{n+Nu}] B_{n-1+min(j,Nu)} \Delta u_{n-1+min(j,Nu)} \end{aligned} \tag{31}$$

Note that to obtain the state prediction at time instance n+j the knowledge of matrix predictions $A_n \dots A_{n+i-1}$ and $B_n \dots B_{n-l+min(j,Nu)}$ is required. The control increments after the control horizon are assumed to be zero.
Next introduce the following notation:

$$\left[\prod_{k=l}^{m} A_{n+k}\right] = \begin{cases} A_{n+m}A_{n+m-1} \dots A_{n+1} & \text{if } l \le m \\ I & \text{if } l > m \end{cases}$$

Then (31) may be represented as:

$$\chi_{n+j} = \left[\prod_{k=0}^{j-1} A_{n+k}\right]\chi_n + \left[\prod_{k=1}^{j-1} A_{n+k}\right] B_n \Delta u_n + \left[\prod_{k=2}^{j-1} A_{n+k}\right] B_{n+1} \Delta u_{n+1}$$

$$+\cdots+\left[\prod_{k=Nu}^{j-1} A_{n+k}\right] B_{n-1+\min(j,Nu)} \Delta u_{n-1+\min(j,Nu)} \tag{32}$$

From (29) and (32) the following equation for the future state predictions vector $X_{n+l,N}$ is obtained:

$$X_{n,Ne} = \Omega_{n,Ne} A_n \chi_n + \Psi_{n,Ne} \Delta u_{n,Nu} \tag{33}$$

Where

$$\Omega_{n,Ne} = \left[\left[\prod_{k=1}^{0} A_{n+k}\right]^T \left[\prod_{k=1}^{1} A_{n+k}\right]^T \cdots \left[\prod_{k=1}^{Ne-1} A_{n+k}\right]^T\right]$$

$$\Psi_{n,Ne} = \begin{bmatrix} \left[\prod_{k=1}^{0} A_{n+k}\right] B_n & 0 & \cdots \\ \left[\prod_{k=1}^{1} A_{n+k}\right] B_n & \left[\prod_{k=2}^{0} A_{n+k}\right] B_{n+1} & \ddots \\ \vdots & \vdots & \ddots \\ \left[\prod_{k=1}^{Ne-1} A_{n+k}\right] B_n & \left[\prod_{k=2}^{Ne-1} A_{n+k}\right] B_{n+1} & \cdots \end{bmatrix} \begin{matrix} 0 \\ \vdots \\ \ddots \\ \left[\prod_{k=Nu}^{Ne-1} A_{n+k}\right] B_{n+Nu-1} \end{matrix}$$

From the output equation (27) it is clear that

$$y_{n+j} = C_{n+j} \chi_{n+j} \tag{34}$$

Combining (29) and (34) the following relationship between vectors $X_{n+l,N}$ and $Y_{n+1,N}$ is obtained:

$$Y_{n+1,Ne} = \Theta_{n,Ne} X_{n+1,Ne} \tag{35}$$

Where:

$$\Theta_{n,Ne} = diag(C_{n+1}, C_{n+2}, \dots, C_{n+Ne})$$

Finally substituting in (35) X_{n+N} by (33) the following equation for output prediction is obtained:

$$Y_{n+1,Ne} = \phi_{n,Ne} A_n \chi_n + S_{n,Ne} \Delta U_{n,Nu} \tag{36}$$

Where:

$$\phi_{n,Ne} = \Theta_{n,Ne} \Omega_{n,Ne} \qquad S_{n,Ne} = \Theta_{n,Ne} \Psi_{n,Ne}$$

Substituting $Y_{n+1,N}$ in the cost function (30) by the equation (36) and performing the static optimization the control minimizing the given cost function is finally derived:

$$\Delta U_{n,Nu} = \left(S_{n,Ne}^T \Lambda_E S_{n,Ne} + \Lambda_u\right)^{-1} S_{n,Ne} \Lambda_E \left(R_{n+1,Ne} - \phi_{n,Ne} A_n \chi_n\right) \tag{37}$$

4. Observer based regulator problem with constraints on the control

4.1 Linearization

Through linearization, the model equations are written in the standard form of state equations, as follows:

$$\begin{cases} \dot{x} = Ax(t) + Bu(t) \\ y(t) = Cx(t) \end{cases} \tag{38}$$

For the model ASM1 simplified trough linearization, the state, input and output vectors are given by the equation (14)-(16):

$$x(t) = [X_{A,nit}(t)\ X_{H,nit}(t)\ S_{S,nit}(t)\ S_{NH,nit}(t)\ S_{NO,nit}(t)\ S_{O,nit}(t)$$

$$X_{A,denit}(t)\ X_{H,denit}(t)\ S_{S,denit}(t)\ S_{NH,denit}(t)\ S_{NO,denit}(t)\ X_{rec}(t)]^T \tag{39}$$

$$y(t) = \left[S_{NH,nit}(t)\ \ S_{NO,nit}(t)\ S_{O,nit}(t)\right]^T \tag{40}$$

$$u(t) = [\,Q_{r1}\ Q_{r2}\ Q_{air}]^T \tag{41}$$

We present the constraint on the control as follows:

$$\begin{cases} -Q_{r1} \le Q_{r1} \le 4Q_{r1} \\ -Q_{r2} \le Q_{r2} \le Q_{r2} \\ -100 \le Q_{air} \le 260 \end{cases}$$

For the steady-state functioning point:

$$x(t) = [69.6\ 623\ 13.5\ 3.2\ 10.4\ 2.4\ 68.9\ 624.6\ 20.9\ 8.9\ 5.3\ 1356.8\ ]^T \tag{42}$$

4.2 Decomposition

Any representation in the state space can be transformed into the equivalent form by the transformation $Z = T_0 x$ [10]:

$$\begin{cases} \dot{Z} = \bar{A}\,Z + \bar{B}u \\ y(t) = \bar{C}\,Z \end{cases} \tag{43}$$

With:

$$\bar{A} = \begin{pmatrix} A_{n0} & A_{12} \\ 0 & A_0 \end{pmatrix};\ \bar{B} = \begin{pmatrix} B_{n0} \\ B_0 \end{pmatrix};\ \bar{C} = (0 \quad C_0);\ Z = \begin{pmatrix} Z_{n0} \\ Z_0 \end{pmatrix}$$

Where $(A_0 \quad C_0)$ is observable but in our case the pair $(A_0 \quad B_0)$ is controllable.
So we obtain the following system of equations:

$$\begin{cases} \dot{Z}_{n0} = A_{n0}Z_{n0} + A_{12}Z_0 + B_{n0} \\ \dot{Z}_0 = A_0 Z_0 + B_0 u \\ y = C_0 Z_0 \end{cases} \tag{44}$$

4.3 Luenberger observer

An observer is a mathematical structure that combines sensor output and plant excitation signals with models of the plant and sensor. An observer provides feedback signals that are superior to the sensor output alone.

When faced with the problem of controlling a system, some scheme must be devised to choose the input vector $x(t)$ so that the system behaves in an acceptable manner. Since the state vector $y(t)$ contains all the essential information about the system, it is reasonable to base the choice of $x(t)$ solely on the values of $y(t)$ and perhaps also t. In other words, x is determined by a relation of the form x(t) = F[y(t), t].
This is, in fact, the approach taken in a large portion of present day control system literature. Several new techniques have been developed to find the function F for special classes of control problems. These techniques include dynamic programming [Labarrere, M., Krief]-[Dutka, A., Ordys, A., Grimble], Pontryagin's maximum principle [K. K. Maitra], and methods based on Lyapunov's theory [J.Oreilly].
In most control situations, however, the state vector is not available for direct measurement. This means that it is not possible to evaluate the function F[y(t), t]. In these cases either the method must be abandoned or a reasonable substitute for the state vector must be found.
In this chapter it is shown how the available system inputs and outputs may be used to construct an estimate of the system state vector. The device which reconstructs the state vector is called an observer. The observer itself as a time-invariant linear system driven by the inputs and outputs of the system it observes.
To observe the system state, sometimes he can go to estimate the entire state vector then part is available as a linear combination of the output [J.Oreilly]. We suppose that we have p linear combinations, we will present the case where one has this information and cannot rebuilt that (n-p) linear combination of system states or

$$\mathrm{z}(.) = TZ_0(.) \tag{45}$$

Is a linear combination, with the matrix T of dimension (n-p, n). The estimated state is then obtained by:

$$\hat{Z}_0 = \begin{pmatrix} C_0 \\ T \end{pmatrix}^{-1} \begin{pmatrix} y(.) \\ \mathrm{z}(.) \end{pmatrix} = \begin{pmatrix} V & P \end{pmatrix} \begin{pmatrix} y(.) \\ \mathrm{z}(.) \end{pmatrix} \tag{46}$$

The matrix T is chosen in such a way that the matrix $\begin{pmatrix} C_0 \\ T \end{pmatrix}$ is invertible. Furthermore the amount $TZ_0(.)$ can be measured which leads us to generate $\mathrm{z}(.)$, from an auxiliary dynamical system as follows:

$$\delta \mathrm{z}(.) = D\mathrm{z}(.) + Ey(.) + Gu(.) \tag{47}$$

Where $\mathrm{z}(.)$ is the state of the observer dynamics. Note here that the matrices V, C_0, T, P, verify

$$VC_0 + PT = II \tag{48}$$

The control problem with constraint via an observer of minimal order may be solved in the following way:
How to choose the state feedback F:

$$u(.) = sat\left(F\hat{Z}_0(.)\right) \tag{49}$$

And matrices D, E and G calculated such that the asymptotic stability and the constraints on inputs are guaranteed
The observation error in this case is given by

$$\epsilon(.) = z(.) - TZ_0(.) \tag{50}$$

We recall that the matrices of the observer of minimal order is given by [11]:

$$D = TA\_0\,P\,,\ E = TA\_0\,V\,,\ G = TB\_0 \tag{51}$$

Which is equivalent to write that the check matrices in the following relation

$$TA_0 - EC_0 = DT \tag{52}$$

Where the matrices T and P are chosen to ensure asymptotic stability of the matrix D, in order to see vanish asymptotically non sampling error, indeed:

$$\begin{aligned}\delta\epsilon(.) &= \delta z(.) - T\delta Z_0\\ &= Dz(.) + Ey(.) + Gu(.) - T\big(A_0 Z_0(.) + B_0 u(.)\big)\\ &= Dz(.) + EC_0 Z_0(.) - TA_0 Z_0(.)\\ &= Dz(.) - DTZ_0(.)\\ &= D\,\epsilon(.)\end{aligned}$$

For the observation error, we define the field $D(II, \epsilon_{max}, \epsilon_{min})$ that give us the limits within which we allow change of error$\epsilon(.)$. The reconstruction error is always given by

$$e(.) = \hat{Z}_0(.) - Z_0(.) \tag{53}$$

Is related to the error of observation:

$$\begin{aligned}e(.) &= Vy(.) + Pz(.) - Z_0(.)\\ &= VC_0 Z_0(.) + Pz(.) - (VC_0 + PT)Z_0(.)\\ &= P\big(z(.) - TZ_0(.)\big)\\ &= P\epsilon(.)\end{aligned}$$

Lemma: The field $D(II, u_{max}, u_{min}) \times D(II, \epsilon_{max}, \epsilon_{min})$ is positively invariant with respect to the system trajectory $\binom{u(.)}{\epsilon(.)}$ only hosts and if so, there exists a matrix $H \in R^{m\times n}$ Such that:

1. $HF = FA_0 + FB_0F$ (54)
2. $\widetilde{M}q_\epsilon \leq 0$

Where:

$$M = \begin{pmatrix} H & L_r \\ 0 & D \end{pmatrix}\ ;\ q_\epsilon = \begin{pmatrix} u_{max} \\ \epsilon_{max} \\ u_{min} \\ \epsilon_{min} \end{pmatrix}\ ;\ L_r = -FVC_0A_0P$$

For every pair $\big(u(0), \epsilon(0)\big) \in D(II, u_{max}, u_{min}) \times D(II, \epsilon_{max}, \epsilon_{min})$

Proof: We start by writing the equation for the evolution of the control u(t) always in the case of a linear behaviour using previous relationship.

$$\begin{aligned}\delta u(.) &= F\delta\hat{Z}_0\\ &= F\delta\big(Pz(.) + VC_0 Z_0(.)\big)\\ &= FP\delta z(.) + FVC_0\delta Z_0(.)\end{aligned}$$

$$
\begin{aligned}
&= FP\big(Dz(.) + Ey(.) + Gu(.)\big) + FVC_0\big(A_0Z_0(.) + B_0u(.)\big) \\
&= FP\big(TA_0Pz(.) + TA_0Vy(.)\big) + (FPTB_0 + FVC_0B_0)u(.) + FVC_0A_0Z_0(.) \\
&= FPTA_0\big(Pz(.) + Vy(.)\big) + F(PT + VC_0)B_0u(.) + FVC_0A_0Z_0(.) \\
&= FPTA_0\hat{Z}_0(.) + FB_0u(.) + FVC_0A_0\left(\hat{Z}_0(.) - e(.)\right) \\
&= F(PT + VC_0)A_0\hat{Z}_0(.) + FB_0F\hat{Z}_0(.) - FVC_0A_0e(.) \\
&= (FA_0 + FB_0F)\hat{Z}_0(.) - FVC_0A_0e(.) \\
&= HF\hat{Z}_0(.) - FVC_0A_0P\epsilon(.) \\
&= Hu(.) + L_r\epsilon(.)
\end{aligned}
$$

Is then augmented system consisting of control u(t) and error $\epsilon(.)$, we get

$$
\begin{pmatrix} \delta u(.) \\ \delta\epsilon(.) \end{pmatrix} = \begin{pmatrix} H & L_r \\ 0 & D \end{pmatrix} \begin{pmatrix} u(.) \\ \epsilon(.) \end{pmatrix}
$$

5. Simulation results

Simulation experiments for the first strategies of control were carried out by numerically integration of the complete model of the biological process. Numerical values of the parameters appearing in the model equations are given in the table I and table II.

Variable	Value	Description
V_{nit}	$1000m^3$	volume of nitrification basin
V_{denit}	$250m^3$	volume of denitrification basin
V_{dec}	$1250m^3$	volume of settler
Q_{in}	$3000m^3/j$	influent flow rate
Q_{r1}	$2955m^3/j$	recycled flow rate
Q_{r2}	$1500m^3/j$	intern recycled flow rate
Q_w	$45m^3/j$	waste flow rate
$X_{A,in}$	$0\ mg/l$	autotrophs in the influent
$X_{H,in}$	$30\ mg/l$	hetertrophs in the influent
$S_{S,in}$	$200\ mg/l$	substrate in the influent
$S_{NH,in}$	$30\ mg/l$	ammonium in the influent
$S_{NO,in}$	$2\ mg/l$	nitrate in the influent
$S_{O,in}$	$0mg/l$	oxygen in the influent

Table I. Process characteristics.

Simulation results are given in figure 3 for the NLGPC strategies. The perturbations pursued on the control variables are due to measurement noises. The output variables evolution that are the global nitrogen and the dissolved oxygen concentrations, and their corresponding reference trajectories are 7 and 3, respectively. The figure 4 presents the results of simulation for the second controller.

Parameter	Value	Description
Y_A	0.24	yield of autotroph mass
Y_H	0.67	yield of heterotroph mass
i_{xb}	0.086	
K_S	20mg/l	affinity constant
$K_{NH,A}$	1mg/l	affinity constant
$K_{NH,H}$	0.05mg/l	affinity constant
K_{NO}	0.5mg/l	affinity constant
$K_{O,A}$	0.4mg/l	affinity constant
$K_{O,H}$	0.2mg/l	affinity constant
μ_{Amax}	0.8l/j	maximum specific growth rate
μ_{Hmax}	0.6l/j	maximum specific growth rate
b_A	0.2l/j	decay coefficient of autotrophs
b_H	0.68l/j	decay coefficient of heterotrophs
η_{NO}	0.8l/j	correction factor for anoxic growth

Table II. Kinetic parameters and stoechimetric coeeficient characteristics

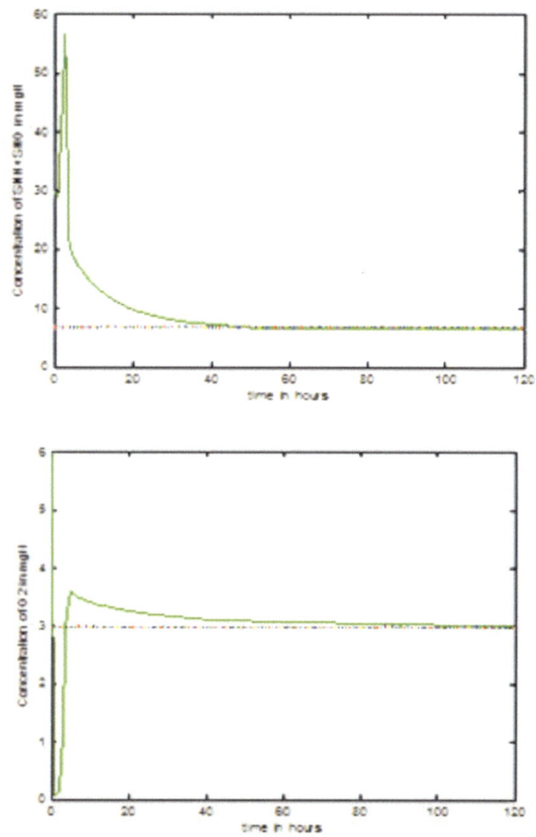

Fig. 3. Evolution of the dissolved oxygen and the global nitrogen concentrations for Non linear system with first controller.

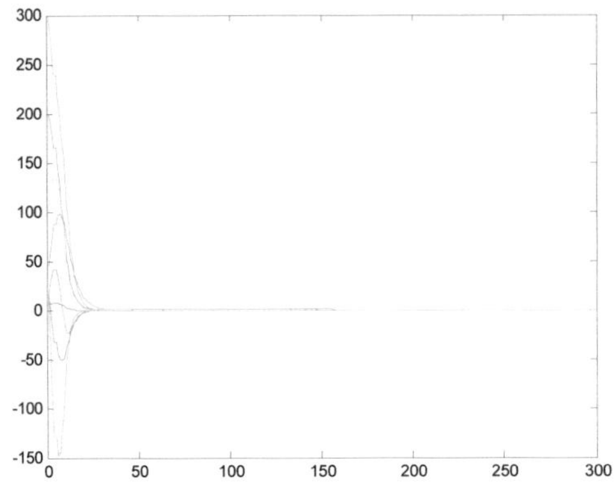

Fig. 4. Evolution of all the states of the linear system with the second controller.

6. Conclusion

Controlling the complex behaviour of the Wastewater Treatment Plant is a challenging mission and requires good control strategies. The process has many variables and presents large time constants. In addition, the process is constantly submitted to significant influent disturbances. These facts make mathematical models and computer simulation to be indispensable in developing new and efficient model based control architecture. This paper presents a part from control, the estimation procedure to compute estimated values of the unavailable state variables of the process, in order to have a more realistic simulation.

In one hand this paper, presents estimation and a predictive non linear controller for a biological nutrient removal have been proposed. The observer performs the twin task of states reconstruction and parameters estimation. The control and estimation techniques developed are based on direct exploitation of the full non-linear IAWQ model Simulation studies show either the efficiency of the non-linear controller in regulation or the effectiveness and the robustness of the estimation scheme, in reconstruction of the unmeasured variables and online estimation of the specific growth rates. The application of estimators such as 'intelligent sensors' to identify important biological variables and parameters with physical meaning constitutes an interesting alternative to the lack of sophisticated instrumentation and provides real time information on the process. In the other hand we introduced the observers in the control loop of a linear system with input constraints. This work is an extension of the theory of control systems with constraints by applying the concept of invariance positive. It addresses the problem of applicability such method in case the states of the systems studied are not measurable or not available at the measure. We presented the case of the observer which part of the information output is used to complete part of the state vector to estimate.

7. Acknowledgment

The authors gratefully acknowledge the support of the Spanish Government through the MICINN project DPI2009-14410-C02-01.

8. References

Clarke, D. W., Montadi C., Tuffs, P. S. (1987). Generalised predictive control - Part 1, The basic algorithm, Part 2, Extensions and interpretations, Automatica, 23(2), pp.137-148.

Dutka, A., Ordys, A., Grimble, M.J. (2003). Nonlinear Predictive Control of a 2dof Helicopter Model, in: Proc. of 42nd IEEE Control and Decision Conference, Maui, Hawaii

Grimble, M. J., Ordys, A. W. (2001). Non-linear Predictive Control for Manufacturing and Robotic Applications, in: Proc: of IEEE Conference on Methods and Models in Automation and Robotics, Miedzyzdroje, Poland

H.El bahja , O.Bakka, P.Vega and F.Mesquine,Modelling and Estimation and Optimal Control Design of a Biological Wastewater Treatment Process, MMAR09,Miedzyzdroje, Poland.

H.El bahja, O.Bakka, P.Vega and F.Mesquine, Non Linear GPC Of a Nuttrient Romoval Biological Plant, ETFA09, Mallorca, Spain.

F.Mesquine, O.Bakka, H.El bahja, and P.Vega , Non Linear GPC Of a Nuttrient Romoval Biological Plant, ETFA10, Bilbao, Spain.

Hajji, S., Farza, M., M'Saad, M., & Kamoun, M. (2008). Observer-based output feedback controller for a class of nonlinear systems. In Proc. of the 17th IFAC world congress.

Hammouri, H., & Farza, M. (2003). Nonlinear observers for locally uniformly observable systems. ESAIM: Control, Optimisation and Calculus of Variations, 9, 353_370.

Shim, H. (2000). A Passivity-based nonlinear observer and a semi-global separation principle. Ph.D. thesis. School of Electrical Engineering, Seoul National University.

Henze, M., Leslie Grady JR., C.P., Gujerm, W., Maraism, G.V.R. and Matsnom, T.,"Activated Sludge Model No.1", I.A.W.Q., scientific and technical Report No.1, 1987.

B. Dahhou, G. Roux, G. Chamilotoris, Modelling and adaptive predictive control of a continuous fermentation process, Appl. Math. Modelling 16 (1992) 545-552.

D. Dochain, Design of adaptive controllers for nonlinear stirred tank bioreactors: extension to the MIMO situation, J. Proc. Cont. 1 (1991) 41-48.

F.Nejjari, Modlisation, Estimation et commande d'un bioprocd de traitement des eaux uses, Thesis Report, Faculty of Sciences, Marrakesh, Morocco, June 1997.

F. Nejjari, A. Benhammou, B. Dahhou, G. Roux, Nonlinear multivariable control of a biological wastewater treatment process, in: Proceedings of ECC 97, Brussels, Belgium, 1-4 July 1997.

Labarrere, M., Krief, J.P. ET Gimonet, B. (1982). Le filtrage et ses applications. Cepadues Edition, Toulouse.

Dutka, A., Ordys, A., Grimble, M.J. (2003). Nonlinear Predictive Control of a 2dof Helicopter Model, in: Proc. of 42nd IEEE Control and Decision Conference, Maui, Hawaii

Fatiha Nejjari and Joseba Quevedo, predictive control of a nutrient removal bio logical plant. Proceeding of the 2004 americain conferance Boston.

R. Bellman and R. Kalaba, "Dynamic programming and feedback control," Proc. of the First IFA C Moscow Congress; 1960.

V. G. Boltyanski, R. V. Gamkrelidze, E. F. Mischenko, and L. S. Pontryagin, "The maximum principle in the theory of optimal processes of control," Proc. of the First IFAC Moscow Congress; 1960.

J. La Salle and S. Lefshetz,"Stability by Liapunov's Direct Method with Applications," Academic Press, New York, N. Y.; 1961.

K. K. Maitra, "An Application of the Theory of Dynamic Programming to Optimal Synthesis of Linear Control Systems," Proc. of Dynamic Programming Workshop, ed, J. E. Gibson, Purdue University, Lafayette,Ind.; 1961.

J.Oreilly. Observer for linear systems. Academic-press edition, 1983. Contributed Works

Part 5

Characterization of Biomass, Pretreatment and Recovery

10

Preparation and Characterization of Bio-Oil from Biomass

Yufu Xu, Xianguo Hu, Wendong Li and Yinyan Shi
Hefei University of Technology
P. R. China

1. Introduction

Bio-oil is a kind of liquid fuel made from biomass materials such as agricultural crops, algal biomass, municipal wastes, and agricultural and forestry by-products via thermo-chemical processes (Demirbas, 2007). As one kind of new inexpensive, clean and green bio-energies, bio-oil is considered as an attractive option instead of conventional fuel in the aspect of reducing environmental pollution.

Currently, biomass crops are distributed abroad in the world and the amount is very large, including woody and herbaceous crops growing in temperate and subtropical regions (Ragauskas, et al., 2006). The annual production is about 2740 Quads (1 Quad = 1016Btu), which is about 8 times the total annual world energy consumption (C. Xu & Lad, 2007). Though, the biomass resource is abundant in the world, the efficiency of utilization is very low. With energy crisis and fuel tension, it is more important to develop new technology in order to use biomass resource efficiently. In addition, biomass resources can also cause less NO_x and SO_x emission due to the less content of nitrogen and sulfur (Sun, et al., 2010).

In recent years, the research on bio-oil has been paid more attention due to the property of sustainable, carbon neutral, and easy to store and transport. Thus, a series of methods are developed to prepare and upgrade bio-oil including fast pyrolysis, liquefaction, gasification, hydrotreatment, and so on. In addition, the characterization of bio-oil is also being focused and got more achievements.

Herein, several selected features concerning the bio-oil are surveyed. In the first part, the preparation of bio-oil is reviewed. The second part will focus on the characterization of bio-oil from biomass.

2. Preparation of bio-oil

Biomass can be converted to bio-oil by the way of fast pyrolysis, liquefaction and gasification processes, and it can also be further obtained high-quality products with the help of upgrading and separation processes. The product properties are different through these approaches, which attribute to the differences in technology as well as equipment.

2.1 Fast pyrolysis

Pyrolysis is defined as a thermo-chemical process of the decomposition to smaller molecules by thermal energy (Demirbas, 2007). Fast pyrolysis is a high temperature process in which

the feedstock is rapidly heated in the absence of oxygen, vaporises and condenses to a dark brown mobile liquid (A. V. Bridgwater & Peacocke, 2000; Q. Zhang, et al., 2006).
The biomass fast pyrolysis is attractive because the biomass can be readily converted into liquid products. These liquids, such as crude bio-oil, have advantages in transport, storage, combustion, retrofitting, and flexibility in production and marketing (Demirbas, 2009).

2.1.1 Raw materials

Biomass raw materials are marked by their tremendous diversity, which include forest products wastes, agricultural residues, organic fractions of municipal solid wastes, paper, cardboard, plastic, food waste, green waste, and other waste (Demirbas, 2009). Nowadays, a lot of materials have been used in the scientific research as raw materials, such as wood leaf (Zabaniotou & Karabelas, 1999), wood industry residues (Garcia-Perez, et al., 2007), rice husks and sawdust (Zheng, et al., 2006), corn stalks (Yi, et al., 2000), and so on.
In the stage of raw materials preparation, drying is usually essential because of the existence of water in the feedstock, which will transfer into liquid product finally. High content of water will reduce the heat value of bio-oil and affect its storage stability. In general, we can make use of by-product gas to dry the feedstock in order to reduce energy consumption (Brammer & Bridgwater, 1999; A. V. Bridgwater & Peacocke, 2000).

2.1.2 Reactors configuration

At present, a variety of reactor configurations have been studied by many organisations, a summary list is shown in Table 1.

Reactor type	Organisation
Fluid bed	Aston University, VTT, NERL, Hamburg university, Oldenberg University, INETI, Dynamotive, etc.
Ablative	NERL, Aston University, ZSW-Stuttgart University
Circulating fluid bed	CRES, CPERI
Entrained flow	GTRI, Egemin
Rotating cone	Twente University, BTG/Schelde/Kara
Transported bed	Ensyn, (at ENEL, Red Arrow, VTT)
Vacuum moving bed	Laval University/Pyrovac

Table 1. Fast pyrolysis reactor types and locations (A. V. Bridgwater & Peacocke, 2000)

However, in all the configurations, fluid bed has been most extensively researched and obtains the huge achievement, which resulting from their ease of operation and ready scale-up. A representational fluid bed fast pyrolysis system is shown in Figure 1. The reactor configurations mainly contain a hopper, two screw feeders, an electric heater, a fluidized-bed reactor, two cyclones, a condenser, and an oil pump, as well as some thermocouples and pressure meters (Zheng, et al., 2006). In the experiment process, the raw materials are fed to the hopper at a certain rate and the hot vapour produced will be quickly cooled into liquid product in the condenser.

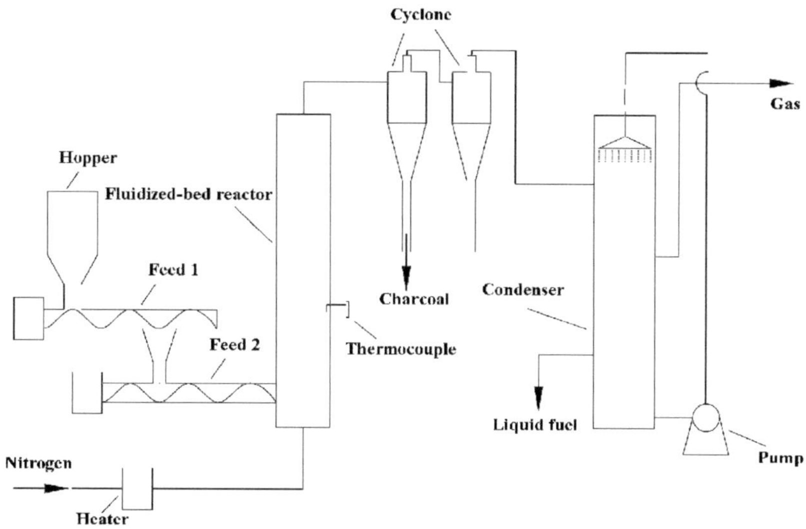

Fig. 1. The experimental device (Zheng, et al., 2006)

As a kind of most popular and ideal configuration, we have reason to believe that fluid bed will achieve greater developments in performance and cost reduction in the near future (A. V. Bridgwater & Peacocke, 2000).

2.1.3 Temperature of reaction

Fast pyrolysis is a high temperature process, thus temperature has tremendous effect to the yield of liquid. The correlation between them is shown in Figure 2 for typical products from fast pyrolysis of wood (Toft, 1996). In the lower temperature, the liquid yield is low due to the less sufficient pyrolysis reaction, which will produce high content of char at the same time. Likewise, the excessive temperature will also lead to liquid yield decreased resulting from the increase of gas product.

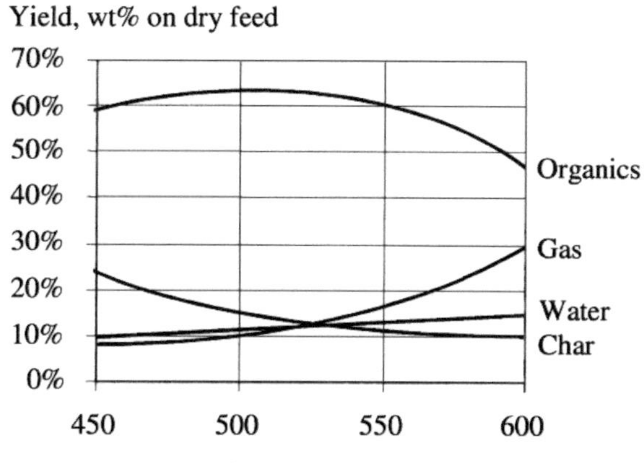

Fig. 2. Typical yields of organic liquid, reaction water, gas and char from fast pyrolysis of wood, wt% on dry feed basis (Toft, 1996)

In order to achieve high liquid yield, the pyrolysis reaction temperature is better to controlled around 500℃ in the vapour phase for most forms of woody biomass (A. V. Bridgwater, et al., 1999b). Of course, different crops may have different maxima yield at different temperatures.

2.1.4 Vapour residence time

Vapour residence time is also important to the liquid yield of pyrolysis reaction. Very short residence times will lead to the incomplete depolymerisation of the lignin, while prolonged residence times can cause further cracking of the primary products (A. V. Bridgwater, et al., 1999b). Too long or short residence time will reduce the organic yield, so it is necessary to select a suitable residence time. In general, the typically vapour residence time is about 1 s.

2.1.5 Liquids collection

The collection of liquids has been a major difficulty in the preparation of fast pyrolysis processes, because the nature of the liquid product is mostly in the form of mist or fume rather than a true vapour, which increases the collection problems (A. V. Bridgwater, et al., 1999b). Furthermore, it is important to choose appropriate condenser and optimum cooling rate; otherwise, some vapour products will take place polymerization and decomposition to produce bitumen (lead to blockage of condenser) and uncondensable gas if cooling time delay. In order to achieve good heat-exchange effect, it is necessary to let the product vapours contact fully with the condensed fluid. Thus, it is regarded as a good method to cool vapour product effectively by using well-sprayed liquid scrub in the bottom of the liquids collection equipment (Zheng, et al., 2006). At present, electrostatic precipitators is widely used by many researchers due to its effectiveness to the liquids collection. However, a kind of very effective method and equipment has not yet to be found by now.

2.2 Liquefaction

Liquefaction is considered as a promising technology to convert biomass to liquefied products through a complex sequence of physical and chemical reactions. In liquefaction process, macromolecular substances are decomposed into small molecules in the condition of heating and the presence of catalyst (Demirbas, 2000a; Demirbas, 2009).

Pyrolysis and liquefaction are both thermo-chemical conversion, but the operating conditions are different as shown in Table 2 (Demirbas, 2000a). Moreover, as two kinds of different transformation method, there are also lots of differences between the liquefaction (Eager, et al., 1983; Hsu & Hixson, 1981) and pyrolysis (Adjaye, et al., 1992; Alen, 1991; Maschio, et al., 1992) mechanisms of biomass.

Process	Temperature(K)	Pressure(MPa)	Drying
Liquefaction	525-600	5-20	Unnecessary
Pyrolysis	650-800	0.5-0.1	Necessary

Table 2. Comparison of liquefaction and pyrolysis (Demirbas, 2000a)

2.2.1 Direct liquefaction

Liquefaction can be divided in two categories, direct liquefaction and indirect liquefaction. Direct liquefaction refers to rapid pyrolysis to produce liquid tars and oils and/or condensable organic vapours, while indirect liquefaction is a kind of condensing process of gas to produce liquid products in the present of catalysts (Demirbas, 2009). In the process of liquefaction, there are lots of reactions occurred such as cracking, hydrogenation, hydrolysis and dehydration, and so on.

The direct liquefaction of Cunninghamia lanceolata in water was investigated, and the maximum heavy oil yield can reach 24% (Qu, et al., 2003). Similar yield of oil (25–34%) are achieved by other researchers through the experiment on the liquefaction of various wood in an autoclave (Demirbas, 2000b). The results show that there are no obvious correlations between the raw materials and bio-oil yields.

2.2.2 Sub/supercritical liquefaction

Supercritical liquefaction is a thermo-chemical process for the conversion of biomass to bio-oil in the presence of supercritical solvents as reaction medium. At present, water, as reaction medium, is attracting widely attention in the aspect of various biomass conversions due to a series of advantages compared with other organic solvents (Sun, et al., 2010). On one hand, water is an economic and environmental friendly solvent, because it will eliminate the costly pretreatment or dying process of wet raw materials and not produce pollution. On the other hand, water possess suitable critical temperature (374℃) and critical pressure (22MPa), and it has a strong solubility for organic compounds derived from biomass in the supercritical condition (C. Xu & Lad, 2007).

There are lots of research works on the aspect of biomass liquefaction in the condition of supercritical condition. For instance, a variety of lignocellulosic materials' conversion at around 350℃ in the presence of CO and $NaCO_3$ at Pittsburgh Energy Technology Center (PETC) (Appell, et al., 1971), woody biomass (Jack pine sawdust) liquefaction in the supercritical water without and with catalysts (alkaline earth and iron ions) at temperatures of 280-380℃ (C. Xu & Lad, 2007), paulownias liquefaction in hot compressed water in a stainless steel autoclave in the conditions of temperature range of 280-360℃, and so on. In general, the yields of liquid through supercritical liquefaction are in the range of 30-50%, which is depend on temperature, pressure, catalysts, etc.

2.2.3 Catalyst

In the process of liquefaction, it is essential to use catalyst in order to achieve higher liquid yield and better quality products. In generally, the common catalysts are used in liquefaction process are alkali salts, such as Na_2CO_3 and KOH, and so on. (Duan & Savage, 2010; Minowa, et al., 1995; Zhou, et al., 2010)

The researcher in university of Michigan produced bio-oils from microalga in the presence of six different heterogeneous catalysts (Pd/C, Pt/C, Ru/C, Ni/SiO_2-Al_2O_3, CoMo/ γ-Al_2O_3 (sulfided), and zeolite) (Duan & Savage, 2010). The bio-oils produced are much lower in oxygen than the original algal biomass feedstock, and their heating values are higher than those of typical petroleum heavy crudes. Moreover, the effects of more catalysts are investigated on the liquefaction, such as Fe, $NaCO_3$ (Sun, et al., 2010), $Ca(OH)_2$, $Ba(OH)_2$, $FeSO_4$ (C. Xu & Lad, 2007), and so on. In summary, the presence of catalyst can decompose macromolecules (including cellulose and hemicellulose) into smaller materials, which will form all kinds of compound through a series of chemical reactions.

2.2.4 Reaction pressure

Hydrogen pressure plays a significant role in the liquefaction of biomass, especially in the condition with extension of reaction times. Yan et al. discuss the effect of hydrogen pressure

to the yield of liquid. The results show that the dependence on H_2 pressure is weak at the early stage of reaction, but the following stage increase the demand to the hydrogen due to the formation of bio-oil accompany with the decomposition reaction of preasphaltene and asphaltene (Yan, et al., 1999).
In addition, the presence of either the hydrogen or the higher pressure in the reaction system will suppress the formation of gas and increase the bio-oil yield. Liquefaction in a high-pressure H_2 environment also led to bio-oil with an increased H content and H/C ratio (Duan & Savage, 2010), which is beneficial to the increase of its heating value in the process of combustion.

2.2.5 Reaction temperature

The yield of bio-oil is depended on the reaction temperature due to differences of reaction type in different temperature periods. Figure 3. reveal the study results on the liquefaction of Cunninghamia lanceolata (Qu, et al., 2003). It is clear that the yield of heavy oil increases firstly and then decreases as the increasing reaction temperature, and reaches the maximum value at around 320℃. The reason might be the competition of hydrolysis and repolymerization. Hydrolysis cause biomass decomposition and then forms small molecule compounds, which rearrange through condensation, cyclization and polymerization to form new compounds. In general, the maximum oil yield is obtained in the temperature range of 525-600K at experimental conditions.

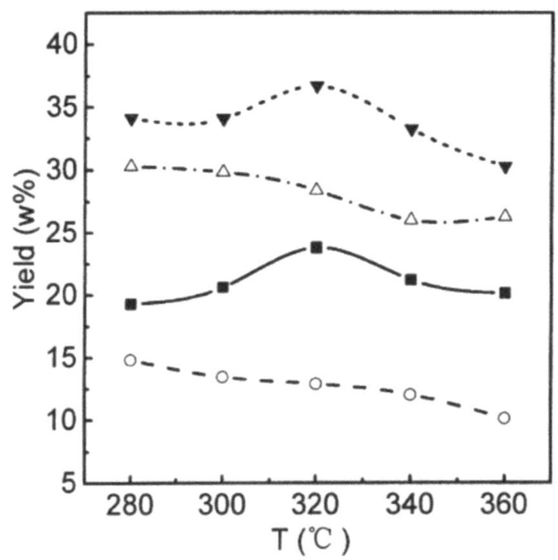

Fig. 3. Effect of reaction temperature on liquefaction of Cunninghamia lanceolata. ■ heavy oil yield; O Organics Dissolved yield; Δ residue yield; ▼ total yield. (Qu, et al., 2003)

2.2.6 Solvent

Bio-oil obtained from liquefaction process is a kind of a very viscous liquid resulting in many problems in the stage of production and storage (Demirbas, 2000a). Therefore, in order to reduce the viscous, it's necessary to add some solvent during the process of liquefaction, such as ethyl acetate, methanol and alcohol due to their high solubility and lower price.

In some conditions, the solvent can play the role of hydrogen-donor solvent in the process of liquefaction. This kind of solvent not only reduce the viscous of products but also increase the yield of liquid, that's because the presence of hydrogen-donor solvent will induce strong destruction of molecular structure of sawdust (Yan, et al., 1999). In summary, it is very important to select a proper solvent for liquefaction of biomass.

2.3 Upgrading and separation

As a renewable energy source, biomass can be convent to bio-oil and has some advantages compared with conventional fossil fuel. Unfortunately, the application range for such oils is limited because of the high acidity (pH~2.5), high viscosity, low volatility, corrosiveness, immiscibility with fossil fuels, thermal instability, tendency to polymerise under exposure to air and the presence of oxygen in a variety of chemical functionalities (Gandarias, et al., 2008; Wildschut, et al., 2009; Q. Zhang, et al., 2007). Hence, upgrading and separation of the oils is required for most applications. The recent upgrading techniques are described as follows.

2.3.1 Catalytic hydrogenation

The catalytic hydrogenation is performed in hydrogen providing solvents activated by the catalysts of Co-Mo, Ni-Mo and their oxides or loaded on Al_2O_3 under pressurized conditions of hydrogen and/or CO. For catalytic hydrogenation, it's important to select a catalyst with higher activity. There's actually been studies show that the Ni-Mo catalyst presented a higher activity than the Ni-W catalyst for the phenol HDO reactions in all the temperature (Gandarias, et al., 2008). Moreover, Senol et al. investigated the elimination of oxygen from carboxylic groups with model compounds in order to understand the reaction mechanism of oxygen-containing functional groups, and obtained three primary paths of producing hydrocarbons through aliphatic methyl esters (Senol, et al., 2005).

In order to improve the properties of pyrolysis liquids and achieve higher liquid yield, A two-stage hydrotreatment process was proposed (Elliott, 2007; Furimsky, 2000). The first stage is to remove the oxygen containing compounds which readily undergo polymerization at high temperature condition. In the second stage, the primary reactants will further convert to other products.

Hydrotreatment is an effective way to convert unsaturated compounds into some more stable ones, but it requires more severe conditions such as higher temperature and hydrogen pressure. Although hydrogenation of bio-oil has made huge progresses, more stable catalysts maybe the largest challenge to make production of the commercial fuels from the bio-oil more attractive.

2.3.2 Catalytic cracking

Catalytic cracking is that oxygen containing bio-oils are catalytically decomposed to hydrocarbons with the removal of oxygen as H_2O, CO_2 or CO. Guo et al. investigated the catalytic cracking of bio-oil in a tubular fixed-bed reactor with HZSM-5 as catalyst. The results show that the yield of organic distillate is about 45%, and that the amount of oxygenated compounds in the bio-oil reduce greatly (Guo, et al., 2003). Moreover, seven mesoporous catalysts were compared in converting the pyrolysis vapours of spruce wood for improving bio-oil properties (Adam, et al., 2006). The experiment results confirmed the advantageous of catalyst usage, and the Al-SBA-15 catalyst performs more balanced among all the catalysts tested.

Catalytic cracking can converting macromolecule oxygenated substances to lighter fractions (Adjaye & Bakhshi, 1995; S. Zhang, et al., 2005). Furthermore, it is considered as a promising method and has drawn wide attention due to the price advantage.

2.3.3 Steam reforming

At present, catalytic steam reforming of bio-oils is a technically to produce hydrogen, which is extremely valuable for the chemical industry. The steam reforming of aqueous fraction from bio-oil is studied at the condition of high temperature (825 and 875℃) using a fixed-bed micro-reactor (Garcia, et al., 2000). The results show that catalytic efficiency is depend on the water-gas shift activity of catalysts. National Renewable Energy Laboratory (NREL) demonstrated reforming of bio-oil in a bench-scale fluidized bed system using several commercial and custom-made catalysts, and hydrogen yield was around 70% (Czernik, et al., 2007). Besides, some researchers also studied the effect of no noble metal-based catalysts for the steam reforming of bio-oil and achieve good results (Rioche, et al., 2005).

A major advantage of producing hydrogen from bio-oil through steam reforming is that bio-oil is much easier and less expensive than other materials.

2.3.4 Emulsification

To combine bio-oil with diesel fuel directly can be carried out through emulsification method by the aid of surfactant. This is a relatively short-term way to use bio-oil. The ratio range of bio-oil/diesel emulsification is very wide, and the viscosity of emulsion is acceptable (D. Chiaramonti, et al., 2003). Zheng studied the emulsification of bio-oil/diesel and obtained many kinds of homogeneous emulsions (Zheng, 2007). The physical properties of emulsions are shown in Table 3, which shows the emulsions have higher heat value, lower pH and lower viscosity compared with bio-oil.

	25% Bio-oil +74%diesel +1% emusifier	50% Bio-oil +49%diesel +1% emusifier	75% Bio-oil +24%diesel +1% emusifier
Viscosity	73	129	192
pH	2.7	2.5	2.2
LHV(MJ/kg)	34.55	29.1	23.65

Table 3. Properties of emulsions (Zheng, 2007)

It is therefore possible to consider bio-oil emulsification as a possible approach to the wide use of these oils reducing the investment in technologies. Nevertheless, high cost and energy consumption input are needed in the transformations. Moreover, the dominant factor is that the corrosion was accelerated by the high velocity turbulent flow in the spray channels in the experiment process.

2.3.5 Distillation

A large amount of water from the raw material is unavoidable in the bio-oil even if it is dry material. The existence of water is bad for the upgrading of the bio-oil, thus water should be removed from the bio-oil. The water in the bio-oil can be removed through azeotropic distillation with toluene (Baker & Elliott, 1988). In addition, the light and weight fractions can also be separated by distillation such as molecular distillation, and the obtained light fraction can be used as the material for upgrading process (Yao, et al., 2008).

2.3.6 Extraction

Bio-oil is a complex mixture, which nearly involves hundreds of compounds, mainly including acids, alcohols, aldehydes, esters, ketones, sugars, phenols, phenol derivatives, and so on. The oil fractions can be separated by the way of water extraction and obtain water-insoluble and water-soluble fractions, which can be separated further (Sipila, et al., 1998). The whole process is shown in Figure 4.

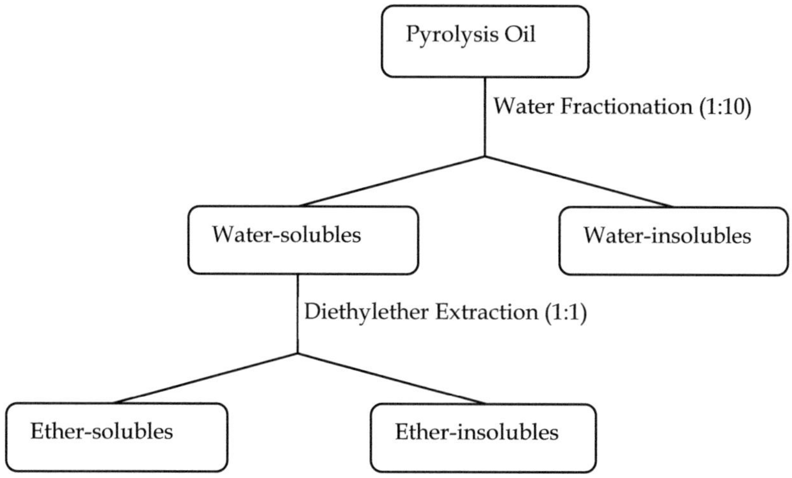

Fig. 4. Fractionation scheme of bio-oil (Sipila, et al., 1998)

There are many substances that can be extracted from bio-oil, including a range of flavourings and essences for the food industry (A. V. Bridgwater, et al., 1999b).

2.3.7 Column chromatography

The composition of the bio-oil is complex and a lot of material properties are similar among them. Thus, it is unrealistic to separate all kinds of fractions by conventional methods such as distillation and extraction. Nevertheless, column chromatography, as a new separation technology, can satisfy the high sensitivity requirement needed by the bio-oil separation. For instance, phthalate esters, which is considered as toxic material to human and being wife, can be separated from bio-oil by the way of column chromatography (Zeng, et al., 2011).

3. Characterization of bio-oil

As well known, the material property depends on its structure and constitute. Bio-oil has poor properties due to the complexity of composition, which causes the limitation of application range. In order to understand the properties and composition of bio-oil so as to use effectively, it's necessary to carry on characterization to bio-oil.

3.1 Physiochemical properties

The bio-oil from biomass is typically a dark-brown liquid with a pungent odour, and the physiochemical properties of the bio-oil are different from conventional fossil fuels. The mainly physiochemical properties contain components, heating value, water content, density, flash point, and so on.

3.1.1 Components

The components of bio-oil are complicated, comprising mainly water, acids, alcohols, aldehydes, esters, ketones, sugars, phenols, phenol derivatives, lignin-derived substances, and so on. The complexity of the bio-oil itself results in the difficult to analyze and characterize (Wildschut, 2009). Gas chromatography-mass spectrometry (GC-MS) has been the technique most widely used in the analyses of the component (Sipila, et al., 1998). The major components of one kind of crude bio-oil based on the GC-MS analyses are shown in Table 4.

Main components	RT/min	Area w/%
formaldehyde	1.42	3.14
aldehyde	1.51	6.52
hydroxyacetaldehyde	1.61	3.14
hydroxypropanone	1.72	2.70
butyric acid	1.82	0.96
acetic acid	2.07	29.76
glyceraldehyde	2.6	3.54
3,4-dihydroxy-dihydro-furan-2-one	2.77	3.27
2,2-dimethoxy-ethanol	2.86	6.83
furfural	3.13	6.56
2,5-dimethoxy-tetrahydro-furan	3.5	3.47
4-hydroxy-butyric acid	4.27	0.43
5H-furan-2-one	4.51	0.74
2,3-dimethyl-cyclohexanol	4.76	1.31
3-methyl-5H-furan-2-one	5.19	0.38
corylon	6.15	1.18
phenol	6.59	1.57
o-cresol	6.8	1.12
m-cresol	7	1.46
2-methoxy-6-methyl-phenol	7.79	1.78
3,4-dimethyl-phenol	8.99	1.14
4-ethyl-phenol	9.7	1.31
3-(2-hydroxy-phenyl)-acrylic acid	10.1	1.53
catechol	10.81	3.53
3-methyl-catechol	11.9	1.36
vanillin	12.7	0.24
4-ethyl-catechol	12.86	0.71
levoglucosan	14.73	9.95
2,3,4-trimethoxy-benzaldehyde	15.5	0.20
3-(4-hydroxy-2-methoxy-phenyl)-propenal	15.8	0.15

Table 4. Components of crude biomass oil (Hu, et al., 2011a)

3.1.2 Heating Value

The standard measurement of the energy content of a fuel is its heating value (HV). HV is divided into lower heating value (LHV) and higher heating value (HHV) depending on the water produced through hydrogen in vapour or liquid phase. Heating value can be determination by the oxygen-bomb colorimeter method (Demirbas, 2009).

The heating value of the pyrolysis oils is affected by the composition of the oil (Sipila, et al., 1998). At present, HHV of bio-oil can be determined directly according to DIN 51900 by the oxygen-bomb colorimeter. In addition, the HHV of the bio-oil is also calculated using the following formula (Milne, et al., 1990).

$$\mathrm{HHV} = 338.2 \times \mathrm{C} + 1442.8 \times (\mathrm{H} - \frac{\mathrm{O}}{8}) \quad (\mathrm{MJ/kg}) \tag{1}$$

The LHV can be determined by the HHV and the total weight percent of hydrogen (from elemental analysis) in the bio-oil according to the formula (Oasmaa, et al., 1997) as shown below.

$$\mathrm{LHV} = \mathrm{HHV} - 218.3 \times H\% \,(\mathrm{wt\%}) \quad (\mathrm{KJ/kg}) \tag{2}$$

Bio-oil is of a lower heating value (15 - 20 MJ/kg), compared to the conventional fossil oil (41 - 43 MJ/kg) (A.V. Bridgwater, et al., 1999a; Wildschut, et al., 2009). That is to say that the energy density of bio-oil is only about half of the fossil oil, which is attribute to the higher water and oxygen contents. In order to improve the heating value of bio-oil so that it can be used in the engine, it is necessary to reduce the contents of water and oxygen by the way of upgrading, as described above.

3.1.3 Water content

The water content in the bio-oil is analyzed by Karl-Fischer titration according to ASTM D 1744. The sample solvent is a mixture of chloroform and methanol (3:1 v/v) (Sipila, et al., 1998), because this solvent can dissolve almost all of the component of bio-oil. In the process of experiment, a small amount of bio-oil (0.03-0.05g) was added to an isolated glass chamber containing Karl Fischer solvent. The titrations were carried out using the Karl Fischer titrant (Wildschut, et al., 2009).

The existence of water in the bio-oil is unavoidable, which is due to moisture in the raw material. In general, the water content of bio-oil is usually in the range of 30-35 wt% (Radlein, 2002), and it is hard to remove from bio-oil resulting from the certain solubility of bio-oil and water. The existence of water has both negative and positive effects on the storage and utilization of bio-oils. On the one hand, it will lessen heating values in combustion, and may cause phase separation in storage. On the other hand, it is beneficial to reduce viscosity and facilitate atomization (Lu, et al., 2009).

3.1.4 Oxygen content

The elemental compositions of the oils (C, H, O and N) can be determined using a CHN-S analyzer according to ASTM D 5373-93. The oxygen content will be calculated by difference (Wildschut, et al., 2009).

The oxygen content of the bio-oil varies in the range of 35-40% (Oasmaa & Czernik, 1999). The presence of high oxygen content is regard as the biggest differences between bio-oil and fossil oil, that's because it lead some bad properties, such as corrosiveness, viscosity, low energy density, thermal instability, and so on (Elliott, et al., 2009). Of course, a certain amount of oxygen in the fuel is beneficial to improve combustion sufficiency. However, it is imperative to removal of oxygen in the bio-oil through hydrodeoxygenation (HDO) and reduction of the oxygen content below 10 wt% by a catalytic hydrotreatment reactions is possible under severe conditions (Wildschut, et al., 2009).

3.1.5 Density

Density can be measured at 15℃ using picnometer by ASTM D 4052 (Sipila, et al., 1998). The density of bio-oil is usually in the range of 1.1-1.3kg/m^3, which is depending on the raw materials and pyrolysis conditions. The density of bio-oil is larger than the gasoline and diesel because of the presence of a large number of water and macromolecule such as cellulose, hemicelluloses, oligomeric phenolic compounds (Oasmaa & Czernik, 1999), and so on.

3.1.6 Ash

Ash is the residue of bio-oil after its combustion, and the ash can be determined according to ASTM D 482. The ash of bio-oil is usually vary in 0.004-0.03 wt% (Oasmaa & Czernik, 1999), which is also relevant to the raw materials and reaction conditions. In general, the ash content is higher for the straw oil than for other oils due to their originally higher amounts in straw than in wood (Sipila, et al., 1998).

The presence of ash in bio-oil can cause erosion, corrosion and kicking problems in the engines and the valves (Q. Zhang, et al., 2007). However, there is no effective way to reduce the content of ash by now.

3.1.7 Mechanical impurities

The mechanical impurities are measure as ethanol insolubles retained by a filter after several washings and vacuum-drying (Sipila, et al., 1998). Generally, the presence of mechanical impurities cannot avoid in the preparation process of the bio-oil. Mechanical impurities mainly contain pyrolysis char, fine sand, materials used in the reactor, and precipitates formed during storage (Oasmaa & Czernik, 1999).

The content of mechanical impurities in different oils are usually varies in 0.01 to 3 wt% with the particle sizes of 1-200 μm (Oasmaa, et al., 1997). The presence of mechanical impurities is harmful to the storage and combustion of bio-oil, resulting in agglomerate and viscosity increases (Lu, et al., 2009). The most economical and efficient method to reduce the content of mechanical impurities would be filtration.

3.1.8 Flash point

The flash point of a volatile liquid is the lowest temperature at which it can vaporize to form an ignitable mixture in air. Flash point is measured using a flash-point analyzer according to ASTM D 93. The test temperature is usually employ increase of 5.5℃/min in the range of 30-80℃ (Wildschut, et al., 2009).

Flash point is influenced by the raw materials and preparation method, because of these will result in the differences in composition and content of the bio-oil from biomass. In general, the bio-oils from hardwood have a high flash point due to the low contents of methanol and evaporation residue of ether soluble (Sipila, et al., 1998).

3.1.9 pH

The bio-oil has amount of diluted water and volatile acids, such as acetic and formic acid, which results in the low pH values varied in 2-3. The presence of acids in the bio-oil is the main reason to account for the property of corrosion to materials in the storage and application processes. Therefore, it requires upgrading to fulfil the requirement of fuels before application through upgrading processes.

3.2 Combustion property

Combustion is the oxidation of the fuel at elevated temperatures, and accompanied by the production of heat and conversion of chemical species.

As a kind of clean and renewable energy, bio-oil has a potential to be used as a conventional fossil fuel substitute. However, the usage of bio-oil has been limited due to some problems during its use in standard equipment constructed for combustion petroleum-derived fuels (Czernik & Bridgwater, 2004). Bio-oil has the low heating values (leading low flame temperature) (Demirbas, 2005) and high water content, which is harmful for ignition. Furthermore, organic acids in the bio-oil are highly corrosive to common construction materials. In addition, the present of solid, high viscosity, coking are also the primary challenge in the process of combustion (Yaman, 2004). Of course, bio-oil has some important advantages such as effectively volatility and combustibility. In the combustion applications, biomass has been fired directly either alone or along with a primary fuel such as diesel, methanol, ethanol, and so on (Demirbas, 2004).

The combustion properties of the bio-oil can be tested by the biomass fuels combustion system, which consists of a droplet generator, a laminar flow reactor, and a video imaging system (Wornat, et al., 1994). The device can observe the combustion behaviors of bio-oil droplets directly. The tests can be performed both a fibre-suspended single droplet and a stream of freefalling mono-dispersed droplets (Lu, et al., 2009).

In the present chapter, we will introduce the combustion property of the bio-oil in standard equipment such as boilers, diesel engines, and gas turbines.

3.2.1 Combustion in boiler

Boiler is a common device used for generate heat and power through burning fuels such as wood, coal, oil, and natural gas. The source of combustion materials for boiler is widespread, but the fuel combustion efficiency is usually less than engines and turbines. It is suitable for bio-oil used in boiler instead of conventional fossil fuel and coal, etc (Czernik & Bridgwater, 2004). Though it is difficult to ignite for bio-oil due to the high content of water, it can burn steadily once ignited, and the observed flame lengths with pyrolysis oils are similar to those of conventional fuel oils (Shaddix & Hardesty, 1999).

The ignition of bio-oil is the key to the combustion in boiler. Some modifications of the existing burner and boiler are better effective method to improve its ignitability and combustion stability. The boiler can be designed in a dual fuel mode, hence the bio-oil can be co-fired with petroleum fuel at different ratios (Gust, 1997).

Emissions of NO_X and SO_X from boilers firing bio-oil are lower than those from residual fuel oil, but emissions of particulate (soot, carbonaceous cenospheres, and ash) are higher from bio-oil resulting from the high content of ash and incomplete combustion of the oil. Generally, Emissions of NO_X and carbon monoxide (CO) from combustion of bio-oil vary in 140-300ppm and 30-50ppm respectively, which are all at acceptable levels (Shaddix & Hardesty, 1999).

3.2.2 Combustion in diesel engine

The diesel engine has the highest thermal efficiency (up to 45%) of any regular internal or external combustion engine due to its very high compression ratio, of course it report a high demand for the fuel quality.

VTT (Technical Research Centre of Finland) investigated the combustion performance of bio-oil in the diesel engine (4.8kW, single-cylinder, high-speed) (Solantausta, et al., 1994). The results showed that bio-oil was not suitable for a conventional diesel engine and produced many problems because of the specific properties. For one thing, bio-oil could not auto-ignition without additives (nitrated alcohol) and it also needs a pilot injection system. For another thing, an amount of coke formed in the process of combustion of bio-oil, which resulting in the periodic clogging of the fuel injector. In addition, severe material wear occurred, which is considered as difficult to avert.

A detailed investigation ignition delay and combustion behavior has been carried out by MIT by comparing with the performance of two bio-oils and No.2 diesel fuel in a direct injection engine (Shihadeh & Hochgreb, 2000). The bio-oil exhibited longer ignition delays due to the relatively slow chemistry process to the diesel fuel.

In recently, more researches about the combustion of bio-oil have been reported, including erosion-corrosion problems to standard materials in UK (A. V. Bridgwater, et al., 2002), selection of optimum operating characteristics (A. V. Bridgwater, et al., 2002; Leech, 1997; Ormrod & Webster, 2000), tests on emulsions of bio-oil in diesel fuel used in different engines (Baglioni, et al., 2001; D. Chiaramonti, et al., 2003), and so on.

3.2.3 Combustion in gas turbine

A gas turbine, also called a combustion turbine, is a rotary engine that produces energy via the flowing combustion gas. Gas turbine is widely used in various aspects, most important of which are driving electric power generators and providing power to aircraft (Czernik & Bridgwater, 2004).

Combustion of bio-oil in has been demonstrated in a 2.5 MWe industrial gas turbine (J69-T-29) at Teledyne CAE (USA) as early as 1980s (Kasper, et al., 1983). The combustion system of the J69 consists of an annular combustor and a centrifugal fuel injector rotating as shaft speed. The test results show that the combustion efficiency of the bio-oil in this gas turbine is over 99%.

The first industrial application of bio-oil in gas turbines combustion was carried out in the year of 1995 (Andrews, et al., 1997; Andrews & Patnaik, 1996). The researchers used a 2.5MWe class-GT2500 turbine engine, which was designed and built by Mashproekt in Ukraine. The fuel of GT2500 turbine is diesel oil rather than its standard fuel (kerosene), and the gas turbine a "silo" type combustion chamber, which can be modified more easily. The results about atomization tests show that both water and bio-oil can generate a wider cone angle than diesel oil, this is because diesel oil has lower viscosity and surface tension and the interaction between primary and secondary flows (David Chiaramonti, et al., 2007).

3.3 Corrosion property

Bio-oil obtained by the fast pyrolysis of straw is an acidic fuel with pH of 3.4–3.5. It contains a large amount of organic acids, phenol and water. For this reason, biomass oils will strongly corrode aluminium, mild steel and nickel based materials, whereas stainless steel, cobalt based materials, brass and various plastics are much more resistant (Oasmaa, et al., 1997).

The corrosion extent of the metal can be determined by the weight increase and variations on the metal surface, which can be analyzed by optical micrography and X-ray photoelectron spectroscopy (XPS). Generally, the corrosion performance of metals are sensitive to materials, temperature condition and bio-oil property. The corrosion in bio-oil of four kinds of metals used frequently in engines (including iron, lead, steel and copper) is studied at different temperatures and for different test durations using a simulation corrosion evaluation apparatus (Figure 5) for internal combustion engine fuel (Hu, et al., 2011b). The results of mass variation rates of four metals at different temperature are summarised in Table 5.

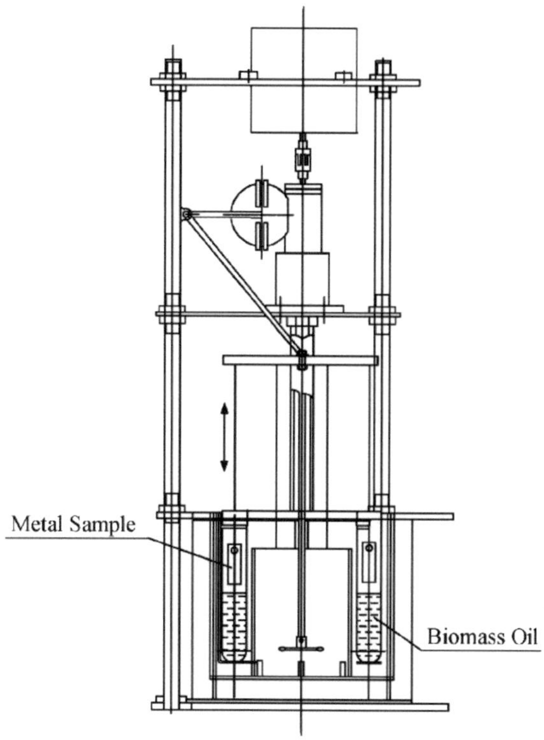

Fig. 5. Schematic diagram of corrosion test apparatus, metal strip was dipped intermittently with frequency of 15 per min (Hu, et al., 2011b)

Metal	25℃		40℃		55℃	
	5h	10h	5h	10h	5h	10h
Iron	10.25	19.89	11.15	21.64	11.81	25.60
Lead	7.53	11.23	16.70	17.28	20.35	25.10
Steel	3.27	8.41	6.11	12.43	7.60	12.91
Copper	0.67	1.56	0.78	1.57	1.19	2.19

Table 5. Weight increase of metals at different temperatures and during different exposure times, g/m^2 (Hu, et al., 2011b)

3.3.1 Cu strip

Corrosion information can be obtained from the weight increase of the metal strips when immersed in the biomass oil. Study shows that the weight increase for copper was the smallest compared with the other metals, which indicated its best anticorrosive ability (Hu, et al., 2011b). The chemisorption of oxygen and other gases in the atmosphere will initially increase the weight of the strips. Furthermore, after contacting with biomass oil, some corrosion products, such as Cu_2O and CuO, are formed on the surface of the metals. These cannot be removed washed by physical methods and result in an increase in weight of the samples. In the case of copper, these corrosion layers do not prevent the underlying metals from further corrosion. However, the corrosion of copper will become slow because of its noble character (Darmstadt, et al., 2004).

3.3.2 Stainless steel

Stainless steel has anti-corrosion ability like Cu strip due to the presence of Cr, which is the mainly anti-corrosion element in the stainless steel. For AISI 1045 steel, the corrosion volumes increased with corrosion time and temperatures. After corrosion, layers of oxide and/or hydroxide are formed on the metal surface. X-ray photoelectron spectroscopy (XPS) results show the presence of Fe_2O_3 and Fe_3O_4, which are mainly corrosion products. However, these layers cannot protect the metal from further oxidation (Hu, et al., 2011b).
For austenitic steel (SS 316), it is not causes corrosion in the experiment condition, which is mainly attribute to the formation of chromium oxide layer that prevents further oxidation (Darmstadt, et al., 2004). Consequently, the stainless steel can be taken into consideration in the selection of construction materials for pyrolysis units and diesel engine.

3.3.3 Lead

The bio-oil corrosiveness to lead is especially severe compared with stainless steel and copper. A significant weight variation was found for lead, which increased with temperature. When lead comes into contact with bio-oil, oxide and/or hydroxide layers are formed on the metal surface. The chief components in this layer are PbO and $Pb(OH)_2$. However, this layer did not protect the underlying metal against further oxidation though the oxide layer is relatively thick (Hu, et al., 2011b).

3.3.4 Iron

Bio-oil is very corrosive to iron compared with stainless steel, which is essentially noncorrosive. There is a oxide layer as the same as stainless steel even the same components (Fe_2O_3 and Fe_3O_4). However, XPS results show that the corrosion product on the steel surface was thicker than on iron (no signal for metallic iron from the substrate). Likewise, the layer cannot protect the metal from further oxidation (Hu, et al., 2011b).

3.4 Tribological performance

As a new type energy fuel, bio-oil is mainly used for combustion heating equipment such as industrial furnace, gas turbine, diesel engine, and so on. However, bio-oil will be able to lead higher friction and wear to the oil pipeline and nozzle in the process of injection, which has very serious effect to the stable combustion even safety performance (Wang, et al., 2008). Therefore, it is necessary to learn about bio-oil tribological properties and its mechanism.

3.4.1 Friction efficiency

Generally, the four-ball tribometer is used to study the tribological performances of bio-oil to obtain friction coefficient, and the wear scar diameter can be measured by digital microscope. Xu et al. studied the tribological performance and explained the lubrication mechanism of the straw based bio-fuel by four-ball tribometer at 1450rpm (Y. Xu, et al., 2007). The experimental results showed that the extreme pressure of the bio-fuel was up to 392 N, and the extreme pressure of diesel oil was 333 N. These results indicated that the straw based bio-oil has a potential lubrication performance than the diesel oil.

The friction coefficient of straw-based bio-oil under different loads suggested that it increased with load (Figure 6), which may be result from the real contact surface distortion increased with the load. The frictional coefficient of bio-oil are varied in 0.08 and 0.11 between 196N and 294N. The wear scar diameter on the ball surface increased with load slowly in 30min (Figure 7).

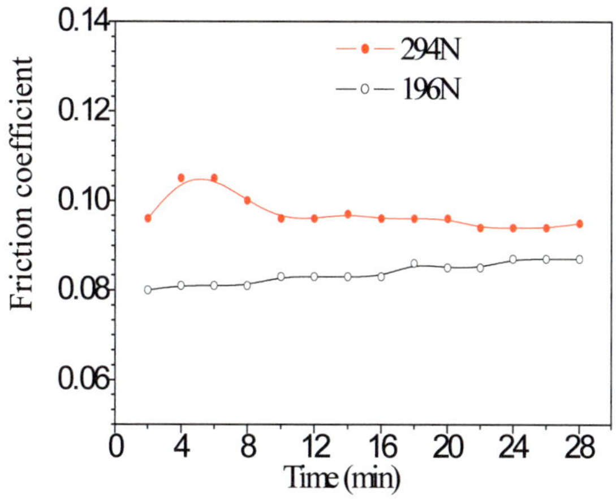

Fig. 6. Variations of friction coefficient of bio-fuel with test duration under different loads (Y. Xu, et al., 2007)

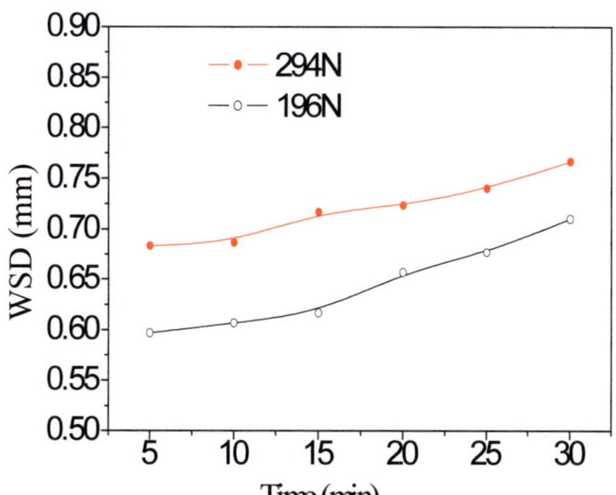

Fig. 7. Variations of wear scar diameter of bio-fuel with test (Y. Xu, et al., 2007)

3.4.2 Wear volume/weight

The weight loss of bio-oil during the process of use can be analyzed by thermo-gravimetric analyze (TGA). In case of used bio-fuel, its weight loss reduced 11% when the temperature was over 530℃ compared with that of fresh bio-oil, because some compounds in bio-oil may reacted during the friction process(Figure 8). (Y. Xu, et al., 2007)

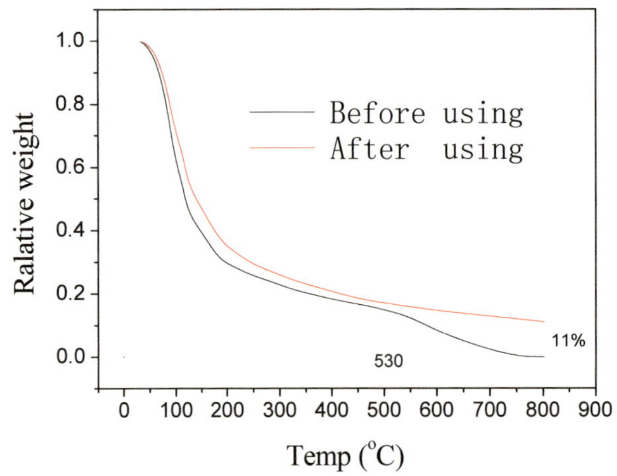

Fig. 8. TGA curves of bio-fuel before and after (Y. Xu, et al., 2007)

3.4.3 Lubricity

As well known, the alternative fuel from biomass cannot be used well in internal combustion engine because of the serious lubrication (Y. Xu, et al., 2007). However, using emulsion technology to mixing bio-oil with diesel is one of the most convenient approaches to use bio-oil reasonable (Ikura, et al., 2003; Qi, et al., 2008).

Xu et al. investigated the lubricity of the bio-oil/diesel emulsion by high frequency reciprocating test rig (Figure 9) (Y. Xu, et al., 2010; Y. Xu, et al., 2009). Table 6 showed that the average friction coefficient of the emulsified bio-oil was 0.130, which was lower than commercial diesel number zero (0.164). This result indicated that the emulsified bio-oil had better lubricity properties than commercial diesel number zero.

Item	Diesel	Emulsified bio-oil
Average friction coefficient	0.164	0.130
Corrected wear scar diameter/ μm	226	284

Table 6. Comparison of friction coefficient and wear resistance between emulsified bio-oil and diesel (Y. Xu, et al., 2010)

The lubrication mechanism of emulsified bio-oil could be attributed to the polar groups and oxygenic compounds. The interaction between them caused the tiny liquid drops deposit on the surface of friction, which generated frictional chemical reaction and led to the better boundary lubrication. However, the existence of oxygen might accelerate the corrosion wear on the rubbing surface.

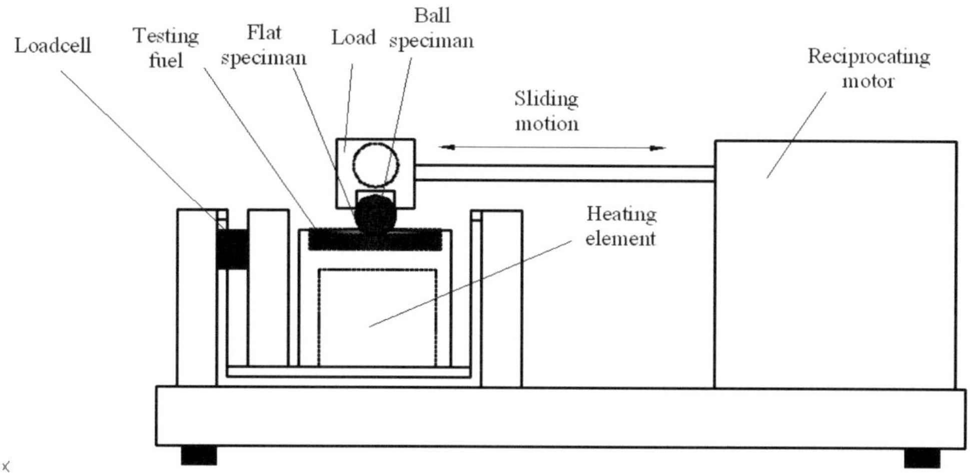

Fig. 9. Schematic diagram of lubricity test by high frequency reciprocating test rig (Y. Xu, et al., 2010)

Hu et al. studied the tribological performance of distilled biomass oil from rice straw by pyrolysis process in a four-ball tribometer. The results showed that the refined biomass oil had certain anti-wear and friction-reducing properties (Hu, et al., 2008b).

3.5 Biodegradability

As the production expanding constantly, bio-oil also caused environmental problems like as fossil fuels. In production, transportation, storage and application processes, bio-oil will destroy local ecological environment if it emissions into the soil and water as a result of the accident or improper management (Hu, et al., 2008a).

Generally, the methods which control oil pollution can be divided into three kinds: physical, chemical, and biological; the former two methods are very expensive and treatment is not completely or cause secondary pollution. However, biological method is economic, efficient and the final product is carbon dioxide and water, without any secondary pollution (Fu, et al., 2009). A mass of research indicate that biological degradation plays an important role in the purification of the oil pollution, but the microbial degradation ability itself restricts the oil pollutant further degradation (Pelletier, et al., 2004).

3.5.1 Degradation properties in soil

The degradation rate of the bio-oil in the soil is responsive to microorganism, temperature, oil content, pH, etc (Hu, et al., 2008a). Hu et al. gained a strain of bio-oil degrading mold (a kind of Aspergill versicoir, named as EL5) through enrichment, separation and purification from sludge collected from a paper mill. The yield of CO_2 was taken as degradation test index. The results showed that the degradation speed of bio-oil was positively correlated to the temperature and negatively correlated to substrate concentration. The degradation rate of the bio-oil in the soil can reach 40% in the suitable temperature (30℃) and neutral pH, compared with only 6% under the same conditions without degrading mold (Hu, et al., 2008a).

3.5.2 Degradation properties in aquatic environment

As the degradation of bio-oil in the soil, the degradation rate of the bio-oil in the aquatic environment is also responsive to microorganism, temperature, oil content, pH, etc. During the acclimation, the biodegradation process of bio-oil is accorded approximately with the first-order reaction by the way of Sturm method which is described by measuring CO_2 volume from the microbes' production (Fu, et al., 2009). A schematic diagram of biodegradation experiment is shown in Figure 10. The whole device was carried out under aerobic conditions. The biodegradation ability could be improved in aqueous culture under neutral and acidic conditions. The optimal temperature for biodegradation of bio-oil is 40℃. The optimal inocula content for the biodegradation of bio-oil was 16%.

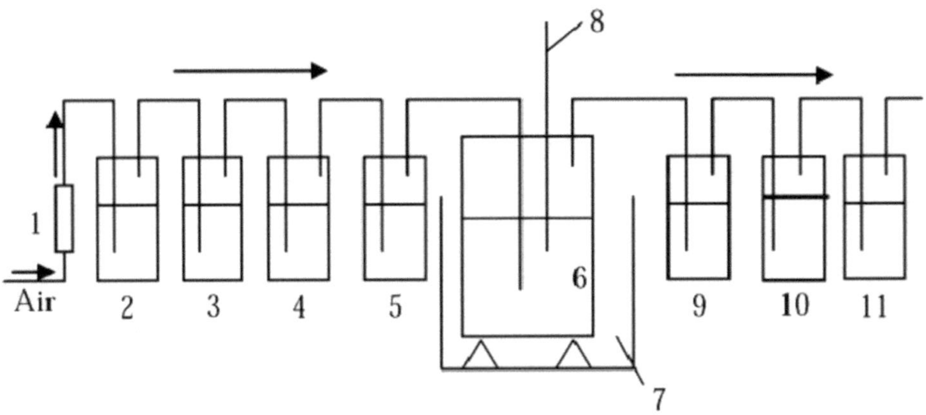

Fig. 10. Schematic diagram of biodegradation experiment
Notes: 1. Flow meter; 2-4. Three bottles for absorbing CO_2 from atmosphere; 5.Bottle for testing the absorbency; 6. Bioreactor; 7. Constant temperature water bath; 8. Thermometer; 9-11. Three bottles for absorbing CO_2 from biodegradation (Fu, et al., 2009)

Blin investigated the biodegradation properties of various pyrolysis oils and EN 590 diesel sample in the Modified Sturm (OECD 301B). The results showed that various bio-oils degraded 41–50% after 28 days, whereas the diesel only has 24% biodegradation. The biodegradation model of bio-oil can be very well described by a first-order kinetic equation (Blin, et al., 2007).

4. Conclusions

This chapter reviewed the preparation methods and characterization of the bio-oil. The bio-oil showed the promising prospects as an alternative renewable energy sources to replace the fossil fuel. However, the bio-oil has high acid value, high oxygen, and low heating values compared with the commercial diesel fuel. It is urgent to investigate the thermo-chemical conversion mechanism of the biomass. What's more, the more effective upgrading methods should be carried out the raw bio-oil because of these disadvantages. The properties such as basic physiochemical property, combustion, corrosion, lubricity and biodegradability of the bio-oil from biomass were also discussed. Furthermore, the chemical components and the quality standard of the bio-oil was needed to be established as soon as possible in order to accelerate the development and application of the bio-oil.

5. Acknowledgements

Financial support from National Natural Science Foundation of China (Grant No. 50875071), Anhui Provincial Natural Science Foundation (Grant No. 11040606Q37), and College Students Innovative Experimental Program Foundation of HFUT (Grant No. cxsy102025) are gratefully acknowledged.

6. References

Adam, J., Antonakou, E., Lappas, A., Stocker, M., Nilsen, M. H., Bouzga, A., Hustad, J. E. & Øye, G. (2006). In Situ Catalytic Upgrading of Biomass derived Fast Pyrolysis Vapours in a Fixed Bed Reactor Using Mesoporous Materials. *Microporous and Mesoporous Materials,* Vol.96, No.1-3, (November 2006), pp. 93-101, ISSN 1387-1811

Adjaye, J. D. & Bakhshi, N. N. (1995). Production of Hydrocarbons by Catalytic Upgrading of a Fast Pyrolysis Bio-oil. Part II: Comparative Catalyst Performance and Reaction Pathways. *Fuel Processing Technology,* Vol.45, No.3, (December 1995), pp. 185-202, ISSN 0378-3820

Adjaye, J. D., Sharma, R. K. & Bakhshi, N. N. (1992). Characterization and Stability Analysis of Wood-derived Bio-oil. *Fuel Processing Technology,* Vol.31, No.3, (1992), pp. 241-256, ISSN 03783820

Alen, R. (1991). Thermochemical Conversion of Some Simple Lignin Model Compounds in the Liquid Phase. *Bioresource Technology,* Vol.35, No.1, (1991), pp. 103-106, ISSN 09608524

Andrews, R. G., Fuleki, D., Zukowski, S. & Patnaik, P. C. (1997). Results of Industrial Gas Turbine Tests Using a Biomass-derived Fuel, In: *Making a Business from Biomass in Energy, Environment, Chemicals, Fibres and Materials,* R. P. Overend and E. Chornet, (Ed.), pp. 425-435, Elsevier Sciences Inc., NewYork, USA

Andrews, R. G. & Patnaik, P. C. (1996). Feasibility of Utilising a Biomass derived Fuel for Industrial Gas Turbine Applications, In: *Bio-oil Production and Utilisation,* A. V. Bridgwater and E. N. Hogan, (Ed.), pp. 236-245, CPL Press, Newbury, UK

Appell, H. R., Fu, Y. C., Friedman, S., Yavorsky, P. M. & Wender, I. (1971). *ConVerting Organic Wastes to Oil.* Report of Investigation No. 7560, U.S. Bureau of Mines: Washington, D.C., 1971.

Baglioni, P., Chiaramonti, D., Bonini, M., Soldaini, I. & Tondi, G. (2001). Bio-Crude-Oil/Diesel oil Emulsification: Main Achievements of the Emulsification Process and Preliminary Results of Tests on Diesel Engine, In: *Progress in Thermochemical Biomass Conversion,* A. V. Bridgwater, (Ed.), pp. 1525-1539, Blackwell Science, Oxford, UK

Baker, E. G. & Elliott, D. C. (1988). Catalytic Hydrotreating of Biomass-derived Oils, In: *Pyrolysis Oils from Biomass,* (Ed.), pp. 228-240, American Chemical Society, ISBN 0-8412-1536-7,

Blin, J., Volle, G., Girard, P., Bridgwater, T. & Meier, D. (2007). Biodegradability of Biomass Pyrolysis Oils: Comparison to Conventional Petroleum Fuels and Alternatives Fuels in Current Use. *Fuel,* Vol.86, No.17-18, (December 2007), pp. 2679-2686, ISSN 0016-2361

Brammer, J. G. & Bridgwater, A. V. (1999). Drying Technologies for an Integrated Gasification Bio-energy Plant. *Renewable and Sustainable Energy Reviews,* Vol.3, No.4, (December 1999), pp. 243-289, ISSN 1364-0321

Bridgwater, A. V., Czernik, S., Diebold, J., Meier, D., Oasmaa, A., Peacocke, C., Piskorz, J. & Radlein, D. (1999a). *Fast Pyrolysis of Biomass: A Handbook.* CPL Scientific Publishing Services Limited, ISBN 1-872691-07-2, Newbury, UK

Bridgwater, A. V., Meier, D. & Radlein, D. (1999b). An Overview of Fast Pyrolysis of Biomass. *Organic Geochemistry,* Vol.30, No.12, (December 1999), pp. 1479-1493, ISSN 0146-6380

Bridgwater, A. V. & Peacocke, G. V. C. (2000). Fast Pyrolysis Processes for Biomass. *Renewable and Sustainable Energy Reviews,* Vol.4, No.1, (March 2000), pp. 1-73, ISSN 1364-0321

Bridgwater, A. V., Toft, A. J. & Brammer, J. G. (2002). A Techno-economic Comparison of Power Production by Biomass Fast Pyrolysis with Gasification and Combustion. *Renewable and Sustainable Energy Reviews,* Vol.6, No.3, (September 2002), pp. 181-246, ISSN 1364-0321

Chiaramonti, D., Bonini, M., Fratini, E., Tondi, G., Gartner, K., Bridgwater, A. V., Grimm, H. P., Soldaini, I., Webster, A. & Baglioni, P. (2003). Development of Emulsions from Biomass Pyrolysis Liquid and Diesel and Their Use in Engines--Part 1 : Emulsion Production. *Biomass and Bioenergy,* Vol.25, No.1, (July 2003), pp. 85-99, ISSN 0961-9534

Chiaramonti, D., Oasmaa, A. & Solantausta, Y. (2007). Power Generation Using Fast Pyrolysis Liquids from Biomass. *Renewable and Sustainable Energy Reviews,* Vol.11, No.6, (August 2007), pp. 1056-1086, ISSN 1364-0321

Czernik, S. & Bridgwater, A. V. (2004). Overview of Applications of Biomass Fast Pyrolysis Oil. *Energy & Fuels,* Vol.18, No.2, (February 2004), pp. 590-598, ISSN 0887-0624

Czernik, S., Evans, R. & French, R. (2007). Hydrogen from Biomass-production by Steam Reforming of Biomass Pyrolysis Oil. *Catalysis Today,* Vol.129, No.3-4, (December 2007), pp. 265-268, ISSN 0920-5861

Darmstadt, H., Garcia-Perez, M., Adnot, A., Chaala, A., Kretschmer, D. & Roy, C. (2004). Corrosion of Metals by Bio-Oil Obtained by Vacuum Pyrolysis of Softwood Bark Residues. An X-ray Photoelectron Spectroscopy and Auger Electron Spectroscopy Study. *Energy & Fuels,* Vol.18, No.5, (July 2004), pp. 1291-1301, ISSN 0887-0624

Demirbas, A. (2000a). Mechanisms of Liquefaction and Pyrolysis Reactions of Biomass. *Energy Conversion and Management,* Vol.41, No.6, (April 2000), pp. 633-646, ISSN 0196-8904

Demirbas, A. (2000b). Effect of Lignin Content on a Queous Liquefaction Products of Biomass. *Energy Conversion and Management,* Vol.41, No.15, (October 2000), pp. 1601-1607, ISSN 0196-8904

Demirbas, A. (2004). Combustion Characteristics of Different Biomass Fuels. *Progress in Energy and Combustion Science,* Vol.30, No.2, (2004), pp. 219-230, ISSN 0360-1285

Demirbas, A. (2005). Potential Applications of Renewable Energy Sources, Biomass Combustion Problems in Boiler Power Systems and Combustion Related Environmental Issues. *Progress in Energy and Combustion Science,* Vol.31, No.2, (2005), pp. 171-192, ISSN 0360-1285

Demirbas, A. (2007). *Biodiesel: A Realistic Fuel Alternative for Diesel Engines.* Springer-Verlag London Limited, ISBN 978-1-84628-994-1, London, UK

Demirbas, A. (2009). *Biofuels: Securing the Planet's Future Energy Needs.* Springer-Verlag London Limited, ISBN 978-1-84882-010-4, Lond, UK

Duan, P. & Savage, P. E. (2010). Hydrothermal Liquefaction of a Microalga with Heterogeneous Catalysts. *Industrial & Engineering Chemistry Research,* Vol.50, No.1, (August 2010), pp. 52-61, ISSN 0888-5885

Eager, R. L., Pepper, J. M., Roy, J. C. & Mathews, J. F. (1983). Chemical Studies on Oils derived from Aspen Poplar Wood, Cellulose, and an Isolated Aspen Poplar Lignin. *Canadian Journal of Chemistry,* Vol.61, No.9, (1983), pp. 2010-2015, ISSN 0008-4042

Elliott, D. C. (2007). Historical Developments in Hydroprocessing Bio-oils. *Energy & Fuels,* Vol.21, No.3, (May 2007), pp. 1792-1815, ISSN 0887-0624

Elliott, D. C., Hart, T. R., Neuenschwander, G. G., Rotness, L. J. & Zacher, A. H. (2009). Catalytic Hydroprocessing of Biomass Fast Pyrolysis Bio-oil to Produce Hydrocarbon Products. *Environmental Progress & Sustainable Energy,* Vol.28, No.3, (October 2009), pp. 441-449, ISSN 1944-7450

Fu, Y., Hu, X., Xu, Y., Zhu, X. & Jiang, S. (2009). Characterization of Biodegradation of Straw-based Biomass-oil in Aqueous Culture Conditions. *Industrial Lubrication and Tribology,* Vol.61, No.5, (2009), pp. 277 - 280, ISSN 0036-8792

Furimsky, E. (2000). Catalytic Hydrodeoxygenation. *Applied Catalysis A: General,* Vol.199, No.2, (June 2000), pp. 147-190, ISSN 0926-860X

Gandarias, I., Barrio, V. L., Requies, J., Arias, P. L., Cambra, J. F. & Guemez, M. B. (2008). From Biomass to Fuels: Hydrotreating of Oxygenated Compounds. *International Journal of Hydrogen Energy,* Vol.33, No.13, (July 2008), pp. 3485-3488, ISSN 0360-3199

Garcia-Perez, M., Chaala, A., Pakdel, H., Kretschmer, D. & Roy, C. (2007). Vacuum Pyrolysis of Softwood and Hardwood Biomass: Comparison between Product Yields and Bio-oil Properties. *Journal of Analytical and Applied Pyrolysis,* Vol.78, No.1, (January 2007), pp. 104-116, ISSN 0165-2370

Garcia, L., French, R., Czernik, S. & Chornet, E. (2000). Catalytic Steam Reforming of Bio-oils for the Production of Hydrogen: Effects of Catalyst Composition. *Applied Catalysis A: General,* Vol.201, No.2, (July 2000), pp. 225-239, ISSN 0926-860X

Guo, X., Yan, Y., Li, T., Ren, Z. & Yuan, C. (2003). Catalytic Cracking of Bio-oil from Biomass Pyrolysis. *The Chinese Journal of Process Engineering,* Vol.3, No.1, (February 2003), pp. 91-95, ISSN 1009-606X

Gust, S. (1997). Combustion Experiences of Flash Pyrolysis Fuel in Intermediate Size Boilers, In: *Developments in Thermochemical Biomass Conversion,* A. V. Bridgwater, (Ed.), pp. 481-488, Blackie Academic & Professional, ISBN 0-7514-0350-4, London, UK

Hsu, C. C. & Hixson, A. N. (1981). C1 to C4 Oxygenated Compounds by Promoted Pyrolysis of Cellulose. *Industrial & Engineering Chemistry Product Research and Development,* Vol.20, No.1, (March 1981), pp. 109-114, ISSN 0196-4321

Hu, X., Ding, H., Xu, Y. & Zhu, X. (2008a). Isolation and Identification of a Strain for the Degradation of Biomass-Oil and Its Degradation Properties in Soil. *Research of Environmental Sciences,* Vol.21, No.6, pp. 182-186, ISSN 1001-6929

Hu, X., Li, C., Xu, Y., Wang, Q. & Zhu, X. (2011a). On the Thermal Oxidation Stability of Pyrolysis Biomass Oil. *International Journal of Renewable Energy Technology,* Vol.2, No.2, (2011), pp. 155-168, ISSN 1757-3971

Hu, X., Xu, Y., Wang, Q., Jiang, S. & Zhu, X. (2008b). Tribological Performance of Distilled Biomass Oil from Rice Straw by Pyrolysis Process. *Journal of Synthetic Lubrication,* Vol.25, No.3, (September 2008), pp. 95-104, ISSN 1557-6841

Hu, X., Zhou, L., Wang, Q., Xu, Y. & Zhu, X. (2011b). On the Corrosion Behavior of Metal in Biomass-oil. *Corrosion Engineering Science and Technology,* (2011), DOI: 10.1179/147842209X12464471864619, ISSN 1478-422X

Ikura, M., Stanciulescu, M. & Hogan, E. (2003). Emulsification of Pyrolysis derived Bio-oil in Diesel Fuel. *Biomass and Bioenergy,* Vol.24, No.3, (March 2003), pp. 221-232, ISSN 0961-9534

Kasper, J. M., Jasas, G. B. & Trauth, R. L. (1983). Use of Pyrolysis-derived Fuel in a Gas Turbine Engine, *Proceedings of 28th ASME International Conference Gas Turbine,* pp. 8, Phoenix, AZ, USA, Mar 27, 1983

Leech, J. (1997). Running a Dual Fuel Engine on Pyrolysis Oil, In: *Biomass Gasification and Pyrolysis, State of the Art and Future Prospects,* M. Kaltschmitt and A. V. Bridgwater, (Ed.), pp. 495-497, CPL Press, Newbury, UK

Lu, Q., Li, W. & Zhu, X. (2009). Overview of Fuel Properties of Biomass Fast Pyrolysis Oils. *Energy Conversion and Management,* Vol.50, No.5, (May 2009), pp. 1376-1383, ISSN 0196-8904

Maschio, G., Koufopanos, C. & Lucchesi, A. (1992). Pyrolysis, a Promising Route for Biomass Utilization. *Bioresource Technology,* Vol.42, No.3, (1992), pp. 219-231, ISSN 09608524

Milne, T. A., Brennan, A. H. & Glenn, B. H. (1990). *Source Book of Methods of Analysis for Biomass and Biomass ConVersion Processes.* Elsevier Applied Science Publishers Ltd, ISBN 1-85166-527-7, London, England

Minowa, T., Yokoyama, S.-y., Kishimoto, M. & Okakura, T. (1995). Oil Production from Algal Cells of Dunaliella Tertiolecta by Direct Thermochemical Liquefaction. *Fuel,* Vol.74, No.12, (December 1995), pp. 1735-1738, ISSN 0016-2361

Oasmaa, A. & Czernik, S. (1999). Fuel Oil Quality of Biomass Pyrolysis Oils State of the Art for the End Users. *Energy & Fuels,* Vol.13, No.4, (April 1999), pp. 914-921, ISSN 0887-0624

Oasmaa, A., Leppamaki, E., Koponen, P., Levander, J. & Tapola, E. (1997). *Physical Characterisation of Biomass-based Pyrolysis Liquids: Application of Standard Fuel Oil Analyses.* VTT Publications 306, ISBN 951-38-5051-X, Espoo, Finland

Ormrod, D. & Webster, A. (2000). Progress in Utilization of Bio-oil in Diesel Engines. *PyNe Newsletter,* Vol.10, (2000), pp. 15, ISSN 1470–3521

Pelletier, E., Delille, D. & Delille, B. (2004). Crude Oil Bioremediation in Sub-Antarctic Intertidal Sediments: Chemistry and Toxicity of Oiled Residues. *Marine Environmental Research,* Vol.57, No.4, (May 2004), pp. 311-327, ISSN 0141-1136

Qi, G., Dong, P., Wang, H. & Tan, H. (2008). Study on Biomass Pyrolysis and Emulsions from Biomass Pyrolysis Oils and Diesel, *Proceedings of the 2nd International Conference on Bioinformatics and Biomedical Engineering,* pp. 4735-4737, ISBN 978-1-4244-1747-6, Shanghai, China, May 16-18, 2008

Qu, Y., Wei, X. & Zhong, C. (2003). Experimental Study on the Direct Liquefaction of Cunninghamia Lanceolata in Water. *Energy,* Vol.28, No.7, (June 2003), pp. 597-606, ISSN 0360-5442

Radlein, D. (2002). Study of Levoglucosan Production-A Review, In: *Fast pyrolysis of Biomass: A Hand Book,* (Ed.), CPL Press, Newbury, England

Ragauskas, A. J., Williams, C. K., Davison, B. H., Britovsek, G., Cairney, J., Eckert, C. A., Frederick, W. J., Hallett, J. P., Leak, D. J., Liotta, C. L., Mielenz, J. R., Murphy, R., Templer, R. & Tschaplinski, T. (2006). The Path Forward for Biofuels and Biomaterials. *Science,* Vol.311, No.5760, (January 2006), pp. 484-489, ISSN 0036-8075

Rioche, C., Kulkarni, S., Meunier, F. C., Breen, J. P. & Burch, R. (2005). Steam Reforming of Model Compounds and Fast Pyrolysis Bio-oil on Supported Noble Metal Catalysts. *Applied Catalysis B: Environmental,* Vol.61, No.1-2, (October 2005), pp. 130-139, ISSN 0926-3373

Senol, O. I., Viljava, T. R. & Krause, A. O. I. (2005). Hydrodeoxygenation of Methyl Esters on Sulphided NiMo/γ-Al2O3 and CoMo/γ-Al2O3 Catalysts. *Catalysis Today,* Vol.100, No.3-4, (2005), pp. 331-335, ISSN 0920-5861

Shaddix, C. R. & Hardesty, D. R. (1999). *Combustion Properties of Biomass Flash Pyrolysis Oils: Final Project Report,* Sandia National Laboratories, pp. 16-20, Albuquerque, NM; Livermore, CA

Shihadeh, A. & Hochgreb, S. (2000). Diesel Engine Combustion of Biomass Pyrolysis Oils. *Energy & Fuels,* Vol.14, No.2, (February 2000), pp. 260-274, ISSN 0887-0624

Sipila, K., Kuoppala, E., Fagernas, L. & Oasmaa, A. (1998). Characterization of Biomass-based Flash Pyrolysis Oils. *Biomass and Bioenergy,* Vol.14, No.2, (March 1998), pp. 103-113, ISSN 0961-9534

Solantausta, Y., Nylund, N.-O. & Gust, S. (1994). Use of Pyrolysis Oil in a Test Diesel Engine to Study the Feasibility of a Diesel Power Plant Concept. *Biomass and Bioenergy,* Vol.7, No.1-6, (1994), pp. 297-306, ISSN 0961-9534

Sun, P., Heng, M., Sun, S. & Chen, J. (2010). Direct Liquefaction of Paulownia in Hot Compressed Water: Influence of Catalysts. *Energy,* Vol.35, No.12, (December 2010), pp. 5421-5429, ISSN 0360-5442

Toft, A. J. (1996). Ph.D, thesis, Aston University, Birmingham, UK.

Wang, Q., Xu, Y., Hu, X. & Zhu, X. (2008). Experimental Study on Friction and Wear Characteristics of Bio-oil. *Transactions of the Chinese Society of Agricultural Engineering[J],* Vol.24, No.9, (September 2008), pp. 188-192, ISSN 1002-6819

Wildschut, J. (2009). Pyrolysis Oil Upgrading to Transportation Fuels by Catalytic Hydrotreatmen, thesis, University of Groningen,

Wildschut, J., Mahfud, F. H., Venderbosch, R. H. & Heeres, H. J. (2009). Hydrotreatment of Fast Pyrolysis Oil Using Heterogeneous Noble-Metal Catalysts. *Industrial & Engineering Chemistry Research,* Vol.48, No.23, (Deccember 2009), pp. 10324-10334, ISSN 0888-5885

Wornat, M. J., Porter, B. G. & Yang, N. Y. C. (1994). Single Droplet Combustion of Biomass Pyrolysis Oils. *Energy & Fuels,* Vol.8, No.5, (September 1994), pp. 1131-1142, ISSN 0887-0624

Xu, C. & Lad, N. (2007). Production of Heavy Oils with High Caloric Values by Direct Liquefaction of Woody Biomass in Sub/Near-critical Water. *Energy & Fuels,* Vol.22, No.1, (November 2007), pp. 635-642, ISSN 0887-0624

Xu, Y., Wang, Q., Hu, X. & Chen, J. (2007). Preliminary Study on Tribological Performance of Straw Based Bio-Fuel, *Proceedings of ASME/STLE 2007 International Joint Tribology Conference,* pp. 81-83, ISBN 0-7918-4810-8, San Diego, California, USA, October 22-24, 2007

Xu, Y., Wang, Q., Hu, X., Li, C. & Zhu, X. (2010). Characterization of the Lubricity of Bio-oil/diesel Fuel Blends by High Frequency Reciprocating Test Rig. *Energy,* Vol.35, No.1, (January 2010), pp. 283-287, ISSN 0360-5442

Xu, Y., Wang, Q., Hu, X. & Zhu, X. (2009). Preparation and Tribological Performance of Micro-emulsified Bio-oil. *Acta Petrolei Sinica(Petroleum Processing Section),* Vol.25, No.z1, (September 2009), pp. 53-56, ISSN 1001-8719

Yaman, S. (2004). Pyrolysis of Biomass to Produce Fuels and Chemical Feedstocks. *Energy Conversion and Management,* Vol.45, No.5, (March 2004), pp. 651-671, ISSN 0196-8904

Yan, Y., Xu, J., Li, T. & Ren, Z. (1999). Liquefaction of Sawdust for Liquid Fuel. *Fuel Processing Technology,* Vol.60, No.2, (July 1999), pp. 135-143, ISSN 0378-3820

Yao, Y., Wang, S., Luo, Z. & Cen, K. (2008). Experimental Research on Catalytic Hydrogenation of Light Fraction of Bio-oil. *Journal of Engineering Thermophysics,* Vol.29, No.4, (Aprial 2008), pp. 715-719, ISSN 0253-231X

Yi, W., Bai, X., He, F. & Yao, F. (2000). Biomass Liqucfaction in a High-Temperature Plasma Jet Flow. *Journal of Shandong Institute of Technology,* Vol.14, No.1, pp. 9-12, ISSN 1672-0040

Zabaniotou, A. A. & Karabelas, A. J. (1999). The Evritania (Greece) Demonstration Plant of Biomass Pyrolysis. *Biomass and Bioenergy,* Vol.16, No.6, (June 1999), pp. 431-445, ISSN 0961-9534

Zeng, F., Liu, W., Jiang, H., Yu, H.-Q., Zeng, R. J. & Guo, Q. (2011). Separation of Phthalate Esters from Bio-oil derived from Rice Husk by a Basification-acidification Process and Column Chromatography. *Bioresource Technology,* Vol.102, No.2, (January 2011), pp. 1982-1987, ISSN 0960-8524

Zhang, Q., Chang, J., Wang & Xu, Y. (2006). Upgrading Bio-oil over Different Solid Catalysts. *Energy & Fuels,* Vol.20, No.6, (October 2006), pp. 2717-2720, ISSN 0887-0624

Zhang, Q., Chang, J., Wang, T. & Xu, Y. (2007). Review of Biomass Pyrolysis Oil Properties and Upgrading Research. *Energy Conversion and Management,* Vol.48, No.1, (January 2007), pp. 87-92, ISSN 0196-8904

Zhang, S., Yan, Y., Li, T. & Ren, Z. (2005). Upgrading of Liquid Fuel from the Pyrolysis of Biomass. *Bioresource Technology,* Vol.96, No.5, (March 2005), pp. 545-550, ISSN 0960-8524

Zheng, J. (2007). Bio-oil from Fast Pyrolysis of Rice Husk: Yields and Related Properties and Improvement of the Pyrolysis System. *Journal of Analytical and Applied Pyrolysis,* Vol.80, No.1, (August 2007), pp. 30-35, ISSN 0165-2370

Zheng, J., Zhu, X., Guo, Q. & Zhu, Q. (2006). Thermal Conversion of Rice Husks and Sawdust to Liquid Fuel. *Waste Management,* Vol.26, No.12, (2006), pp. 1430-1435, ISSN 0956-053X

Zhou, D., Zhang, L., Zhang, S., Fu, H. & Chen, J. (2010). Hydrothermal Liquefaction of Macroalgae Enteromorpha Prolifera to Bio-oil. *Energy & Fuels,* Vol.24, No.7, (June 2010), pp. 4054-4061, ISSN 0887-0624

11

Combined Microwave - Acid Pretreatment of the Biomass

Adina-Elena Segneanu, Corina Amalia Macarie,
Raluca Oana Pop and Ionel Balcu
National Institute of Research and Development for Electrochemistry and Condensed Matter,Timisoara Romania

1. Introduction

Bioethanol represents an important alternative for the fossil fuels. The limited fossil fuel stock, the growth of the energy necessary all over the world and the environmental safety lead to an increasing interest in alternative fuels [Balat et al., 2008]. One of the most important renewable energy sources is the lignocellulosic biomass, including wood and crop residues, and that may have applications in the energetic field (both thermal energy and biofuels). There are four main steps in the conversion process of lignocellulosic biomass to ethanol: pretreatment, enzymatic hydrolysis, fermentation and separation [Petersen et al., 2009]. One of the key factors that influence the obtaining of bioethanol is the pretreatment stage. Biomass composition consists in 70-85% cellulosic materials (cellulose and hemicelluloses) and 15-30% lignins. For a corresponding capitalization of biomass, the removal of the lignin content and the transformation of cellulose and its derivatives in sugars are required.

Pretreatment of the lignocellulosic biomass is an important preliminary step that is performed in order to improve the yield of the hydrolysis reaction of cellulosic derivatives in fermentable sugars. The goal of the pretreatment stage consists in changes that are made in the lignocellulosic materials structure, in order to facilitate the access of enzymes in the hydrolysis reaction (Soccol, 2010). A corresponding pretreatment stage must fulfill the following conditions (Balat et al., 2008; Del Campo, 2006; Balat, 2010):

- to improve the sugar formation or the capacity to subsequently obtain sugars by hydrolysis
- -to prevent degradation or the loss of carbohydrates
- to prevent the obtaining of possible inhibitory by-products in the hydrolysis and fermentation stages
- costs efficiency
- to avoid the destroy of cellulose and hemicelluloses
- the use of a minimum amount of chemical products

The above-mentioned characteristics represent the basis for the comparisons among various pretreatment methods that are used in the bioethanol industry. A number of different methodologies have been developed in order to accomplish the first stage of the lignocellulosic biomass to ethanol, namely the pretreatment of the biomass.

2. Pretreatment methods of the lignocellulosic biomass

2.1 Acid pretreatment

The main objective of the acid pretreatment is the solubilization of the hemicellulosic fraction of the biomass, in order to increase the accessibility of the enzymes in the enzymatic hydrolysis reaction (Alvira et al., 2010). Inorganic acids like H_2SO_4, HCl and H_3PO_4 have been used for the pretreatment of the lignocellulosic biomass, in order to improve the enzymatic hydrolysis. There may be used both concentrated and diluted inorganic acids. Pretreatment of the biomass with concentrated acids, at ambient temperature, will lead to higher yields of fermentable sugars and to the hydrolysis of both cellulose and hemicelluloses. There are frequently used acids like H_2SO_4 72%, HCl 41% and trifluoroacetic acid 100%. In this case, a necessary step is the recovery of the acid, in order to lower the economic costs of the process (Girio et al, 2010). The method has the advantage not to use enzymes for saccharification in the further stage, but there are also a number of drawbacks: energy consumption, the use of equipment that is resistant to corrosion, a longer reaction time and the necessary operation of acid recovery (Talebnia et al., 2010).

Pretreatment with diluted acids presents many advantages for an industrial use and it may be applied to different types of biomass. The pretreatment stage may occur at higher temperatures (180°C) for a shorter time, or at lower temperatures (120°C) and a longer residence time. Pretreatment with dilute acid shows the advantage of hemicellulose solubilization, but also of the conversion of the solubilized hemicellulose in fermentable sugars. Pretreatment with diluted acids leads to the obtaining of a fewer degradation products than the preteatment with concentrated acids (Alvira et al., 2010). The highest yields of the hydrolysis reaction have been recorded after treating the lignocellulosic material with dilute sulfuric acid. Usually, sulfuric acid concentrations are in the range 0.5-1.5%, and the working temperatures are 120-160°C (Alvira et al., 2010).

Organic acids (fumaric acid, maleic acid) appear as alternatives for the improving of the hydrolysis yield of the lignocellulosic biomass. Maleic acid is proven to be more efficient than the fumaric acid, and has the advantage to lead to the obtaining of lower amounts of furfural (compared to the dilute sulfuric acid) (Kootstra et al., 2009). Another pretreatment method with dilute acids uses H_2CO_3 (obtained through the absorption of CO_2 in aqueous solutions) (van Walsum and Shi, 2004).

2.2 Alkaline pretreatment

Pretreatment with alkaline solutions increases the digestibility of cellulose and favors the solubilization of lignins (Alvira, 2010). It may occur at room temperature and reaction time may vary from seconds to days. It leads to a smaller degradation of sugars than in the case of acid pretreatment, but is proven to be more efficient for crop residues than for lignocellulosic biomass (Kumar and Wyman, 2009). For the optimization of the pretreatment conditions, the possibility of losing the fermentable sugars and the formation of some inhibitory compounds must be taken into account.

Reagents that are frequently used for the alkaline pretreatment are NaOH, KOH, $Ca(OH)_2$, $(NH_4)_2OH$. Amond them, most widely used is NaOH. For example, pretreatment with NaOH solutions leads to swelling and the increasing of internal surface of cellulose (Alvira et al., 2010). The same authors mentioned that pretreatment with NaOH of hardwood increases the digestibility by the decreasing to 20% of the lignin content.

Although alkaline pretreatments show great efficiency as regards the lignin solubilization, they are less efficient concerning the solubilization of cellulose and hemicelluloses (Girio, 2010).
A widely spread alkaline pretreatment method of the biomass is represented by the ARP (Ammonia Recycle Percolation) procedure (Wu and Lee, 1997). It consists in the use of aqueous ammonia at temperatures around 170°C (Kim and Lee, 2005). The solubilization of hemicelluloses in an oligomeric form occurs within 40-60% range (Girio et al., 2010). The cellulosic fraction is hardly degraded, but in the following steps of the hydrolysis the yields are closed to the theoretic ones (Kim and Lee 2005, Kim et al. 2008). The mechanism of the reaction with aqueous ammonia is very similar to the pretreatment with $Ca(OH)_2$ and NaOH, especially as regarding the swelling of biomass and the breakdown of the ester and ether bonds of the carbohydrates that exist in lignin (Girio et al, 2010). The advantages of the use of NH_3 are: swelling of the lignocellulosic material, a selective reaction for the removal of lignin, low interaction with carbohydrates, high volatility. One of the known reactions of aqueous NH_3 with lignin is represented by the breakdown of the C-O-C bonds from lignin, as well as of the etheric and esteric bonds from the complex lignin-carbohydrates (Stavrinides et al., 2010). As a result of ARP pretreatment, 60-85% from the entire lignin content is removed (Kim et al., 2008).
Another procedure that uses ammonia for the pretreatment of biomass is the AFEX (Ammonia Fiber Explosion) process. It consists in the contact of biomass with liquid ammonia at elevated temperatures and under pressure for a certain time, followed by a fast decompression (Zheng et al., 2009). The method proved to be less efficient in the case of hardwood and softwood residues (Zheng et al., 2009)

2.3 Organosolv pretreatment

In the Organosolv process, there are used a number of organic or aqueous solvents (methanol, ethanol, acetone, ethylene glycol) in order to solubilize the lignin and to obtain a corresponding treated cellulose for the hydrolysis process (Chum et al., 1988). The advantage of the Organosolv procedure consists in the recovery of lignin as secondary product. The maximum working temperature is 205°C, regarding the used solvent. The economicity of the process depends on the recovery of the organic solvent (Zhao et al., 2009).
The main advantages of the Organosolv pretreatment are: organic solvents can be easily removed by distillation and they can be reused; lignin may be isolated as solid materials (solids) and the carbohydrates are isolated as syrup (Zhao et al., 2009; Kim et al., 2008).
Disadvantages: the pretreated solids need to be initially washed with organic solvents in order to prevent reprecipitation of the dissolved lignin. Also, the process must be strictly controlled, due to the volatility of the organic solvents (Zhao et al., 2009).
Regarding the economy of the process, recovery of the solvents is necessary, even though high amounts of energy are needed. Due to these considerations, Organosolv pretreatment has no applications at industrial level.
The Organosolv pretreatment undergoes both in the presence or absence of a catalyst, at temperatures in the range 185-210°C. The yields od delignification process are improved if mineral acids like HCl, H_2SO_4 or H_3PO_4 or organic acids like formic, oxalic or acetylsalicylic acid (Sun and Cheng, 2002) are used. After pretreatment with Organosolv, three fractions are obtained: dry lignin, an aqueous hemicellulosic phase and a cellulosic fraction (Duff and Murray, 1996).
The most frequently used is the Organosolv pretreatment with aliphatic alcohols (especially methanol and ethanol), mostly due to their low price. Among the alcohols with higher

boiling points, mostly used are polyhydroxylic alcohols like ethylene glycol and glycerol. The main advantage is the fact that the process could occur at atmospheric pressure. Pretreatment with aqueous glycerol leads to the removal of the lignin, but also to a significant loss of cellulose (Kucuk, 2005).

2.4 Pretreatment with solid superacids

The solid acid catalysts appeared as a consequence of the developing of a new, eco-friendly process for the obtaining of bioethanol. Particles of solid acid can be separated by the liquid products through decantation or filtration, and the catalyst may be reused without further processing stages to be necessary.

Solid superacids are made from a solid medium treated with Lewis or Bronsted acids (Zhao et al., 2009). They have the great advantage of being non-toxic, non-corrosive and safe for the environment. They are better donors than pure sulfuric acid and show a higher selectivity in the hydrolysis reaction and require low temperatures and atmospheric pressure (Zhao et al.,2009; Yamaguchi and Hara). Some of the superacids used in the process of the obtaining of bioethanol are: niobic acid (Nb_2O_5-nH_2O), zeolite, Amberlyst-15, amorphous C that contains SO_3H, COOH and OH groups (Zhao et al., 2009).

Another superacid used for the selective conversion of cellulose to glucose is the heteropoly acid $H_3PW_{12}O_{40}$ (Tian et al., 2010). The selectivity of the pretreatment method is very high (around 90%) and requires mild reaction conditions (160-180°C). Another advantage of this method is the possibility to reuse the catalyst, which can be recycled by extraction with diethyl ether (Tian et al., 2010).

2.5 Ionic liquids

The main advantage of using ionic liquids for the bioethanol production is represented by the possibility of a complete solubilization of the lignocellulosic biomass. Swatloski et al. suggested that solubilization is due to the breakdown of the H bonds of the polysaccharides by the anion of the ionic liquids. In the present, the process cannot be applied at industrial level due to the high costs of the ionic liquids (Swatloski et al., 2002).

A variant of the pretreatment with ionic liquids is represented by the microwave-assisted pretreatment of lignocellulosics in ionic liquids (Zhang and Zhao, 2009; Zhu et al., 2006). The method is characterized by shorter reaction time (due to the microwave irradiation) and a better solubilization of the biomass. According to Zhu et al., the raw lignocellulosic material is directly solubilized in the ionic liquid in the presence of microwaves and cellulose is precipitated by adding water. The other organic compounds (like lignins) remain in solution. Experimental results (Zhu et al., 2006) showed that the yields in ethanol are very similar to the ones obtained through steam explosion or chemical pretreatment.

2.6 Hydrothermal methods of pretreatment

The hydrothermal reactions for the pretreatment of biomass are new, eco-friendly pretreatment methods. They consist in the contact of the lignocellulosic materials with water at elevated temperature and pressure. During the process, hemicelluloses are hydrolyzed to sugars. The reaction time is very short (seconds) in order to avoid degradation of the sugars (http://www.ecn.nl/units/bkm/biomass-and-coal/transportation-fuels-and-chemicals/transportation-fuels/biomass-pre-treatment-fractionation/).

A variant of the hydrothermal pretreatment consists in the use of catalytic hydrothermal reaction that uses a solid catalyst (for example, amorphous carbon that contains -SO_3H groups) and results in higher amounts of fermentable sugars (Onda et al., 2009).

2.7 Ozonolysis

Pretreatment with ozone occurs in mild conditions (room temperature, atmospheric pressure) and results in a strong delignification of the biomass (Sun and Cheng, 2002). The major drawback of the process is represented by the high costs, due to the large quantity of ozone that is needed during the pretreatment (Sun and Cheng, 2002).

2.8 Combined methods of pretreatment

2.8.1 Pretreatment with alkaline peroxides, followed by steam explosion

The procedure combines the advantages of alkaline pretreatment and steam explosion. It will lead to an efficient delignification and to the chemical swelling of the lignocellulosics fibers (Zhao et al., 2009). Use of a combined process (steam explosion and NaOH 10%) led to a significant increase of the free sugars concentration towards the pretreatment with H_2O_2 1% and NaOH 1% (Chen and Qiu, 2010).

2.8.2 Pretreatment with ionic liquids coupled with steam explosion

Pretreatment of the lignocellulosics biomass with ionic liquids coupled with steam explosion led to the degradation of hemicelluloses in fermentable sugars (Chen and Qiu, 2010). Lignin with high molecular mass is insoluble in ionic liquids, so it can be separated from cellulose.

2.8.3 Biological pretreatment

For the biological pretreatment of the lignocellulosic biomass there are used both microorganisms (fungi and bacteria) and enzymes (Mtui, 2009; Balat, 2011). There are used white, brown and soft-rot fungi for the solubilization of hemicelluloses and also for the lignin degradation (Mtui, 2009; Balat, 2011). For the enzymatic pretreatment of the biomass, different cellulases (endoglucanases, exoglucanases and β-glucosidases) are used (Sun and Cheng, 2002).

3. Studies regarding the determination of the optimum parameters of the microwave-assisted dilute acid pretreatment of lignocellulosic biomass

Lignocellulosics biomass has three main components: cellulose (40-50%), hemicelluloses (25-35%), lignin (15-20%) and also small amounts of proteins, lipids, acids, mineral salts. As it was mentioned before, the aim of the pretreatment stage is the removal of hemicelluloses and lignin. Also, the cellulose structure is altered in order to facilitate the enzymatic attack.

From all the pretreatment methods presented in the former chapter, pretreatment with dilute mineral acids (especially H_2SO_4) combined with microwave irradiation has been chosen. The advantages of this process are the reaction conditions (that does not involve corrosion problems, or volatility or very high temperatures issues) and the low economic costs. Also, the use of microwave irradiation leads to shorter reaction time and also provides a uniform heating of the reaction mixture.

Experimental part: three types of sawdust (hardwood (oak) and softwood (fir) essences and herbs (hemp)) were treated with dilute sulfuric acid (for different concentrations: 0.55, 0.82, 1.23 and 1.64%) and heated (in the presence of microwaves) at three different temperatures: 120, 140 and 160°C, for 15 and 30 minutes, in order to perform an extensive study on the pretreatment in acid medium. The concentration in sugars (expressed as free glucose) of the solutions obtained after the hydrolysis reactions was considered in order to establish the best pretreatment method.

After cooling, the suspension was neutralized with $CaCO_3$ until a pH value of 5.5-6, for the removal of sulfates. Pretreated sawdust were filtered and washed with water, in order to remove the entire amount of sugars.

Determination of the total amount of carbon hydrates after performing dilute acid pretreatment on different types of sawdust was made by the colorimetric method with 3,5-dinitrosalicylic acid. 5 milliliters from the solution obtained after pretreatment were treated with DNS 1%, boiled for 15 minutes on a water bath and then cooled. Extinction was measured (against blank) at 575 nm.

Results of the pretreatment method with dilute sulfuric acid (H_2SO_4 0.55%) at 120°C, for 15 and 30 minutes, are presented in the table below:

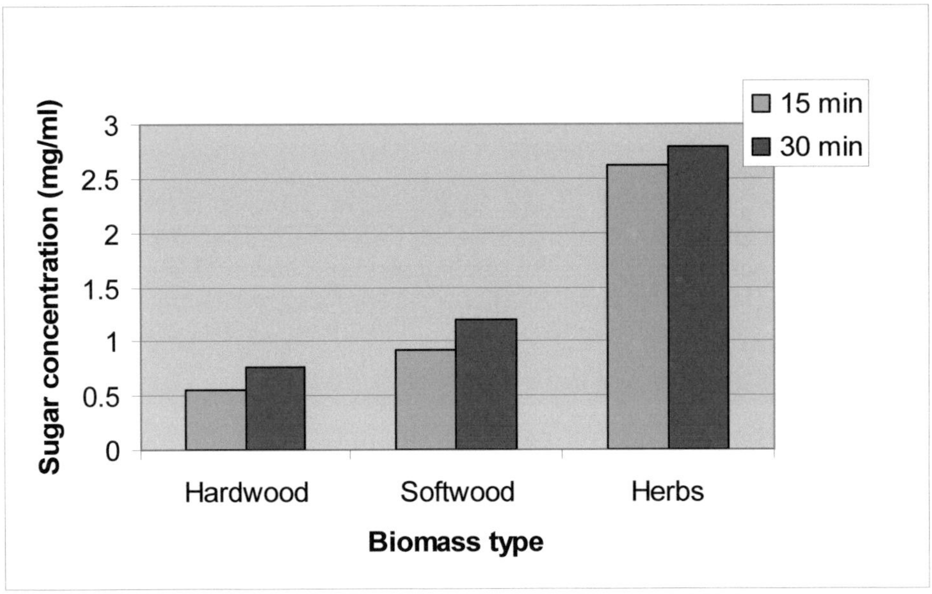

Fig. 1. Pretreatment of the biomass with H_2SO_4 0.55% at 120°C

As it may be seen, best results are obtained for the sawdust from herbaceous plants (in our case, hemp). The amount of sugars (expressed as free glucose) obtained after pretreatment is almost three times higher in the case of hemp sawdust than in the case of hardwood sawdust.

Pretreatment with the same acid solution (H_2SO_4 0.55%) at 140 and 160°C, respectively, led to the following results:

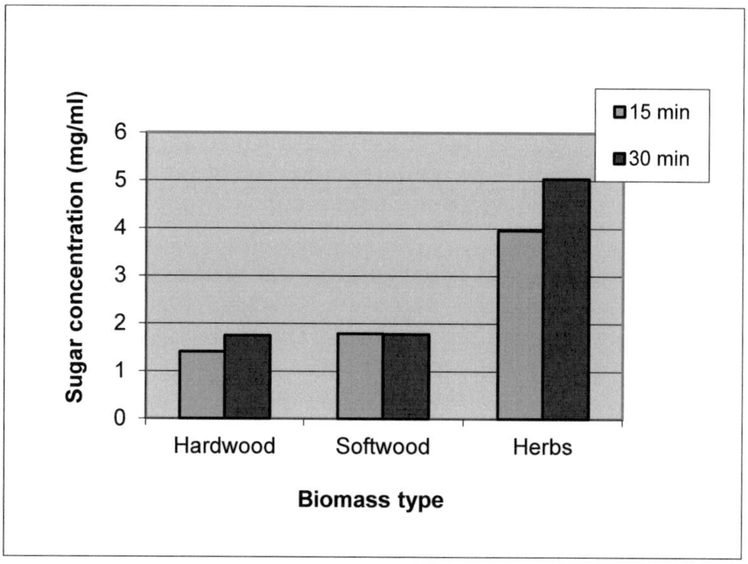

Fig. 2. Pretreatment of the biomass with H_2SO_4 0.55% at 140°C

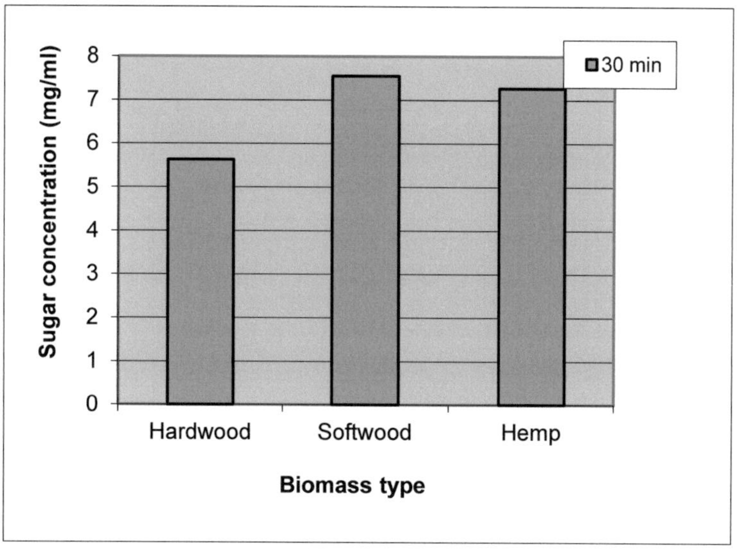

Fig. 3. Pretreatment of the biomass with H_2SO_4 0.55% at 160°C

In the case of the pretreatment with H_2SO_4 0.55% at 140°C, an increase of the reaction (pretreatment) time has significant consequences only in the case of hemp sawdust, when higher concentration of free sugars are obtained when the pretreatment time is 30 minutes instead of 15 minutes. For the hardwood (oak) and softwood (fir) sawdust, an increase of the pretreatment time does not lead to a significant improvement of the free sugars yield.

In the case of pretreatment with dilute acid at 160°C, our previous studies showed that there is no difference between the results of the pretreatment process at 15 or 30 min. Taking into account that in the other pretreatment methods best results have been obtained when the pretreatment lasted 30 minutes, the same period was chosen for the hydrolysis with H_2SO_4 0.55% at 160°C.

All the presented results show that, best results are obtained when pretreatment at 160°C is performed. The highest yields in free sugars are obtained for softwood and herbaceous sawdust, respectively, so it may be said that the softwood and herbaceous sawdust structure is more easily attacked than the hardwood sawdust structure during the acid hydrolysis.

The same pretreatment method with dilute sulfuric acid (0.82%) combined with microwave irradiation was used for the same types of sawdust (hardwood-oak, softwood-fir, herbaceous-hemp) at three different temperatures. The experiments were carried out in the same conditions as mentioned before, the only change being the different concentration of the acid. The aim of the study was to establish if an increase of the acid concentration leads to an increase of the amount of obtained sugars in the same temperatures conditions or, as a result, much of the already formed sugars will be degraded. The results are presented in the figures below:

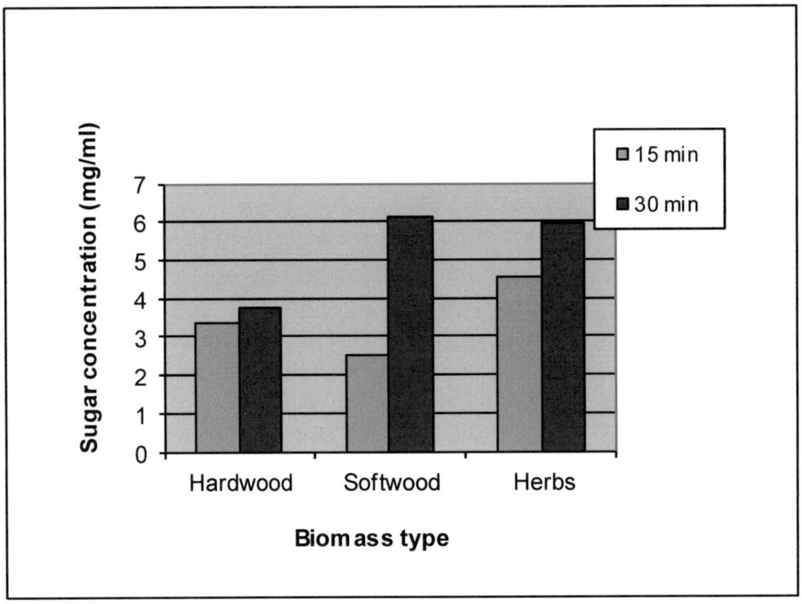

Fig. 4. Pretreatment of the biomass with H_2SO_4 0.82% at 120°C

According to these results, a slight concentrated solution of sulfuric acid has better results regarding the concentration in fermentable sugars of the solutions obtained after pretreatment. Good results are obtained especially for the fir sawdust, the level of sugars is almost 5 times higher when treated with H_2SO_4 0.82% at 120°C for 30 minutes than with H_2SO_4 0.55% for an identical time and temperature. Also the results of hardwood sawdust pretreatment are improved, the concentration of final solutions after pretreatment in free sugars is almost three times higher than in the case when H_2SO_4 0.55% was used. The results of the pretreatment are much poorer for the oak (hardwood) sawdust than for the fir (softwood) and herbaceous (hemp) sawdust.

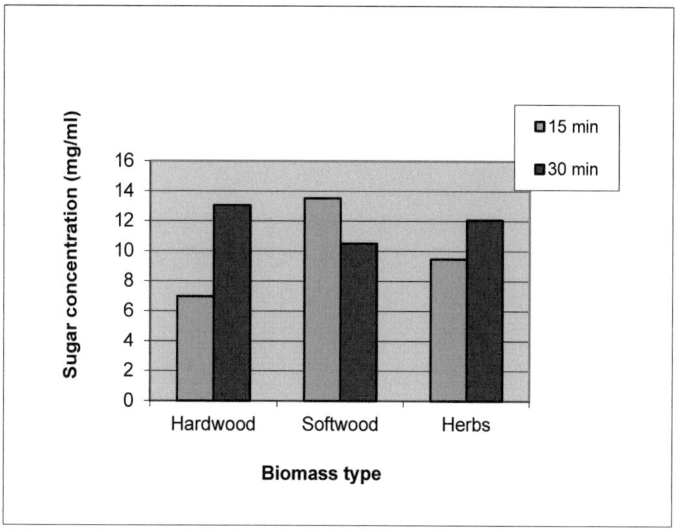

Fig. 5. Pretreatment of the biomass with H_2SO_4 0.82% at 140°C

Pretreatment with sulfuric acid 0.82% at 140°C led to the obtaining of very similar results for all the sawdust types used in the study. Except the softwood sawdust, when best results were obtained for a shorter reaction time (15 minutes), pretreatment with H_2SO_4 0.82% at 140°C for 30 minutes is more efficient than the similar one with H_2SO_4 0.55%.

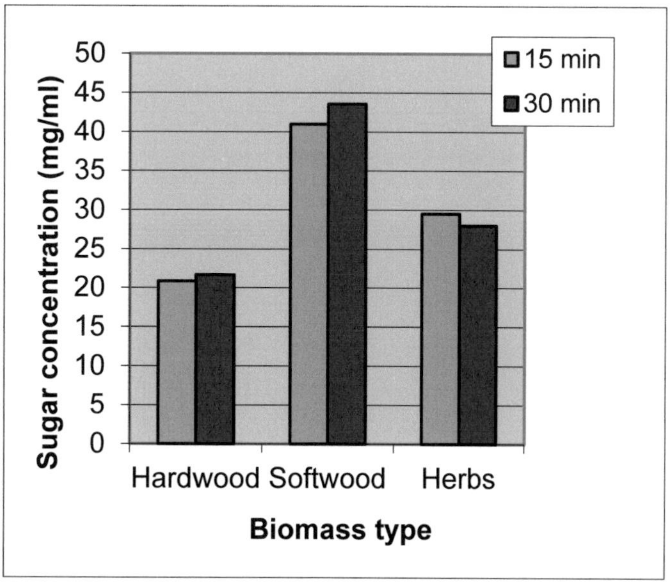

Fig. 6. Pretreatment of the biomass with H_2SO_4 0.82% at 160°C

When temperature is increased to 160°C, much higher concentrations of fermentable sugars are obtained. It may be observed that, at this temperature, there are almost no differences between the results of the 15 minutes and 30 minutes pretreatment. The pretreatment method shows its efficiency especially as regards the fir sawdust, followed by the hemp sawdust. As happened in all of the previous cases, poorer concentrations in fermentable sugars are obtained for the oak sawdust.

Same pretreatment method was used for the three types of sawdust, but in this case a solution of H_2SO_4 1.23% was used. The results are presented below in a graphic form:

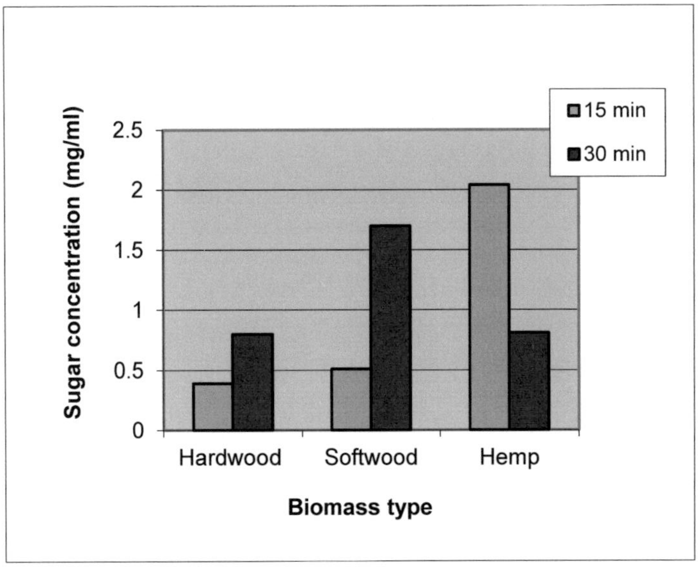

Fig. 7. Pretreatment of the biomass with H_2SO_4 1.23% at 120°C

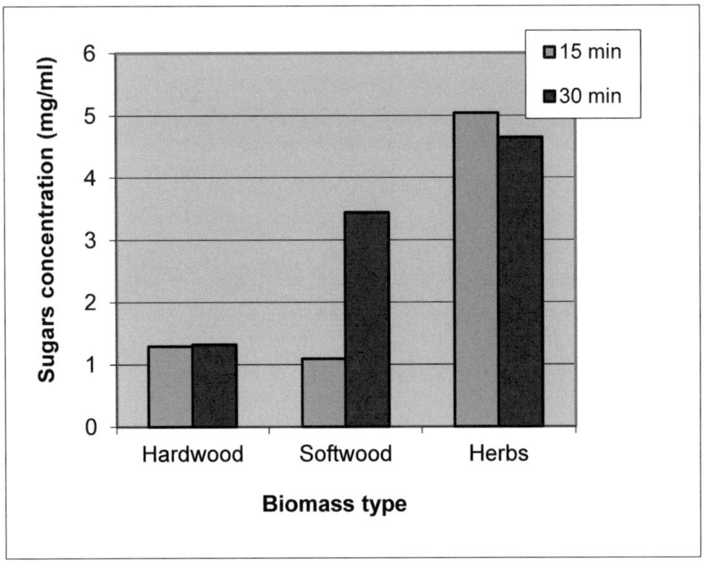

Fig. 8. Pretreatment of the biomass with H_2SO_4 1.23% at 140°C

It may be seen that the results of the pretreatment with a solution of sulfuric acid 1.23% in the same conditions of temperature and residence time result in much poorer results than in the above-mentioned case, when sulfuric acid 0.82% was used. A possible explanation consists in the fact that, at higher concentrations of acidic solution, the already formed sugars to be destroyed and degraded.
Taking into account the similarity of the results of the pretreatment with H_2SO_4 0.82% at 160°C for 15 and 30 minutes respectively, reaction of the sawdust with H_2SO_4 1.23% at 160°C was carried out only for 30 minutes. The results are presented below:

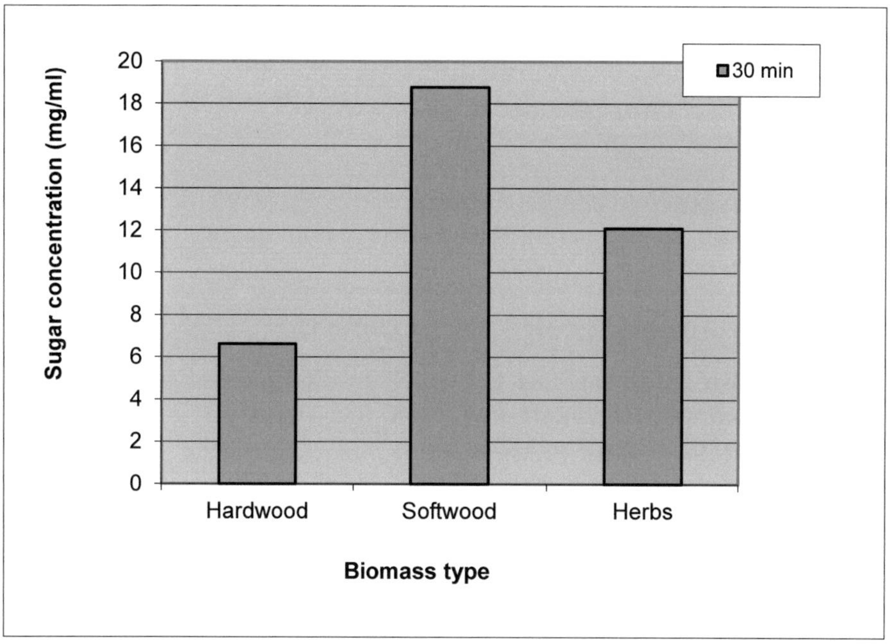

Fig. 9. Pretreatment of the biomass with H_2SO_4 1.23% at 160°C

Unlike the pretreatment with H_2SO_4 0.55%, it may be observed that in the case of herbaceous sawdust (hemp), an increased reaction time leads to smaller amounts of fermentable sugars. A stronger acid and a longer pretreatment time have better results only for the softwood (fir) sawdust, while as regarding the herbaceous sawdust it appears than a shorter reaction time leads to an increase yield in fermentable sugars. Data presented in Figures… show that the best results are obtained for the fir sawdust, and, as in the previous case (H_2SO_4 0.55%), the pretreatment method gives the poorer results for the hardwood sawdust. It appears that a prolonged acid pretreatment, with a slight acidic solution (than the concentrations of H_2SO_4 used before, namely 0.55% and 0.82%) is not benefic for the herbaceous sawdust, being possible that a great part of the already formed fermentable sugars to be simultaneously degraded during the pretreatment time.
In order to see if a more concentrated acid has a positive influence on the acid hydrolysis of the lingnocellulosic materials, a solution of H_2SO_4 1.64% was employed for the pretreatment of the three types of sawdust, at the same temperatures (120, 140 and 160°C) and 15 and 30 minutes reaction time, respectively. The results are the following:

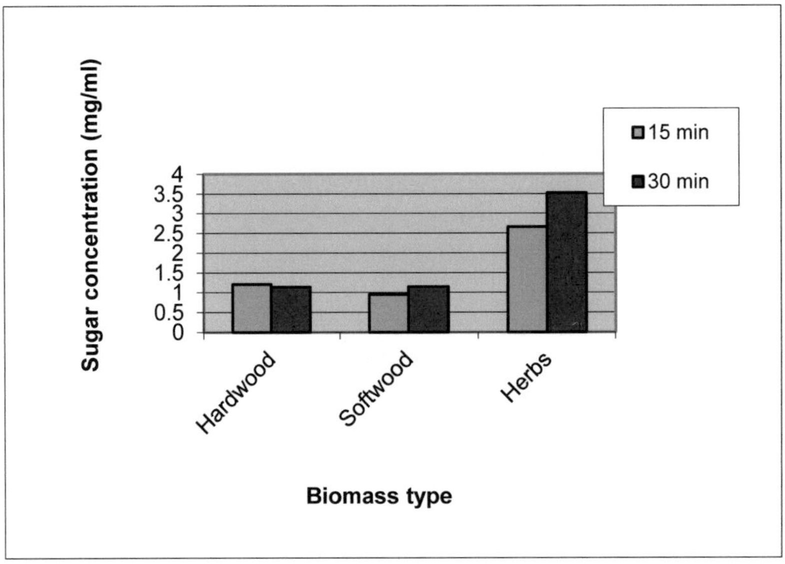

Fig. 10. Pretreatment of the biomass with H_2SO_4 1.64% at 120°C

The results show that hemp sawdust is favored by this pretreatment method, but the concentrations in fermentable sugars are lower than the ones obtained in the same conditions, but when H_2SO_4 0.82% was used.

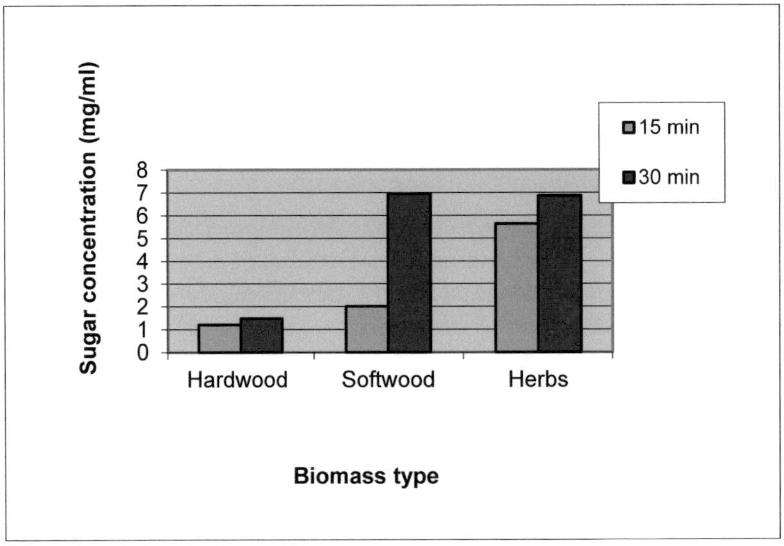

Fig. 11. Pretreatment of the biomass with H_2SO_4 1.64% at 140°C

An increase of the temperature leads to a higher concentrations in free sugars, but only for fir and hemp sawdust, respectively. Elevated residence time led to considerably improved results, especially as regarding the hemp sawdust.

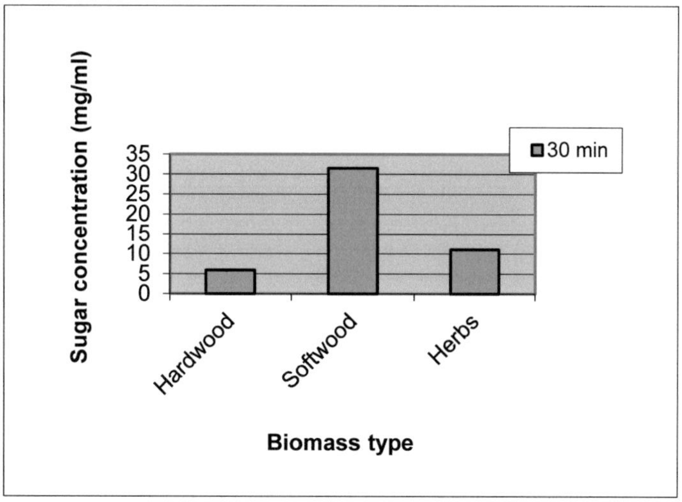

Fig. 12. Pretreatment of the biomass with H_2SO_4 1.64% at 160°C

The profile of the results is, somewhat, similar to the pretreatment with H_2SO_4 0.82% in the same conditions. It may be observed that, quantitatively, pretreatment at higher temperatures and longer time leads to better results. The amount of fermentable sugars increases with the acid concentration and with the residence time. Best results are obtained for the fir sawdust, when pretreated with H_2SO_4 1.64% at 160°C. Poorer results are obtained for the herbaceous sawdust (hemp) and hardwood sawdust, respectively. It appears that harsh conditions are required for a corresponding pretreatment in the case of fir sawdust (30 minutes residence time and 140 or 160°C).

Best results are obtained for the fir sawdust, when pretreated with H_2SO_4 0.82% at 160°C, with no significant difference due to the residence time (15 or 30 minutes).

As regarding the hemp sawdust, the best results are obtained when pretreatment with H_2SO_4 0.82% at 160°C for 15 minutes is employed. It can be said that a corresponding hydrolysis of the lignocellulosics from herbaceous sawdust requires less harsh conditions than the acid hydrolysis of softwood sawdust.

Concerning the hardwood sawdust, it may be said that pretreatment with dilute acids at temperatures in the range 120-160°C is not suitable. In all of the cases, only small amounts of free, fermentable sugars are obtained after the pretreatment. From all the pretreatment variant presented, it appears that the most suitable is the method that uses H_2SO_4 0.82% at 160°C for 15 minutes (the differences are very small between results of the 15 minutes and 30 minutes pretreatment, respectively.

It may be said that a corresponding microwave-assisted pretreatment of oak, fir and hemp sawdust is achieved by means of dilute sulfuric acid (0.82%) at 160°C, for 15 minutes.

In order to determine the pretreatment severity, the combined severity factor (CSF) that includes acid concentration, temperature and pretreatment time was used (Hsu et al., 2010).

$$CSF = \log\left\{t \cdot \exp\left[\left(T_H - T_R\right)14.75\right]\right\} - pH$$

Where: t - time (minutes), TH – temperature of the process, TR – reference temperature (100°C), pH – pH of the dilute sulfuric acid.

Pretreatment conditions	Acid concentration (%)	CSF
120°C, 15′	0.55	0.65
	0.82	0.80
	1.23	0.95
	1.64	1.10
120°C, 30′	0.55	0.95
	0.82	1.10
	1.23	1.25
	1.64	1.40
140°C, 15′	0.55	1.25
	0.82	1.40
	1.23	1.55
	1.64	1.65
140°C, 30′	0.55	1.55
	0.82	1.70
	1.23	1.85
	1.64	1.95
160°C, 30′	0.55	2.10
	0.82	2.30
	1.23	2.45
	1.64	2.55

Table 1. The combined severity factor (CSF) of the different variants of the microwave-assisted dilute acid hydrolysis process

4. A study concerning the possibility of using lyophilization as an efficient pretreatment method of the lignocellulosic residues

Experimental part: a suspension of sawdust and NaOH 1% and H_2SO_4 1% solution (1:10 w/v) was lyophilized at -52°C for 24 hours. The pretreated suspensions were filtered, washed with ultrapure water and the filtrate was neutralized with a solution of H_2SO_4 0.82% (the alkaline ones) and with $CaCO_3$ (the acid ones). The concentration in free, fermentable sugars was determined using the colorimetric method with 3,5-dinitrosalicylic acid.

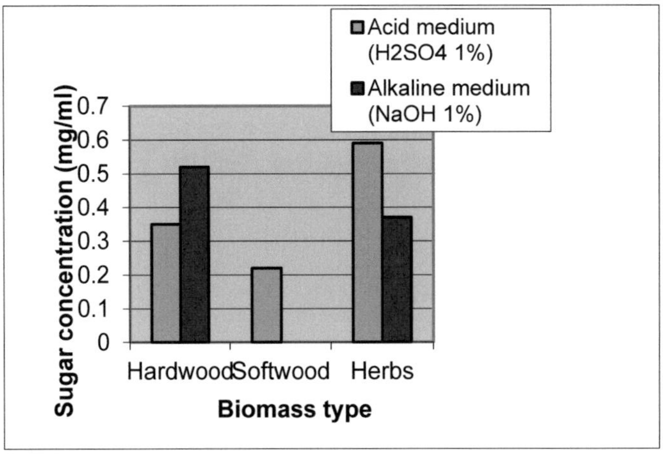

Fig. 13. Results of the alkaline and acid lyophilization pretreatment

The concentrations of free sugars are much poorer compared to the ones obtained after the combined pretreatment of microwave irradiation and dilute acid hydrolysis. No detectable concentrations of fermentable sugars were obtained for fir sawdust, when treated with an alkaline solution. A comparison between the two proposed methods is clearly in the favor of the microwave-assisted acid hydrolysis, which requires much less time and lower economic costs.

5. Conclusions

The results of the microwave-assisted acid pretreatment of the lignocellulosic biomass show that for good results in free sugars concentration there are not necessary elevated temperatures and high acid concentration. As results from the performed study, very efficient seems to be the pretreatment with sulfuric acid 0.82% at a temperature of 140°C, conditions that are characterized by a combined severity factor of 1.7. As regarding the possibility of using lyophilization in acid or alkaline medium, the obtained results are very poor and do not stand for the use of lyophilization as a viable pretreatment method.

6. References

Alvira, P., Tomas-Pejo, E., Ballesteros, M., Negro, M. J. (2010). Pretreatment technologies for an efficient bioethanol production process based on enzymatic hydrolysis: A review. *Bioresource Technology,* Vol. 101, pp. 4851-4861

Balat, M. (2011). Production of bioethanol from lignocellulosic materials via the biochemical pathway: A review. *Energy Conversion and Management,* Vol. 52, pp. 858-875

Balat, M., Balat, H., Oz, C. (2008). Progress in bioethanol processing. *Progress in Energy and Combustion Science,* Vol. 34, pp. 551-573

Chen, H., Qiu, W. (2010). Key technologies for bioethanol production from lignocellulose. *Biotechnology Advances,* Vol.28, pp. 556-562

Chum, H. L., Johnson, D. K., Black, S., Baker, J., Grohmann, K., Sarkanen, K. V., Wallace, K., Schroeder, H. A. (1988). Organosolv pretreatment for enzymatic hydrolysis of poplars: I. Enzyme hydrolysis of cellulosic residues. *Biotechnology and Bioengineering,* Vol. 31, pp. 643-649

Del Campo, I. et al. (2006). Diluted acid hydrolysis pretreatment of agri-food wastes for bioethanol production. *Industrial Crops and Products,* Vol. 24, pp. 214-221

Duff, S. J. B., Murray, W. D. (1996). Bioconversion of forest products industry waste cellulosics to fuel ethanol: A review. *Bioresource Technology,* Vol. 55, pp. 1-33

Girio, F. M., Fonseca, C., Carvalheiro, F., Duarte, L. C., Marques, S., Bogel-Lucasik, R. (2010). Hemicelluloses for fuel ethanol: A review. *Bioresource Technology,* Vol. 101, pp. 4775-4800

Hsu, T.-C., Guo, G.-L., Chen, W.-H., Hwang, W.-S. (2010). Effect of dilute acid pretreatment of rice straw on structural properties and enzymatic hydrolysis. *Bioresource Technology,* Vol. 101, pp. 4907-4913

Inoue, H., Yano, S., Endo, T., Sakaki, T., Sawayama, S. (2008). Combining hot-compressed water and ball milling pretreatments to improve the efficiency of the enzymatic hydrolysis of eucalyptus. *Biotechnology for Biofuels,* 1:2

Kim, J.-S., Kim, H., Lee, J.-S., Lee, J.-P., Park, S.-C. (2008). Pretreatment characteristics of waste oak wood by ammonia percolation. *Appl. Biochem. Biotechnol.,* Vol. 148, pp. 15-22

Kim, T. H., Lee, Y. Y. (2005). Pretreatment and fractionation of corn stover by ammonia recycle percolation *Process. Bioresource Technology,* Vol. 96, No. 18, pp. 2007-2013

Kootstra, A. M. J., Beeftink, H. H., Scott, E. L., Sanders, J. P. M. (2009). Optimization of the dilute maleic acid pretreatment of wheat straw. *Biotechnology for Biofuels*, Vol. 2, No. 31

Kucuk, M. M. (2005). Delignification of biomass using alkaline glycerol. *Energ. Source*, Vol. 27, pp.1245–1255

Kumar, R., Wyman, C. E. (2009) Does change in accesibility with conversion depend on both the substrate and pretreatment technology? *Bioresource Technology*, Vol. 100, pp. 4193-4202

Mtui, G. Y. S. (2009). Recent advances in pretreatment of lignocellulosic wastes and production of value added products. *African J. of Biotechnology*, Vol.8, No.8, pp. 1398-1415

Onda, A., Ochi, T., Yanagisawa, K. (2009). Hydrolysis of cellulose selectively into glucose over sulfonated activated-carbon catalyst under hydrothermal conditions. *Top Catal.*, Vol. 52, pp. 801-807

Soccol, C. R. et al. (2010). Bioethanol from lignocelluloses: Status and perspectives in Brazil. *Bioresource Technology*, Vol. 101, pp. 4820-4825

Stavrinides, A. J., Phipps, D. A., Al-Shamma'a, A. (2010). Review: Current and developing lignocellulosic pretreatment methods for bioethanol production, Available from: www.ljmu.ac.uk/.../Amended_PROCEEDINGS_BEAN_2010_WEB_VERSION_24.pdf

Swatloski, R. P., Spear, S. K., Holbrey, J. D., Rogers, R. D.(2002). Dissolution of cellose with ionic liquids. *J. Am. Chem. Soc.*, Vol. 124, pp. 4974-4975

Sun, Y., Cheng, J. (2002). Hydrolysis of lignocellulosic materials for ethanol production: a review. *Bioresource Technology*, Vol. 83, pp. 1-11

Talebnia, F., Karakashev, D., Angelidaki, I. (2010). Production of bioethanol from wheat straw: An overview on pretreatment, hydrolysis and fermentation. *Bioresource Technology*, Vol. 101, pp. 4744-4753

Tian, J., Wang, J., Zhao, S., Jiang, C., Zhang, X., Wang, X. (2010). Hydrolysis of cellulose by the heteropoly acid $H_3PW_{12}O_{40}$. *Cellulose*, Vol. 17, pp. 587-594

Van Walsum, G. P., Shi, H. (2004). Carbonic acid enhancement of hydrolysis in aquoeus pretreatment of corn stover. *Bioresource Technology*, Vol. 93, No. 3, pp. 217-226

Wu, Z., Lee, Y. Y. (1997). Ammonia recycled percolation as a complementary pretreatment to the dilute-acid process. *Applied Biochemistry and Biotechnology*, Vol. 63-65, No. 1, pp. 21-34

Yamaguchi, D., Hara, M. (2010). Optimization of hydrolysis of cellulosic materials by a solid acid catalyst. Available from: www.iiis.org/CDs2010/CD2010IMC/ICEME_2010/.../FB297UD.pdf

Zhao, X., Cheng, K., Liu, D. (2009). Organosolv pretreatment of lignocellulosic biomass for enzymatic hydrolysis. *Appl. Microbiol. Biotechnol.*, Vol. 82, pp. 815-827

Zhang, Z., Zhao, Z. K. (2010). Solid acid and microwave-assisted hydrolysis of cellulose in ionic liquid. *Carbohydrate Research*, Vol. 344, pp. 2069-2072

Zheng, Y., Pan, Z., Zhang, R. (2009). Overview of biomass pretreatment for cellulosic ethanol production. *Int. J. Agric. & Biol. Eng.*, Vol. 2, No. 3, pp. 51

Zhu, S. et al. (2006). Dissolution of cellulose with ionic liquids and its application: a mini-review. *Green Chem.*, Vol. 8, pp. 325-327

http://www.ecn.nl/units/bkm/biomass-and-coal/transportation-fuels-and-chemicals/transportation-fuels/biomass-pre-treatment-fractionation/

12

Relationship between Microbial C, Microbial N and Microbial DNA Extracts During Municipal Solid Waste Composting Process

Bouzaiane Olfa, Saidi Neila, Ben Ayed Leila,
Jedidi Naceur and Hassen Abdennaceur
*Centre de Recherche et des Technologie des Eaux (CERTE),
Laboratoire Traitement et Recyclage des Eaux, Cité Mahrajène, Tunis,
Tunisie*

1. Introduction

The municipal solid waste composting process has been defined as a controlled aerobic microbial process widely used to decompose organic matter to obtain a stable product consisting of a humus-like substance (Michel *et al.*, 1995). The end product or compost is available for agricultural use. However, the main requirement for the safe use or application of compost to agricultural lands is its degree of stability, which implies stable organic matter content (Castaldi *et al.*, 2004, 2008; Mondini *et al.*, 2004). This practice is becoming one of the most promising ways for the reclamation of degraded soils in semiarid areas of the Mediterranean countries like Tunisia (Bouzaiane *et al.*, 2007 a). Optimization of the composting process depends on optimization of environmental conditions that promote the development and activity of microbial communities. In fact the microbial biomass (MB) amount plays an important role on the biochemical transformations, on the optimization and on the quality of the end product (Mondini *et al.*, 2002; Jedidi *et al.*, 2004).
The chloroform- fumigation–extraction (CFE) is currently the most common method used to quantify the microbial biomass in soil samples (Vance *et al.*, 1987). Some authors have applied the CFE technique on compost substrates (De Nobili *et al.*, 1996, Hellmann *et al.*, 1997, Mondini *et al.*, 1997; Ben Ayed *et al.*, 2007).
On the other hand, the application of molecular methods to study the composting process and the microbial communities governing the transformation of the organic matter presents some unique challenges. One such challenge is the dynamic nature of the process, characterized by rapid changes in microbial population, temperature and oxygen gradients, and the availability of nutrients for microorganisms. The analysis of nucleic acids extracted from environmental samples allows researchers to study natural microbial communities without the need for cultivation (Peters *et al.*, 2000; Dees and Ghiorse, 2001). Although there have been many published studies on methods for the extraction DNA from environmental samples, very few have focused upon the extraction of DNA from compost. Compost samples may also contain 10–100 times greater humic acid concentrations than mineral soils (Pfaller *et al.*, 1994). Humic acids co-purify with DNA during many purification steps (Ogram *et al.*, 1987). These factors combine to make DNA quantification in compost

exceptionally difficult. Methods designed to extract DNA from soils and sediments have been adapted to obtain DNA from composts (Blanc *et al.*, 1999; Kowalchuk *et al.*, 1999). However, the relative effectiveness of extraction and purification methods for isolating compost DNA of sufficient purity has not been examined. Also, potential bias introduced by different extraction protocols has not been investigated yet. In this paper, we adopted the Fast DNA Kit for Soil DNA extraction and purification procedures to extract and purify DNA from compost.

In the present study, we attempted to evaluate (i) the evolution of microbiological parameters such as microbial biomass C, N and DNA content during municipal solid waste composting process and (ii) the relationship between microbial biomass C, N and DNA concentration during municipal solid waste composting process and possibly use these parameters to find out the compost stability.

2. Materials and methods

2.1 Composting process

The compost was prepared at the pilot composting station of Beja City, situated about 100 km to the west of Tunis. At the entry of the composting station, the wastes were stocked on big pile with a pyramidal shape (3.0 m length x 2.5 m width x 1.5 m x high) during 2 months without any previous treatment. The non-biodegradable coarse wastes (mostly plastics and glasses) were manually removed; therefore the remaining wastes were subsequently crushed and sieved to 40 mm in order to decrease the waste heterogeneity. Sawdust and green wastes were added to the wastes and these wastes were stocked on new pile during 3 months for stabilization.

Temperature and humidity were controlled daily, and pile was turned and watered (humidity regularly adjusted to 50%) as soon as the inner temperature of the pile reached or exceed 65°C. These operations of turning and watering were performed almost twice per month on an average.

2.2 Sampling of organic wastes during the composting process

Ten samples (approximately 5 kg each) were collected every 15 days from day 5 to day 139 from ten randomly selected locations in the pile by digging a small pit to 1 m depth with a shovel. At each sampling time, samples were mixed thoroughly and three portions of 1 kg each were separated. The first portion was stored at -20°C to constitute a collection of samples, the second was for pH determination, and the third was for microbiological analyses.

2.3 Temperature and pH determination during the composting process

Temperature inside the windrows was measured, every day during the composting period, with a special sensing device stuck introduced to 60 cm depth in randomly selected points. For pH, 400 g of compost were placed in an Erlenmeyer flask containing 2 l of distilled water and stirred for 3-5 min. The mixture was allowed to settle for 5 min and the pH was measured using a pH meter. For dry weight, 400 g of fresh compost was dried at 105 °C until the weight remained constant.

2.4 Determination of microbial biomass C and N

Microbial biomass C and N were determined by the CFE method, according to Vance *et al.* (1987) and Brookes (1995), respectively. Twenty grams were fumigated with ethanol-free

$CHCl_3$ for 24 h at 25°C in a dessicator. After removing the fumigant the samples were extracted for 60 min with 80 ml 0.5 mol l^{-1} K_2SO_4 solutions (1/4, w/v) and then filtered through a Whatman filter paper. Non-fumigated samples were extracted as above at the time the fumigation started. The amounts of soluble C in the fumigated and non-fumigated compost extract are used to determine biomass C. Organic C was quantified by the potassium dichromate oxidation method (Jenkinson and Powlson, 1976) and subsequent back-titration of the unreduced dichromate. The sample microbial biomass C (MBC) was estimated using the following equation (Jenkinson and Powlson, 1976):

$$MBC = CE/0.35$$

Where CE was the difference between organic C extracted from fumigated and non-fumigated treated samples.
Total N in the extracts was determined according to the Kjeldahl methods as described by Brookes *et al.*, 1985.
The microbial biomass N was estimated using the following equation:

$$MBN = NE/0.68$$

Where NE was the difference between total N extracted from fumigated and non fumigated samples. Amounts of microbial biomass C or N were expressed (mg C or N kg-1 dry weight) on air-dry soil basis and represent the average of three determinations (repeated three times on a single sample).

2.5 DNA extraction

About 0.5g of compost was weighed into DNA extraction matrix tubes using the Bio 101 Fast DNA Kit for Soil (Biogene, France). All extraction steps were carried out according to the manufacturer's instructions. DNA was eluted in 100µl of elution buffer. Purified DNA was quantified by spectrophotometer (Bio-RAD Smart Spec TM Plus, France) (Leckie *et al.*, 2004). Reserve aliquots were stored at - 20°C and working stocks at 8°C.
The spectrophotometric *A260 /A280* and *A260 /A230* ratios were determined to evaluate levels of protein and humic acid impurities, respectively, in the extracted DNA (Ogram *et al.*, 1987; Steffan *et al.*, 1988).

2.6 Statistical analysis

The ANOVA analysis was carried out using the SPSS statistical program for Windows (SPSS Inc., Chicago, IL). The means were compared according to the Newman and Keuls multiple range-test at $P < 0.05$.

3. Results

3.1 Physico-chemical parameters of composting process

The physicochemical characteristics evolution obtained during the municipal solid waste composting process was presented in Table 1.
In this study the temperature progress vary according the two phases of composting process, digestion and maturation (Fig. 1). The phase of digestion starts with a mesophilic phase in which the temperature reached 42°C. During this mesophilic step, the humidity rate was up to 45%. After 20 days of composting, the temperature reached 65°C and the

thermophilic step started. In this step the humidity decreased significantly. Then, the temperature decreased gradually to reach 40°C. At the 62 days, and after the addition of sawdust and green wastes in order to enhance the microbial activity, the maturation phase took place. In this phase, like in the digestion phase, the temperature increased gradually to reach 50°C, stabilised for a short period then decreased. In this phase there was also mesophilic, thermophilic and cooling steps.

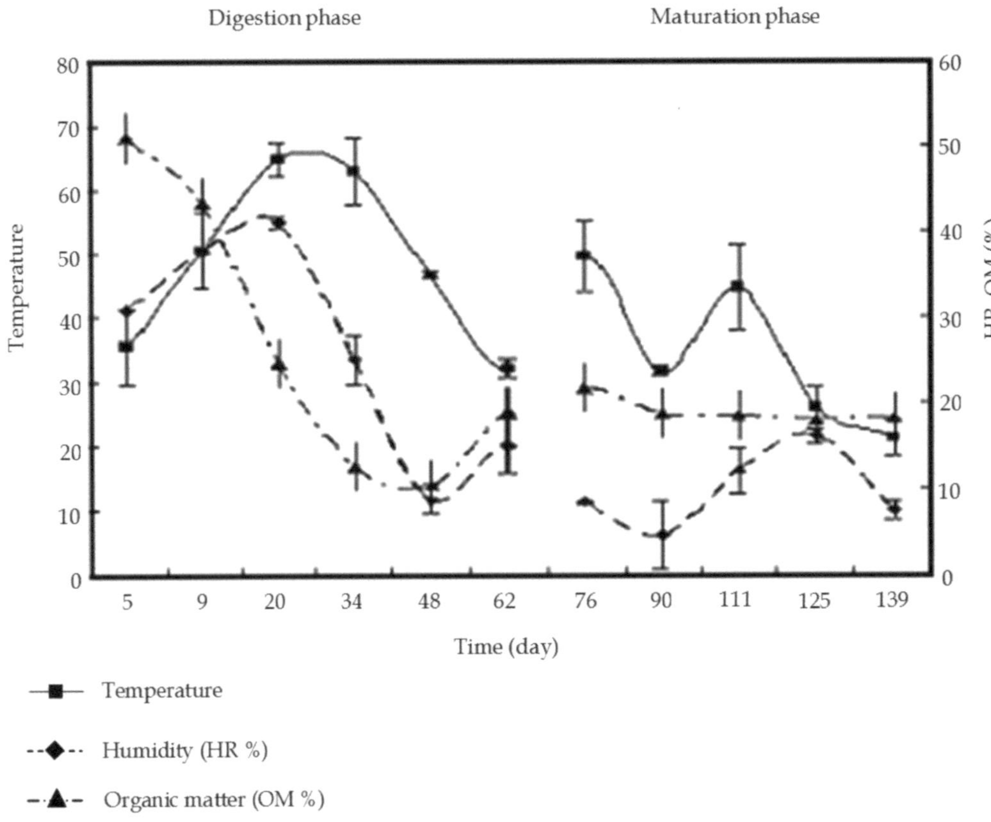

Fig. 1. Progress of temperature, humidity and organic matter during composting process

3.2 Evolution of microbial biomass C, microbial N and microbial DNA extracts during composting process

The progress of microbial biomasses (BC and BN) over time marked a real variation, particularly with a decrease of BC, BN and DNA concentration registered during the digestion and maturation phases (Figure 2). During the digestion phase of composting process microbial biomass C (*BC*) and microbial biomass N (*BN*) ranged from 4.86 to 1 µg kg^{-1} and from 1.472 µg kg^{-1} to 0.65, respectively. During the maturation phase these values decreased to reach 0.44 mg kg^{-1} for BC and 0.26 mg kg^{-1} BN. DNA content evolution ranged from 51.9 to 39 µg g^{-1} of dry matter in digestion phase and this content decrease to reach 18.5 µg g^{-1} of dry matter in the end product.

The BC/BN values registered in digestion phase indicate the dominance of three types of microbial communities. Homogeneous microbial community was found during mesophilic and thermophilic steps of municipal solid waste process was found particularly with BC/BN values of 3.3. Heterogeneous microbial communities were found particularly with BC/BN values of 7.92 and 1.54 (Table 1).

The BC/BN values registered in maturation phase indicate the dominance of two types of microbial communities. Heterogeneous microbial communities were found particularly with BC/BN values of 2.3 and 1.6.

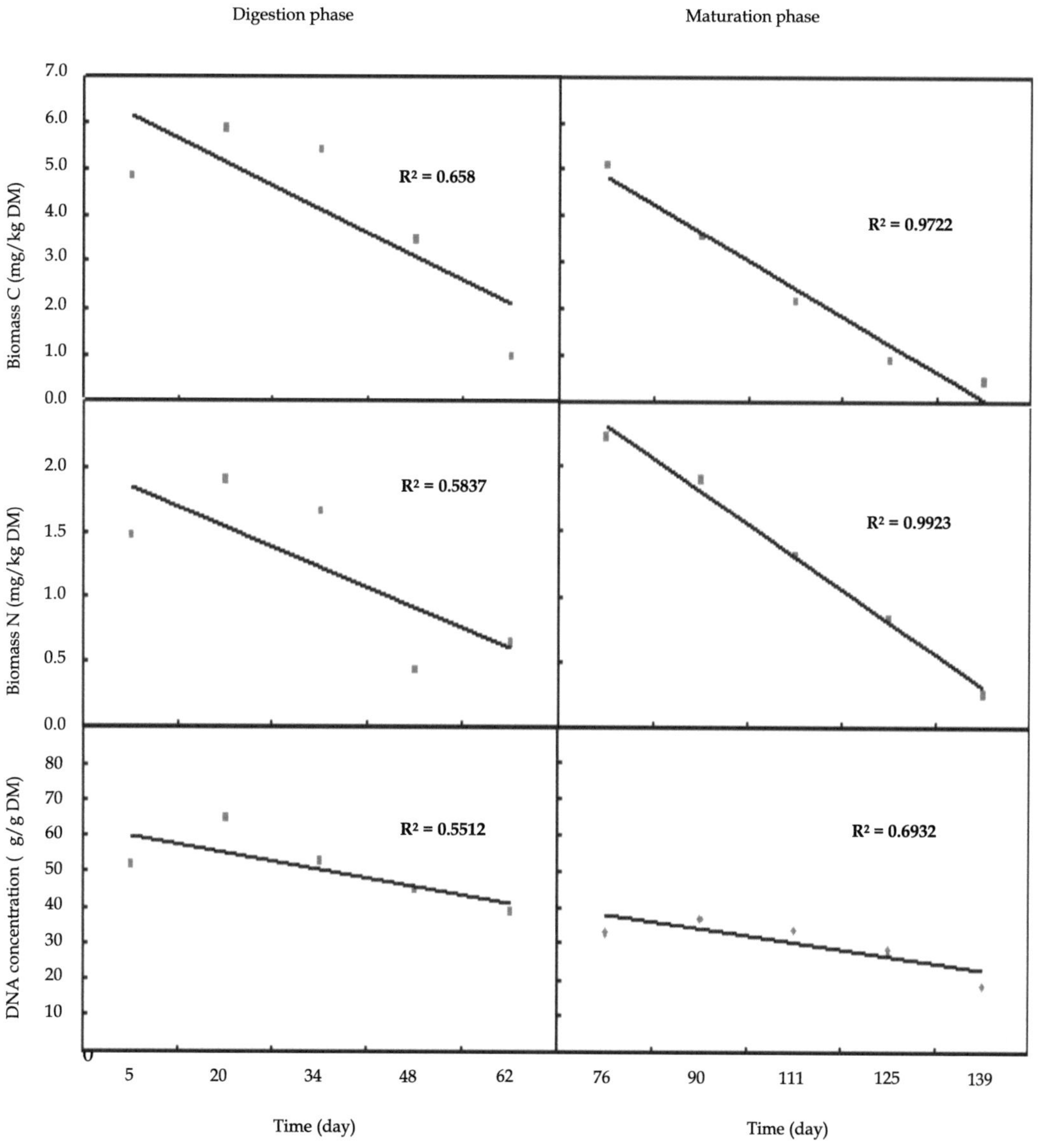

Fig. 2. Progress of microbial biomass C, microbial biomass N and microbial DNA extracts during composting process

The addition of sawdust and green wastes is considered to be a source of organic matter that stimulates microbial biomass. In fact, the addition of sawdust and green wastes affect the structure and composition of the microbial communities that colonize the municipal solid waste.

Days of composting	pH	TOC (mg kg-1DM)	TN (mg kg-1DM)	C/N	BC/BN
5	6.50	31.67	1.14	27.78	3.299
20	6.97	22.10	1.22	18.11	3.077
34	7.81	18.05	1.33	13.57	3.274
48	7.04	14.00	1.78	7.86	7.923
62	7.19	30.32	2..30	13.18	1.540
Addition of sawdust and green wastes					
76	7.41	14.18	1.14	12.44	2.301
90	7.60	15.79	1.21	13.05	1.889
111	7.32	16.54	1.41	11.73	1.662
125	7.52	15.51	1.54	10.07	1.080
139	7.65	10.36	1.14	9.09	1.698

TOC, Total organic carbon; TN, total nitrogen; C/N, carbon: nitrogen ratio; DM: dry matter

Table 1. Physicochemical properties obtained during municipal solid waste composting process

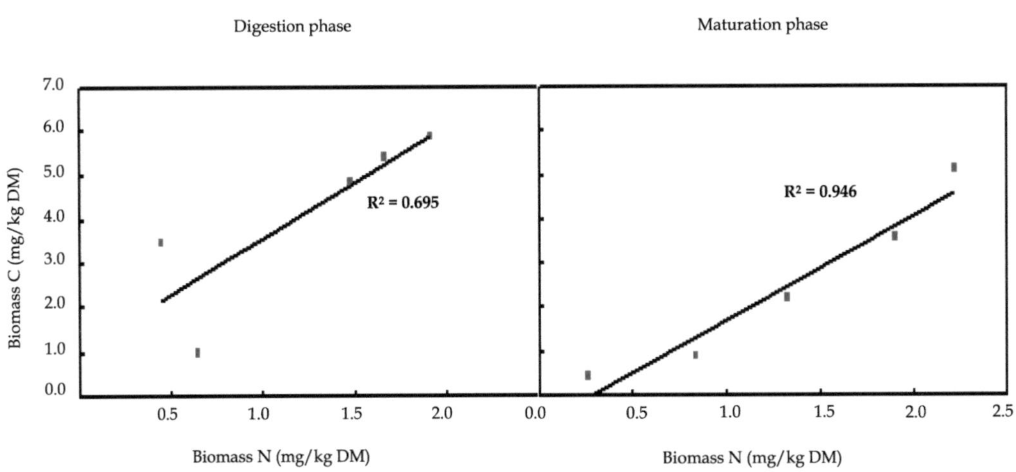

Fig. 3. Relationship between biomass N and biomass C during digestion and maturation phases of composting process.

3.3 Relationship between microbial biomasses BC and BN and DNA content

A good linear relation between microbial BC and BN was found during the digestion and maturation phases, with *r* coefficients of 0.69 and 0.94, respectively (Figure 3). The result showed clearly (*r* coefficients) that the microbial biomasses BC and BN obtained in the digestion phase were higher in comparison with those obtained during the maturation phase.

A linear relationship between biomass C and DNA concentration was found (Fig. 4A and B). DNA concentrations and BC were highly correlated during the digestion phase of municipal solid waste composting process with *r* coefficients of 0.80 (Fig. 4A).

On the other hand there is a linear relationship between biomass N and DNA concentration (Fig. 4C and D). DNA concentrations and BN were highly correlated during the digestion and maturation phases of municipal solid waste composting process with *r* coefficients of 0.78 and 0.76, respectively (Fig. 4B). Nevertheless, the DNA concentration was generally proportional to the BC or to BN and both methods seemed to give reliable values of compost microbial biomass. Our results indicate that BC and BN and DNA contents of the compost can be related with biological and chemical parameters in a combined way.

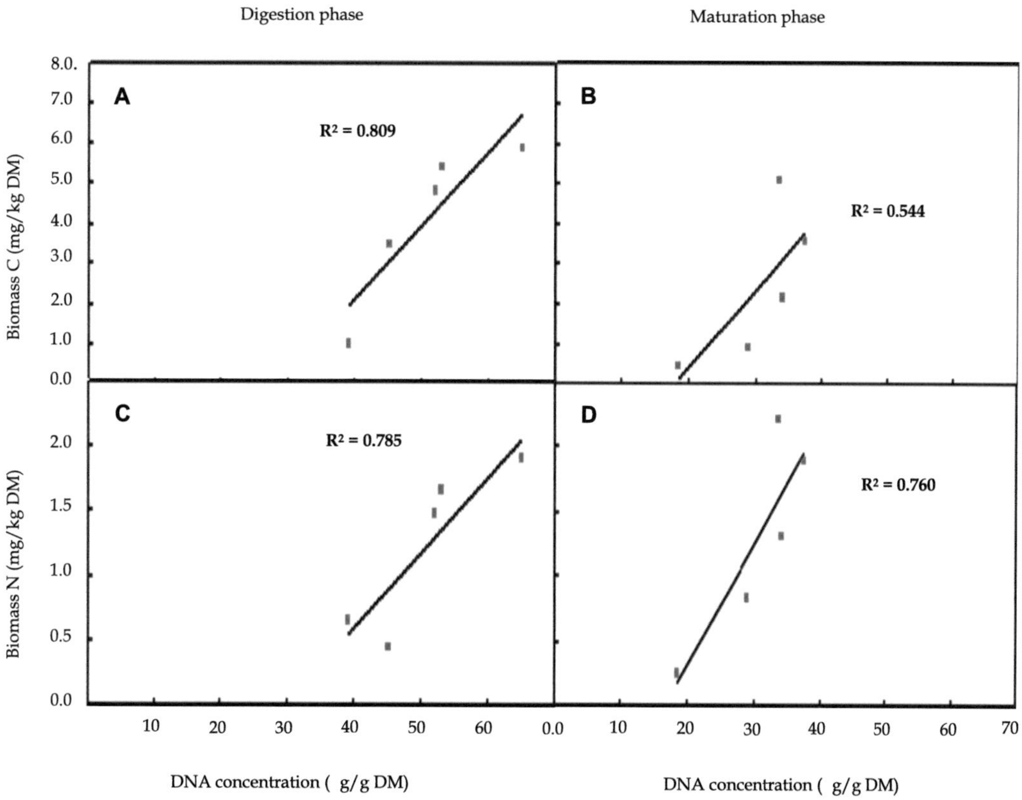

Fig. 4. Relationship between DNA concentration and biomass C (A and B) and biomass N (C and D) during digestion and maturation phases of composting process

3.4 Humic acid and protein impurities during composting

The *A260 /A230* and *A260 /A280* ratios for compost DNA were significantly lower than the ratios for DNA solutions from pure cultures showing that compost DNA was coextracted with humic compounds (Table 2).

DNA extracts from the cooling stage of maturation phase showed the lowest ratio *A260/A280* and *A260/A230* ratios than those obtained with the other stage of composting which may due to the high proportion of humic acids with the composting progress. Accordingly, the decrease in the microbial biomass DNA concentration in the cooling stage of composting could be explained by the DNA binding to compost humic acids and the formation of humic-DNA complexes.

The extracted DNA with low *A260 /A230* or unsuitable *A260 /A280* ratio decreases the efficiency of PCR amplification.

The extraction method will be suitable for the DNA purity. The purity will determine the extent to which the microbial DNA template can be amplified by PCR during the composting analysis. However, in this study the humic acid content could not interfere with PCR. Then the PCR products were successfully used for DGGE analysis (data not shown).

The DNA extract was thus suitable to be used for molecular studies on the microbial communities in municipal solid waste composting process.

Days of composting	DNA yield (μg DNA g^{-1} DM)	A260/ A280 ratio	A260/ A230 ratio
5	51.90	1.33 (0.04) d	1.02 (0.01) d
20	65.00	1.31(0.05) d	1.03 (0.04) d
34	53.00	1.32 (0.02) d	1.02 (0.02) d
48	45,00	1.22 (0.02) c	0.94 (0.04) c
62	39.00	1.22 (0.03) c	0.95 (0.03) c
Addition of sawdust and green wastes			
76	33.50	0.98 (0.03) b	0.71 (0.01) b
90	37.40	0.98 (0.01) b	0.73 (0.03) b
111	34.00	0.89 (0.03) a	0.72 (0.04) b
125	28.70	0.87 (0.04) a	0.66 (0.04) a
139	18,50	0.88 (0.02) a	0.65 (0.02) a
Pure culture		1.89	1.57

Pure culture: DNA from Gram positive bacteria. n = 3 determined by spectrophotometry at 260 nm (A260), 280 nm (A280) and 230 nm (A230); (In brackets): standard deviation; within a column different letter after bracket means that the value is significantly difference according to Student-Newmann-Keuls test at $P < 0.05$; DM: dry matter

Table 2. Comparison of compost DNA yields and purity

5. Discussion

5.1 Physico-chemical parameters of composting process

The composting process at the microbial level involves several interrelated factors, namely temperature, ventilation (O2 imputed), moisture content and available nutrients. Based on temperature, the process of aerobic composting can be divided into three major steps, a mesophilic-heating step, a thermophilic step and a cooling step (Mustin, 1987). During the mesophilic step, the temperature and the water content increased as a consequence of biodegradation of organic compounds. The temperature increment is the consequence of the organic matter oxidation (Hassen *et al.*, 2001). The mesophilic step is followed by the thermophilic step. The latter step occurred between days 20 and days 34 of the composting process. As mentioned by Hachicha *et al.*, 1992 and Marrug *et al.*, 1993, a temperature above 60 °C seriously affect the decomposition rate of the organic waste as a result of a reduction in microbiological activity. The temperature started to decrease after 48 days, and then increased again after the addition of fresh organic matter. A second decrease of the temperature then occurred after 111 days of the process, this decrease led to the depletion of organic matters and the carbon/azotes (C/N) ratio tended to stabilize. By the end of the composting process, the average temperature inside the windrow showed a decrement and reached approximately 30 °C at the end of the process (Ben Ayed *et al.*, 2007).

Composting is a self-heating, aerobic, solid phase, useful way of transforming organic wastes into valuable organic matter for use as an organic amendment for soils. The composting process can provide stable and valuable substrates through the bio-oxidation of the organic fraction deriving from different waste matrices (Castaldi *et al.*, 2004, 2008). Many tests have been considered as maturity indices for compost, and most of them focus on the chemical and physical properties of compost. The most common parameters include compost temperature, pH, cation exchange capacity, dissolved organic C, C/N ratio, humification index, plant growth bioassay, spectroscopic methods, etc. (Garcia *et al.*, 1992; Castaldi *et al.*, 2004).

5.2 Evolution of microbial biomass C, microbial N and microbial DNA extracts during composting process

The evolution of microbial biomass C, microbial N and microbial DNA extracts during composting process is probably related to the availability of readily decomposable substrates; in fact when organisms are presented with a substrate they normally multiply rapidly until the substrate is nearly exhausted, when numbers reach a maximum (Joergensen *et al.*, 1990; Ben Ayed *et al.*, 2007). Thereafter, with the exhaustion of these substances caused by the intense microbial activity and by ongoing humification, the microbial biomass decreased. The BC and BN decreased possibly due to the degradation of the depletion of organic substrates available for micro-organisms growth (Manuael *et al.*, 2009).

With the progress of the process the DNA content decrease and the extraction and purification method yielded 18.5 µg DNA/g of dry compost in the end of the process. Howeler *et al.*, 2003 found 18.2 µg DNA/g of wet compost yielded by extraction and purification method from compost.

This result could be explained by (i) the microbial DNases degradation or by (ii) the protection of the DNA by binding to compost humic acids. The formation of humic-DNA complexes should be considered as a process related to the changes in compost matrix, i.e. formation of humic like substances, which is one of the main purposes for the composting process.

Biological parameters such as microbial biomass are useful indicators of biological activity in ecosystems (Benitez *et al.*, 1999). Since, during the composting process microbial biomass C, microbial biomass N and DNA contents could indicate compost stability, defined as the degree of decomposition of the readily biodegradable organic matter.

5.3 Relationship between microbial BC and BN and DNA

A good linear relationship between microbial BC and BN, during the different stages of composting process. The same result was found during three consecutive years of compost amendment at the level of the upper and deep horizon of non cultivated soil (Bouzaiane *et al.*, 2007 a). Jedidi *et al.*, 2004 found the same linear relationship between BC and BN in amended soil and in laboratory conditions. Franzluebbers *et al.*, 1995 found the same linear relationship between BC and BN with $r = 0.86$.
In the composting process the humification and mineralization of organic substances occurs simultaneously. The DNA content, BC and BN could be related to the humification index and degree of polymerisation evolutions.
In the digestion phase we think that the micro-organisms diversity is due to the incorporation of extra-cellular DNA from degrade microbial in to bacterial genome as possible source of genetic instructions (transformation, conjugation and transduction).
Similar results were obtained by Bouzaiane *et al.*, 2007b who found a strong relationship between BC, BN estimated by CFE, and extracted DNA in cultivated-compost-amended soil. Marstorp *et al.*, 2000 found also a strong relationship between BC, estimated by CFE, and extracted DNA in a mineral soil. They suggested that DNA could be used as a measure of microbial biomass in agricultural soils with low organic matter content. Tejada *et al.*, 2009 were found a strong correlation between biological and chemical parameters during municipal solid waste composting process.
Tejada *et al.*, 2009 suggested that humification index (HI) and degree of polymerisation (DP) of the compost can be related with biological and chemical parameters in a combined way.

5.4 Humic acid and protein impurities during composting

The humic acid increased during municipal solid waste composting process. Also Tejada *et al.*, 2009 showed that the humic index and degree of polymerisation parameters, both increased during composting process (66% and 41%, respectively at the end of the composting process when compared to values at 0 days).
Composting DNA was often contaminated with humic acid or proteins that interfered with accurate quantification of DNA by UV absorbance at 260 nm (Tebbe and Vahjen, 1993; Kuske *et al.*, 1998). In this work, we used Fast DNA Kit for Soil DNA extraction and UV absorbance at 260 nm to detect very low DNA concentrations in diluted samples (typically 100 to 1000 fold), so that the effect of humic acid contamination could be ignored. UV absorbance at 260 nm was an excellent method for DNA quantification of samples extracted from environmental sources containing high levels of humic acids. A simple and accurate method of humic acid quantification (e.g. absorbance) should also be used to determine the correct dilution required for DNA quantification and to measure the progress of humic acid.

6. Conclusion

It can be concluded that the microbial biomass C and N and DNA content during the municipal solid waste composting process can be of great use in understanding the compost stability state. This fact does not mean that the study of these biological properties diminishes the study of the chemical properties, but rather, both types of properties can be combined to indicate the compost stability. In fact the linear regression analysis developed in this work indicates a strong relationship between the biological properties. On the other hand the commercial method for extraction DNA was suitable for PCR-DNA amplification of microbial analysis during the composting of municipal solid waste and of the end product such as the compost that could be used for the detection of microbial pathogens.

7. Acknowledgements

Special thanks to all who helped in the water treatment and recycling laboratory of CERTE (Centre de Recherche et des Technologie des Eaux).

8. References

Ben Ayed, L.; Hassen, A.; Jedidi, N.; SaidiI, N.; Bouzaiane, O. & Murano, F. (2007). Microbial C and N dynamics during composting process of urban solid waste. *Waste Management and Research,* Vol. 25, No.1, (February 2007), pp. 24-29, ISSN 0734–242X.

Benitez, E.; Nogales, R.; Elvira, C.; Masciandaro, G. & Ceccanti B. (1999). Enzyme activities as indicators of the stabilization of sewage sludges composting with Eisenia foetida. *Bioresource Technology,* Vol. 67, No. 3, (March 1999), pp. 297-303, ISSN 0960-8524.

Bouzaiane, O.; Cherif, H.; Saidi, N.; Jedidi, N. & Hassen, A. (2007a). Effects of municipal solid waste compost application on the microbial biomass of cultivated and non-cultivated soil in a semi-arid zone. *Waste Management and Research,* Vol. 25, No.4, (August 2007), pp. 334-342, ISSN 0734–242X.

Bouzaiane, O.; Cherif, H.; Saidi, N.; Hassen, A. & Jedidi, N. (2007b). Municipal solid waste compost dose effects on soil microbial biomass determined by chloroform fumigation-extraction and DNA methods. *Annals of Microbiology,* Vol. 57, No.4, pp. 681-686, ISSN 1590-4261.

Blanc, M.; Marilley, L.; Beffa, T. & Aragno, M. (1999). Thermophilic bacterial communities in hot composts as revealed by most probable number counts and molecular (16S rDNA) methods. *FEMS Microbiology Ecology,* Vol. 28, No. 2, (February 1999), pp. 141-149, ISSN 0168-6496.

Brookes, P.C. (1995). The use of microbial parameters in monitoring soil pollution by heavy metals. *Biology and Fertility of Soils,* Vol. 19, No. 4, (March 1995), pp. 269-279, ISSN 0178-2762.

Brookes, P.C.; Landman, A.; Pruden, G. & Jenkinson, D.S. (1985). Chloroform fumigation and the release of soil nitrogen: a rapid direct extraction method to measure microbial biomass nitrogen in soil. *Soil Biology and Biochemistry,* Vol. 17, No. 6, (1985), pp. 837-842 ISSN 0038-0717.

Castaldi, P.; Garau, G. & Melis P. (2004). Influence of compost from sea weeds on heavy metal dynamics in the soil–plant system. *Fresenius Environment Bulletin,* Vol. 13, pp. 1322-1328, ISSN 1018-4619.

Castaldi, P.; Garau, G. & Melis, P. (2008). Maturity assessment of compost from municipal solid waste through the study of enzyme activities and water-soluble fractions. *Waste Management*, Vol. 28, No. 3, (March 2008), pp. 534-540, ISSN 0956-053X.

Dees, P.M. & Ghiorse W.C. (2001). Microbial diversity in hot synthetic compost as revealed by PCR-amplified rRNA sequences from cultivated isolates and extracted DNA. *FEMS Microbiology Ecology*, Vol. 35, No. 2, (April 2001), pp. 207–216, ISSN 0168-6496.

De Nobili, M.; Baca, M. T.; Fornassier F. & Mondini C. (1996). Ninhydrinreactive nitrogen of CHCl3 fumigated and non fumigated compost extracts as a parameter to evaluate compost stability. In: De Bertoldi, M., Sequi, P., Lemmes, P. & Papi, T. (eds): *The Science of Composting*, pp. 255-261. Blackie Academic & Professional, Glasgow, Scotland.

Franzluebbers, A.T.; Hons, F.M. &Zuberer, D.A. (1995). Soil organic carbon, microbial biomass and miniralizable carbon and nitrogen in sorghum. *Soil Science Society of America Journal*, Vol. 59, No. 2, (March, April 1995), pp. 460-466, ISSN 0361-5995.

Garcia C., HernЈndez T., Costa F., Ceccanti B., & Ciardi C. (1992). Changes in ATP content, enzyme activity and inorganic nitrogen species during composting of organic wastes. *Canadian Journal of Soil Science*, Vol. 72, No. 3, (August 1992), p 243-253, ISSN 0008-4271.

Hachicha R., Hassen A., Jedidi N., & Kallali H. (1992). Optimal conditions for municipal solid waste. *Biocycle* J Waste Recyc, Vol. 33, No. 6, No. pp. 76-77.

Hassen A., Belguith K., Jedidi N., Cherif A., Cherif M., & Boudabbous A. (2001). Microbial characterization during composting of municipal solid waste. *Bioresource and Technology*, Vol. 80, No. 3, (December 2001), pp. 185-192, ISSN 0960-8524.

Hellmann B., Zelles L., Palojarvi A. & Quingyun B. (1997). Emission of climate relevant trace gases and succession of microbial communities during open-window composting. *Applied and Environmental Microbiology*, Vol. 63, No. 3, (March 1997), pp. 1011-1018, ISSN 0099-2240.

Howeler M., Ghiorse W. C., & Walker L. P. (2003). A quantitative analysis of DNA extraction and purification from compost. *Journal of Microbiol and Methods*, Vol. 54, No. 1, (July 2003), pp. 37-45, ISSN 0167-7012.

Jedidi N., Hassen A., Van Cleemput O., & M'hiri A. (2004). Microbial biomass in soil amended with different types of organic wastes. *Waste Management and Research*, Vol. 22, No. 2, (April 1999), pp. 93-99, ISSN 0734–242X.

Jenkinson D.S., & Powlson D.S. (1976). The effects of biocidal treatments on metabolism in soil-I. Fumigation with chloroform. *Soil Biology and Biochemistry*, Vol. 8, No. 3, (1976), pp. 167-177, ISSN 0038-0717.

Joergensen, R. G. & Brooks, P. C. (1990). Ninhydrin-reactive nitrogen measurements of microbial biomass in 0.5 M K2SO4 soil extracts. *Soil Biology and Biochemistry*, Vol. 22, No.8, (1990), pp. 1023-1027, ISSN 0038-0717.

Kowalchuk, G.A.; Naoumenko, Z.S.; Derikx, P.J.L.; Felske, A.; Stephen, J.R. & Arkhipchenko I.A. (1999). Molecular analysis of ammonia-oxidizing bacteria of the beta subdivision of the class proteobacteria in compost and composted materials. *Applied and Environmental Microbiology*, Vol. 65, No. 2, (February 1999), pp. 396–403, ISSN 0099-2240.

Kuske C.R., Banton K.L., Adorada D.L., Stark P.C., Hill K.K., & Jackson P.J., (1998). Small-scale DNA sample preparation method for field PCR detection of microbial cells and spores in soil. *Applied and Environmental Microbiology*, Vol. 64, No. 7, (July 1998), 2463-2472, ISSN 0099-2240.

Leckie S.E., Prescott C.E., Grayston S.J., Neufeld J.D., & Mohn W.W. (2004). Comparison of chloroform fumigation-extraction, phospholipid fatty acid, and DNA methods to determine microbial biomass in forest humus. *Soil Biology and Biochemistry*, Vol.36, No. 3, (March 2004), pp. 529-532, ISSN 0038-0717.

Marstorp H., Guan X., & Gong P. (2000). Relationship between dsDNA, chloroform labile C and ergosterol in soils of different organic matter contents and pH. *Soil Biology and Biochemistry*, Vol. 32, No. 6, (June 2000), pp. 879-882, ISSN 0038-0717.

Marrug C., Grebus M., Hassen R.C., Keener H.M., & Hoitink H.A.J. (1993). A kinetic model of yard waste composting process. *Composting Science and Utilization*, Vol. 1, pp. 38-51, ISSN 1065-657X.

Michel F.C., Reddy C.A., & Forney L.J. (1995). Microbial-degradation and humification of the lawn care pesticide 2, 4-dichlorophenoxyacetic acid during the composting of yard trimmings. *Applied and Environmental Microbiology*, Vol. 61, No. 7, (July 1995), pp. 2566-2571, ISSN 0099-2240.

Mondini C., Sanchez Monedero M. A., Leita L., Bragato G., & De Nobili M. (1997). Carbon and ninhydrin reactive nitrogen of the microbial biomass in rewetted compost samples. *Communications in Soil Science and Plant Analysis*, Vol. 28, No. 1-2, (January, Febuary1997), pp. 113–122, ISSN 0010-3624.

Mondini C., Contin M., Leita L., & De Nobili M. (2002). Response of microbial biomass to air-drying and rewetting in soils and compost. *Geoderma.*, Vol. 105, No. 1-2, (January 1999), pp. 111–124, ISSN 0016-7061.

Mondini C., Fornasier F., & Sinicco T. (2004). Enzymatic activity as a parameter for the characterization of the composting processes. *Soil Biology and Biochemistry*, Vol. 36, No. 10, (October 2004), pp. 1587–1594, ISSN 0038-0717.

Mustin M. (1987). *Le Compost: Gestion de la Matière Organique*. (Editions François Dubusc), ISBN 9782864720089, Paris 954p.

Ogram A., Sayler G.S., & Barkay T. (1987). The extraction and purification of microbial DNA from sediments. *Journal of Microbiological and Methods*, Vol. 7, No. 2-3, (December1987), pp. 57-66, ISSN 0167-7012.

Peters S., Koschinsky S., Schwieger F., & Tebbe C.C. (2000). Succession of microbial communities during hot composting as detected by PCR-single-strandconformation polymorphism-based genetic profiles of small-subunit rRNA genes. *Applied and Environmental Microbiology*, Vol. 66, No. 3, (March 2000), pp. 930–936, ISSN 0099-2240.

Pfaller S.L., Vesper S.J., & Moreno H. (1994). The use of PCR to detect a pathogen in compost. *Composting Science and Utilization*, Vol. 2, pp. 48-54, ISSN 1065-657X.

Steffan R.J., Goksoyr J., Bej A.K., & Atlas R.M. (1988). Recovery of DNA from soils and sediments. *Applied and Environmental Microbiology*, Vol. 54, No.12, (December 1988), pp. 2908-2915, ISSN 0099-2240.

Tebbe C.C., & Vahjen W. (1993). Interference of humic acids and DNA extracted directly from soil in detection and transformation of recombinant DNA from bacteria and yeast. *Applied and Environmental Microbiology*, Vol. 59, No.8, (August 1993), pp. 2657-2665, ISSN 0099-2240.

Tejada M., Garcia-Marinez A. M., & Parrado J., (2009). Relationships between biological and chemical parameters on the composting of a municipal solid waste. Bioresource Technology, Vol. 100, No. 17, (September 2009), pp. 4062-4065, ISSN 0960-8524.

Vance E.D., Brookes P.C., & Jenkinson D.S. (1987). Microbial biomass measurements in forest soils: determination of KC values and tests of hypotheses to explain the failure of the chloroform fumigation-incubation method in acid soils. *Soil Biology and Biochemistry*, Vol. 19, No.6, (1987), pp. 689-696, ISSN 0038-0717.

13

Characterization of Activated Carbons Produced from Oleaster Stones

Hale Sütcü
Zonguldak Karaelmas University
Turkey

1. Introduction

Activated carbon has a porous structure surrounded by carbon atoms and therefore is a material with adsorbent capability. The most important parameter that is put into consideration to investigate its chemical characterization is porosity. Pore size determines how adsorption takes place in pores (Marsh & Reinoso, 2006). In accordance with IUPAC, pores are classified into three different sizes. Pores less than 2,0 nm are classified as micropores, those in the range of 2,0-50 nm mesopores and those greater than 50 nm macropores (IUPAC).

The selection of raw material for the production of activated carbon is made on the basis of carbon amount, mineral matter and sulfur content, availability, cost, and shelf life (Kroschwitz,1992). Raw material may be of vegetable, animal and mineral origin and the production can be carried out by means of physical and chemical activation depending on the type of raw material.

The physical activation method generally involves carbonization and activation stages (Singh, 2001). In the activation stage oxidizing agents are used such as carbondioxide and steam and thus form pores and canals (Jankowska et al., 1991).

Chemical activation involves a carbonization stage where a chemical activating agent that is in the form of a solution or dry is blended with the raw material. Chemicals employed in chemical activation (potassium hydroxide, phosphoric acid, zinc chloride etc.) are effective at decomposing the structure of the raw material and forming micropores (Marsh & Reinoso, 2006).

The literature has many articles dealing with activated carbons produced from raw material using both the chemical and physical activation methods. Materials frequently used as raw material of vegetable origin include corncobs (Sun et al., 2007; Aworn et al., 2009; Preethi et al., 2006), hazelnuts (Demiral et al., 2008; Soleimani & Kaghazchi, 2007), olives (Yavuz et al., 2010), nuts (Yeganeh et al., 2006; Aygun et al., 2003), peaches (Kim, 2004), loquat stones (Sütcü & Demiral, 2009), wood (Ould-Idriss et al., 2011; Sun & Jiang, 2010) and bamboo (Ip et al., 2008), those of animal origin bones (Moreno-Pirajan et al., 2010) and hide waste (Demiral & Demiral, 2008), and those of mineral origin coal (Alcaniz-Monge et al., 2010; Cuhadaroglu & Uygun, 2008; Liu et al., 2007; Sütcü & Dural, 2007), petroleum coke (Lu et al., 2010) and rubber (Gupta et al., 2011; Nabais et al., 2010).

In this study I produced activated carbons from chars obtained through the carbonization of oleaster stones by physical, chemical and chemical+physical activation, and performed their surface characterization.

2. Experimental

2.1 Material and its structural characterization

The oleasters used in this study were obtained from a green grocer and their stones removed. The stones were washed clean and dried at 105°C for 24 hours.

The structural analysis of the oleaster stones were carried out by proximate and ultimate analyses, thermogravimetric analysis (TG), fourier transform infrared spectroscopy (FTIR) and scanning electron microscopy (SEM). The results regarding the proximate and ultimate analyses are given in Table 1.

The TG analysis was performed using a PL 1500TGA apparatus from ambient to 800°C at a heating rate of 10 °C/min and a nitrogen flow rate of 100 ml/min.

The FTIR spectrum was taken by means of a Perkin Elmer Spectrum One apparatus at wavelengths ranging from 4000 to 650 cm^{-1}.

The SEM image was obtained using a JEOL JSM model 5410 LV scanning electron microscope.

Ash^a	Volatile Matter^a	Fixed Carbon^a	C^b	H^b	N^b	S^b
0.57	74.27	25.16	48.16	0.66	3.44	0.29

Table 1. The results of proximate and ultimate analyses of oleaster stones (a. on dry basis, %, b. on dry and ash free basis, %)

2.2 Production of chars and their structural characterization

The stones were subjected to carbonization at a heating rate of 10 °C/min, a carbonization temperature of 600 °C and a nitrogen flow rate of 100 ml/min, and held at that temperature for 1 h. The carbonization was performed in a tube furnace of internal diameter 6 cm and length 110 cm. The chars were reduced to a size range of 0.5-1.0 mm to make them ready for the production of activated carbon.

The structural analysis of the chars were conducted by proximate and ultimate analyses, TG analysis, FTIR spectroscopy and SEM. Table 2 gives the results from the proximate and ultimate analyses undertaken.

The FTIR spectrum was taken using a Perkin Elmer Spectrum One apparatus within a wavelength range of 4000-650 cm^{-1}.

The SEM image was obtained using a JEOL JSM model 5410 LV scanning electron microscope .

Ash^a	Volatile Matter^a	Fixed Carbon^a	C^b	H^b	N^b
2.30	10.40	87.30	62.60	2.45	0.63

Table 2. Results from analyses of chars (a. on dry basis, %, b. on dry and ash free basis, %)

2.3 Activated carbon production

2.3.1 Physical Activation (PH)

The production of activated carbon from chars by physical activation was conducted in a tube furnace at carbonization temperatures of 650°C, 750°C and 850°C. The chars were heated up to the above-mentioned temperatures at a nitrogen atmosphere in a flow rate of 100 ml/min and a heating rate of 10 °C/min, and subjected to a CO_2 atmosphere with a flow rate of 100 ml/min. The chars thus obtained were kept in an desiccator. The chars produced by means of this method were designated as PH650, PH750 and PH850, respectively.

2.3.2 Chemical Activation (CH)

The production of chemical activation from chars was carried out using the chemical KOH at carbonization temperatures of 650°C, 750°C and 850°C. The mixture prepared in such a way that the char/KOH ratio would be 1/1 (mass ratio) was mixed with water of 10ml and held in a drying oven at 50 °C for 24 hours. The mixture was then heated up to the aforementioned temperatures at a heating rate of 10 °C/min and a nitrogen flow rate of 100 ml/min, and kept at that temperature for 1 hour. The activated carbons produced were boiled with 0.5 N HCI for 30 minutes and washed with distilled water until their pH was 6.5. Finally, they were dried in a vacuum drying oven and kept in an desiccator. The activated carbons thus obtained were designated as CH650, CH750 and CH850, respectively.

2.3.3 Sequential Activation (Chemical+Physical, CHPH)

The chars were blended with 10ml of distilled water in such a way that the char/KOH ratio would be 1/1 (mass ratio) and held in a drying oven for 24 hours. After that, this mixture was heated up to 650 °C, 750 °C and 850 °C at a nitrogen flow rate of 100 ml/min and a heating rate of 10 °C/min and was held at these temperatures under CO_2 with a flow rate of 100 ml/min. After the activated carbons produced were boiled with 0,5 N HCI for 30 minutes, they were washed with distilled water until a pH value of 6.5 was achieved, dried in a vacuum drying oven at 105 °C for 24 hours and kept in an desiccator. The activated obtained through this method are denoted by CHPH650, CHPH750 and CHPH850, respectively.

2.4 Strucural characterization of activated carbons

Structural characterization of the activated carbons was carried out by FTIR spectroscopy, SEM and a Quantachrome Autosorb Automated Gas Sorption System.
The FTIR spectra of the activated carbons were taken by means of a Perkin Elmer Spectrum One apparatus at wavelengths in the range of 4000 to 650 cm^{-1}.
The SEM images were obtained using a JEOL JSM model 5410 LV scanning electron microscope.
The iodine number of the activated carbons was determined in accordance with ASTM D 4607-94.
Surface analyses were performed by nitrogen adsorption at -196°C using a Quantachrome Autosorb Automated Gas Sorption System. Prior to adsorption, the activated carbons were outgassed under vacuum conditions at 250°C for 3 hours. Adsorption isotherms were obtained at pressures in the range of 10^{-5}-1.0. The surface areas and pore volumes were determined by means of Brunauer-Emmett-Teller (BET) and t-pilot software and pore size distribution using density functional theory (DFT) software.

3. Results and discussion

3.1 Structural analysis of oleaster stones and chars

Figure 1 gives the results from TG analysis carried out on oleaster stones. The decomposition of oleaster stones takes place in three stages. The first stage, which occurs at temperatures ranging from 30 °C to 140 °C involves moisture loss (Popescu et al., 20111). The other stages are related to the release of volatiles resulting from the decomposition of hemicellulose, cellulose and lignin (Tongpoothorn et al., 2011; Luangkiattikhun et al., 2008; Antal, 1982). In the second stage occuring at 140 °C-245 °C, hemicellulose decomposes as well as cellulose which also starts to disintegrate. Within this temperature range, the maximum decomposition temperature and rate were established to be 222 °C and 1,50%, respectively. The last stage, which takes place within a temperature range of 245 °C and 600 °C, is characterized by the decomposition of cellulose and lignin. The maximum decomposition temperature in this stage was found to be 333 °C and the maximum decomposition rate 6.71%.

The amount of char remaining as a result of TG analysis of oleaster stones in nitrogen atmosphere is 25.57%.

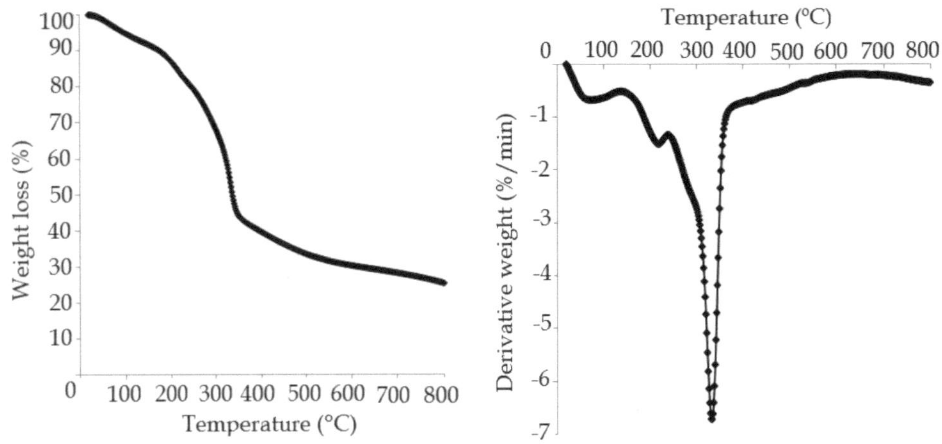

Fig. 1. Graph depicting the results from thermogravimetric and differential thermogravimetric analyses of oleaster stones.

Figure 2 gives the FTIR spectra of oleaster stones and chars obtained from them. An interpretation of the FTIR spectra reveals the existence of functional groups occurring in the structure.

The band observed at 3600-3200 cm^{-1} is indicative of the -OH stretching peak and existence of phenol, alcohol and carboxylic containing hydroxyl groups. This band, which is present in oleaster stones, do not exist in chars. This can be attributed to the decomposition of the structure and removal of the groups containing hydroxyl groups.

The band at 3000-2800 cm^{-1} indicates the presence of an aliphatic -CH stretching. This band is visible in oleaster stones but not in chars.

The band at around 1700 cm^{-1} denotes the existence of carbonyl/carboxyl groups and can be observed in oleaster stones.

The 1600-1500 cm^{-1} band, which is visible in both oleaster stones and chars, indicates the presence of an aromatic C=C ring stretching.

The bands at 1450-1300 cm^{-1} denotes the existence of C-H vibrating alkene groups. This band which exist in oleaster stones occurs in chars more densely.

The bands observed at 1240-1000 cm^{-1} indicates the existence of phenolic and alcoholic groups, and were identified in the FTIR spectra of oleaster stones and chars. The bands at 900-600 cm^{-1} denotes the presence of aromatic ring structures and are visible in both oleaster stones and chars.

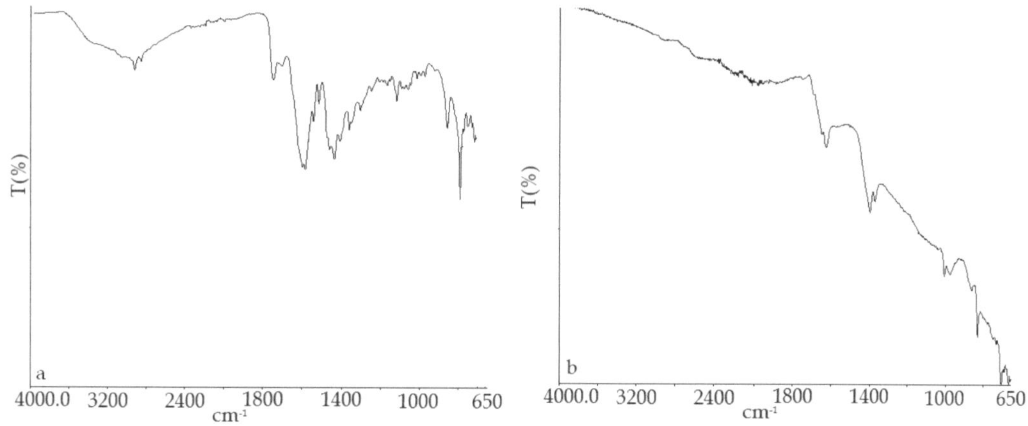

Fig. 2. FTIR spectra of (a) oleaster stones and (b) chars produced from them

Figure 3 gives SEM micrographs of oleaster stones and chars obtained from them.
It is clear from the SEM micrograph of oleaster stones that they have a fibrous structure. Chars produced at a carbonization temperature of 600°C were also determined to have a fibrous structure, a heterogeneous size and pores without any homogeneous distribution.

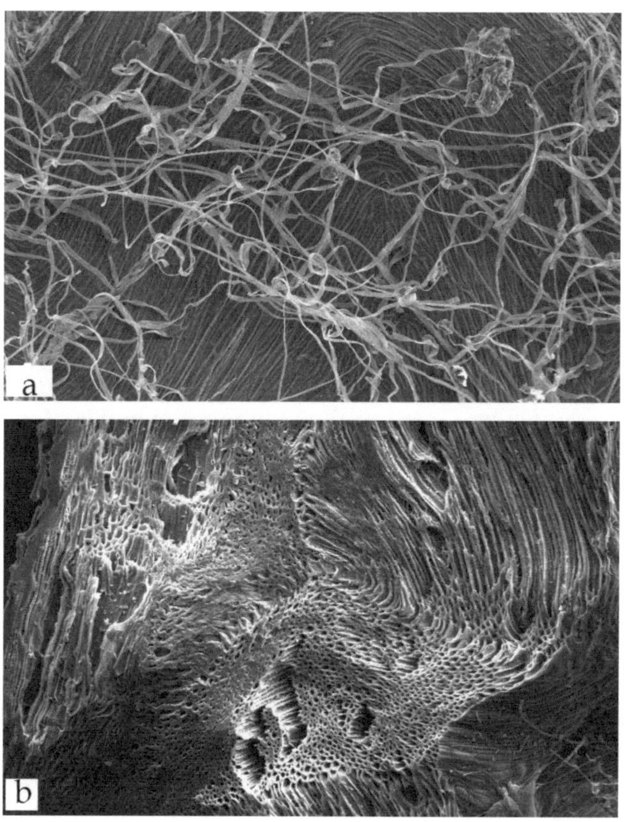

Fig. 3. SEM micrographs of (a) oleaster stones and (b) chars produced from them

3.2 Activated carbon yields

Figure 4 illustrates variations in the yield of activated carbon produced at varying temperatures and conditions. It is evident from the graph that activated carbon yields are affected by the activation method and carbonization temperature. With increasing temperature the yield of activated carbons produced by physical, chemical and sequential activation exhibits a downward trend. The yields obtained through sequential activation were found to be significantly low.

As the process of sequential activation involves the use of both potassium hydroxide and carbondioxide, there is an increase in the decomposition of the structure. In other words, with increasing decomposition more volatiles are released, which leads to a lower yield.

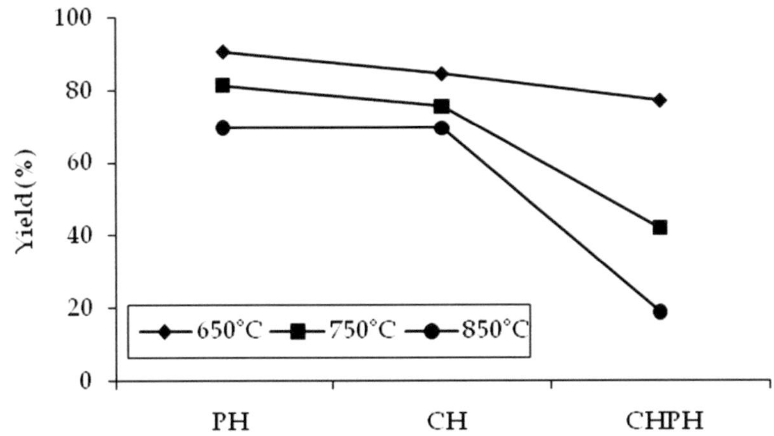

Fig. 4. Variations in activated carbon yields in relation to conditions for the production of activated carbon

3.3 Structural characterization of activated carbons

3.3.1 Isotherms

Figure 5 gives the nitrogen adsorption isotherms at 77K of activated carbons produced at three different temperatures by means of three different methods. An investigation of the adsorption isotherms found them to be isotherms (Type 1) in accordance with IUPAC classification except for activated carbon PH650. Based on this, we can speak of high microporosity (Sing et al., 1985-IUPAC Recommendations).

Adsorption of activated carbons produced at 650°C, 750°C and 850°C displays an upward trend from the lowest to the highest depending on physical, chemical and sequential activation methods in their respective order. Moreover, for each activation method, as temperature increases, so does the adsorption of activated carbons.

The experiments carried out at 650°C revealed that chemically produced activated carbons have a higher adsorption tendency compared to that of activated carbons produced by physical and sequential methods. There was an increase in the adsorption tendency of activated carbons obtained at 750°C and 850°C using all three methods. Activated carbons produced by sequential activation at both temperatures were established to have a comparatively higher adsorption tendency.

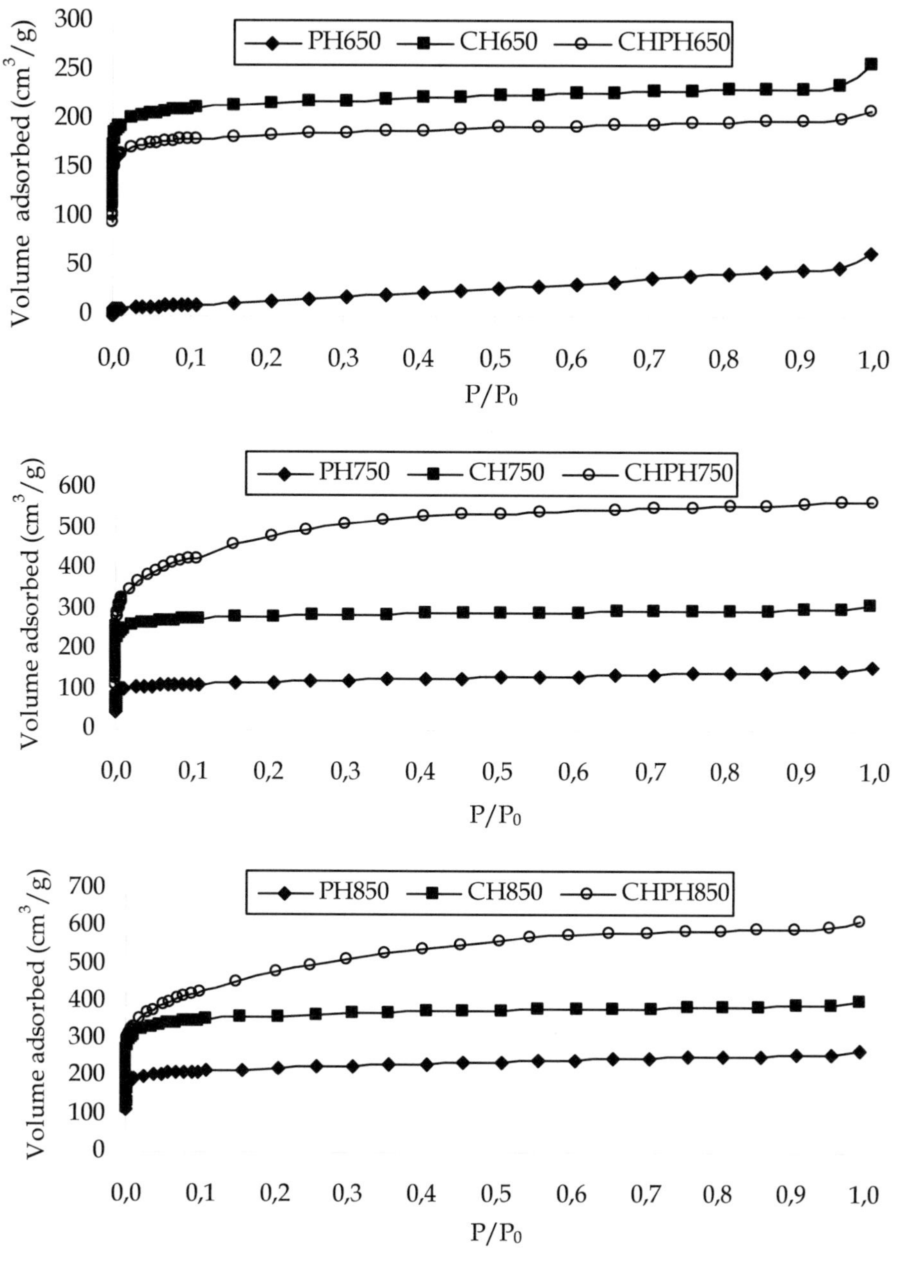

Fig. 5. Nitrogen adsorption isotherms

3.3.2 Surface area

Figure 6 illustrates variations in BET and micropore surface areas of activated carbons produced under three different activation conditions and at three different temperatures. The graph shows that BET and micropore surface areas exhibit variations depending on the activation method and temperature.

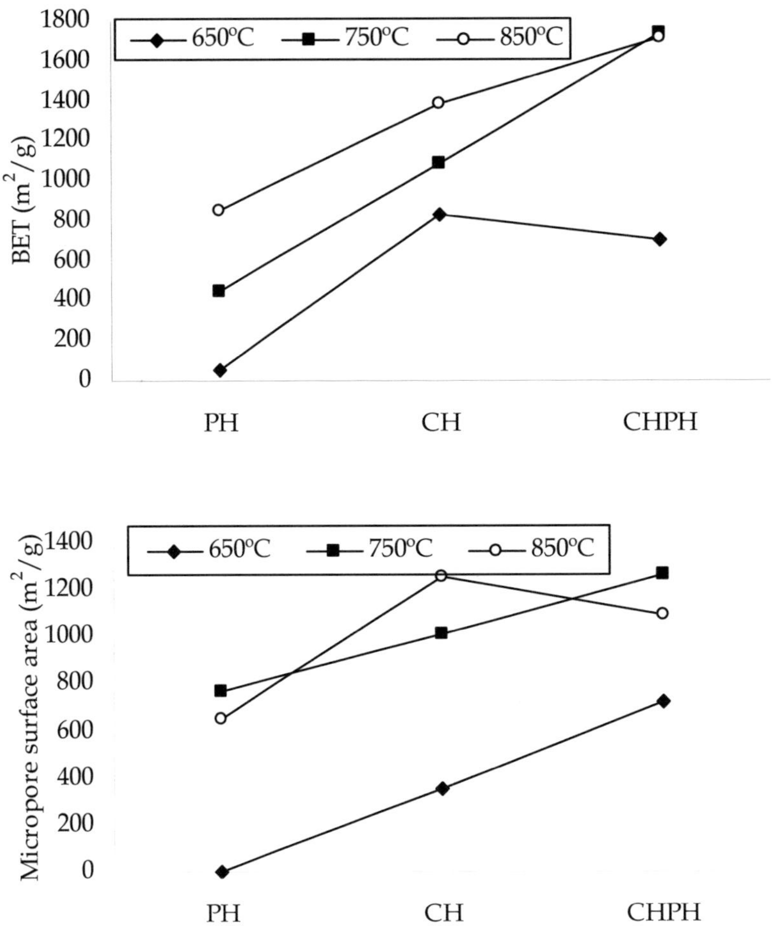

Fig. 6. Variations in BET and micropore surface areas in relation to activation method and temperature

The highest BET and micropore surface area were achieved at a carbonization temperature of 650°C through the production of activated carbons by chemical activation. Activated carbons PH650, CH650 and CHPH650 were found to have BET values of 53 m^2/g, 830 m^2/g and 707 m^2/g, respectively. The micropore surface areas of activated carbons PH650, CH650 and CHPH650 were established to be 0 m^2/g, 765 m^2/g and 650 m^2/g, respectively. The BET surface area for PH650 obtained was found to be low and no pores were observed in the microstructure. It can be stated that physical activation is not effective at this carbonization temperature but chemical activation is suitable. The micropore percentage of activated carbons produced through chemical and sequential activation is 92%.

It was found that activated carbons obtained at 750 °C have a comparatively higher surface area than those produced at 650 °C. The BET values of activated carbons PH750, CH750 and CHPH750 were determined to be 447 m^2/g, 1084 m^2/g and 1733 m^2/g, respectively. The same activated carbons were found to have micropore surface areas of 356 m^2/g, 1008 m^2/g

and 1254 m^2/g, respectively. The percentage of the micropore surface area for PH750, CH750 and CHPH750 were established to be 79%, 93% and 72%, respectively. It is clear that the chemical and sequential methods at the same carbonization temperature are suitable for producing activated carbons with a high BET and microporosity. However, it was found that sequential activation is more effective at obtaining a higher BET surface area as compared to chemical activation, which is capable of producing structures with micropores.

As for activated carbons produced at a carbonization temperature of 850 °C, their surface areas were found to be higher than those produced at the other two temperatures. Activated carbons produced at this temperature by physical activation, chemical activation and sequential activation were found to have BET values of 849 m^2/g, 1387 m^2/g and 1713 m^2/g, respectively. The micropore surface areas of carbons produced by the same methods were established to be 721 m^2/g, 1261 m^2/g and 1094 m^2/g, respectively. The percentage of micropore surface area of activated carbons produced by means of physical, chemical and sequential activation were determined to be 85%, 91% and 64%, respectively. The BET surface areas were observed to display an upward trend in the order of physical, chemical and sequential activation. In contrast, sequential activation yields a lower micropore surface area. This decrease is attributable to the fact that micropores decompose to become larger.

A comparison of each carbonization temperature reveals that activated carbons produced by chemical activation have higher BET values. BET values of activated carbons obtained through sequential activation are higher compared to those of activated carbons produced by means of both physical and chemical activation.

Figure 7 illustrates how total pore and micropore volumes vary depending on the carbonization temperature and activation method employed.

The highest total pore volume (0,4001 cm^3/g) was achieved through chemical activation employed in experiments carried out at a carbonization temperature of 650 °C. At the same carbonization temperature, physical activation and sequential activation yielded total pore volumes of 0,1014 cm^3/g and 0,3273 cm^3/g, respectively. Micropore volume displays variation similar to that observed in total pore volume. It was determined that physical activation does not lead to the formation of micropores. Total pore volume obtained through chemical activation and sequential activation were calculated to be 77% and 79%, respectively. Sequential activation at the same carbonization temperature results in micropore volume increasing.

At 750 °C total pore volume was observed to increase during physical, chemical and sequential activation. For these activation methods, total pore volumes were found to be 0,2441 cm^3/g, 0,4820 cm^3/g and 0,9529 cm^3/g, respectively. For the same activation methods, the micropore volume percentages have values of 59%, 84% and 55%, respectively. At this temperature, micropore volume obtained by means of chemical activation was determined to be higher compared to that achieved by means of the other methods.

Total pore volume achieved at 850 °C was established to be higher than that obtained at the other carbonization temperatures. Physical, chemical and sequential activation at this temperature yielded total pore volumes of 0,4285 cm^3/g, 0,6294 cm^3/g and 0,9557 cm^3/g, respectively. The micropore volume percentages were calculated to be, in the same order of activation methods employed, 68%, 80% and 49%, respectively. Chemical activation produced a higher micropore volume, whereas micropore volume obtained through sequential activation proved to be comparatively lower.

The densest micropore structure was achieved in activated carbons produced through chemical activation at carbonization temperatures of 750ºC and 850ºC. During chemical activation at three cabonization temperatures, KOH reacts with carbon to form an alkali metal carbonate. This, in turn, decomposes at high temperatures, and the resultant carbon dioxide leads to new pores being formed and the micropores becoming larger (Alcanz-Monge & Illan-Gomez, 2008; Nabais et al., 2008; Tseng et al., 2008). As the sequential activation method involved using both KOH and CO_2, the micropores and new pores become larger. With the physical activation method, carbon dioxide proved to be ineffective at forming new pores.

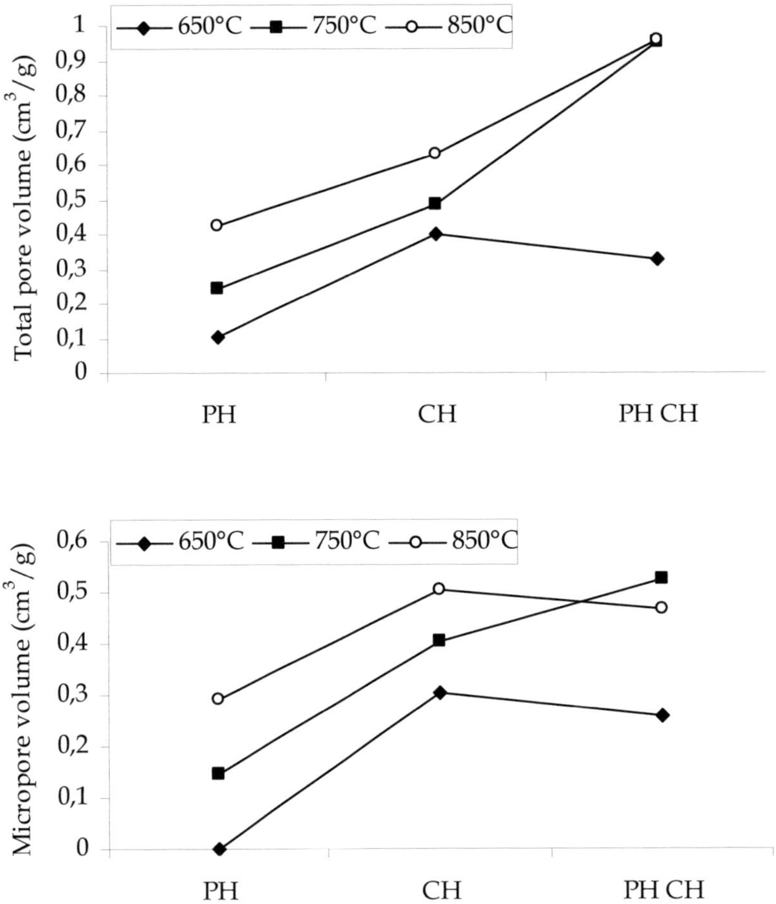

Fig. 7. Variations of total pore and micropore volumes in relation to carbonization temperature and activation method

3.3.3 Pore size distribution

Figure 8 gives variations of pore size distribution calculated based on the DFT method depending on carbonization temperature and the activation method employed.

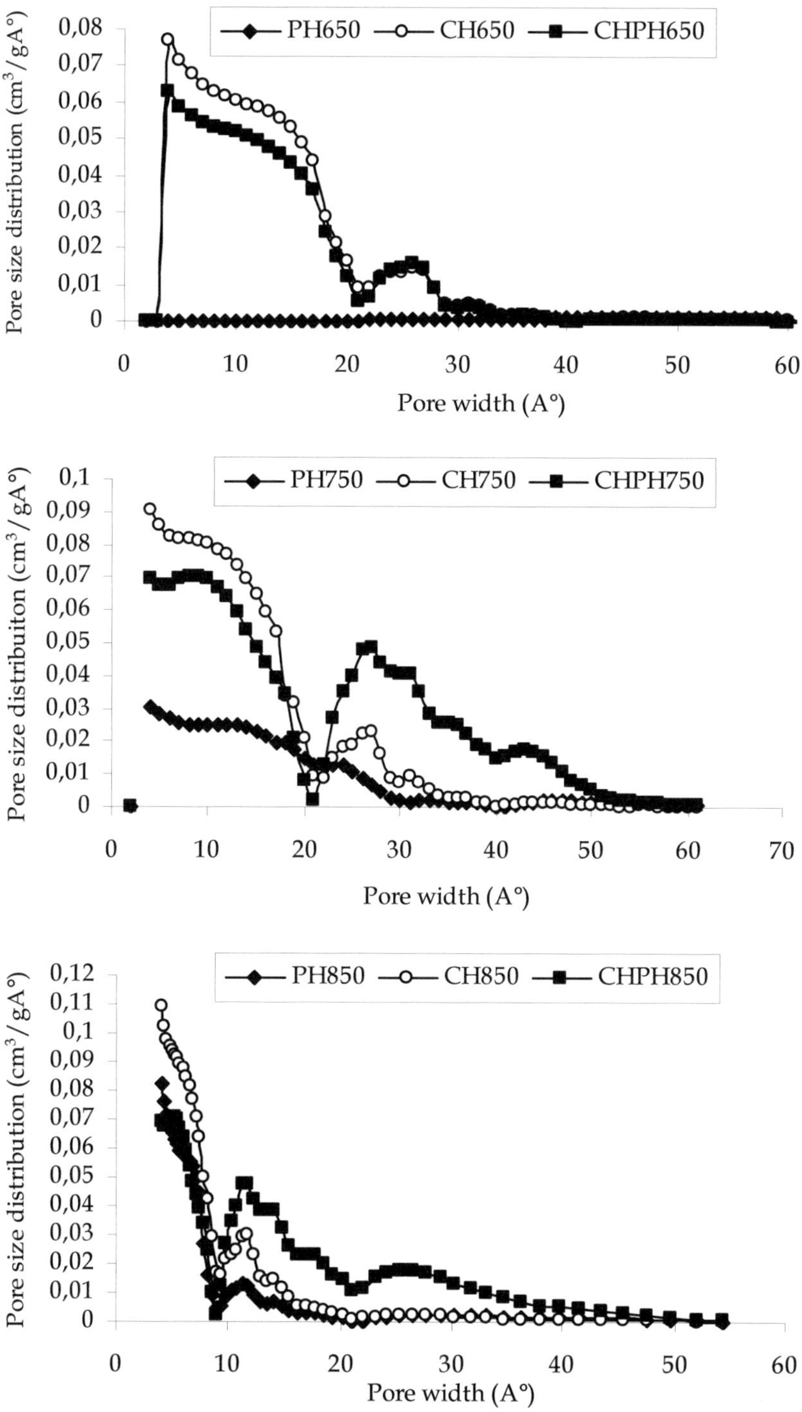

Fig. 8. Variations in pore size distribution in relation to carbonization temperature and activation method

The pore size of activated carbons produced by physical activation at a carbonization temperature of 650 °C is in the range of 4-55 A°. Moreover, this activated carbon has a very low BET surface area (53 m^2/g) and its micropore surface area could not be determined. The pore size distribution of activated carbons produced through chemical and sequential activation methods is observed to be in the ranges of 2-20 A° and 20-35 A°, respectively. This indicates that activated carbons have, along with mesopores, a more dense micropore contents.

A carbonization temperature of 750°C is observed to lead to both micro- and mesopores forming. Physical activation yielded a pore size distribution in the ranges of 4-20 A° and 20-30 A°, chemical activation a pore size distribution in the ranges 4-21 A° and 21-34 A°, and sequential activation led to a pore size distribution within the ranges of 4-20 A° and 20-51 A°. Chemical activation made it possible for micropores to become more dense at this temperature. As for sequential activation, it was observed to bring about an increase in mesopore density.

It was observed that micropores decrease and mesopores increase even more at a carbonization temperature of 850 °C. At this temperature, the decomposition of the structure displays an upward trend. Physical activation produced pore size distribution in the ranges of 4-9 A° and 9-19 A°, chemical activation led to a pore size distribution ranging from 4 to 9 A° and from 9 to 19 A°, and the pore size distribution achieved through sequential activation was within the ranges of 4-9 A°, 9-12 A° and 9-19 A°. At this temperature, new micropores are formed and the existing and new micropores decompose to form mesopores. The densest micropore structure was achieved in activated carbons produced through chemical activation at carbonization temperatures of 750 °C and 850 °C.

3.3.4 FTIR spectra

Figure 9 gives FTIR spectra of activated carbons obtained at three different carbonization temperatures using three different activation methods.

The band observed at 3600-3200 cm^{-1} is not present in chars but visible in the spectra of activated carbons produced using the three activation methods. This is because chemical activation and physical activation applied caused oxygen compounds to enter the structure. The aliphatic groups in the structure of activated carbons are observed at 3000-2800 cm^{-1} . Aromatic structures associated with the band observed 1600-1500 cm^{-1} is not visible in the spectra of activated carbons produced by sequential activation.

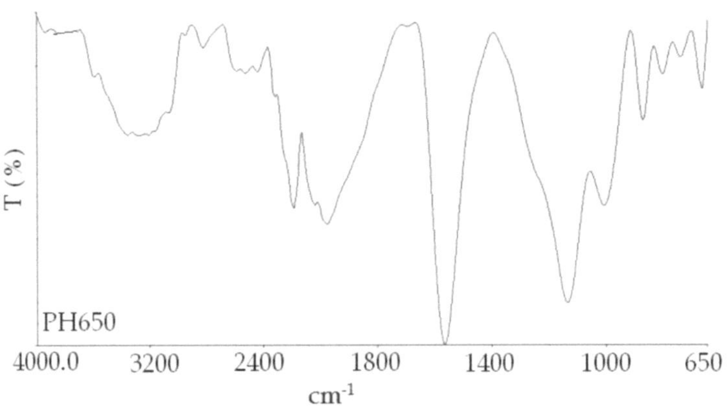

Fig. 9. FITR spectra of activated carbons

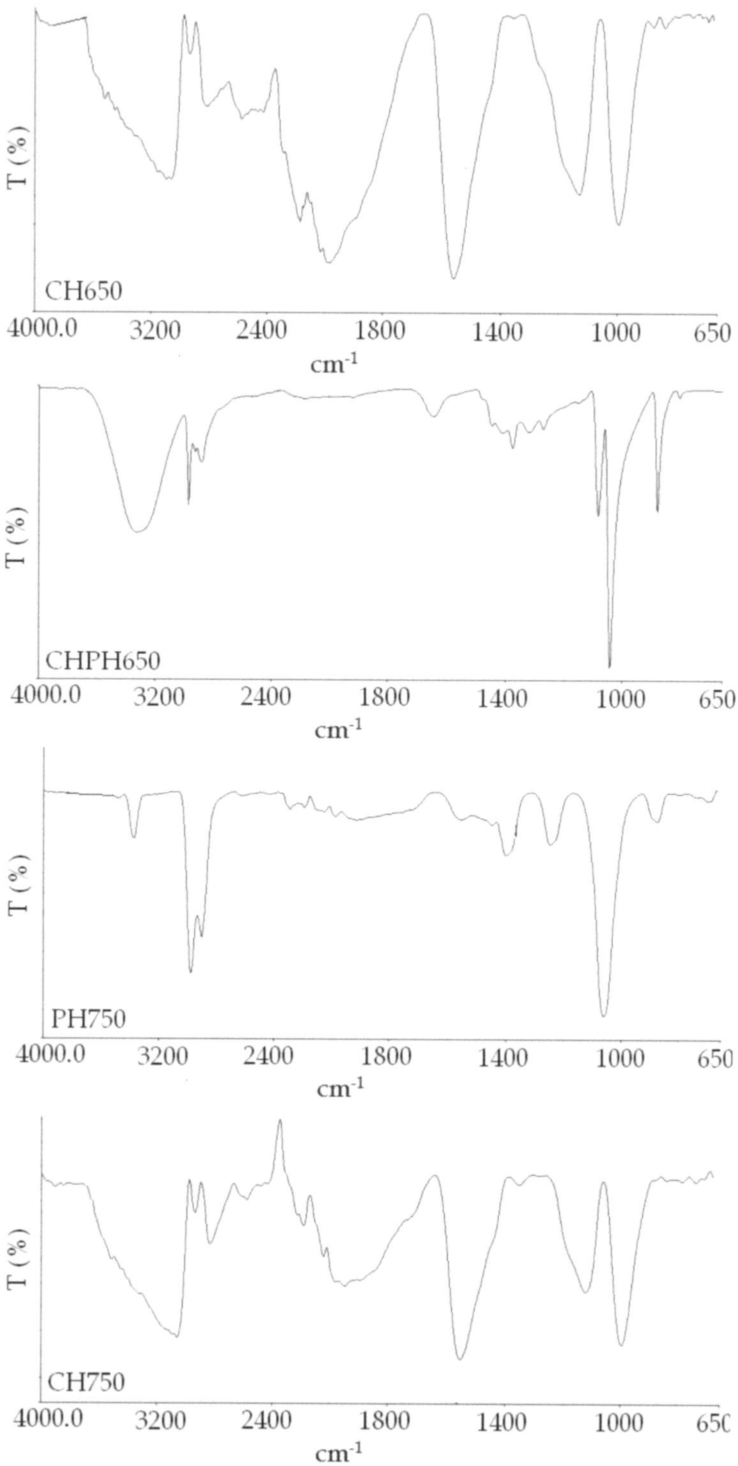

Fig. 9. Continued

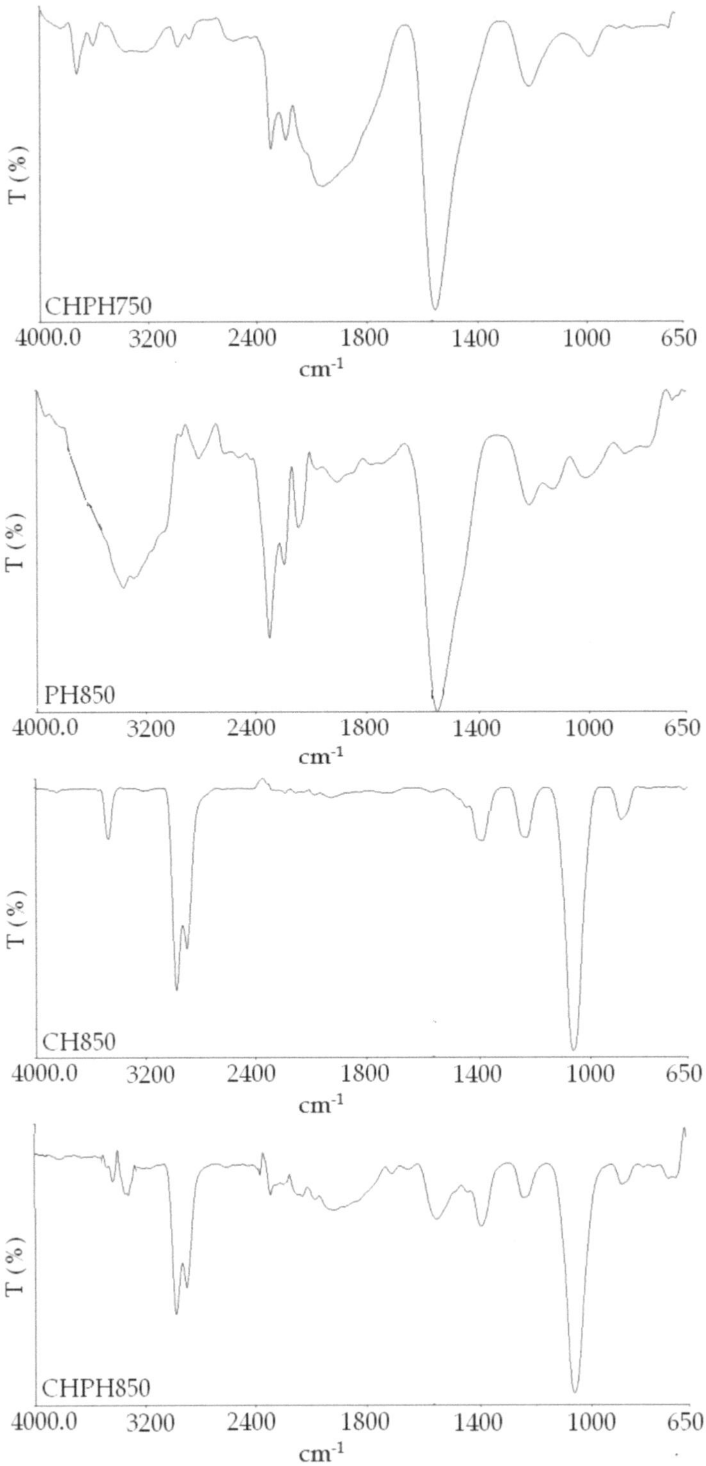

Fig. 9. Continued

Alkene groups at 1450-1300 cm^{-1} are observed as a multiple peak in activated carbons produced using the sequential activation method.
The bands (1240-1000 cm^{-1}) indicative of phenolic and alcoholic structures also occur in activated carbons.
It is evident from the FTIR spectra that functional groups present in oleaster stones decreased, disapeared or became smaller in their chars. Functional groups occurring in the structure of activated carbons produced by physical, chemical and sequential activation at 650°C, 750°C and 850°C exhibited variations as opposed to functional groups in chars. It is evident from the FTIR spectra that the structure of activated carbons was found to contain aromatic, aliphatic and oxygen-containing functional groups.

3.3.5 SEM micrographs

Figure 10 depicts SEM micrographs of activated carbons obtained at three different carbonization temperatures by means of three activation methods.
It can be concluded from SEM micrographs taken during experiments performed at a carbonization temperature of 650°C that the fibers disintegrated and no porous structure was formed. This proves that the value of surface area is low. It is observed that chemical and sequential activation lead to the formation of pores but, do not provide a homogenous distribution.
Physical activation at a carbonization temperature of 750°C was observed to lead to the formation of pores. The chemical and activation methods not only maintained the fibrous structure, but made it possible for pore distribution to be homogenous as well.
Physical activation at a carbonization temperature of 850°C made the porous structure of the activated carbon produced even clearer. In contrast, the chemical and sequential activation methods resulted in the pores decomposing.

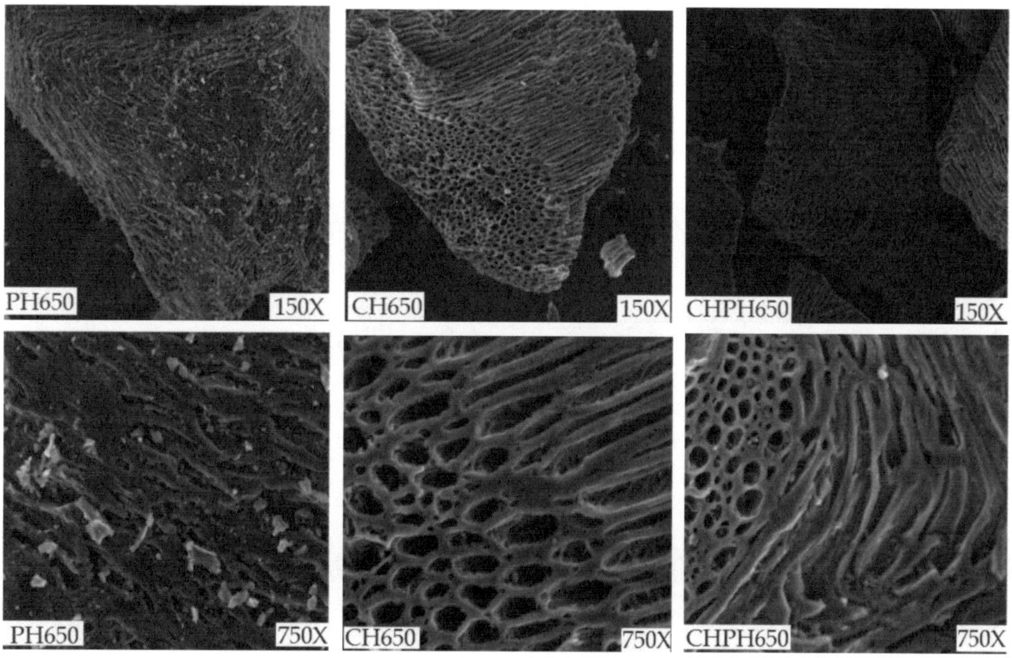

Fig. 10. SEM micrographs of activated carbons (150X and 750X)

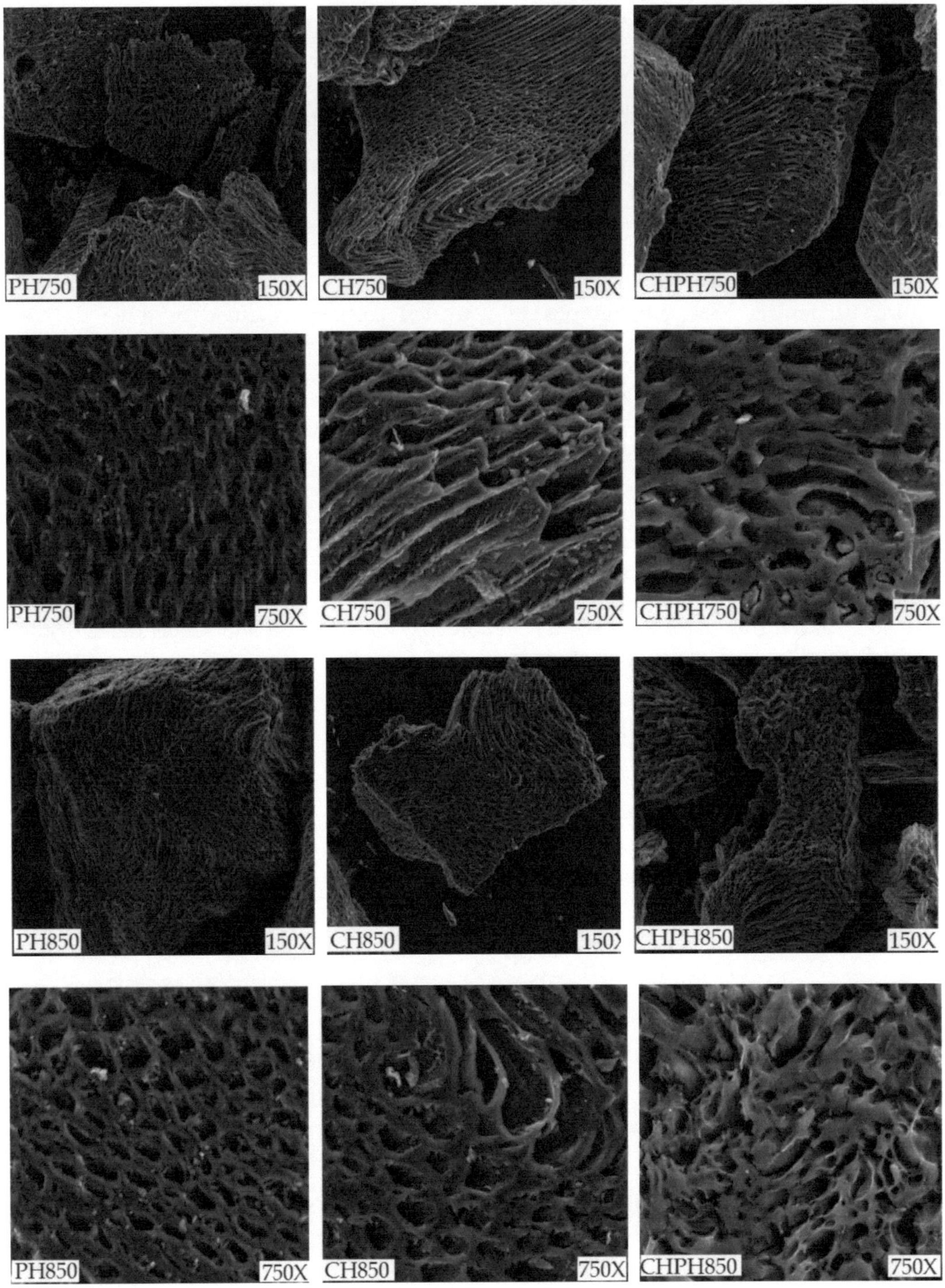

Fig. 10. Continued

3.3.6 Iodine number

The iodine number is a technique employed by producers, sellers, researchers etc. in order to determine the adsorption capacity of activated carbons. The iodine number is the amount of iodine adsorbed by 1g of carbon at the mg level. The iodine value is a measure of porosity for activated carbons. However, no relationship can be established between the iodine number and surface area (ASTM D4607, 2006; Qui&Guo, 2010). The iodine number displays variation depending on the raw material, production conditions and the distribution of the pore volume (ASTM D4607, 2006).

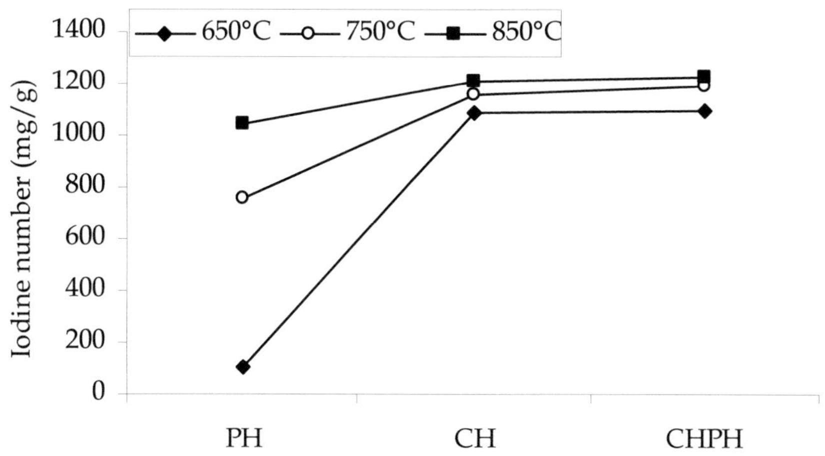

Fig. 11. Variations in the iodine number in relation to the activation method employed and carbonization temperature

Variations in the iodine numbers of activated carbons are given in Table 11. The iodine number is affected by both carbonization temperature and the activation method applied.

4. Conclusion

In this study, I sought to produce activated carbons by physical, chemical and sequential (chemical+physical) activation at carbonizaton temperatures of 650 °C, 750 °C and 850 °C.
It has been established that the porous structure parameters of the activated carbons produced are affected by both carbonization temperature and the activation method employed. The chemical and sequential activation methods led to the formation of activated carbons with a relatively higher BET and micropore surface starting from a carbonization temperature of 750 °C in particular.
Activated carbon produced by means of the sequential activation method at a cabonization temperature of 750 °C yielded the highest BET surface area of 1733 m^2/g. The highest micropore surface area was achieved through chemical activation at a carbonization temperature of 850 °C. By contrast, the highest percentage of micropore surface area with 93% was obtained by means of chemical activation at a carbonization temperature of 750 °C.
The iodine number was also affected by both carbonization temperature and the activation methods employed. Activated carbon obtained at a carbonization temperature of 850 °C using the sequential activation method yielded the highest iodine number.

Also, the FTIR spectra and SEM micrographs taken confirm that, due to their structural characterization, oleaster stones are a suitable material for activated carbon production, and accordingly, use as adsorbents.

5. Acknowledgement

The writer would like to express her gratitude to the Management Zonguldak Karaelmas University Scientific Resaerch Fund (Project No.2008-70-01-01) for their financial assistance at the project level.

6. References

Alcaniz-Monge, J. & Illan-Gomez, M.J. (2008). Insight into hydroxides-activated coals: Chemical or physical activation?, Journal of Colloid Interface Science, Vol. 318, Issue:1, pp. 35-41.

Alcaniz-Monge, J., Perez-Cadenas, M. & Lozano-Castello, D. (2010). Effect of the preoxidation of coals in the preparation of chemically activated carbon pellets, *Energy & Fuels,* Vol. 24, pp. 3385-3393.

Antal, M.J. (1982). Biomass pyrolysis: a review of the literature, Part I carbohydrate pyrolysis, *In Advances in Solar Energy,* (Boer, K.W., Duffie, J.A., Eds.), American Solar Energy Society: Boulder.

ASTM D4607-94 (2006). Standard test method for determination of iodine number of activated carbon.

Aworn, A., Thiravetyan, P. & Nakbanpote, W. (2009) Preparation of CO_2 activated carbon from corncob for monoethylene glycol adsorption, *Colloids And Surfaces A-Physicochemical And Engineering Aspects,* Vol. 333, Issue: 1-3, pp. 19-25.

Aygun, A., Yenisoy-Karakas, S. & Duman, I. (2003). Production of granular activated carbon from fruit stones and nutshells and evaluation of their physical, chemical and adsorption properties, *Microporous and Mesoporous Materials*, Vol. 66, Issue:2-3-, pp. 189-195.

Cuhadaroglu, D. & Uygun, O.A. (2008). Production and characterization of activated carbon from a bituminous coal by chemical activation, *African Journal of Biotechnology*, Vol. 7, Issue: 20, pp. 3703-3710.

Demiral, H., Demiral, I., Tumsek, F. & Karabacakoğlu, B. (2008). Pore structure of activated carbon prepared from hazelnut bagasse by chemical activation, *Surface and Interface Analysis*, Vol. 40, Issue:3-4,pp.616-619.

Demiral, H. & Demiral, I. (2008). Surface properties of activated carbon prepared from wastes, *Surface and Interface Analysis*, Vol. 40, Issue, 3-4, pp.612-615.

Gupta, V.K., Gupta, B., Rastogi, A., Agarwal, S. & Nayak, A. (2011). A comparative investigation on adsorption performances of mesoporous activated carbon prepared from waste rubber tire and activated carbon for a hazardous azo dye-Acid Blue 113, *Journal of Hazardous Materials*, Vol. 186, Issue: 1, pp. 891-901.

Ip, A.W.M., Barford, J.P. & Mckay, G. (2008). Production and comparison of high surface area bamboo derived active carbons, *Bioresource Technology*, Vol. 99, Issue:18, pp. 8909-8916.

IUPAC Manual of symbols and terminology for physicochemical quantities and units. (1972). Butterworths, London.

Jankowska, H. Swiatkowski, A. & Choma, J. (1991). *Active Carbon*, Ellis Horwood.

Kim, D.S. (2004), Activated carbon from peach stones using phosphoric acid activation at medium temperatures, *Journal of Environmental Science and Health Part A-Toxic/Hazardous Substances & Environmental Engineering*, Vol. 39, Issue: 5, pp. 1301-1318.

Kroschwitz, J.I. (Ed.) (1992). *Kirk-Othmer Encyclopeia of Chemical Technology*, John Wiley & Sons.

Liu, L.S., Liu, Z.Y., Yang, J., Huang, Z. & Liu, Z. (2007). Effect of preparation conditions on the properties of a coal-derived activated carbon honeycomb monolith, *Carbon*, Vol. 45, Issue: 14, pp. 2836-2842.

Lu, C., Xu, S. & Liu, C. (2010). The role of K_2CO_3 during the chemical activation of petroleum coke with KOH, *Journal of Analytical and Applied Pyrolysis*, Vol. 87, Issue:2, 282-287.

Luangkiattikhun, P., Tangsathitkulchai, C. & Tangsathitkulchai, M. (2008). Non-isothermal thermogravimetric analysis of oil-palm solid wastes, *Bioresource Technology*, Vol. 99. pp. 986–997.

Marsh, H. & Rodriguez-Reinoso, F. (2006). *Activated Carbon*, Elsevier.

Moreno-Pirajan, J.C., Gomez-Cruz, R., Garcia-Cuello, V.S. & Giraldo, L. (2010). Binary system Cu(II)/Pb(II) adsorption on activated carbon obtained by pyrolysis of cow bone study, *Journal of Analytical and Applied Pyrolysis*, Vol. 89, Issue:1, pp.122-128.

Nabais, J.M., Laginhas, C., Carrott, P.J.M., Ribeiro Carrott, M.M.L., Galacho, C., Amoros, J.E.C. & Gisbert, A.V.N. (2010). Characterization of the surface of activated carbons produced from tire residues, *Materials Science Forum*, Vol. 636-637, pp. 1383-1388.

Nabais, J.V., Carrott, P., Ribeiro Carrott, M.M.L., Luz, V. & Ortiz, A.L. (2008). Influence of preparation conditions in the textural and chemical properties of activated carbons from a novel biomass precursor: The coffee endocarp, *Bioresource Technology*, Vol. 99, No:15, pp.7224-7231.

Ould-Idriss, A., Stitou, M., Cuerda-Correa, E.M., Fernandez-Gonzalez, C., Macias-Garcia, A., Alexandre-Franco, M.F. & Gomez-Serrano, V. (2011). Preparation of activated carbons from olive-tree wood revisited. II. Physical activation with air, *Fuel Processing Technology*, Vol. 92, Issue: 2, pp. 266-270.

Popescu, M.A., Popescu, C.M., Lisa, G. & Sakata, Y. (2011). Evaluation of morphological and chemical aspects of different wood species by spectroscopy and thermal methods, *Journal of Molecular Structure*, Vol. 988, pp. 65–72.

Preethi, S., Sivasamy, A., Sivanesan S, Ramamurthi, V. & Swaminathan, G. (2006). Removal of safranin basic dye from aqueous solutions by adsorption onto corncob activated carbon, *Industrial & Engineering Chemistry Research*, Vol. 45, Issue: 22, pp. 7627-7632.

Sing, K.S.W., Everett, D.H., Haul, R.A.W., Moscou, L., Pierotti, R.A., Rouquerol, J. & Siemieniewska, T. (1985). Reporting physisorption data for gas/solid systems with special reference to the determination of surface area and porosity,, IUPAC Recommendations (1984), *Pure Applied Chemistry*, Vol. 57, No: 4, pp. 603-619.

Singh, Y. (2011). Wealth from waste, Activated Carbon. http://www.wealthywaste.com/activated-carbon

Sun, K. & Jiang, J.C. (2010) Preparation and characterization of activated carbon from rubber-seed shell by physical activation with steam, *Biomass & Bioenergy*, Vol. 34, Issue: 4, pp. 539-544.

Sun, Y., Zhang J.P., Yang, G. & Li, Z.H. (2007). Production of activated carbon by H_3PO_4 activation treatment of corncob and its performance in removing nitrobenzene from water, *Environmental Progress,* Vol. 26, No.1, pp. 78-85.

Soleimani, M. & Kaghazchi, T. (2007) Agricultural waste conversion to activated carbon by chemical activation with phosphoric acid, *Chemical Engineering & Technology,* Vol. 30, Issue: 5, pp. 649-654.

Sutcu, H. & Demiral, H. (2009) Production of granular activated carbons from loquat stones by chemical activation, *Journal of Analytical and Applied Pyrolysis*, Vol. 84, Issue: 1, pp. 47-52.

Sutcu, H. & Dural, A. (2007). The adsorption of lead, copper and nickel ions from aqueous solutions on activated carbon produced from bituminous coal, *Fresenius Environmental Bulletin*, Vol. 16, ıssue: 3, pp. 235-241.

Qiu G. & Guo, M. (2010). Quality of poultry litter-derived granular activated carbon, *Bioresource Technology*, Vol. 101, pp.379-386.

Tongpoothorn, W., Sriuttha, M., Homchan, P. Chanthai, S. & Ruangviriyachai, C. (2011). Preparation of activated carbon derived from Jatropha curcas fruit shell by simple thermo-chemical activation and characterization of their physico-chemical properties, *Chemical Engineering Research and Design*, Vol. 89, pp. 335–340.

Tseng, R.L., Tseng, S.K., Wu, F.C., Hu, C.C. & Wang, C.C. (2008). Effects of micropore development on the physicochemical properties of KOH-activated carbons, *Journal of Chinese Institute of Chemical Engineers,* Vol. 39, Issue: 1, pp.37-47.

Yavuz, R., Akyildiz, H., Karatepe, N. & Çetinkaya, E. (2010). Influence of preparation conditions on porous structures of olive stone activated by H_3PO_4, *Fuel Processing Technology,* Vol. 91, Issue: 1, pp. 80-87.

Yeganeh, M.M., Kaghazchi, T. & Soleimani, M. (2006). Effect of raw materials on properties of activated carbons, *Chemical Engineering & Technology*, Vol. 29, Issue: 10, pp. 1247-1251.

14

Effect of the Presence of Subtituted Urea and also Ammonia as Nitrogen Source in Cultivied Medium on *Chlorella*'s Lipid Content

Anondho Wijanarko
Department of Chemical Engineering, Universitas Indonesia,
Jalan Prof. Fuad Hasan, Kampus UI,
Indonesia

1. Introduction

Global warming has become one of the most serious environment problems. The main cause of this is because of the increasing of CO_2 level in the atmosphere. In recent years, many attempts have been done to reduce the quantity of CO_2 in the atmosphere. Studies on photosynthesis, CO_2 fixation and utilization of micro algae biomass has been carried out. Similar to another *Chlorella* strain, *Chlorella vulgaris* Buitenzorg is known widely of its high valued potential substances such as chlorophyll, CGF, carotene, and protein, and it can be used as potential biomass albeit the function of CO_2 fixation and also possible content long chain un-saturated fatty acid potencies biodiesel as a renewable fuel stock. These characteristics suggest that *Chlorella* is potential for removal and utilization of CO_2 to minimize the accumulation carbon dioxide emitted from industrial plant as a solution to GHG problem.

For its growth, CO_2 that was also enriched by a little content of unburned hydro carbon (PAH), NO_x, SO_x, CO in flue gas (Wijanarko & Dianursanti, 2009; Dianursanti et al, 2010), *Chlorella* needs light energy that was converted to chemical energy in the form of ATP to be used in photosynthesis, metabolism, growth and cell division. It also need substrates such bi-phosphoric salt as phosphor source that was functioned in phosphoric linkage of RNA and DNA structure; urea, nitrate salt or mono ethanol amine as nitrogen source that is an important factor for protein synthesis and cellular growth (Ohtaguchi & Wijanarko, 2002). Based on previous work using *Chlorella*, this work uses a large flat surface photo bioreactors as a part of scale up design for large scale biomass production by using NO_x enriched flue gas utilization as carbon source and also using ammonia or urea as substitution nitrate salt content in its substrate medium as simulated waste contaminated water.

2. Materials & methods

Chlorella vulgaris Buitenzorg is taken from Depok Fresh Water Fishery Research Center that was grown in Benneck medium. This strain grows in 18.0 dm^3 of culture medium in bubble column photo bioreactor that have sizing of (38.5 cm x 10 cm x 60 cm). Experimental apparatus used in the experiment is shown on **Figure 1**.

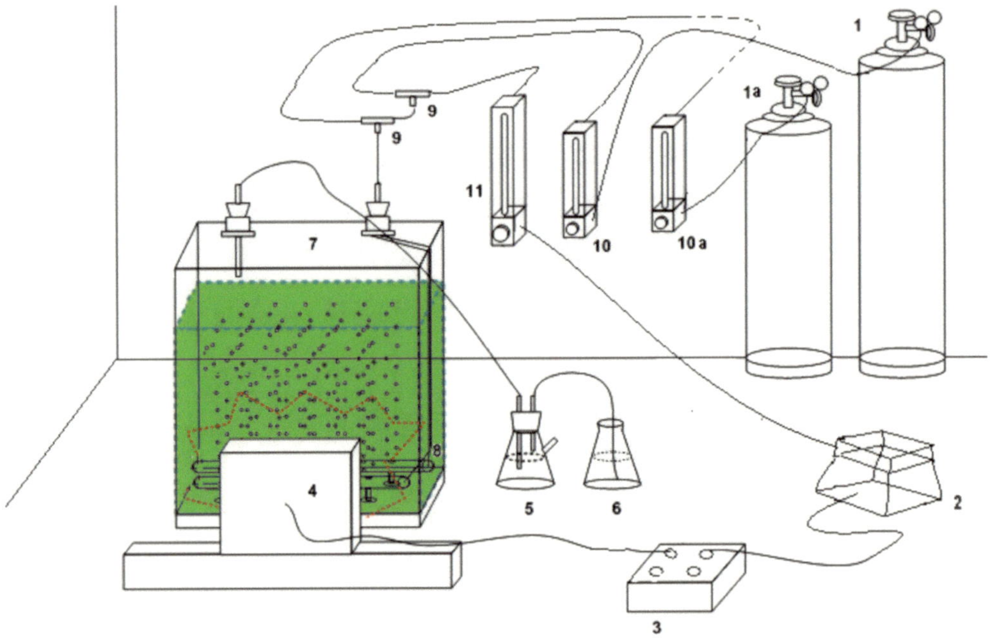

1. CO_2 storage bomb; 1a. N_2O storage bomb; 2. Air blower; 3. Electric switch; 4. Light source; 5. CO_2 trapped erlenmeyer; 6. CO_2 Erlenmeyer discharge; 7. Flat bubble column photo – bioreactor; 8. Gas bubbler; 9. T – Septum; 10. CO_2 flow-meter; 10a. N_2O flow-meter; 11. Air Flow – meter.

Fig. 1. Experimental apparatus

Conditions were defined as following. Temperature (T) was set at 29.0 °C (302 K), Pressure (P) was set at ambient pressure (1 atm.; 101 kPa), Light intensity (I) was set at 3.0 Klx, superficial gas velocity (U_G) was set at 15.7 m/h and CO_2 concentration (y_{CO2i}) in blown bubble air was set around 5.0%. Before cultivation, this strain was grown with pre-culture condition that was set by blowing bubble fresh air with U_G 1.0 vvm with similar operation condition. These photo bioreactors are illuminated by 4 (four) lamps [*Philips Halogen* lamp 20W/12V/50Hz].

Culture biomass content (OD_{600} method) was measured at 600 nm using UV-Vis Spectrophotometer (Labo-Med Inc.); Ammonia was measured at 425 nm using Spectrofotometer and calculated by Nessler method; Lipid content is analysis by Bligh-Dyer Method [Manirakizal et al, 2001); extracted fatty acid content is analyzed using GCMS; protein was measured by Lowry method; elemental analysis is done by XRD and CHNS analyzer; CO_2 inlet and outlet is measured using TCD Gas Chromatography; Chlorophyl a and carotene contents are assayed and calculated by pigment assay procedure (Richmond, 2004; Wijanarko et al, 2006a. 2006b).

3. Results & discussion

For industrial application purposes, utilization of waste water that was analyzed rich of nitrogen source such as urea CO $(NH_2)_2$, ammonia NH_3 or other excess nitrogen substance make biomass production more economically and important cause of a prediction of it's biomass contain more un-saturated fatty acid.

Figure 2 tend a determination of proper diluted nitrogen nutrients for *Chlorella* growth that it varied into control experiment that existed at the Benneck Medium (500 mg/L $NaNO_3$), deficiency diluted nitrogen (250 mg/L $NaNO_3$), excess diluted nitrogen (750 mg/L $NaNO_3$), and different diluted nitrogen sources (500 mg/L CO $(NH_2)_2$). At excess diluted nitrogen source that was shown at medium content 750 mg/L $NaNO_3$ and 500 mg/L CO $(NH_2)_2$, *Chlorella*'s growth result tend lower although growth result in medium content urea more higher than result on excess nitrate salt.

Based on our previous result that was known $CH_{3.3}N_{0.203}O_{0.322}P_{0.041}$ as biomass compound and was constructed from elemental analysis result of dry biomass of *Chlorell vulgaris* Buitenzorg, in presence of nitrate salt in cultivation media, whole chemical reaction of biomass cultivation (Dianursanti et al, 2010) could be shown as below:

$$CH_{3.3}N_{0.203}O_{0.322}P_{0.041} + 1.11\ H_2O + HCO_3^- + 0.041\ H_2PO_4^- + 0.203\ NO_3^- \rightarrow 2CH_{3.3}N_{0.203}O_{0.322}P_{0.041} + 2.03\ O_2 \quad (1)$$

Meanwhile, in case of presence of different diluted nitrogen sources such as $CO(NH_2)_2)$, whole chemical reaction of biomass cultivation could be changed as below:

$$CH_{3.3}N_{0.203}O_{0.322}P_{0.041} + 0.984\ H_2O + 0.898\ HCO_3^- + 0.041\ H_2PO_4^- + 0.102\ CO\ (NH_2)_2 \rightarrow 2CH_{3.3}N_{0.203}O_{0.322}P_{0.041} + 1.81\ O_2 \quad (2)$$

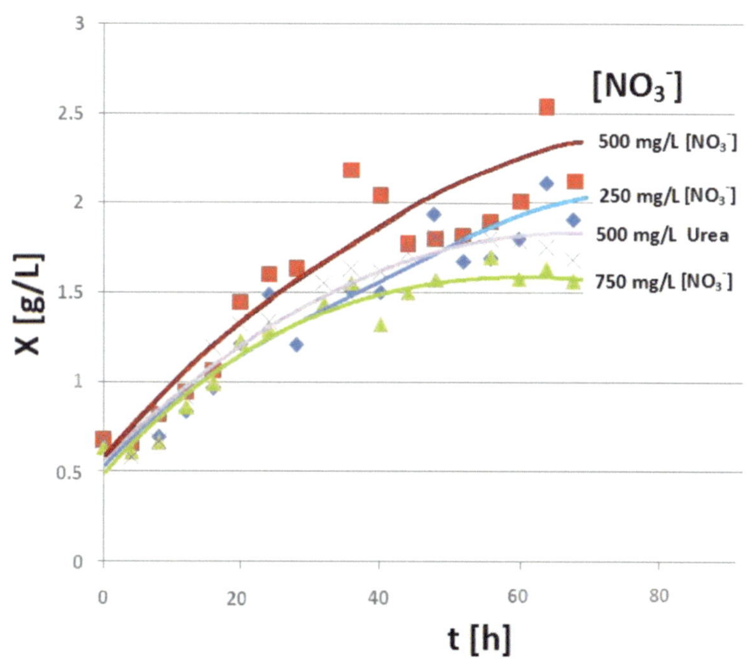

Fig. 2. Effect of composition nitrogen source on *Chlorella*'s growth at beginning 72 hours cultivation

It could be understood, presence of 500 mg/L CO $(NH_2)_2$ that was equivalent to two times concentration compare to diluted nitrate salt in cultivation media making nitrogen source concentration excess around 40% and then it change to form ammonium ion that was easily and freely to metabolize for making essential amino acid, protein and chlorophyll, cause of intracellular conversion of urea could be change to ammonium ion easily using urease (urea amidohydrolase) or urea amidolyase that was reacted together with ATP. Both of enzymes was commonly present in unicellular algae (Leftley & Syrett, 1973).

urea amidohydrolase pathway

$$CO(NH_2)_2 + H_2O \rightarrow CO_2 + 2NH_3 \quad (3)$$

urea amidolyase pathway

$$CO(NH_2)_2 + ATP + HCO_3^- \xleftrightarrow{Mg^{2+}K^+} \text{allophanate} + ADP + P_i \quad (4)$$

$$\text{allophanate} \rightarrow 2NH_3 + 2CO_2 \quad (5)$$

In case of nitrate assimilate reaction, intercellular conversion of nitrate ion was performed via nitrate reduction pathway need NADH that was also needed for intracellular lipid, protein and chlorophyll formation and it directly influence to cellular growth.

Nitrate Reduction pathway

$$NO_3^- + NADH + H^+ \xrightarrow{NR} NO_2^- + NAD^+ + H_2O \quad (6)$$

$$NO_2^- + 3H_2O + 2H^+ + h\upsilon \rightarrow NH_4^+ + 1.5O_2 + 2H_2O \quad (7)$$

Meanwhile, excess of intracellular ammonium ion or ammonia could be inhibited formation ATP in chloroplast [9] and it could be understood that optimum condition for *Chlorella*'s growth was around 500 mg/L $NaNO_3$ that existed at the Benneck Medium. This phenomenon could be impressed that Chlorella's growth was followed subtrate activation and inhibition model (Sallisbury & Ross, 1992).

Determination of proper diluted nitrogen nutrients for *Chlorella* growth shown that diluted nitrogen concentration in the Benneck medium (control) there is the most optimal nutrition to produce lipids up to 0.42 g / g biomass for biodiesel utilizing purpose [**Figure 3**].

Cause of intracellular conversion of urea could be change to ammonium ion more easily using both of intracellular algal's urease (urea amidohydrolase) or urea amidolyase, it could be understood why algal's lipid content of alga that was cultivated in diluted urea tend more high [0.3 g/g biomass] at beginning and hereafter shown relatively constant. Urea metabolism was not consumed NADH which was also necessary for intracellular lipid formation. In the meantime, composition of diluted nitrate ion as nitrogen source, at excess diluted nitrogen source that was shown at medium content 750 mg/L $NaNO_3$, algal's cellular produce lipid up to 0.40 g/g biomass but similar to experimental result that was held by Yanqun, as consequence of substrate activation and inhibition growth model, this lipid formation could be happen only at stationer phase of cellular growth (Bailey & Ollis, 1986).

Although cellular growth was decrease around 30%, presence of urea as nitrogen source, diluted urea in cultivation media is the most appropriate nutrients to produce protein until it reaches 0.54 g / g biomass [**Figure 4**]. This protein content is attractable for food supplement development purpose and it was around one and half times increasing compare to result on control experiment. The evidence of intracellular protein formation was closed similar to the reason of lipid formation. Urea metabolism was not consumed NADH which was also necessary for intracellular protein formation and produced ammonium was easily to metabolize for making essential amino acid and also protein (Leftley & Syrett, 1973; Sallisbury & Ross, 1992)

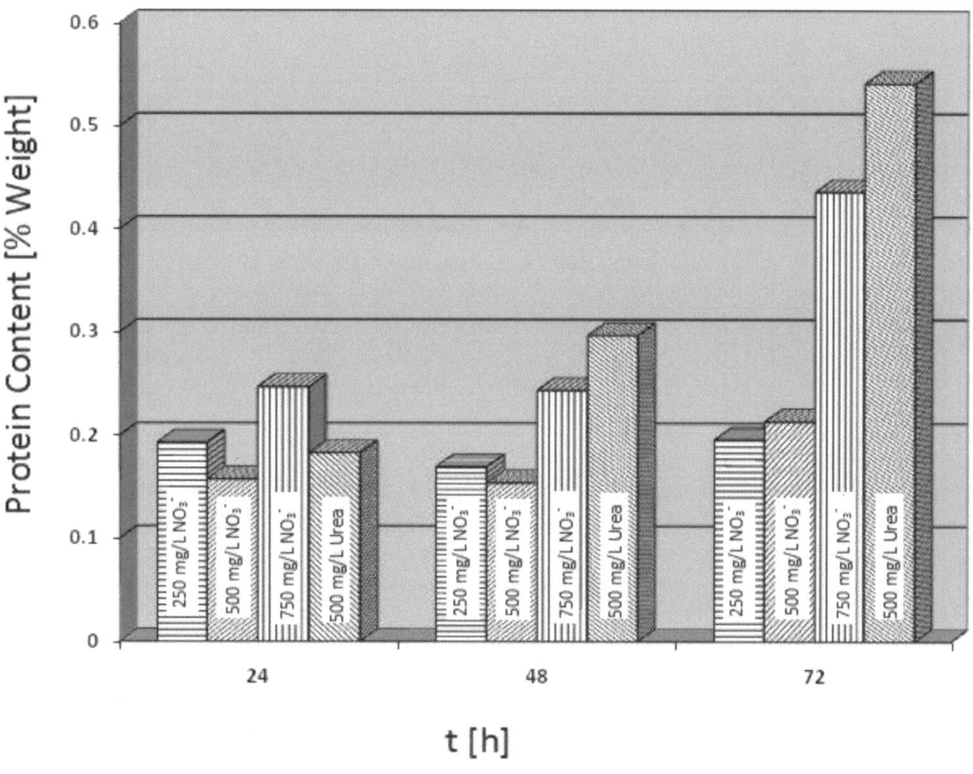

Fig. 3. Effect of composition nitrogen source on *Chlorella*'s lipid content at beginning 72 hours cultivation

Whereas, in excess diluted nitrogen (750 mg/L $NaNO_3$), cell growth produced relatively high protein content on its intracellular around 0.24 g / g biomass at the beginning and increasing to 0.43 g / g biomass at 72 h cultivation and it was closed to result in media contain urea as nitrogen source [**Figure 5**]. Cause of growth relatively lower than both of control experiment that existed at the Benneck Medium (500 mg/L $NaNO_3$) and deficiency diluted nitrogen (250 mg/L $NaNO_3$), increasing of ammonium as conversion produced of excess nitrate via nitrate reduction pathway, together with carbon metabolite product spontaneously could be metabolize for making essential amino acid and then also protein (Sallisbury & Ross, 1992).

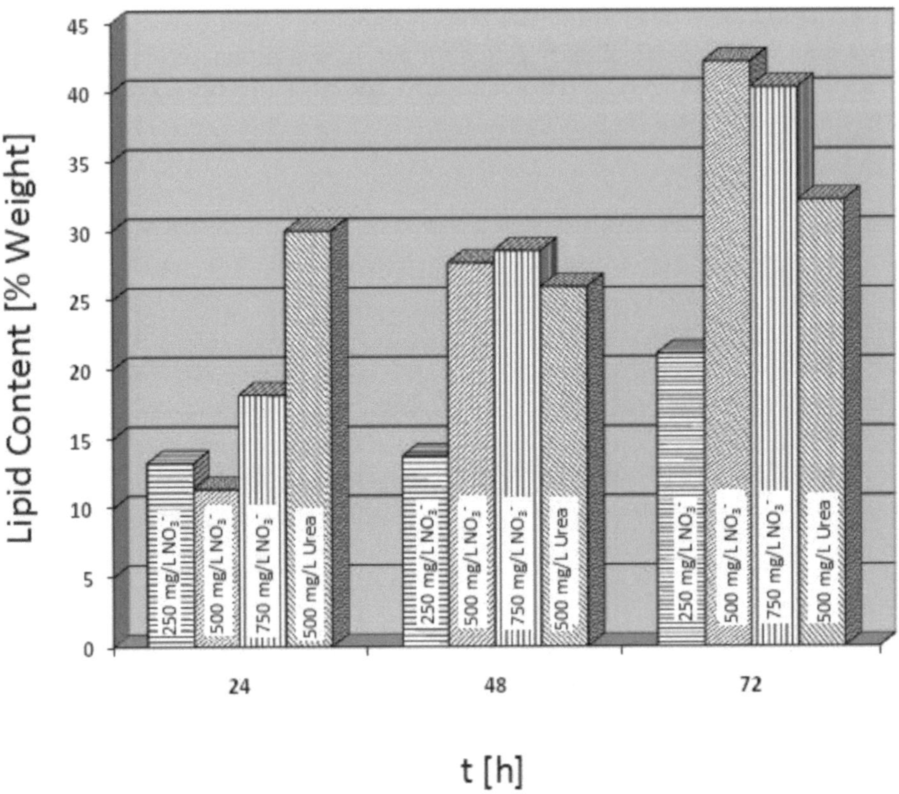

Fig. 4. Effect of composition nitrogen source on *Chlorella*'s protein content at beginning 72 hours cultivation

Furthermore, medium that excess diluted nitrogen is the most appropriate nutrients to produce chlorophyll and it reach 4.9 g/100g biomass at beginning 48 hours [**Figure 6**]. Similar to explanation in above, increasing of ammonium as conversion product from media contain excess nitrate via nitrate reduction pathway, beside making essential amino acid and then also protein, together with carbon metabolite product spontaneously could be metabolize for intracellular chlorophyll (Sallisbury & Ross, 1992). Meanwhile, presence of urea as nitrogen source, as consequence of its high cellular protein producing, algal's growth produce small amount of cellular chlorophyll.

Henceforth, presence of urea as nitrogen source, drastically change intracellular fatty acid content [**Table 1**]. It is shown that presence of urea as substitution species of nitrate salt in Benneck medium, was converted fatty acid C_{16} species (around 30.4 % C_{16} in Benneck) to be fatty acid C_{18} species significantly (around 77.0 % C_{18} in presence of urea) that was guessed by presence of additional carbonyl group in urea structure that was already absorbed into cytoplasm and carry out in cellular metabolizing and converting significantly 16:0 fatty acid to be 18:0 fatty acid and also other species un-significantly 18:1, 18:2 fatty acids.

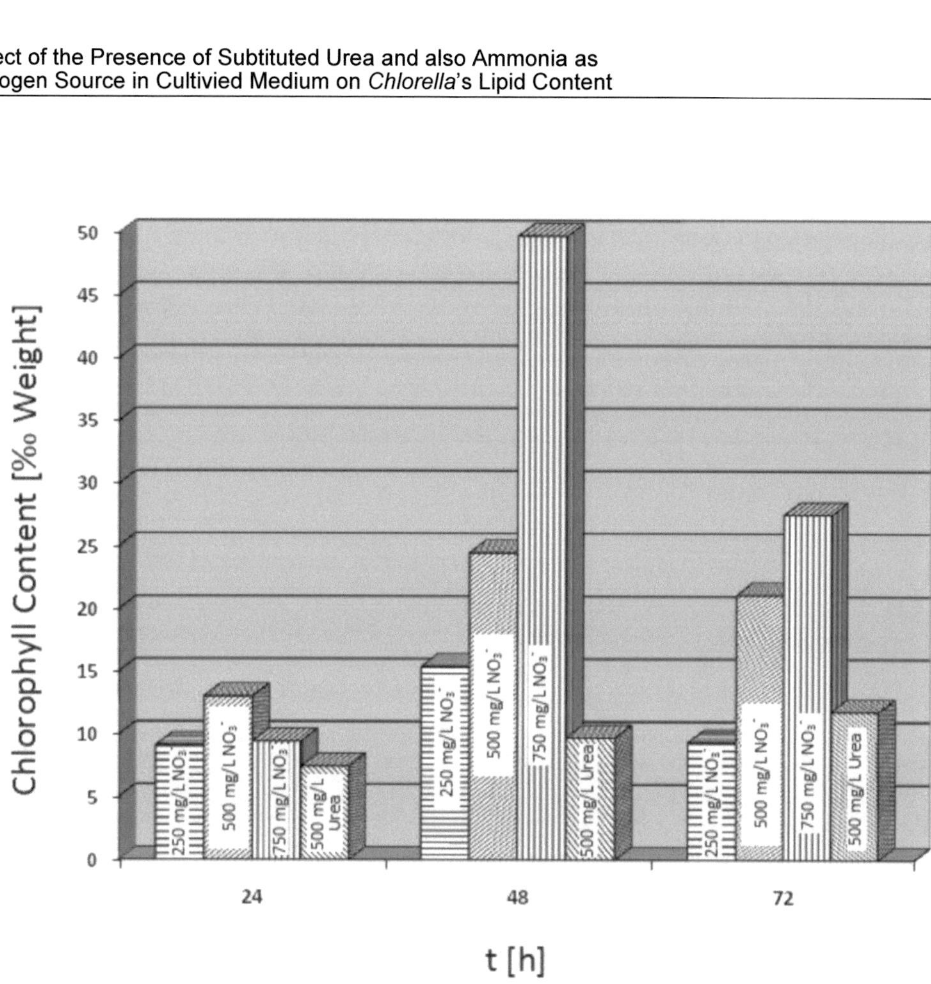

Fig. 5. Effect of composition nitrogen source on *Chlorella*'s chlorophyll content at beginning 72 hours cultivation

Fatty Acid	% Content	
	Appropriate diluted Nitrate Salt (Benneck)	Diluted Urea Media
08 : 0	0.48	0.65
12 : 0	5.50	4.93
14 : 0	3.15	8.60
16 : 0	30.04	0.55
16 : 1	0.33	1.63
18 : 0	9.53	18.04
18 : 1	34.23	40.91
18 : 2	16.74	18.04
20 : 0	0.0	0.60

Table 1. *Chlorella*'s fatty acid content that was cultivated in media contain urea or nitrate salt as nitrogen source.

Determination of proper ammonia nutrients from diluted domestic waste water by 1 : 15 for *Chlorella* growth and compare to appropriate nitrate ion concentration in the Benneck medium (control, 500 mg/L $NaNO_3$) was shown in **Figure 6.** This diluted domestic waste water contain 4.7 mg/L NH3, 330.8 Chemical Oxygen Demand, 78.8 mg/L phosphate salt and pH 8.67. This comparison was done for elaborate effect of substitution nitrate salt in cultivation media with more cheaply and acceptable consumed chemical substance which was contained in waste water such as ammonia to maximize producing of cellular lipids for biodiesel development purpose.

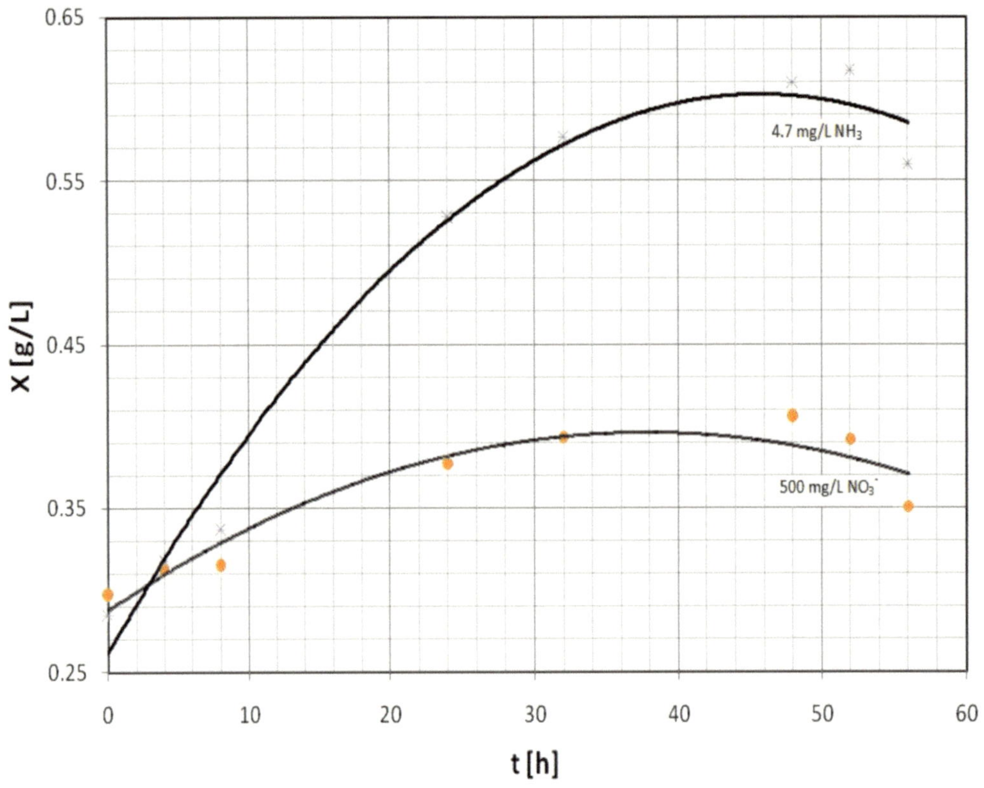

Fig. 6. Effect of replacement diluted domestic waste water 1 : 15 which contained NH_3 as nitrogen source on *Chlorella*'s growth at beginning 56 hours cultivation

At diluted domestic waste water that was measured 4.7 mg/L NH_3 as nitrogen source shown that chlorella's growth result tend near 60% higher than cultivated biomass production in commonly growth media contained appropriate nitrate salt content. It could be understood, in diluted waste water, contained ammonium ion could be directly metabolized for making essential amino acid, protein and chlorophyll that directly related to microbial growth. Composition of free ammonia and ammonium ion in diluted waste water was found 1.05 and 3.65 g/L, as a notification, presence free ammonia could be inhibited cellular growth. Although free ammonia in cultivation media was inhibited algal's growth but in this waste water, presence only 1.05 g/L free ammonia and it was lower than *Chlorella*'s tolerance limit that was found around 6 g/L free ammonia(Strauss et al, 2010).

Compare to intercellular growth in nitrate salt contained media that must be converted to ammonium species at beginning step, presence of ammonium ion in this waste water make it more quickly utilized and of course increasing its biomass production significantly. This phenomenon was similar to previous result on cellular growth of *Chlorella pyrenesoide* which was already done (Ogbonna & Tanaka, 1996). During 48 hours hours cultivation in waste water, ammonia could be decreased to 1.6 mg/L and it is around 66% ammonia nitrogen removal. Furthermore, intracellular lipid formation in algal's growth in waste water, was un-significantly higher than in appropriate nitrate content in Benneck media. **Table 2** shown that change nitrate salt to ammonia as nitrogen source could be increased around 15% in algal's lipid formation. Beside it, chlorophyll formation was also increasing significantly, it was around 55% increasing.

Media	Lipid Content (% weight)	Chlorophyill content (mg/L)
Diluted waste water	57.1	12.1
Benneck	48.7	7.8

Table 2. *Chlorella*'s fatty acid content that was cultivated in diluted waste water and Benneck media

Finally, as a conclusion remarks, compare to result on utilization urea as nitrogen source, substitution nitrate salt in cultivation media with ammonia that was more cheaply cause it presence in domestic waste water, is more significantly for maximizing producing of cellular lipids for biodiesel development purpose.

4. Conclusion

For biodiesel utilizing purpose, diluted nitrogen concentration in the Benneck medium (control) is the most optimal nutrition to produce lipids up to 0.42 g / g biomass. In another case, although cellular growth was decreased around 30%, presence of urea as substituted nitrogen source is the most appropriate nutrients to produce protein up to 0.54 g / g biomass that is necessary for food supplement purpose. Beside that, for producing chlorophyll, medium that excess diluted nitrogen is the most appropriate nutrients to reach up to 49 $^{o}/_{oo}$ weight. Furthermore, presence of urea, drastically change intracellular fatty acid content and it is shown that presence of urea as substitution species of nitrate salt in Benneck medium, was converted fatty acid C_{16} species (around 30.4 % C_{16} in Benneck) to be fatty acid C_{18} species significantly (around 77.0 % C_{18} in presence of urea) that was guessed by presence of additional carbonyl group in urea structure that was already absorbed into cytoplasm and carry out in cellular metabolizing. Finally, compared to result on utilization urea as nitrogen source, substitution nitrate salt in cultivation media with ammonia which was used to minimizing operation cost cause it more cheaply and commonly presence in domestic waste water. Utilization of ammonia for maximizing producing of biomass and cellular lipids is more interesting for biodiesel development purpose. It makes around 55 - 60 % increasing in both *Chlorella*'s growth and cellular lipid formation.

5. Acknowledgement

The author would like to thanks to Dianursanti, Fadli Yusandi and Fitri Kurniati for their technical assistance.

6. References

Wijanarko, A. & Dianursanti. 2009. Simulated flue gas fixation for large-scale biomass production of *Chlorella vulgaris Buitenzorg*. *International Journal for Algae*, 11: 351-358

Dianursanti; Nasikin, M. & Wijanarko, A. 2010. NOx enriched flue gas fixation for biomass production of *Chlorella vulgaris Buitenzorg*. *Asian Journal of Chemical Engineering*, 10: 24-30

Ohtaguchi, K. and Wijanarko, A. 2002. Elevation of the efficiency of cyanobacterial carbon dioxide removal by mono ethanol amine solution. *Technology*, 8:. 267 - 286

Manirakizal, P.; Covaci, A. & Schepens, P. 2001. *Comparative* Study on Total Lipid Determination using Soxhlet, Roese Gottlieb, Bligh Dyer, and Modified Bligh Dyer Extraction Method. *Journal of Food Composition and Analysis*, 14: 93 - 100

Richmond A. [Ed.]. 2004. *Handbook of Microalgal Culture: Biotechnology and Applied Phycology*. Jhon Wiley & Son, New York: 40 - 54

Wijanarko, A.; Dianursanti; Heidi; Soemantojo, R W. and Ohtaguchi, K. 2006. *Effect* of Light Illumination alteration on *Chlorella vulgaris* Buitenzorg's CO_2 fixation in bubble column photobioreactor. *International Journal for Algae*, 8: 53-60

Wijanarko, A.; Dianursanti; Gozan, M.; Andika, S. M. K.; Widiastuti, P.; Hermansyah, H.; Witarto, A. B.; Asami, K.; Soemantojo, R. W.; Ohtaguchi, K. & Song, S. K. 2006. Enhancement of carbon dioxide fixation by alteration of illumination during *Chlorella vulgaris* Buitenzorg's growth. *Biotechnology and Bioprocess Engineering*, 11: 484-488

Leftley, J.W. & Syrett, P.J. (1973). Urease and ATP: Urea Amidolyase Activity in Unicellular Algae. *Journal of General Microbiology*, 77: 109-115

Salisbury, F. B. & Ross. C. W. (1992). *Plant Physiology*, 4th ed., Wadsworth Publishing Co., Colorado

Bailey, J. E. & Ollis, D. F. (1986). *Biochemical Engineering Fundamentals*, 2nd Ed., McGraw Hill Book Co., Singapore

Yanqun, L.; Horsman, M.; Wang, B.; Wu, N. & Lan, C. Q. (2008). Effects of nitrogen sources on cell growth and lipid accumulation of green alga *Neochloris oleoabundans*. *Applied Microbiology and Biotechnology*, 81, pp:629-636

Strauss, M.; Larmie, S. A. & Montenegro, H. A. (2010). Treating Faecal Sludges in Ponds. Available from: *www.eawag.ch/forschung/sandec/.../treating_FS_in_Ponds_Strauss_IWA.pdf*

Ogbonna, J. C. & Tanaka, H. (1996). Night biomass loss and changes in biochemical composition of cells during light/dark cycle culture of *Chlorella pyrenesoide*. *Journal of Fermentation and Bioengineering*, 82: 558 - 564

15

Recovery of Ammonia and Ketones from Biomass Wastes

Eri Fumoto[1], Teruoki Tago[2] and Takao Masuda[2]
[1]*Energy Technology Research Institute, National Institute of Advanced Industrial Science and Technology, 16-1, Onogawa, Tsukuba*
[2]*Division of Chemical Process Engineering, Faculty of Engineering, Hokkaido University, Sapporo, Japan*

1. Introduction

Huge amounts of biomass wastes, such as animal waste and sewage sludge, are produced continuously in farms and disposal plants. The most common method for treating these wastes is to use landfill and/or incineration methods that consume large amounts of energy and cause environmental problems such as air and soil pollution. Because biomass wastes contain nitrogen compounds and various hydrocarbons, a new alternative process to convert the wastes into useful chemicals is desirable.

Ammonia, one such chemical, has been used as a fertilizer, and increasing interest has focused on it as a hydrogen carrier. Ammonia is a liquid around 0.8 MPa at room temperature and offers significant hydrogen storage capacity (17.7 wt% hydrogen in ammonia). Hydrogen has been produced by the decomposition of ammonia with catalysts, such as ruthenium and nickel (Ganley et al., 2004; Liu et al., 2008; Wang et al., 2004; Yin et al., 2004, 2006; Zhen et al., 2008). Hence, the recovery of ammonia from biomass wastes is demanded. After the treatment for ammonia recovery, the remaining liquid wastes containing lower ammonia concentrations could be used as liquid fertilizer, whereas the high concentration of ammonia in raw biomass wastes causes eutrophication of the soil.

Biomass wastes also contain various hydrocarbons, and several methods exist, such as thermal cracking and fermentation, to convert these wastes into useful chemicals. Methane and hydrogen have been produced by the gasification of biomass wastes above 1000 K with the addition of steam or air (Gross et al., 2008; Nipattummakul et al., 2010). Supercritical water gasification is a method conducted under high pressure to produce hydrogen (Guo et al., 2010a). Fuel oil has been produced by the treatment of biomass wastes at relatively low temperatures of between 673 and 823 K (Shen et al., 2005). Anaerobic fermentation has produced methane (Guo et al., 2010b). The treatment of biomass wastes under moderate conditions is desirable because of the high moisture content of the wastes. Biomass wastes contain various oxygen-containing hydrocarbons, and thus the conversion of these hydrocarbons into useful chemicals, such as ketones, appears to be a promising approach. Acetone is used as a raw material for plastics, such as poly(methyl methacrylate) (PMMA) and polycarbonate (PC).

This chapter describes a new method, shown in Fig. 1, to recover useful chemicals, such as ammonia and ketones, from biomass wastes. Ammonia is recovered by the adsorption of nitrogen compounds in the waste, and oxygen-containing hydrocarbons in the waste are catalytically cracked to produce ketones.

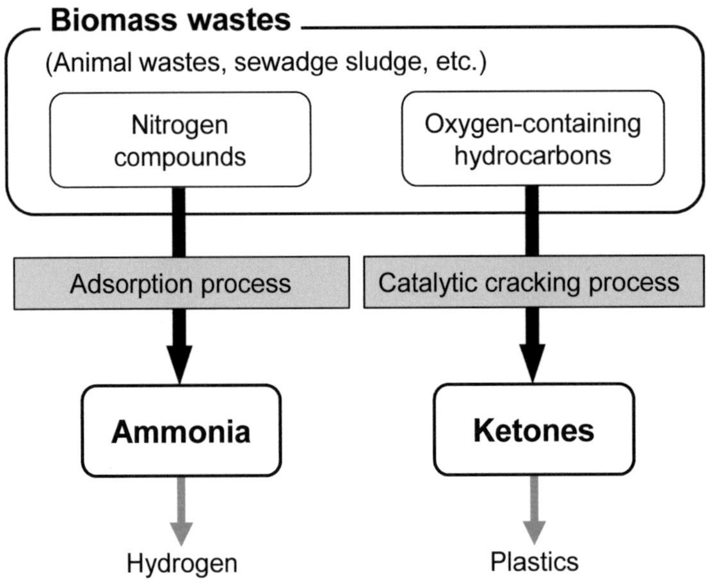

Fig. 1. Recovery method of ammonia and ketones from biomass wastes.

2. Recovery of ammonia

A promising method of ammonia recovery from biomass wastes includes two processes: the recovery of gaseous ammonia, which is generated by aeration of the biomass wastes, and the recovery of aqueous ammonium ions. Adequate adsorbents are required in both processes.

Some adsorbents, such as zeolite, sepiolite, and activated carbon, have been used to recover ammonia gas (Bernal and Lopez-Real, 1993; Park et al., 2005). The maximum amounts of ammonia adsorption on zeolite and sepiolite were approximately 0.8 mol-N/kg-zeolite and 0.3 mol-N/kg-sepiolite (Bernal and Lopez-Real, 1993). Zeolite and sepiolite have also been used to recover ammonium ions in liquid phase (Balci, 2004; Bernal and Lopez-Real, 1993; Yusofa et al., 2010). The maximum adsorption of ammonium ions on zeolite Y was approximately 2.4 mol-N/kg-zeolite (Yusofa et al., 2010).

The precipitation of magnesium ammonium phosphate ($MgNH_4PO_4 \cdot 6H_2O$, MAP) is a useful process for removing ammonium ions in liquid phase (Chimenos et al., 2003; Diwania et al., 2007; Nelson et al., 2003; Stratful et al., 2001). MAP can be precipitated by adding magnesium and phosphate to ammonium solution at a pH above 7. Sugiyama et al. (2005, 2007) reported that an adsorbent derived from MAP was useful for the recovery of ammonium ions from aqueous solution. Ammonia was removed from MAP by thermal treatment above 353 K, yielding a solid, which is the adsorbent for the recovery of aqueous ammonium ions.

The application of MAP-derived adsorbents to both the adsorption process of gaseous ammonia and aqueous ammonium ions could be a promising approach to recover ammonia from biomass wastes. The recovery process of ammonia from biomass wastes in liquid or gas phase is depicted in Fig. 2. Thermal treating of MAP produces ammonia, water, and the adsorbent, which is $Mg(NH_3)_{1-X}HPO_4$. The behaviors of adsorption of gaseous ammonia and aqueous ammonium ions and desorption of ammonia are discussed in this section.

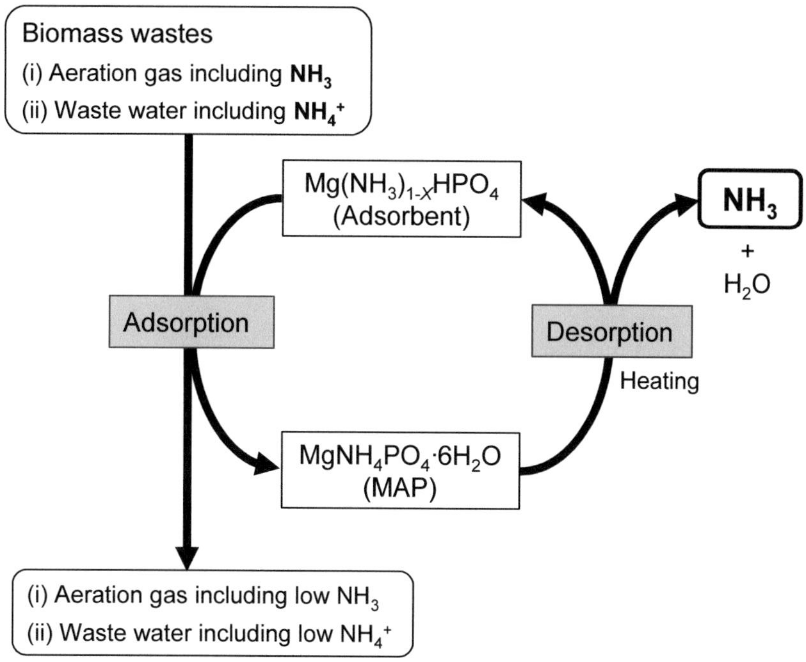

Fig. 2. Ammonia recovery process from biomass wastes using MAP-derived adsorbents.

2.1 Adsorbents derived from MAP

Ammonia can be removed from MAP by heating. Sugiyama et al. (2005) reported that the weight of MAP decreased drastically in the temperature range of 350–400 K due to the elimination of ammonia and water when MAP was heated. The nitrogen content in MAP was reduced by heating, and ammonia was largely eliminated from MAP in the temperature range of 340–360 K (Fumoto et al., 2009). Table 1 shows the remaining nitrogen content in the solids treated at 378 K and 573 K for 24 h in a thermostatic oven. Approximately 70% and 90% of ammonia was eliminated from MAP by thermal treatment at 378 K and 573 K, respectively (Fumoto et al., 2009). The remaining nitrogen content and the amount of weight loss indicate that the adsorption capacity of ammonia onto the solids treated at 378 K and 573 K was 3.6 and 6.0 mol-N/kg-solid, respectively.

Treatment temperature [K]	Remaining nitrogen content [mol-N/mol-Mg]	Surface area [m^2/g]
378	0.30	204
573	0.090	111

Table 1. Remaining nitrogen content and BET surface area of solids obtained by thermal treatment of MAP (Fumoto et al., 2009).

Figures 3 and 4 illustrate the nitrogen sorption isotherms and pore volume distributions of the solids obtained by treating MAP at 378 K and 573 K. The Brunauer-Emmett-Teller (BET) surface area of the solids was calculated and is given in Table 1. The sorption isotherms exhibited hysteresis, indicating that the solids have pores. The solid treated at 378 K had several nanopores, and the surface area of this solid was larger than that of the solid treated at 573 K (Fumoto et al., 2009). These results suggest that the solid treated at 378 K may be a suitable adsorbent for recovering gaseous ammonia and aqueous ammonium ions.

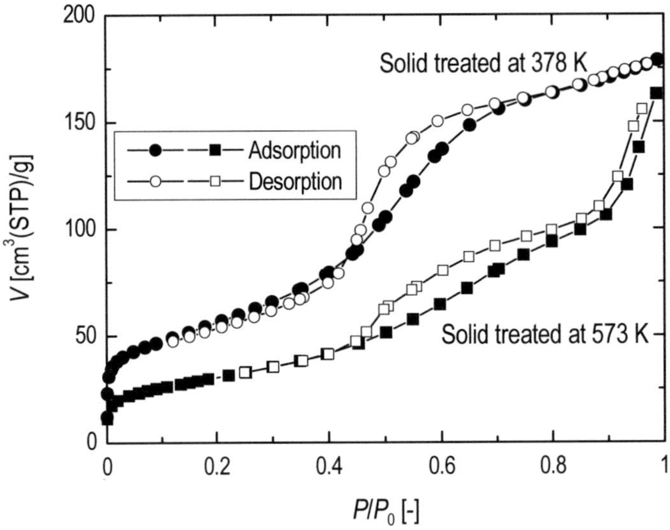

Fig. 3. Nitrogen sorption isotherms of solids obtained by treating MAP at 378 K and 573 K (Fumoto et al., 2009).

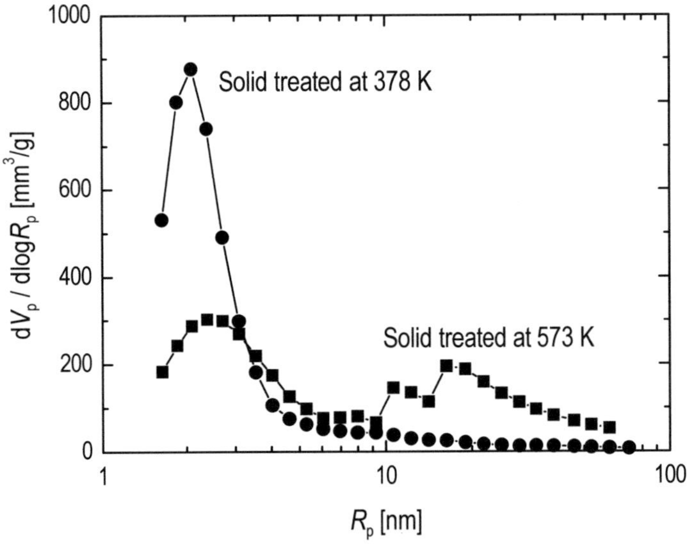

Fig. 4. Pore volume distribution of solids obtained by treating MAP at 378 K and 573 K (Fumoto et al., 2009).

2.2 Gas phase adsorption of ammonia on MAP-derived adsorbents

The adsorption of gaseous ammonia on the adsorbent obtained by treating MAP at 378 K was investigated. The adsorbent, loaded in a stainless steel column, was controlled at 313–353 K, and the experiment of ammonia adsorption was conducted by introducing a mixture of ammonia, hydrogen, and argon. The concentration of ammonia in the inlet gas, C_0, was 2.4 mol/m^3. The outlet gas, including ammonia, hydrogen, and argon, was monitored with a quadrupole mass spectrometer (Q-MS). The mass numbers were chosen as 15, 2, and 40, to detect ammonia, hydrogen, and argon, respectively. Hydrogen was introduced to determine the travel time of the gas from the inlet to the Q-MS. In a preliminary experiment, hydrogen and argon were confirmed to not be adsorbed on the adsorbent.

Figure 5 depicts the effect of temperature on the amount of adsorption of gaseous ammonia on the adsorbent obtained by treating MAP at 378 K. Breakthrough curves of ammonia adsorption were obtained from the measured ammonia concentration in the outlet gas, C_t. The amount of ammonia adsorption, q, was calculated according to Eq. (1).

$$q = \frac{v \cdot C_0}{W} \int_0^{\infty} (1 - C_t / C_0) dt, \tag{1}$$

where v is the flow rate and W is the weight of the adsorbent. The lower the temperature is, the larger the amount of ammonia is adsorbed. The maximum adsorption amount was 2.56 mol-N/kg-adsorbent (Fumoto et al., 2009), which is much larger than that on zeolite (0.8 mol-N/kg-zeolite; Bernal and Lopez-Real, 1993).

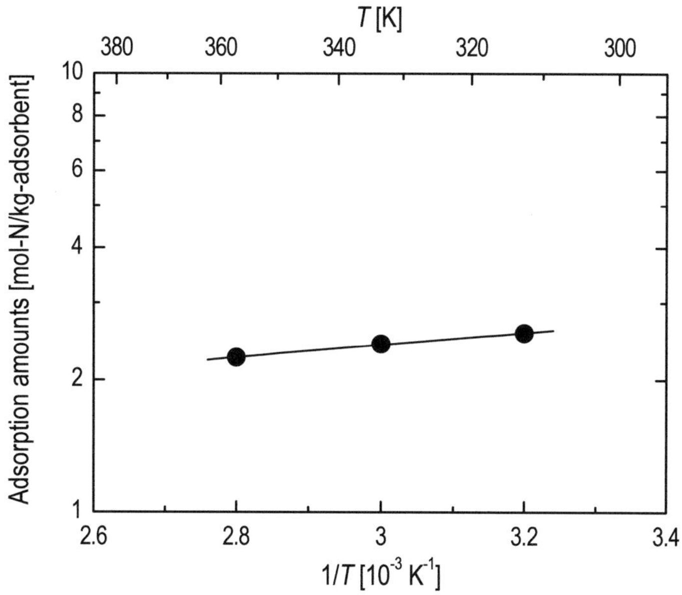

Fig. 5. Effect of temperature on the amount of gaseous ammonia adsorbed on the adsorbent obtained by treating MAP at 378 K (Fumoto et al., 2009).

Figure 6 presents an adsorption isotherm of gaseous ammonia at 313 K. The adsorbent was obtained by thermal treatment of MAP at 378 K. The amount of adsorption of ammonia was proportional to the ammonia concentration, indicating Henry-type adsorption (Fumoto et al., 2009). The adsorption energy, calculated from the data of Arrhenius plots in Fig. 5, was low (-3.0 kJ/mol). These results suggest that gaseous ammonia was physically adsorbed on the adsorbent.

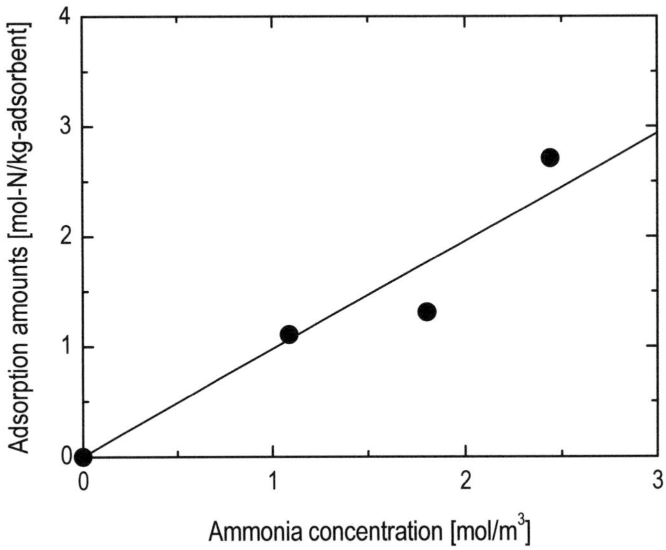

Fig. 6. Adsorption isotherm of gaseous ammonia at 313 K on the adsorbent obtained by treating MAP at 378 K (Fumoto et al., 2009).

2.3 Liquid phase adsorption of ammonium ions on MAP-derived adsorbents

The adsorption of ammonium ions on MAP-derived adsorbents from ammonia water was investigated. The ammonium concentration was 500–12000 ppm and the pH of the ammonia water was adjusted to 11 by adding sodium hydroxide. The adsorbent obtained by treating MAP at 378 K and 573 K was added to the ammonia water at a weight ratio of adsorbents to ammonia water of 1:100, and the ammonium concentration was analyzed after 1 h of stirring at room temperature.

Figure 7 depicts the adsorption isotherms of ammonium ions on the adsorbent from ammonia water at room temperature; the calculated adsorption capacity of ammonium ions is also shown. A large amount of ammonium ions became adsorbed on the adsorbent treated at 378 K (Fumoto et al., 2009), whereas the maximum adsorption amount on zeolite Y was approximately 2.4 mol-N/kg-zeolite (Yusofa et al., 2010). The experimental value of the adsorbed ammonium ions on the adsorbent treated at 378 K was larger than the calculated value. An adsorption isotherm of ammonium ions on the adsorbent treated at 378 K shows Langmuir-type adsorption, indicating chemical adsorption. Figure 8 shows X-ray diffraction (XRD) patterns of MAP and the adsorbent treated at 378 K before and after the adsorption of ammonium ions. The pattern of the adsorbent treated at 378 K shows peaks corresponding to $MgNH_4PO_4 \cdot H_2O$. Sugiyama et al. (2005) reported that MAP was converted to amorphous $MgHPO_4$ by thermal treatment below 773 K. These results suggest that the adsorbent consisted of amorphous $MgHPO_4$ and $MgNH_4PO_4 \cdot H_2O$. The adsorbent after the adsorption of ammonium ions exhibited peaks similar to those of the MAP (Fumoto et al., 2009). Hence, ammonium ions were adsorbed on the site of $MgHPO_4$ of the adsorbent, and the MAP was re-formed in the presence of water.

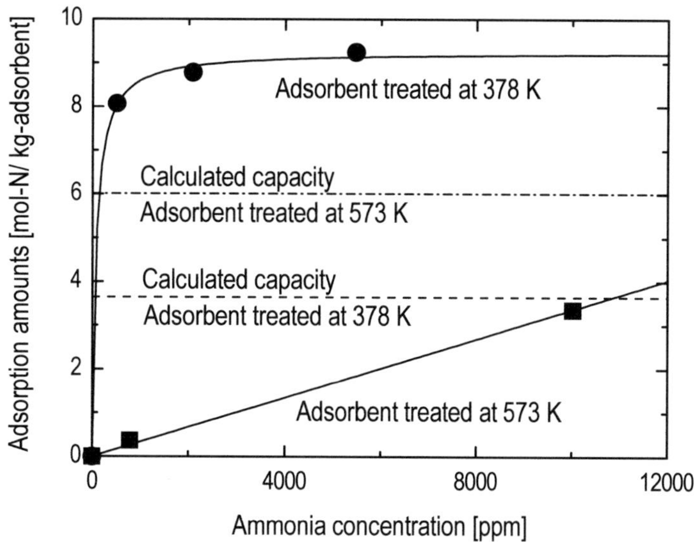

Fig. 7. Adsorption isotherms of ammonium ions at room temperature on the adsorbent obtained by treating MAP at 378 K and 573 K (Fumoto et al., 2009).

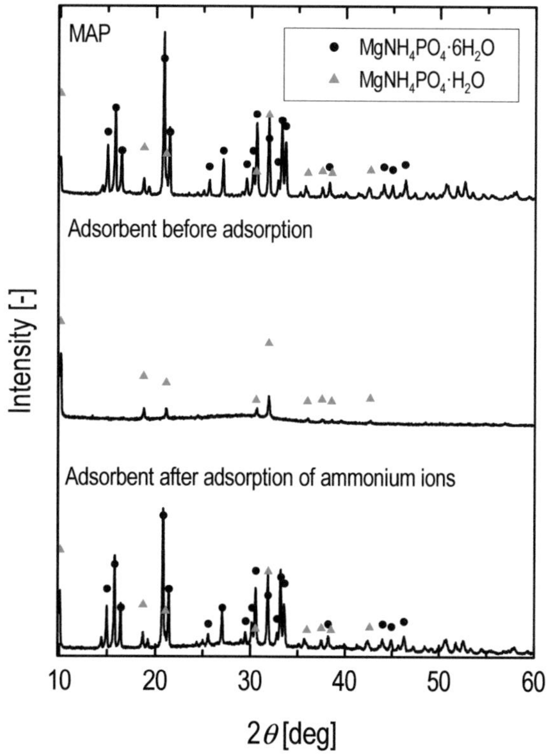

Fig. 8. XRD patterns of MAP and adsorbent treated at 378 K before and after the adsorption of ammonium ions (Fumoto et al., 2009).

The amount of ammonium ions adsorbed on the adsorbent treated at 573 K was significantly less than that of the adsorbent treated at 378 K, as shown in Fig. 7. Furthermore, the experimental value was less than the calculated capacity in the case of the adsorbent treated at 573 K. The fewer nanopores and smaller surface area of the adsorbent treated at 573 K caused the lower adsorption of ammonium ions. The surface chemical properties of the adsorbent may be different between the adsorbents treated at 378 K and 573 K. Consequently, the adsorbent obtained by treating MAP at 378 K was more suitable for the adsorption of ammonium ions.

2.4 Recovery of ammonium ions from animal wastes

The feasibility of recovering ammonia from biomass wastes was demonstrated using cow urine. The urine was pretreated under a hydrothermal condition at 573 K for 1 h to convert nitrogen compounds in the urine into ammonium ions. The pH was adjusted to 10.5 by adding sodium hydroxide and the adsorbent treated at 378 K was added to the pretreated urine at an adsorbent to urine weight ratio of 1:10. The nitrogen concentration was analyzed after 1 h of stirring.

	Recovery yield [mol%-N]	Impurities deposition [mol/mol]	
		C/N	S/N
Pretreated urine	56.9	0.103	0
Untreated urine	65.2	0.486	0.0196

Table 2. Nitrogen recovery yield and impurities deposited on the adsorbent from urine solution (Fumoto et al., 2009).

Table 2 lists the nitrogen recovery yield and the impurities deposited on the adsorbent from the urine; the results obtained using untreated urine are also shown. More than 50% of the nitrogen was recovered from the urine using the adsorbent obtained by treating MAP at 378 K (Fumoto et al., 2009). The nitrogen concentration of the urine decreased to 2000 ppm after the recovery experiment, and the remaining liquid wastes could be used as liquid fertilizer because the liquid contained a low concentration of ammonia.

The nitrogen recovered from pretreated urine corresponded well with ammonium ions because the carbon deposition on the adsorbent was small, as shown in Table 2. In contrast, some carbon was deposited on the adsorbent from the untreated urine, indicating that most of the nitrogen adsorbed on the adsorbent was urea. Furthermore, no sulfur was deposited on the adsorbent from the pretreated urine, which contained sulfur. Therefore, large amounts of ammonia were recovered from the biomass wastes using this method without impurities.

2.5 Desorption of ammonia from solids adsorbing ammonia

The recovery of ammonia by thermal treatment of the solids adsorbing gaseous ammonia and aqueous ammonium ions was examined. The MAP structure was re-formed after the adsorption of ammonium ions in liquid phase. Hence, the solid adsorbing gaseous ammonia and MAP were loaded in the stainless column, followed by heating the column at a rate of 1 K/min in an argon stream. The solid, which was obtained by treating MAP at 378 K, was used after the adsorption of gaseous ammonia. The ammonia and steam eliminated from the solid and MAP was measured by Q-MS. The mass numbers were chosen as 15, 18, and 40 to detect ammonia, steam, and argon, respectively.

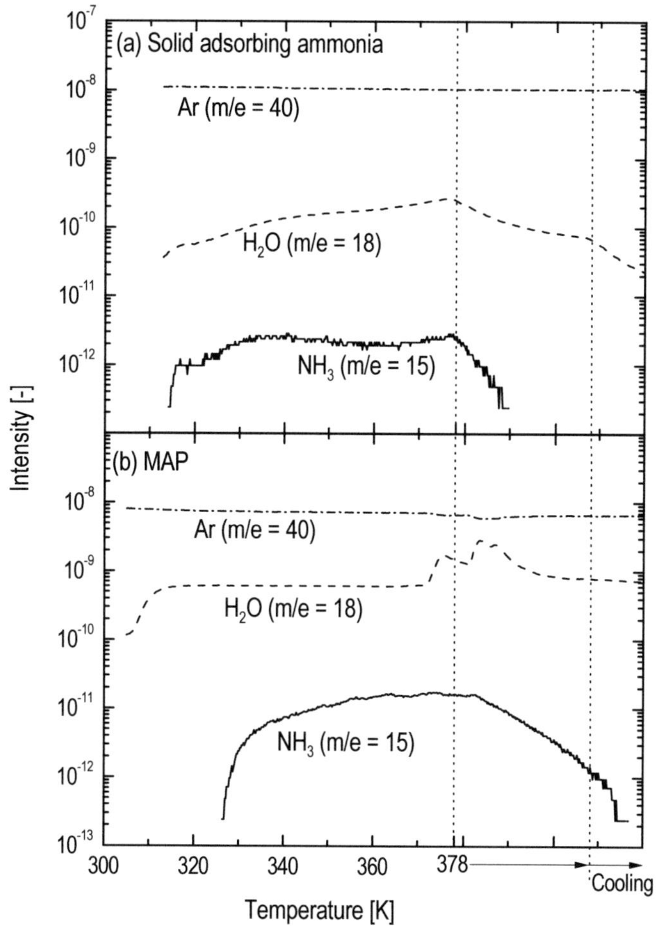

Fig. 9. Gas fractions generated from the solid adsorbing gaseous ammonia and MAP.

Figure 9 describes the gas fractions eliminated from the solids adsorbing gaseous ammonia and MAP. Ammonia was eliminated when these samples were heated. The solid adsorbing gaseous ammonia released ammonia at a relatively lower temperature compared with the MAP, suggesting the physical adsorption of gaseous ammonia. Steam may be desorbed from moisture adsorbed on the surface of the solids and crystallization water of MAP. These results indicate that ammonia could be recovered by thermal treatment of the solids after the adsorption of gaseous ammonia and ammonium ions. Hence, the adsorbent derived from MAP could be used repeatedly.

2.6 Stability of adsorbents for repeated use

The adsorbents derived from MAP are expected to be reused for the recycling process of adsorption and desorption of ammonia. Sugiyama et al. (2005) reported that the removal of ammonium ions in the second run was about 80% of that in the first run when an ammonium removal experiment from aqueous ammonium ions was conducted using adsorbent derived from MAP. The stability of the adsorbents was investigated for repeated use in gaseous ammonium adsorption.

Figure 10 illustrates the change in amounts of adsorbed ammonia on the adsorbents when the sequence of ammonia adsorption and desorption was repeated. After the adsorption of gaseous ammonia at 313 K on the adsorbent obtained by treating MAP at 378 K, the adsorbent was heated to 378 K to eliminate the ammonia, and it was used repeatedly for the adsorption experiment. The amount of ammonia hardly changed in the adsorption/desorption sequence. The pore structure of the adsorbent was almost maintained. Accordingly, this adsorbent is useful for the recovery of ammonia with repeated sequences of adsorption and desorption.

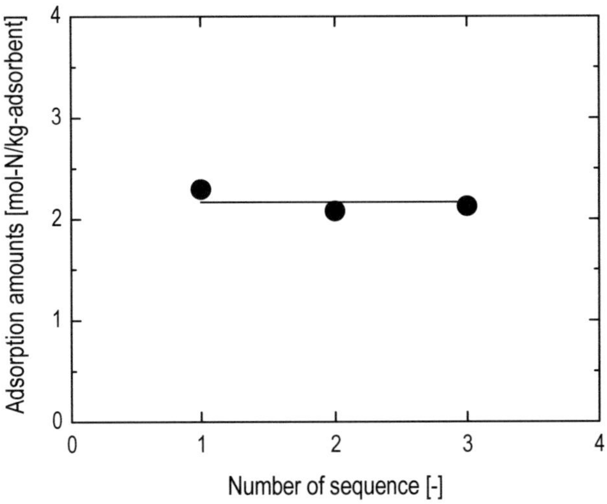

Fig. 10. Change in the amount of adsorbed gaseous ammonia with repeated sequences of ammonia adsorption and desorption.

3. Recovery of ketones

The conversion of hydrocarbons in biomass wastes into useful chemicals is also a promising method. Figure 11 depicts the recovery process of ketones from biomass wastes. To solubilize the solid biomass wastes, such as sewage sludge, the wastes are hydrothermally treated, producing black water. The obtained black water consists of oxygen-containing hydrocarbons and a large amount of water. Some impurities, such as nitrogen and sulfur, are contained in the black water. The conversion of black water into useful chemicals requires catalysts having the following properties: a strong ability to decompose the hydrocarbons in the black water, stable activity in the presence of water, and resistance to the deposition of impurities contained in the black water.

Zirconia-supporting iron oxide catalysts are effective for the decomposition of oil palm waste (Masuda et al., 2001) and petroleum residual oil (Fumoto et al., 2004) in a steam atmosphere. Oil palm waste can be converted to a mixture containing phenol, acetone, and butanone using the catalyst. Hydrocarbons in oil palm waste and petroleum residual oil react with active oxygen species generated from steam on the iron oxide catalyst. Zirconia promotes the generation of the active oxygen species from steam.

The production of ketones from sewage-derived black water was investigated. Figure 12 presents the conversion of oxygen-containing hydrocarbons to ketones with the zirconia-supporting iron oxide catalysts. The active oxygen species generated from steam could react with the hydrocarbons.

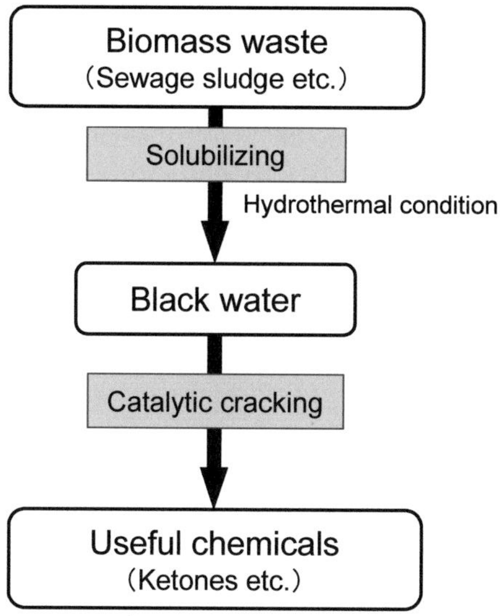

Fig. 11. Recovery of ketones from biomass wastes.

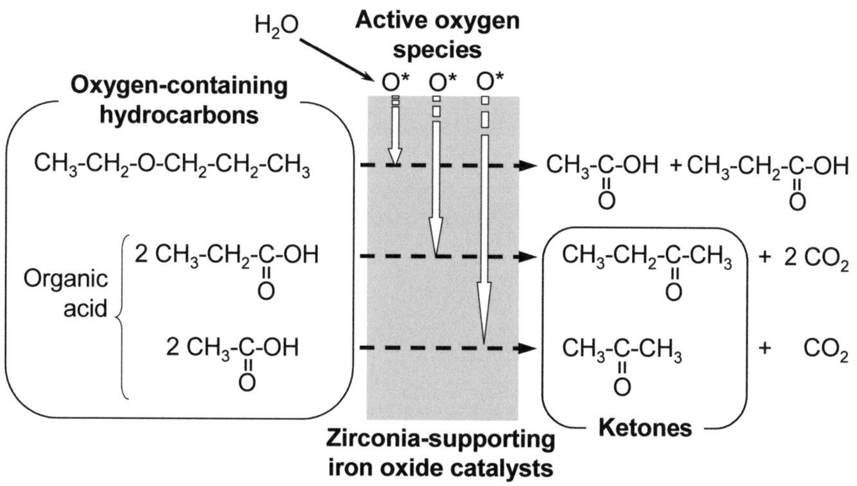

Fig. 12. Reaction mechanism of oxygen-containing hydrocarbons with zirconia-supporting iron oxide catalysts.

3.1 Production of ketones from sewage sludge

Catalytic cracking of sewage-derived black water was investigated under superheating steam conditions. The black water was obtained by the hydrothermal treatment of digested sewage sludge at 573 K. The moisture content of the black water was 98 wt%. The zirconia-supporting iron oxide catalyst was prepared by a coprecipitation method using $FeCl_3 \cdot 6H_2O$ and $ZrOCl_3 \cdot 8H_2O$, yielding the catalyst denoted as Zr(*Y*)-FeO_X, where *Y* is the amount of the supported zirconia by weight percent. The catalytic cracking of sewage-derived black water was carried out at 523 K under 2 MPa for 2 h using a batch autoclave reactor loaded with 0.2 g of catalyst and 3.2 g of black water. The product was analyzed by gas chromatography (GC).

Figure 13 illustrates the product yield after the reaction of black water with Zr(Y)-FeO_X catalysts. The catalysts were active for producing acetone from black water (Fumoto et al., 2006a). The yield of acetone produced from black water increased with increasing zirconia content and reached the maximum value at 7.7 wt% zirconia content. Figure 14 shows the desorption rate of hydrogen generated by the decomposition of steam when the catalysts were heated after the pre-adsorption of steam on the catalysts. The catalyst supporting zirconia exhibited higher steam decomposition activity, even at lower temperatures, producing hydrogen (Masuda et al., 2001). Simultaneously, active oxygen species were generated from steam. These oxygen species spill over to the surface of iron oxide, and oxygen-containing hydrocarbons in black water react with the active oxygen species on the iron oxide. The yield of acetone produced in the reaction with the Zr(15.8)- FeO_X catalyst was less than that in case of the Zr(7.7)-FeO_X catalyst. The active sites on the iron oxide may be covered with the excessively supported zirconia. Consequently, the largest amount of acetone was produced by the reaction of sewage-derived black water with the Zr(7.7)-FeO_X catalyst.

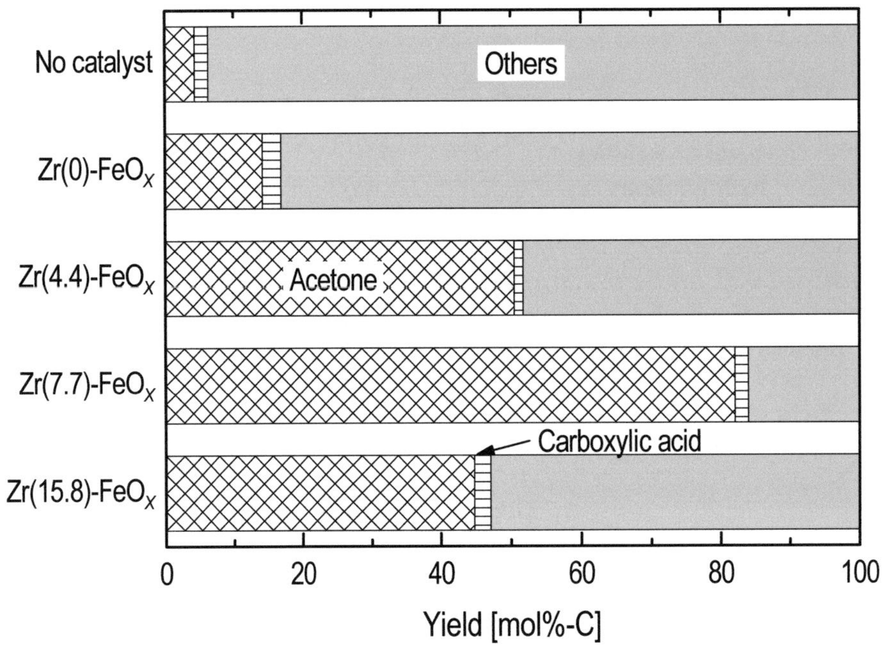

Fig. 13. Product yield of the reaction of black water derived from sewage sludge with Zr(Y)-FeO_X catalysts (Fumoto et al., 2006a).

3.2 Durability of zirconia-supporting iron oxide catalysts

High durability of the catalysts is demanded for their long-term use. The black water contains impurities, such as nitrogen and sulfur, which have the potential of poisoning the catalysts. Nitrogen compounds could be removed by adsorption using the MAP-derived adsorbent. To examine the durability of the catalysts, an accelerated deterioration test using petroleum residual oil, which contained sulfur, was conducted.

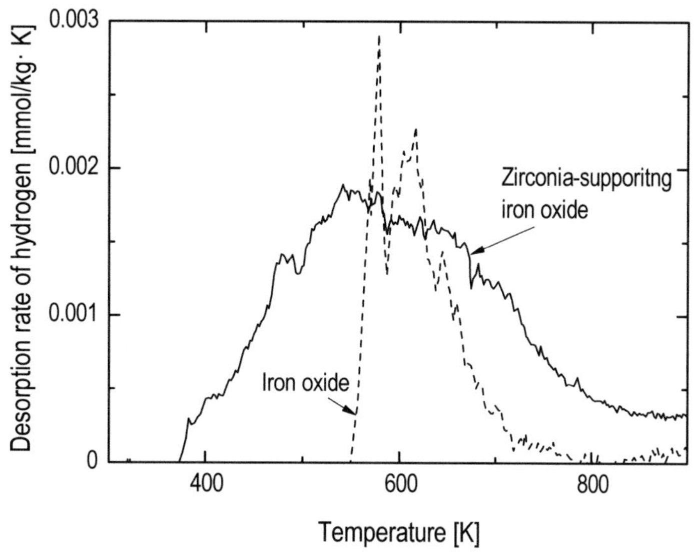

Fig. 14. Desorption rate of hydrogen from steam when the catalyst was heated after the pre-adsorption of steam on the catalysts (Masuda et al., 2001).

Three types of catalysts, Zr/FeO_X, $Zr/Al\text{-}FeO_X$, and $Zr\text{-}Al\text{-}FeO_X$, were prepared. Zirconia was supported on the iron oxide, which was generated from the treatment of α-FeOOH with steam, by impregnation using $ZrOCl_3 \cdot 8H_2O$, yielding the Zr/FeO_X catalyst. The complex metal oxide of aluminum and iron was obtained by a coprecipitation method using $FeCl_3 \cdot 6H_2O$ and $Al_2(SO_4)_3 \cdot 14\text{-}18H_2O$, and zirconia was supported on the complex metal oxide by impregnation, yielding the $Zr/Al\text{-}FeO_X$ catalyst. The $Zr\text{-}Al\text{-}FeO_X$ catalyst was prepared by coprecipitation using $FeCl_3 \cdot 6H_2O$, $Al_2(SO_4)_3 \cdot 14\text{-}18H_2O$, and $ZrOCl_3 \cdot 8H_2O$. The loaded amount of zirconia was 7.7 wt% and the atomic fraction of Al in $Al\text{-}FeO_X$ was 0.079. The catalytic cracking of atmospheric residual oil was conducted in a steam atmosphere at 773 K under atmospheric pressure using a fixed bed reactor loaded with the catalyst. The product oil was analyzed by GC and gel permeation chromatography (GPC).

Figure 15 depicts the change in catalytic activity for the decomposition of heavy oil after the sequence of reaction of residual oil and regeneration of the catalyst. The reaction rate constant, k, was calculated according to Eq. (2):

$$\frac{df_{C30+}}{d(W/F_R)} = -k \cdot f_{C30+}^{\;2}, \tag{2}$$

where f_{C30+} represents the weight fraction of heavy oil (carbon number above 30), and W/F_R is the time factor corresponding to the ratio of the weight of catalyst to the flow rate of residual oil. The activity of the Zr/FeO_X catalyst decreased when the sequence of reaction and regeneration was repeated (Fumoto et al., 2006b). The peeling of zirconia from iron oxide due to structural changes of the iron oxide catalyst caused the deactivation. The $Zr/Al\text{-}FeO_X$ catalyst was not deactivated after the reaction and regeneration sequence. The addition of alumina prevented the structural change of iron oxide. When the reaction was repeated without regeneration, the $Zr\text{-}Al\text{-}FeO_X$ catalyst maintained high activity (Fumoto et al., 2006c), whereas the activity of the $Zr/Al\text{-}FeO_X$ catalyst decreased without the regeneration. The lattice oxygen of iron oxide was consumed during the reaction, causing a phase change of the iron oxide of Zr/FeO_X and Zr/Al-FeO_X catalysts from hematite to magnetite. Hence, the catalyst was regenerated by calcinations. In contrast, the hematite of the Zr-Al-FeO_X catalyst was maintained after the reaction, leading to stable activity without regeneration. No correlation was observed between the activity of the catalyst and the deposition of impurities from residual oil. Accordingly, the Zr-Al-FeO_X catalyst could be useful for long-term application in the conversion process of biomass wastes.

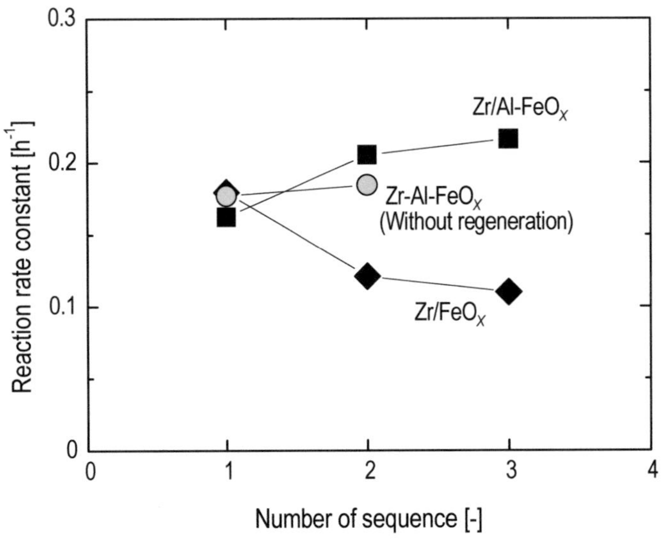

Fig. 15. Change in catalytic activity for the decomposition of heavy oil with a repeated sequence of reaction and regeneration (Fumoto et al., 2006b, 2006c).

4. Conclusion

New methods for recovering ammonia and ketones from biomass wastes were investigated. The gaseous ammonia and aqueous ammonium ions were adsorbed effectively on the adsorbent obtained by treating MAP at 378 K. The adsorption of gaseous ammonia and aqueous ammonium ions was physical and chemical adsorption, respectively. The ammonia could be recovered by thermal treating of the adsorbent after the adsorption of ammonia and ammonium ions, suggesting that the adsorbent is useful for repeated use of the ammonia adsorption/desorption sequence. Large amounts of ammonia were recovered from hydrothermally treated cow urine using the adsorbent, without impurities contained in the urine. Biomass wastes also contain various hydrocarbons. The solid wastes, such as sewage sludge, were solubilized by hydrothermal treatment, producing black water, and catalytic cracking of the black water was conducted. As a result, large amounts of acetone were produced with the zirconia-supporting iron oxide catalyst. Oxygen-containing hydrocarbons reacted with the active oxygen species generated from steam on the iron oxide catalyst. Supported zirconia promoted the generation of the active species. Hence, the yield of acetone increased with the increasing zirconia content in the catalyst. Furthermore, the complex metal oxide catalyst of iron, zirconium, and aluminum showed stable activity for the decomposition of heavy oil. Accordingly, the catalyst may be suitable for the catalytic cracking of biomass wastes.

5. References

Balci, S. (2004). Nature of Ammonium Ion Adsorption by Sepiolite: Analysis of Equilibrium Data with Several Isotherms. *Water Res.*, Vol.38, No.5, (March 2004), pp. 1129-1138, ISSN 0043-1354

Bernal, M. P. & Lopez-Real, J. M. (1993). Natural Zeolites and Sepiolite as Ammonium and Ammonia Adsorbent Materials. *Bioresour. Technol.*, Vol.43, No.1, (1993), pp. 27-33, ISSN 0960-8524

Chimenos, J. M., Fernandez, A. I., Villalba, G., Segarra, M., Urruticoechea, A., Artaza, B. & Espiell, F. (2003). Removal of Ammonium and Phosphates from Wastewater Resulting from the Process of Cochineal Extraction using MgO-Containing By-Product. *Water Res.*, Vol.37, No.7, (April 2003), pp. 1601-1607, ISSN 0043-1354

Diwania, G. E., Rafiea, S. E., Ibiaria, N. N. E. & Ailab, H. I. E. (2007). Recovery of Ammonia Nitrogen from Industrial Wastewater Treatment as Struvite Slow Releasing Fertilizer. *Desalin.*, Vol.214, No.1-3, (August 2007), pp. 200–214, ISSN 0011-9164

Fumoto, E., Tago, T., Tsuji, T. & Masuda, T. (2004). Recovery of Useful Hydrocarbons from Petroleum Residual Oil by Catalytic Cracking with Steam over Zirconia-Supporting Iron Oxide Catalyst. *Energy Fuels*, Vol.18, No.6, (November-December 2004), pp. 1770-1774, ISSN 0887-0624

Fumoto, E., Mizutani, Y., Tago, T. & Masuda, T. (2006a). Production of Ketones from Sewage Sludge over Zirconia-Supporting Iron Oxide Catalysts in a Steam Atmosphere. *Appl. Catal. B*, Vol.68, No.3-4, (November 2006), pp. 154-159, ISSN 0926-3373

Fumoto, E., Tago, T. & Masuda, T. (2006b). Production of Lighter Fuels by Cracking Petroleum Residual Oils with Steam over Zirconia-Supporting Iron Oxide Catalysts. *Energy Fuels*, Vol.20, No.1, (January-February 2006), pp. 1-6, ISSN 0887-0624

Fumoto, E., Tago, T. & Masuda, T. (2006c). Recovery of Lighter Fuels from Petroleum Residual Oil by Oxidative Cracking with Steam over Zr-Al-FeOx Catalyst. *Chem. Lett.*, Vol.35, No.9, (September 2006), pp. 998-999, ISSN 0366-7022

Fumoto, E., Tago, T. & Masuda, T. (2009). Recovery of Ammonia from Biomass Waste by Adsorption on Magnesium Phosphate Derived from Magnesium Ammonium Phosphate. *J. Chem. Eng. Jpn.*, Vol.42, No.3, (2009), pp.184-190, ISSN 0021-9592

Ganley, J. C., Thomas, F. S., Seebauer, E. G. & Masel, R. I. (2004). A Priori Catalytic Activity Correlations the Difficult Case of Hydrogen Production from Ammonia. *Catal. Lett.*, Vol.96, No.3-4, (July 2004), pp. 117-122, ISSN 1011-372X

Gross, B., Eder, C., Grziwa, P., Horst, J. & Kimmerle, K. (2008). Energy Recovery from Sewage Sludge by Means of Fluidised Bed Gasification. *Waste Manage.*, Vol.28, No.10, (2008), pp. 1819-1826, ISSN 0956-053X

Guo, Y., Wang, S. Z., Xu, D. H., Gong, Y. M., Ma, H. H. & Tang, X. Y. (2010a). Review of Catalytic Supercritical Water Gasification for Hydrogen Production from Biomass. *Renewable Sustainable Energy Rev.*, Vol.14, No.1, (January 2010), pp. 334–343, ISSN 1364-0321

Guo, X. M., Trably, E., Latrille, E., Carrère, H. & Steyer, J. P. (2010b). Hydrogen Production from Agricultural Waste by Dark Fermentation: A Review. *Int. J. Hydrogen Energy*, Vol.35, No.19, (October 2010), pp. 10660-10673, ISSN 0360-3199

Liu, H. C., Wang, H., Shen, J. G., Sun, Y, & Liu, Z. M. (2008). Preparation, Characterization and Activities of the Nano-Sized Ni/SBA-15 Catalyst for Producing COx-Free

Hydrogen from Ammonia. *Appl. Catal. A*, Vol.337, No. 2, (March 2008), pp. 138-147, ISSN 0926-860X

Masuda, T., Kondo, Y., Miwa, M., Shimotori, T., Mukai, S. R., Hashimoto, K., Takano, M., Kawasaki, S. & Yoshida, S. (2001). Recovery of Useful Hydrocarbons from Oil Palm Waste Using ZrO_2 Supporting FeOOH Catalyst. *Chem. Eng. Sci.*, Vol.56, No.3, (February 2001), pp. 897-904, ISSN 0009-2509

Nelson, N. O., Mikkelsen, R. L. & Hesterberg, D. L. (2003). Struvite Precipitation in Anaerobic Swine Lagoon Liquid: Effect of pH and Mg : P Ratio and Determination of Rate Constant. *Bioresour. Technol.*, Vol.89, No.3, (September 2003), pp. 229-236, ISSN 0960-8524

Nipattummakul, N., Ahmed, I. I., Kerdsuwan, S. & Gupta, A. K. (2010). Hydrogen and Syngas Production from Sewage Sludge via Steam Gasification. *Int. J. Hydrogen Energy*, Vol.35, No.21, (November 2010), pp. 11738-11745, ISSN 0360-3199

Park, S. J. & Kim, B. J. (2005). Ammonia Removal of Activated Carbon Fibers Produced by Oxyfluorination. *J. Colloid Interface Sci.*, Vol.291, No.2, (November 2005), pp. 597-599, ISSN 0021-9797

Shen, L. & Zhang, D. K. (2005). Low-Temperature Pyrolysis of Sewage Sludge and Putrescible Garbage for Fuel Oil Production. *Fuel*, Vol.84, No.7-8, (May 2005), pp. 809-815, ISSN 0016-2361

Stratful, I., Scrimshaw, M. D. & Lester, J. N. (2001). Conditions Influencing the Precipitation of Magnesium Ammonium Phosphate. *Water Res.*, Vol.35, No.17, (December 2001), pp. 4191-4199, ISSN 0043-1354

Sugiyama, S., Yokoyama, M., Ishizuka, H., Sotowa, K., Tomida, T. & Shigemoto, N. (2005). Removal of Aqueous Ammonium with Magnesium Phosphates Obtained from the Ammonium-Elimination of Magnesium Ammonium Phosphate. *J. Colloid Interface Sci.*, Vol.292, No.1, (December 2005), pp. 133-138, ISSN 0021-9797

Sugiyama, S., Yokoyama, M., Fujii, M., Seyama, K. & Sotowa, K. (2007). Recycling of Thin-Layer of Magnesium Hydrogenphosphate for Removal and Recovery of Aqueous Ammonium. *J. Chem. Eng. Jpn.*, Vol.40, No.2, (February 2007), pp. 198-201., ISSN 0021-9592

Wang, S. J., Yin, S. F., Li, L., Xu, B. Q., Ng, C. F. & Au, C. T. (2004). Investigation on Modification of Ru/CNTs Catalyst for the Generation of COx-Free Hydrogen from Ammonia. *Appl. Catal. B*, Vol.52, No.4, (October 2004), pp. 287-299, ISSN 0926-3373

Yin, S. F., Xu, B. Q., Zhou, X. P. & Au, C. T. (2004). A Mini-Review on Ammonia Decomposition Catalysts for On-Site Generation of Hydrogen for Fuel Cell Applications. *Appl. Catal. A*, Vol.277, No.1-2, (December 2004), pp. 1-9, ISSN 0926-860X

Yin, S. F., Xu, B. Q., Wang, S. J. & Au, C. T. (2006). Nanosized Ru on High-Surface-Area Superbasic ZrO_2-KOH for Efficient Generation of Hydrogen via Ammonia Decomposition. *Appl. Catal. A*, Vol.301, No.2, (February 2006), pp. 202-210, ISSN 0926-860X

Yusofa, A. M., Keata, L. K., Ibrahimb, Z., Majida, Z. A. & Nizamb, N. A. (2010). Kinetic and Equilibrium Studies of the Removal of Ammonium Ions from Aqueous Solution by Rice Husk Ash-Synthesized Zeolite Y and Powdered and Granulated Forms of Mordenite. *J. Hazard. Mater.*, Vol.174, No.1-3, (February 2010), pp. 380-385, ISSN 0304-3894

Zheng, W. Q., Zhang, J., Ge, Q. J., Xu, H. Y. & Li, W. Z. (2008). Effects of CeO_2 Addition on Ni/Al_2O_3 Catalysts for the Reaction of Ammonia Decomposition to Hydrogen. *Appl. Catal. B*, Vol.80, No.1-2, (April 2008), pp. 98-105, ISSN 0926-3373

16

Characterization of Biomass as Non Conventional Fuels by Thermal Techniques

Osvalda Senneca
Consiglio Nazionale delle Ricerche (C.N.R.),
Istituto di Ricerche sulla Combustione
Italy

1. Introduction

In the last decades the problem of CO_2 emission in the atmosphere has driven the industry of power generation towards an increasing use of biomass fuels in addition to conventional fuels.

Figure 1 reports the well known Van Krevelen diagram for a wide variety of solid fuels. It can be seen that biomasses are in general characterized by larger O/C and H/C ratios compared to fossil fuels such as coals. They stand, instead, close to RDFs (refuse derived fuels). As a matter of fact it is not easy to draw a clear demarcation line between biomass and RDFs. Biomass itself is a broad category of materials ranging from raw vegetal materials to solid refuses of industrial and civil origin (wood and agricultural residues, residues of paper, food and dairy industry, sludge of civil origin etc) .

A further element of despair in this already very broad category of fuels lies in the content of inorganics and/or metals, which are present in some biomasses at levels distinctively higher than in traditional fuels. Under this respect biomasses appear even more close to industrial wastes. The presence of metals and inorganic matter may produce unusual effects in terms of both energetic and environmental performance.

It has been shown for a variety of solid fuels that the process and reactor design, in particular the temperature level and the inert/oxidizing nature of the gaseous atmosphere, determine the reaction path and affects severely the fate of the organic matter [1-2] but is also expected to determine the fate of inorganic matter and metals.

As far as the organic content is concerned, upon heating under inert atmosphere this undergoes a combination of thermal cracking and condensation reactions, called pyrolysis, producing a gas, a liquid (tar) and a solid product (char). Gaseous species generally include hydrogen, carbon monoxide, methane, carbon dioxide and other incondensable hydrocarbons; tar consists of chemicals, such as methanol, acetone, acetic acid etc. liquid at room temperature; char is a carbonaceous type solid containing mainly carbon but also the residual inorganic matter.

As shown by Senneca et al. [3] heating of a solid fuel in the presence of oxygen may result in two types of processes depending on the fuel properties and on the process conditions (oxygen concentration, temperature and heating rate). At low temperature (if under isothermal conditions), or at low heating rate (if under non isothermal conditions) thermal cracking and condensation reactions are assisted and enhanced by parallel oxidative and combustion reactions. Char and tar combustion occur in parallel with thermal cracking as shown by fig. 2A and the resulting gas is rich in CO and CO_2. For high enough temperature (under isothermal conditions) or for large particle heating rates (under non isothermal conditions), purely thermally activated pyrolysis overtakes direct combustion and the reaction follows the most typical pattern: pyrolysis occurs first, followed by heterogeneous combustion of the tar and char. The corresponding reduced network is represented in Fig. 2B. It must be noted that volatile matter emission and the formation of an attached or detached volatile flame further contribute to preventing the occurrence of heterogeneous oxidation in this case.

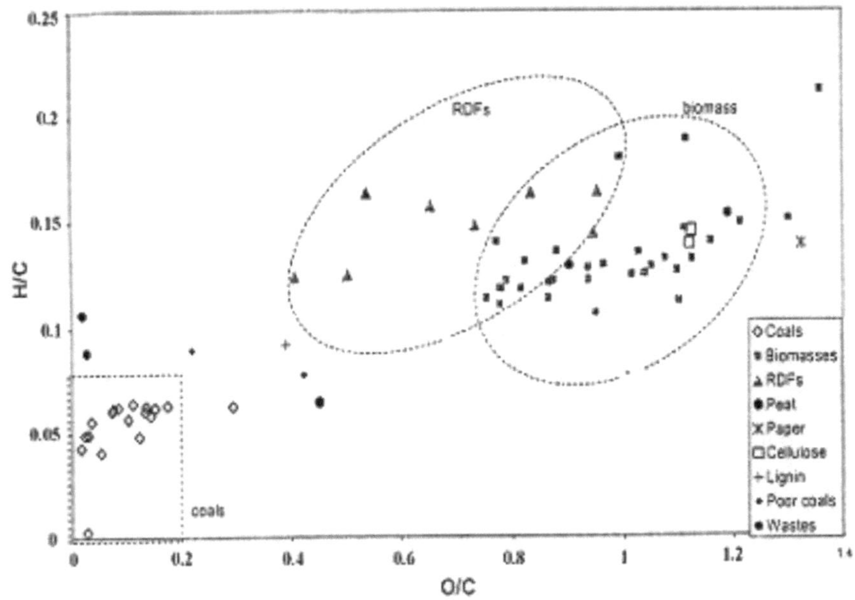

Fig. 1. Van Krevelen diagram

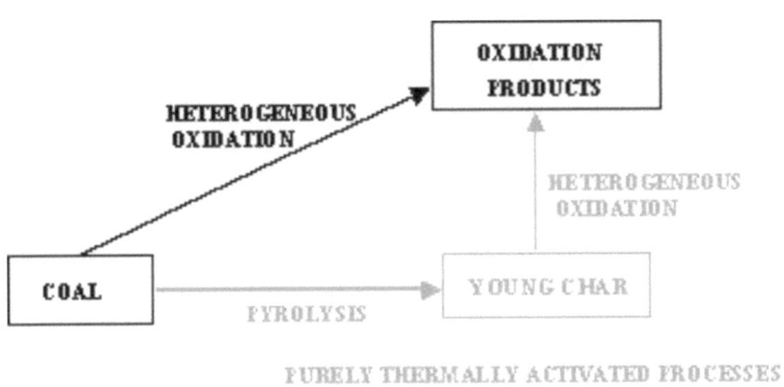

Fig. 2. A Reaction path of oxidative pyrolysis

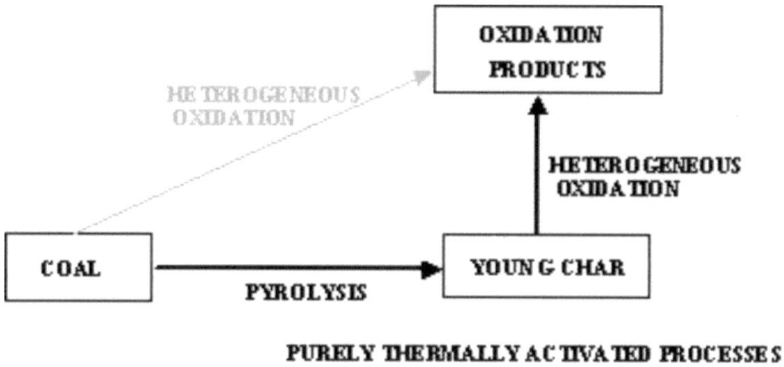

Fig. 2. B Reaction path of pyrolysis-char combustion

The presence of metals and inorganic matter in biomass further complicates the scenario and makes it difficult to predict whether a reaction pathway of type A or B would be active. A first consequence is that the yields and the chemical composition of gaseous, liquid and solid products cannot be predicted a-priori and require appropriate consideration of the process conditions. A second consequence concerns the fate of inorganics and metals themselves, which is a matter of utmost importance for environmental reasons. The potential hazard of emission of volatile metals during the pyrolysis/combustion process and of leaching upon disposal of the final residue is indeed a problem that has already been underlined for a number of wastes of industrial origin, such as sludges and wastes obtained from the reclamation of metal from insulated wires and electronic equipments and automobile wastes [4-9], but also for some wastes that may be included in the category of biomass, for instance meat and bone meal [10] and residues of the pulp and paper industry [11].

In conclusions the design of thermal processes aiming at the exploitation of biomass as solid fuels requires a more comprehensive understanding of how process conditions and reactor design affect the efficiency in terms of energy conversion, yields and chemical composition of gaseous, liquid and solid products as well as the fate of inorganic matter. In other words the exploitation of biomass fuels in thermal processes requires biased experimental investigation of its pyrolysis and combustion behaviour. To this end a diversity of techniques at the laboratory scale can be used. The present paper discusses the problems related with the standard laboratory techniques and presents a comprehensive experimental protocol for the characterization of biomass fuels based on thermal analysis and lab-scale reactors. Examples of selected fuels are presented to demonstrate and clarify the issue.

2. Conventional experimental techniques

The most commonly used lab scale technique for the study of thermal processes involving biomass it thermal analysis, because of it apparent simplicity. Thermal analysis is definitely the easiest and most accurate tool to perform proximate analysis but its natural and most valuable goal is the kinetic study.

Today it is well known that the most reliable kinetic methods for the analysis of non isothermal TG experiments are the Friedman plot [12,13], the Kissinger±Akahira± Sunose plot [13-15] and the Ozawa-Flynn-Wall method [16,17]. A very important point is that this analysis is easy and reliable in the case of single power law reactions but is more complicated in the case of parallel reactions. Thermal processes of biomass in fact have often

been described using power law kinetic expressions, for a single reaction when one major event of weigh loss is distinguished, for two or more parallel reactions when two or more stages of weight loss are observed. This choice is made for sake of simplicity and also because the method for kinetic analysis of TG curves is well consolidated. In the case of multiple/competitive reactions in series/parallel some methods for kinetic analysis have been proposed, but there are few examples of their application.
In any case it must be clear that thermogravimetric analysis can be used confidently to predict the thermal life of a fuel only at relatively low temperature and heating rate. Outsiders may misunderstand there are serious problems to apply the results of thermogravimetric analysis to practical operating conditions of pyrolysers/combustors, where temperature and heating rates are quite different from those of thermogravimetric analysis.
The potential of thermogravimetric analysis in the study of thermal processes of biomass is considerably enhanced by the introduction of simultaneous DSC or DTA and analysis of evolved gas (EGA) by FT-IR and mass-spectrometry. The former technique reveals the presence of transitions, particularly important for biomasses rich in minerals and metals, moreover it gives information on the endothermic/exothermic nature of the processes, thus contributing significantly to interprete the weight changes events detected by the TG curves. Again outsiders should not be tempted to use the DSC data obtained during simultaneous TG/DSC experiments of biomass for a quantitative measure of its heat of pyrolysis/combustion.
Analysis of gaseous species by FT-IR and MS is also very useful to obtain information on the type of gaseous species evolved throughout a thermal process and to understand the reaction paths, but also in this case results must be regarded as qualitative more than quantitative and caution is needed to extend them to real situation. Examples of this type of equipment are shown in Fig. 3.
For the study of the yields of biomass pyrolysis the most common experimental approach is the recourse to purposely made lab furnaces equipped for the collection of tar and the analysis and tar and gases. Different configurations and different collection systems have been proposed. An example of this type of equipment is shown in Fig. 4. Typically the sample is located inside a pyrolysis reactor which is heated by an external electrical furnace with heating rates in the order of 5-50°C/min. The product is conveyed to a set of consecutive traps for tar condensation at progressively lower temperature. Tar is analysed off-line typically by Gass Cromatography. Uncondensables are analysed either of line or online by different analythical tools, such as GC (off-line) or FT-IR or MS (on-line).
This type of experiments is able to give quantitative data on the yield of biomass pyrolysis, however the extrapolation of these results to reaction conditions far from those of the experiment would again be ingenuous.

3. Experimental protocol

The experimental protocol proposed for biomass fuels couples experiments in a thermobalance with experiments in lab scale reactors and tests of physico-chemical characterization of the fuels themselves and of their solid products. It therefore includes three activities.

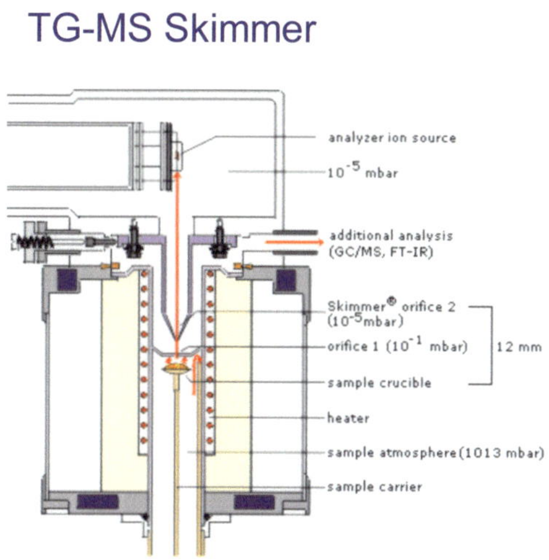

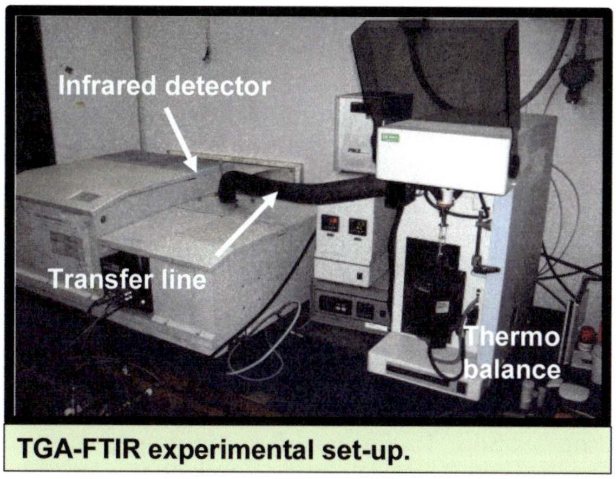

Fig. 3. TG-MS and TF-FTIR apparati.

1. Physico-chemical characterization of the solid

This includes proximate and ultimate analysis, SEM-EDX, ICP, XRD, Porosimetry by Hg and/or gas adsorption, Granulometric analysis.

The same set of analysis is applied to the raw sample and to samples of char and ash. The char is obtained in the necessary amount by pyrolysis in a tubular furnace or in a fluidized bed reactor at temperatures in the order of 600-800°C in a flow of nitrogen.

Ashes are obtained from complete burn-off of the material in lab scale reactors such as tubular furnaces or fluidized bed reactors, in air at temperatures in the order of 800°C.

2. Thermogravimetric analysis

This includes three sets of experiments:

TG-IP.	pyrolysis under inert conditions;
TG-OP.	oxidative pyrolysis;
TG-C.	combustion of char.

Thermal analysis is carried out in a TG system, possibly coupled with a DSC/EGA equipment for on-line analysis of the gaseous products. It is important that such devices are designed to minimize condensation and secondary reactions in the gas phase.

Approximately 10mg of sample are loaded in the pan in each test. Notably the particle size of the sample must be reduced when possible to 100-200 o µm to minimize heat gradients inside the particle and mass transfer limitations. An upward flow of gas of 100-200mL/min is used.

In pyrolysis experiments (TG-IP and TG-OP) the temperature is raised from 25°C to 110°C and held at 110°C for 5-10min to release moisture. The sample is then further heated up to 850-900°C at a constant heating rate. Heating rates in the range 5-20°C/min are scanned. During the ramp, 100% He or Ar or N_2 or a mixture of 0.01-21% oxygen in He/Ar/N_2 are used. The sample is finally held at 850-900°C for 30min, while the gas is switched to 21%O_2 in He/Ar/N_2 to burn the residual char.

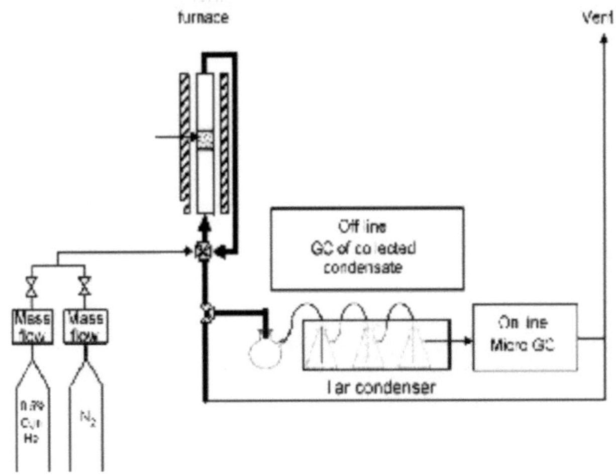

Fig. 4. Lab scale pyrolysis reactor.

In experiments of char combustion (TG-C), the char can be prepared in the thermobalance immediately prior to the combustion test or externally in a lab scale reactor. The char can be then heated in the thermobalance up to 850-900°C at a constant heating rate in a the desired mixture of 0.01-21% oxygen. Alternatively the char is heated in He/Ar/N_2 up to a desired temperature in the range 350-600°C. The gas is then switched to the desired mixture of 0.01-21% oxygen O_2 to burn the char isothermally.

It must be noted that the conditions chosen for the thermogravimetric experiments have been used in past experimental campaigns of pyrolysis and combustion of a wide range of solid fuels. In most cases such conditions proved successful to avoid internal gradients of heat and gas concentration as well as particle overheating and guaranteed that reactions took place under kinetic control. However such precautions may result insufficient to guarantee kinetic control in some cases.

The mass recorded during experiments of pyrolysis and oxidative pyrolysis is worked out in order to obtain TG plots of m/m_o versus T and DTG plots of $\frac{dm}{dT}\frac{1}{m_o} = -\frac{df}{dT}\frac{m_o - m_\infty}{m_o}$ versus T.

where where m, m_o and m_∞ are the actual weight of the sample, the initial weight of sample (after the dehumidification stage in pyrolysis and oxidative pyrolysis experiments) and the weight of the sample residue at the end of the experiment, respectively.
Results were fitted to a power law expression:

$$-\frac{1}{m_o}\frac{dm}{dt} = k_o \exp\left(\frac{-E}{RT}\right)\left(1-\frac{m_\infty}{m_o}\right)^\alpha \tag{1}$$

The kinetic parameters of equation (1) can be obtained by non linear regression analysis of the DTG curves according to the Friedman and Kissinger methods using general-purpose regression tools. Data from experiments at heating rate (H_R) below 20°C/min are used.
The mass loss recorded during experiments of char combustion is further worked out to calculate:

- the carbon conversion degree f =(mo-m)/(mo-m_∞)

m, mo and m_∞ being the actual weight of the sample, the initial weight of sample and the weight of the sample residue at the end of the experiment;

- the instantaneous rate of carbon conversion df/dt

Assuming that a power law kinetic expression of the type

$$\frac{df}{dt} = A(f)\cdot k_o \exp\left(\frac{-E}{RT}\right)p_g^n \tag{2}$$

is a good approximation in most cases, where p_g is the partial pressure of the oxygen and A(f) describes the evolution of instantaneous conversion rate along burn-off.
Accordingly the time $\tau_{0.5}$ required to achieve 50% conversion reads:

$$\tau_{0.5} = \frac{1}{k_o}\exp\left(\frac{E}{RT}\right)p_g^{-n}\int_0^{0.5}\frac{df}{A(f)} \tag{3}$$

and the reaction rate averaged over the first 50% conversion:

$$R_{0.5} = \frac{0.5}{\tau_{0.5}} = k_o' \exp\left(\frac{-E}{RT}\right)p_g^n \tag{4}$$

kinetic parameters of equation (4) can be obtained by non linear regression analysis of average reaction rate over the conversion interval f=[0, 0.5] at different temperature and different values of oxygen partial pressure. Alternatively an average over a larger conversion interval can be adopted.

3. Experiments in lab scale reactors

These include:

TR-IP-SH	Experiments of inert pyrolysis with slow heating.
TR-OP-SH	Experiments of oxidative pyrolysis with slow heating.
TR-IP-I	Experiments of inert pyrolysis under isothermal conditions.
TR-OP-I	Experiments of oxidative pyrolysis under isothermal conditions
TR-CC-SH	Experiments of char combustion with slow heating
TR-CC-I	Experiments of char combustion under isothermal conditions

In experiments of slow pyrolysis (TR-IP-SH and TR-OP-SH) typically tubular reactors are used heated externally by electric furnaces at 5-10°C/min. The vessel with the sample is placed inside the reactor from the very beginning of the experiment and heated accordingly. In experiments of pyrolysis under isothermal conditions (TR-IP-I and TR-OP-I) the sample is fed to the already hot reactor at a given temperature, typically in the range 600-850°C.
Inert pyrolysis is carried out using helium, while for oxidative pyrolysis inert gas is mixed with a small quantity (0.1-5%) of O_2. The reaction products are quickly cooled down as they flow through bubblers held at 0°C and -12°C respectively. Tar captured by the bubblers are characterized off line by GC or simulated distillation. The gas which passes through the bubblers is sent directly to a gas analysis system, possibly a micro-GC in order to analyse the gaseous products on line. These experiments allow to measure the overall yield in gas-tar and solid products. Further data concern the composition of the tar cumulatively produced during the test and the profiles of gaseous species evolved as a function of time/temperature.
In experiments of char combustion at slow heating rate (TR-CC-SH) the same tubular reactor and experimental procedure as for experiments of slow pyrolysis can be used. In experiments of char combustion under isothermal conditions (TR-CC-I) the reactor used for experiments of isothermal pyrolysis can be used or alternatively small scale fluidized bed reactors. In fluidized bed reactors a bed of inert material such as quarzite can be used with particle size typically between 300-400 μm. Particles are fed from the top of the reactor at a fixed temperature (between 500-900°C). During char combustion experiments the gas is initially nitrogen. After pyrolysis is complete, the gas is switched from nitrogen to an O_2/N_2 mixture (with O_2 at values between 4-15%). The profiles of CO and CO_2 evolved as a function of time can be worked out to evaluate char combustion rate according to the following expressions:

$$R = \frac{f}{t_R} = \frac{\dfrac{\int_{to}^{t_R} (c_{CO} + c_{CO2}) Q dt}{n_c}}{t_R} \quad 1/s \qquad (5)$$

f:	carbon conversion degree
t_R, t_o:	reaction time,time when oxygen feed started, s
c_{CO}, c_{CO2}:	concentration of CO e CO_2, mol/l
Q:	gas flow rate, l/s
n_C:	moles of carbon fed with the solid fuel, mol

4. Examples of sample preparation and physico-chemical characyterization tests

In order to explain the experimental protocol proposed in the previous paragraph, results will be presented here for a set of different biomasses as well as for other carbon rich materials. The examples have been selected so as to show typical and problematic cases.
As a first example the case of meat and bone meal (MBM) has been chosen, from a previously published paper [10].
MBM char was prepared in an electrically heated tubular furnace at 650°C for 5min in a flow of nitrogen. Ashes of MBM were produced in the same electrically heated tubular furnace at 800°C in a flow of air.

Elemental analysis, SEM, ICP and granulometric analysis have been carried out on the above samples The following instruments have been used: LECO CHN 2000 and Perkin Elmer CHNOS elemental analysers, a Philips XL30 SEM equipped for EDAX analysis, an Agilent 7500 CE ICP-MS, a Mastersizer 2000 granulometer of Malvern Instruments.
Results are reported in Table 1 and in Fig. 5.
The granulometric analysis of MBM indicate that the sample has a quite dispersed particle size distribution with average particle diameter of 250μm. In the SEM picture of MBM some smooth and roughly cylindrical particles can be recognized within the bulk of the material. The EDAX analysis reveals large amounts of C, O, Ca, P. In comparison the roughly cylindrical particles are poor in Ca and P and quite rich in C and S.

Fig. 5. SEM picture of MBM

The ICP analyses indicates that raw MBM contains large amounts of Na and Ca, followed by K, Mg and by small amounts of Fe, Zn, Al, Sr with traces of Ba, Mn, Cr, Co, Pb. The same metals are found in ashes of MBM produced in the electrical furnace at 800°C, however upon ashing the amounts of Ca, Mg increase by a factor of 3, those of Al, Na, Fe, Zn by a factor of 2; K significantly decreases. XRD of MBM reveals that the only crystalline substance present in MBM is Apatite ($Ca10(PO4)6(OH)2$).
In order to provide a good example of the tests of characterization of the microstructural properties the case of three biomass materials, investigated in ref. [18] will also be reported: namely, wood chips (Pinus radiata), pine seed shells and exhausted olive husk. Porosimetric analysis was carried out on the raw materials, on chars and on partially reacted chars.
Char samples were prepared in a bubbling fluidised bed reactor operated with nitrogen at 850°C for 5min. A selection of char particles prepared in the fluidised bed reactor were embedded in a in epoxy resin and cut. Cross-sections were observed under a scanning electron microscope (Philips XL30 with LaB6 filament) at magnifications up to 50 times. Some char samples were ground and sieved to particle size <300μm and further reacted with air or up to 10% carbon conversion in an electrically heated tubular furnace operated at 440°C in air.

Proximate analysis of MBM			
Moisture (as received w%)		6	
Ash (as received w%)		20	
Fixed carbon (as received w%)		10	
Volatile Matter (as received w%)		64	
Ultimate analysis			
	MBM (as received, w%)	*MBM char (w%)*	
C	43.4	31.1	
H	6.4	1.7	
N	9.2	5.1	
S	0.4	n.d.	
Cl	0.3	n.d.	
P	n.d.	n.d.	
Heating Value of MBM (*d.b. %w*)			
HHV (MJ/kg)		15.50	
LHV (MJ/kg)		14.47	
ICP analysis of raw and ashed material			
	MBM (as received, ppm)	*Ash (ppm)*	
Al	57	108	
Na	11422	19498	
Fe	138	331	
Ca	19832	58541	
K	3910	808	
Mg	1777	5150	
Ba	11	78	
Mn	8	31	
Sr	37	140	
Cr	1	17	
Va	0	0	
Ni	0	8	
Zn	70	139	
Ce	0	0	
Co	2	9	
La	0	0	
Pb	10	9	
Granulometric analysis			
	MBM	EP1	EP3
d (0.1) μm	6	4	1
d (0.5) μm	124	44	8
d (0.9) μm	706	162	51
Mean d (Surface weighted), μm	16	9	3
Mean d (Volume weighted), μm	252	73	18

Table 1. Characterization of MBM

The analysis included mercury intrusion porosimetry, adsorption of N_2 at 77K and of CO_2 at 273K. The porosimetric station consisted in a high-pressure mercury porosimeter Carlo Erba 2000 equipped with a Macropore unit and a Carlo Erba Sorptomatic 1900. Mercury porosimetry allowed to evaluate the pore size distribution of char in the size range of 200μm>dp>75Å and the % porosity, ε. Nitrogen adsorption results allowed to evaluate BET surface areas. Data of carbon dioxide adsorption were analysed according to Dubinin Radishkevich method to evaluate micropore volumes.

Figures 6 A-C show the cross-sections of char particles of wood chips, pine seed shells and olive husk observed under the scanning electron microscope. The micrographs show that char from wood chips and pine seed shells has a highly anisotropic pore structure characterized by parallel channels running in the axial direction (orthogonal to the paper sheet). This is a consequence of the fibrous structure of the parent biomass. Large pores and cavities are also evident in the case of olive husk char, but the orientation appears to be random. A comparison of the three micrographs shows that the solid matrix of the char from wood chips is the most porous, while that of pine seed shell char is the most compact.

The cumulative pore size distribution on volume basis for the chars of the three biomass fuels is reported in Figure 7. Table 2 reports the overall char porosity and density calculated from porosimetric data. Table 3 reports the BET surface area and the micropore volume of unconverted char samples and of char reacted with air or with carbon dioxide up to 10% carbon conversion.

It can be observed that wood chip char is characterized by the lowest density and the largest porosity, which consists predominantly of macropores (>1μm). Wood chip char has also the smallest micropore volume of the three chars (0.17 cm^3/g). Moreover micropore volume of wood chip chars is scarcely affected by partial conversion both with air and with carbon dioxide. BET area of wood chip char is negligible after pyrolysis, it increases to 300m2/g after 10% combustion. Noteworthy the increase in BET surface with the progress of carbon consumption can be related with the opening up and development of mesopores, while the increase of micropore volume can be related with the evolution of microporosity [19]. The observed results therefore suggest that reaction of wood chip char with oxygen opens up larger pores (macro and mesopores). The extent and the role of microporosity is very limited in wood chip char.

Olive husk generates a char that is denser than wood chips char and relatively less macro-porous. The pore size distribution is indeed shifted toward smaller pore sizes. Micropore volume of the unconverted char is comparable with that of wood chips (0.18 cm3/g) but increases by 40% after combustion. BET surface increases up to 320m^2/g after combustion Altogether results of porosimetric analysis suggest that olive husk char possess a more extensive network of mesopores compared to wood chip char and again quite modest microporosity. Moreover mesoporosity develops along with reaction with oxygen.

The char obtained from pyrolysis of pine seed shells has the smallest pore size distribution and the highest density of the three biomass chars investigated. Its micropore volume is 0.23 cm^3/g and increases by 32% after combustion indicating a considerable activation of small pores especially by carbon dioxide. BET area reaches 580 m^2/g after combustion suggesting that mesoporosity is significantly developed by oxygen. Altogether results indicate that pine seed shell char contains a large portion of micro and mesopores prone to be activated by the reaction.

	Average pore diameter [μm]	ε %	Particle density [kg/m3]
Wood chips char	17	91	170
Olive husk char	7.5	80	400
Pine seed shells char	17.5	70	490

Table 2. Results of Hg-porosimetry on three biomass chars

	BET area (N2) [m2/g]	Micropore volume (CO2) [cm3/g]
Unreacted char		
Wood chips char	< 1	0.165
Olive husk char	< 1	0.183
Pine seed shells char	< 1	0.232
Char reacted with air up to 10% conversion		
Wood chips char	296	0.175
Olive husk char	320	0.256
Pine seed shells char	579	0.307

Table 3. Results of gas adsorption on unreacted and partially reacted biomass chars

A last example is reported to demonstrate the study of the fate of metals by SEM, ICP and XRD analysis. The case reported here refers to a bitumen like refuse of the oil industry, particularly riched in Mo and V. Although this is not a biomass fuel, it is presented here because it is particularly instructive of the problematic related to the presence of metals.

In addition to the raw material , also char, ashes, a sample of leached material and a sample of char at intermediate burn-off have been characterized The char was prepared in a tubular reactor at 600°C in a flow of nitrogen. Ashes were obtained from complete burn-off of the material in the same reactor in the excess of 800°C. Partial conversion of the char was accomplished at 600°C in air. Additionally a sample was obtained by overnight leaching of the raw material in pentane.

SEM and ICP analysis were carried out using a Philips XL30 SEM equipped for EDAX analysis and an Agilent 7500 CE ICP-MS. XRD measurements were made with a Brucker D8 ADVANCE diffractometer in reflection mode from 3°(2θ) to 70°(2θ) with a step size of 0.03°(2θ) with an energy dispersive detector Sol-X. Porosimetric analysis was carried out by nitrogen absorption at 77K with a Carlo Erba Sorptomatic.

Results are reported in Tables 4-5 and in Fig 8. Notably the raw material has a very high carbon content and good calorific value (PCS 34050 kJ/kg). It contains also non-negligible contents of selected heteroatoms and several impurities, such as S, Cl, Ca, V, Fe, Ni, Mo. These metals, identified also by XRD, form different crystalline phases: $V_{1.87}FeS_4$, $S_8V_{5.44}$, V_2NiS_4, $V_2Fe_{0.67}S_4$, V_3S_4 and V_2MoS_4. XRD reveals also the presence of a sharp peak at $2\theta=26°$ indicative of the presence of graphitic carbon, probably resulting from catalytic graphitization. The BET area is $200 m^2/g$.

Notably during vacuum treatment prior to nitrogen adsorption tests the sample released a large quantity of sticky and intensely odorous volatile matter. This sticky matter removed under vacuum could also be removed by mild heat treatment up to 150°C or alternatively by leaching the sample with organic solvents such as pentane, as already explained. The proximate and ultimate analyses and the ICP analysis confirm that such pre-treatments removed mainly volatile organic matter with very low boiling point which impregnated the raw sample, while metals remained in the sample.

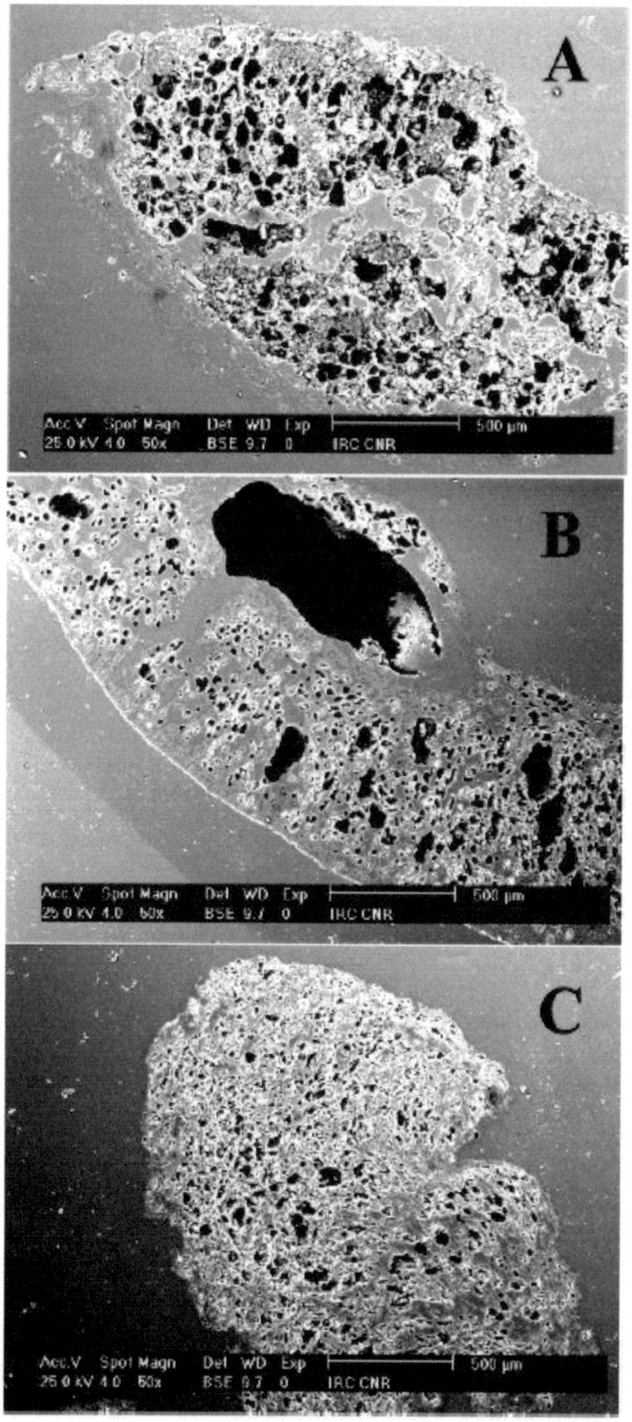

Fig. 6. SEM picture of cross-section of biomass chars. A. Wood chips; B. Pine seed shells; C. Olive husk

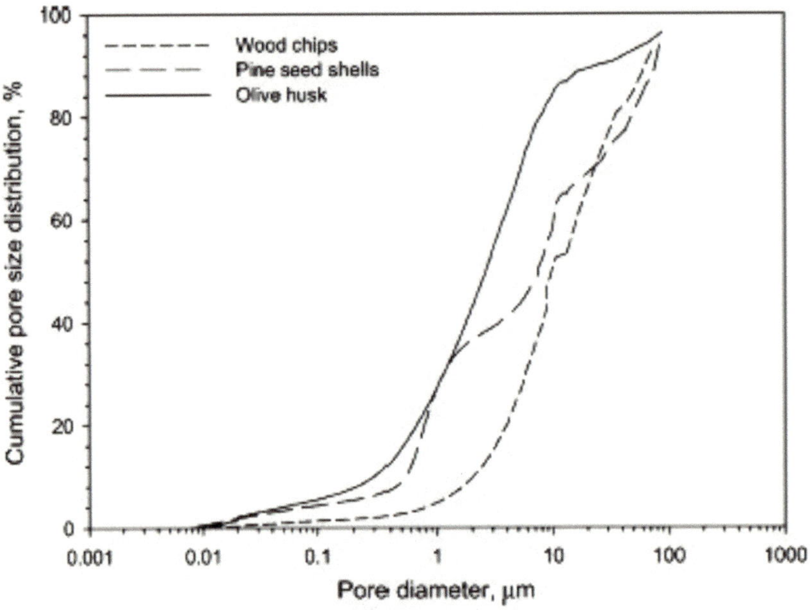

Fig. 7. Cumulative pore size distribution of three biomass chars from Hg porosimetry

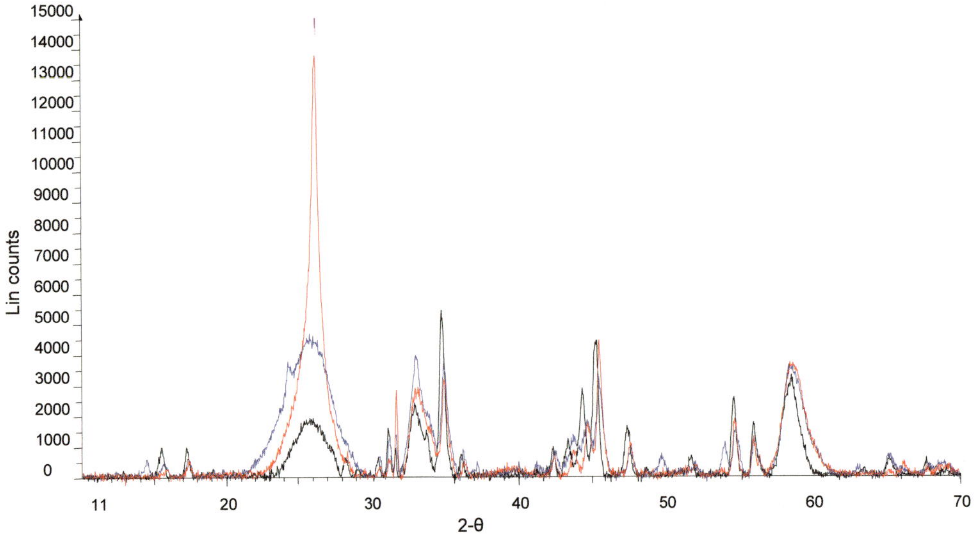

Fig. 8. Results of XRD analysis for a residue of the oil industry: raw sample (red), char produced under inert conditions (black) and partially burned sample (blue).

Upon pyrolysis in nitrogen at 600°C volatile organic matter is further lost whereas metals mainly remain in the solid residue. Results from XRD characterization of the char surprisingly show that the sample becomes less graphitic in nature. When the char is burnt with air in the excess of 800°C the carbon content gradually decreases and the concentration of metals increases. XRD reveals the appearance of vanadium oxides (VO_2 and V_2O_3) and a renewed increase in graphitic order.

The ash composition has been characterized by ICP, and results are reported in Table 5. If one considers that ash residue remaining after complete burn off of the raw fuel represents about 10% of the original sample mass, one would expect that the content of metals in the ash residue should be nearly ten times the corresponding amount in the raw sample. Inspection of Table 5 suggests that this is not the case. To better appreciate the partitioning of metals between the solid residue and the leachate (for samples leached with pentane) or the gas phase (for char remaining after pyrolysis and for the ash residue remaining after combustion), a partitioning factor α has been reported for all but the raw samples and for each metal. Based on an ash-tracing concept, the partitioning factor α has been defined as:

$$\alpha = \frac{\dfrac{w_{i,k}}{w_{i,raw}}}{\dfrac{w_{ref,k}}{w_{ref,raw}}}$$

where $w_{i,k}$ represents the amount of metal i in sample k (k=pentane-leached sample, char, ash) and $w_{i,raw}$ the amount of the same metal in the raw sample. Similarly, $w_{ref,k}$ and $w_{ref,raw}$ represent the amounts of a reference metal in sample k and in the raw sample, respectively. The reference metal was selected so as to meet two constraints: stability upon both heat treatment and combustion, abundance so as to minimize uncertainties associated with its quantification. After consideration of different candidates, Nickel proved to be the better suited reference metal.

Analysis of the partitioning factors provides a clear picture of the relative stability of the different metals upon pentane-leaching, fuel pyrolysis and combustion, which can be related to the departure of α from unity. Most metals are relatively stable upon pyrolysis (with possible exceptions of sodium and lead). More pronounced is the effect of combustion on selected metals: extensive depletion of Se, Sb, Cd and Hg is observed. The more pronounced effect is no doubt that associated with Mo, whose abundant content in the raw residue is only marginally retained in the ash residue after combustion, possibly because of the large volatility of this metal in the oxidized state.

	Raw sample	Pentane-leached sample
Moisture	0.1-0.3	0
Volatiles	22.1-23.8	15.1
Ashes	8.7-10.9	15.9
Fixed carbon	66.7-67.5	69.0

	Raw sample	Pentane-leached sample	Char
C	78.3	78.2	77.2
H	4.6	4.1	1.8
N	0.8	1.0	1.0
S	6.5	n.d.	n.d.
O_2 (by difference)	0-1.1	n.d.	n.d.

Table 4. Analysis (a.r. w%) of refuses of the oil industry. *Minimum and maximum values.

	Raw sample	Pentane-leached sample		Char		Ash	
	ppm	ppm	α	ppm	α	ppm	α
Na	3896	3757	0.97	2980	0.62	29904	0.82
Mg	105	99	0.95	142	1.09	1046	1.07
Al	355	320	0.91	414	0.94	2804	0.84
P	19	10	0.53	17	0.72	152	0.86
K	77	70	0.92	95	0.99	493	0.68
Ca	586	538	0.93	1394	1.92	5193	0.95
Ti	17	15	0.89	22	1.04	118	0.74
V	12720	12360	0.98	15340	0.97	85450	0.72
Cr	24	23	0.97	32	1.07	233	1.04
Mn	17	16	0.95	21	0.99	155	0.97
Fe	2048	1823	0.90	2542	1.00	18546	0.97
Co	39	39	1.01	49	1.01	336	0.92
Ni	4006	3970	1.00	4975	1.00	37469	1.00
Cu	58	56	0.97	76	1.06	381	0.70
Zn	62	61	0.99	63	0.82	553	0.95
Ga	10	10	1.01	13	1.05	78	0.83
As	5	5	1.01	6	0.97	24	0.51
Se	4	4	1.01	3.7	0.74	0.1	0.00
Sr	11	11	1.01	15	1.10	82	0.80
Mo	30818	31298	1.02	40068	1.05	21309	0.07
Cd	51	49	0.97	62	0.98	36	0.08

Table 5. ICP analysis of inorganics and metal partitioning factors of refuses of the oil industry

5. Examples of thermogravimetric analysis

TG-IP/TG-OP/TG-C results for typical biomass

The results of thermal analysis carried out on some of the previously mentioned fuels will be reported in this section in order to provide examples of typical as well as more problematic results as regards TG-IP, TG-OP and TG-C experiments.

Notably the TG analysis was carried out in a Netzsch STA 409 CD TG/DSC equipped with a special Skimmer device and a quadrupole Mass Spectrometer Balzers QMG422 (0-300 a.m.u.). This device, described in detail in ref [20] and shown in Fig. 3, enables on-line analysis of gaseous products while minimizing condensation and secondary reactions in the gas phase.

Pine seed shells provide a good example of the typical biomass behaviour during TG-IP and TG-OP experiments. The TG-DTG curves are reported in Fig. 9. The DTG curves exhibit a single peak anticipated by a shoulder. The peak temperature increases with the heating rate. The char residue is around 25% at all heating rates.

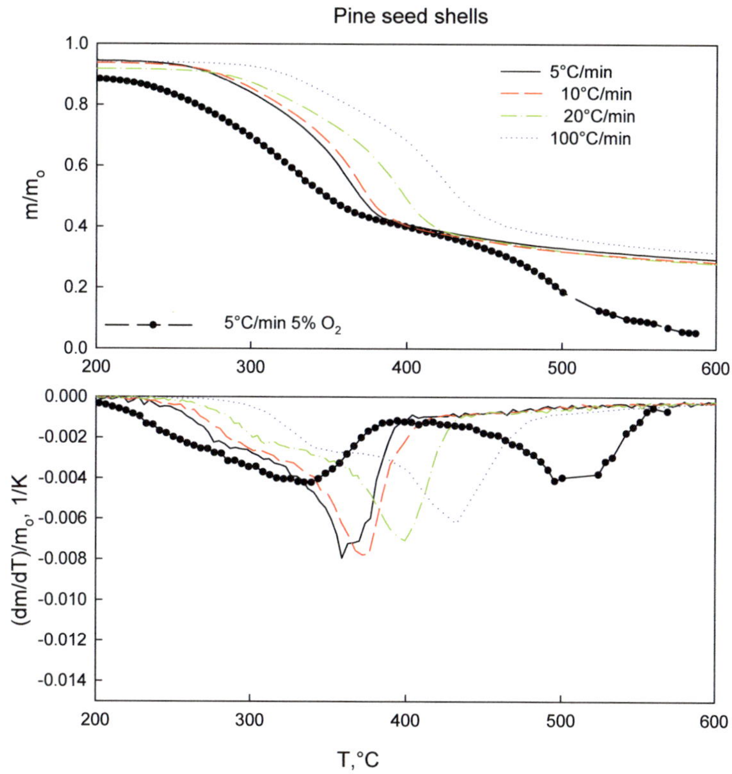

Fig. 9. TG-DTG curves from experiments TG-IP and TG-OP on a typical biomass.

With 5% oxygen at 5°C/min two stages of weight loss are observed: the first stage, accounts for the loss of volatile matter, the second one accounts for conversion of char, leaving behind only residual ash. The first DTG peak is anticipated by 25°C compared to the case of inert pyrolysis. Altogether results of pyrolysis under inert and oxidative conditions indicate a reaction path of the type descried in Fig. 2B.

Completely different results have been obtained for coals [2] which, upon oxidative pyrolysis experience a single stage of weight loss with peak at quite higher temperatures (around 670°C at 5°C/min), according to the reaction path of reported in fig. 2A.

DTG curves of the type shown can be analysed easily according to the Friedmann and the Kissinger methods in order to obtain kinetic parameters. Notably large values of the parameter α are commonly obtained when a single reaction model of the type of eqn. (1) is applied to a set of multiple parallel reactions with a broad distribution of activation energies. For ligno-cellulosic materials in fact a wide range of kinetic schemes have been used, including two parallel reactions, nucleation models, discrete activation energy models. The adoption of a single reaction model is a relatively good simplification in the case of pine seed shells, it may be an oversimplification in the case of other biomasses, nevertheless it remains useful for calculations within a first order approximation.

Fig. 10-11 report results of TG-C experiments on pine seed shell char and other biomasses. In particular the procedure followed to assess the kinetics of char-combustion reaction is exemplified assuming a power law expression of the type in eqn. (2).

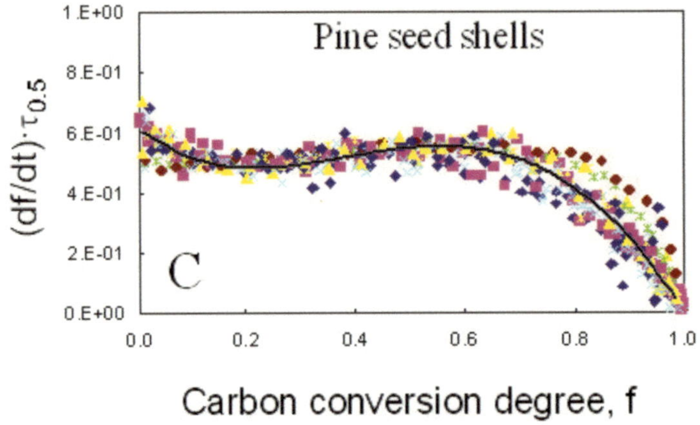

Fig. 10. Instantaneous char combustion rate vs burn-off for a typical biomass.

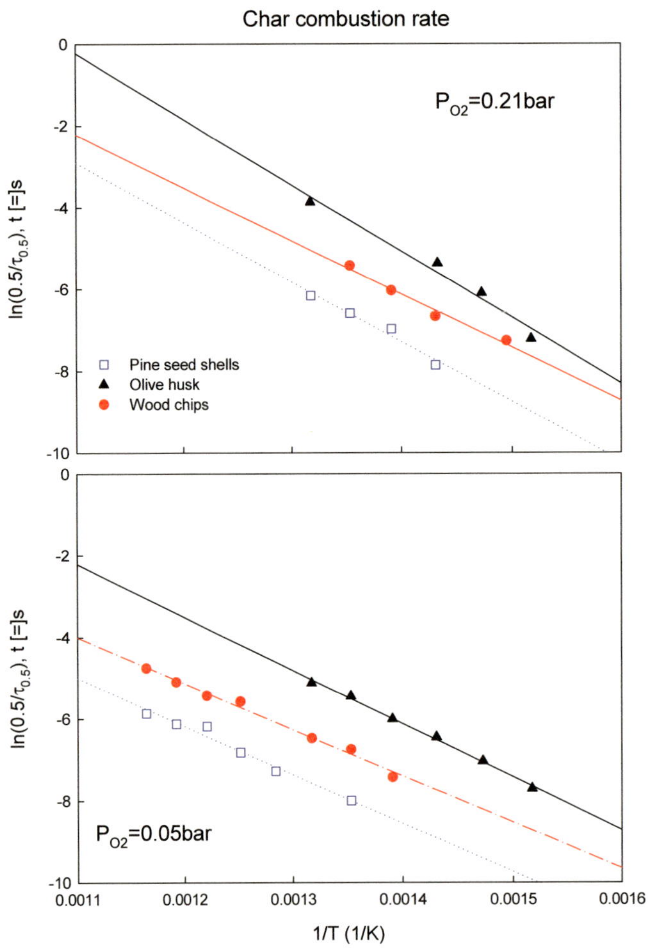

Fig. 11. Arrhenius plots for combustion of three biomass chars.

The instantaneous carbon conversion rate has been normalised with respect to the time $\tau_{0.5}$, By this procedure data points at different temperature and oxygen partial pressure overlay and a regression curve can be drawn, which gives the variation of the rate of carbon conversion along burnoff (A(f)).
It is important to observe that the Arrhenius plots in the temperature range investigated are reasonably linear. This allows to obtain the kinetic parameters for char combustion from regression over all data points in the figure. At higher temperature the slope of the Arrhenius plot might decrease as a consequence of the onset of mass transfer limitations to the rate of combustion. In this case only the linear portion of the Arrhenius plot should be used to obtain kinetic data.
An additional information which can be readily obtained from Fig. 11 is the scale of reactivity of different chars: in the case shown it is evident that olive husk char is the most reactive of the three samples.

TG-IP/TG-OP/TG-C results for metal rich materials

Olive husk and MBM provide examples of atypical TG results that can be obtained for fuels with high inorganic or metal content as regards the effect of oxygen during pyrolysis. As shown in Fig.12 the presence of oxygen during pyrolysis of this type of fuels does not enhance the mass loss, on the contrary the pyrolytic scission seems to be delaied.

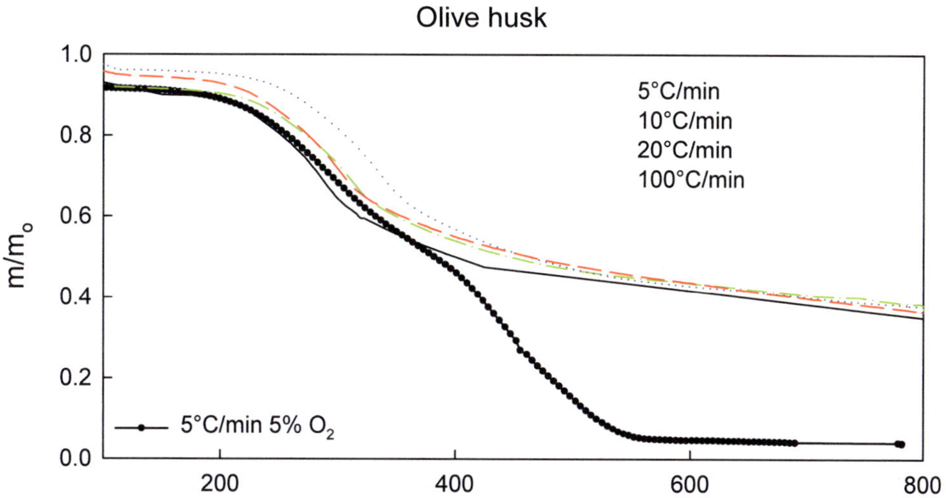

Fig. 12. TG curves of olive husk during TG-IP and TG-OP experiments

This apparently anomalous trend can be related to the high metal and inorganic content of these biomasses which promote formation of oxidized complexes. The uptake of oxygen partly compensates the mass loss due to pyrolysis at moderate temperature. At higher temperature complexes are released and eventually lead to complete burnout above 500°C.
The case of the bitumen like residue, particularly rich in Mo and V, will also be presented here, because it is considered very useful to explain this phenomenology. The results of TG-IP and TG-OP tests for this material are reported in Figs. 13-14 TG-DTG-DSC curves are complemented by MS curves.

Under inert conditions between 150 and 300°C distinct peaks can be recognized in MS profiles corresponding to M/e equal to 29, 43, 41, 57, 56. Although it is not possible to achieve a full identification of the chemical species corresponding to these values of M/e, it is likely that alkanes (propane M/e 29, butane M/e 43,29; pentane M/e 43, 57) and butene (M/e 41, 56) are produced in this temperature range. The complete spectrum at 200°C reveals the release also of species with high molecular mass: M/e 81, 95-98, 104, 118, 140, 154, 160, 164, 176, 180. In the temperature range 350-800°C a large MS peak identified as hydrogen is observed, which may be due to dehydrogenation/aromatization reactions and breakage/condensation of higher molecular weight chains.
When mild oxidizing conditions are established (p_{O2}=0.001bar) the pattern does not change significantly up to 400°C, although above 450°C the weight loss increases from 2 to 9%. When strongly oxidizing conditions are established (p_{O2}=0.21 bar) a totally different behaviour is observed: up to 350°C the weight loss is depressed. Combustion takes off above 450°C with a marked heat release resulting in 90% weight loss. MS curves are not reported, but it is CO and CO_2 are the major products in this case.
The presence of solid state reactions under oxidizing conditions at moderate temperature at the expense of metals is supported by the evidence during experiment DTG-OP of a small though noticeable DSC peak at 350°C and by results of XRD that show the formation of vanadium and molibdenum oxides.

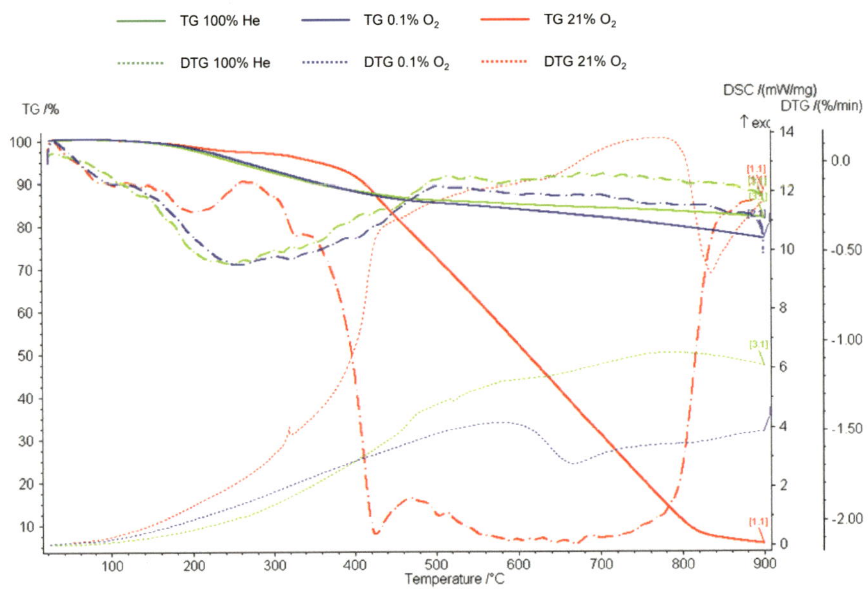

Fig. 13. TG-DTG-DSC curves from experiments TG-IP and TG-OP of a residue of the oil industry.

6. Examples of lab scale experiments

Pyrolysis (TR-IP-SH, TR-OP-SH, TR-IP-I, TR-OP-I)

Examples of results of pyrolysis in ab scale reactors with the specific aim of studying the yields in tar and gas under different pyrolysis conditions are reported for the bitumen like

residue of oil residue already mentioned before. In particular, the effect of inert vs mild oxidizing conditions and the effect of slow vs fast heating are presented.
Pyrolysis and oxidative pyrolysis experiments have been carried out in the tubular reactor described in Fig. 4. The reaction products were quickly cooled down as they flowed through 250ml bubblers held at 0°C and -12°C respectively. Tar captured by the bubblers has been characterized off line by means of simulated distillation. The gas which passed through the bubblers was sent directly to a micro-GC Agilent 3000° equipped with four columns (Molesieve MS5A, Poraplot U, Poraplot allumina and OV1) in order to analyse the gaseous products on line.

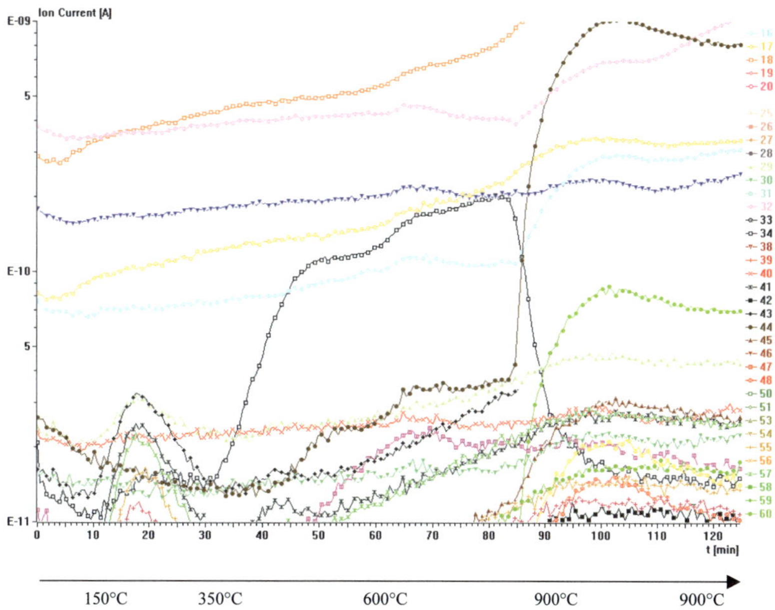

Fig. 14. MS curves from experiments TG-IP of a residue of the oil industry.

The overall char yield was between 19-22% in all the tests. The tar yield was around 10% but turned out to be rather scattered. The analysis of the tars collected by the bubblers is reported in Tab. 6. The weight fractions corresponding to different boiling points are reported. It can be observed that tar produced from slow pyrolysis under inert conditions has a minor fraction of components with boiling point between 170-300°C, a 60% weight fraction has boiling point in the range 350-500°C and 30% above 500°C. These figures are consistent with the weight loss measured by TGA. Tar obtained by fast pyrolysis under inert conditions and by slow pyrolysis with a mild oxidizing atmosphere both contain a larger fraction with boiling point below 350°C.
The composition of the gas leaving the bubblers during an experiment of slow pyrolysis in He are reported in Fig. 15. It can be observed that hydrocarbons with more than two carbon atoms are released in two stages. The first, more pronounced one, occurs between 150-400°C, the second between 400 and 600°C. Methane is instead released over the entire temperature range of the experiment. Under moderately oxidizing conditions similar profiles are obtained up to 300°C, but at higher temperatures CO_2 is produced at the expense of methane and other hydrocarbons.

Figure 16 reports the cumulative yields of different gaseous species throughout pyrolysis in the tubular reactor under different conditions. It can be observed that during slow heating rate pyrolysis in helium the product gas contains mainly CH_4 (90%) and small percentages of CO, CO_2, C_2H_6 (3-5% each). The presence of oxygen in the pyrolysis atmosphere at low concentration levels (0.1%) produces a gas with 50% di CO_2 and 40% CH_4. Upon fast heating pyrolysis rate under inert conditions produces a rather different gas, with a marked increase in C_2H_4 , which becomes the most abundant species, followed by CH_4, C_2H_6, CO, CO_2.

	Experiment TR-IP-SH w%		Experiment TR-OP-SH w%		Experiment TR-OP-I w%
	1st bubbler	2nd bubbler	1st bubbler		1st bubbler
<170	0	0	0	<170	0
170-350	4.7	12.3	18	170-350	4.7
350-500	61.5	55.8	58	350-500	61.5
500-800	33.8	31.9	24	500-800	33.8

Table 6. Boiling points of tar collected during pyrolysis in lab scale reactor of a residue of the oil industry

Char combustion (TR-CC-SH, TR-CC-I)

Char combustion experiments have been carried out in a fluidized bed reactor (FB-C) consisting of a 1.1 m long quartz tube with 20 mm id.. The tube is heated by a vertical electrical furnace with 110 mm ID and length 750 mm. Gas flows bottom up and passes through a distributor positioned at the centre of the tube. The gas flow rate is 100NL/h. A bed of 20mm quarzite is used with particle size between 300-400 μm. Exhaust gas is analysed on line by ABB IR analysers. In each test initially the bed is fluidized by nitrogen. One single particle of approximately 5mm diameter is fed from the top of the reactor at a fixed temperature (between 500-600°C). After pyrolysis is complete, the gas is switched from nitrogen to an O_2/N_2 mixture (with O_2 at values between 4-15%).

Figure 17 shows typical results of a fluidized bed experiment. In the example reported in this figure the particle was fed at t=100s under inert conditions. The bed was at 600°C. The progress of pyrolysis can be followed from the profile of CH_4. The time of pyrolysis in this experiment was 58s. At time t=800s oxygen was let into the reactor at the desired level of concentration (15% in the example), this produced a fast increase of combustion products. The CO and CO_2 profiles obtained during this stage are reported in the figure and show that char combustion took 430s. Notably in all the experiments devolatilization took roughly 60s. Pyrolysis time was indeed not affected by the operating conditions, in the range investigated, suggesting that the process was dominated by heat and mass transfer effects.

The char combustion time increased from 430s to 1500s when the temperature was lowered from 600 to 500°C at a value of oxygen concentration of 15% and from 430s to 1700s when oxygen concentration was lowered from 15 to 4 % at the temperature of 600°C. A regression of data of average rate of char combustion at different temperature and oxygen concentration allows to estimate the values of kinetic parameters.

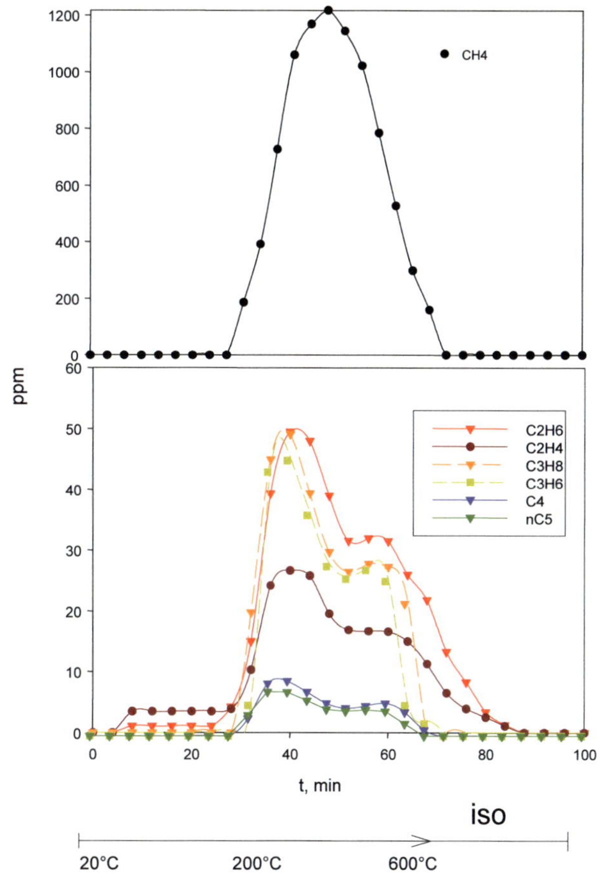

Fig. 15. Gas evolved during TR-IP experiment of a residue of the oil industry

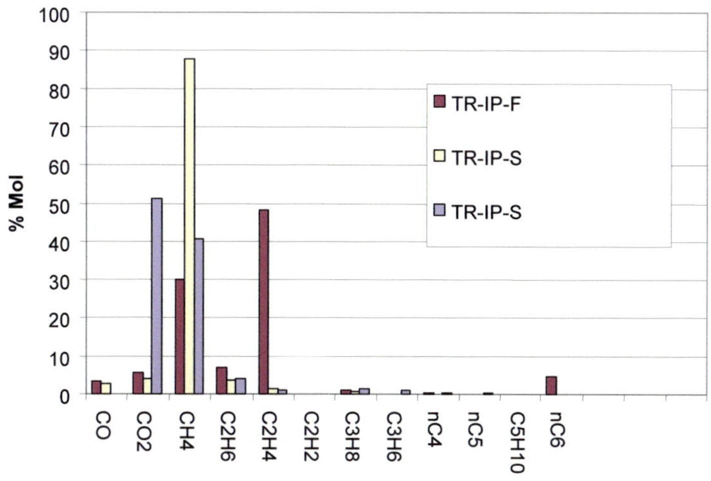

Fig. 16. Analysis of gas evolved during TR-IP experiment of a residue of the oil industry.

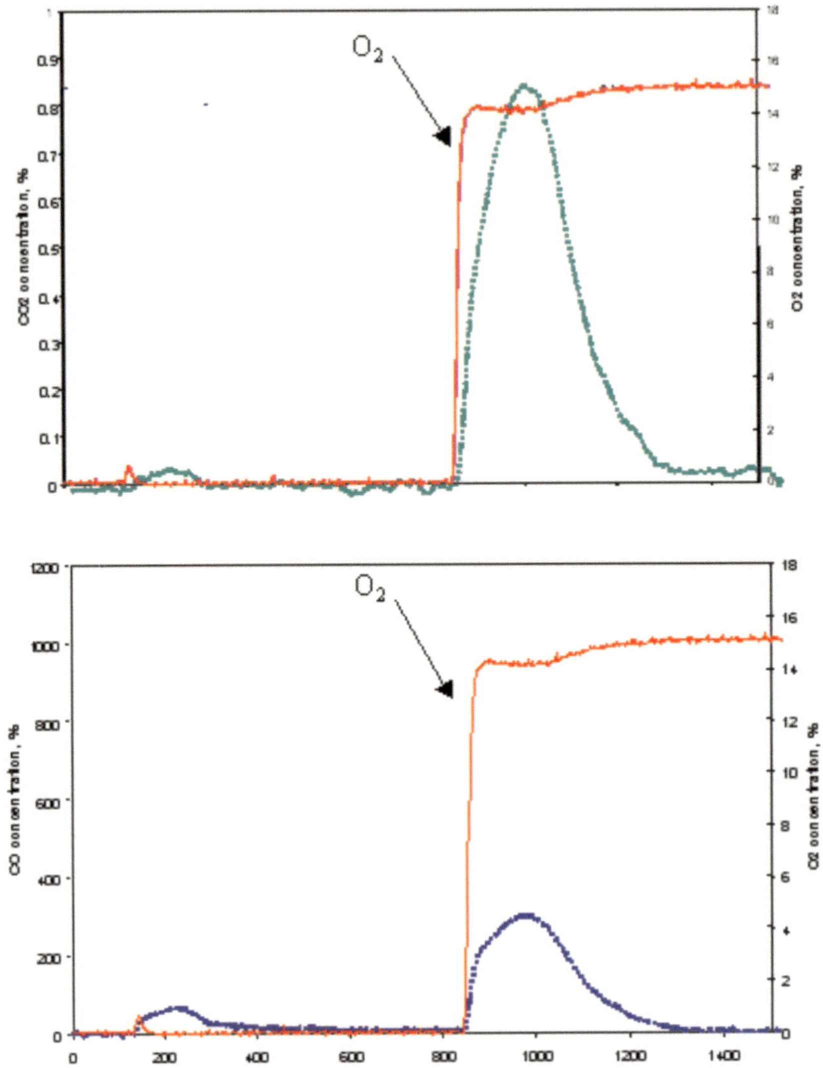

Fig. 17. Profiles of O_2 CO and CO_2 released during an experiment of TRCCI at 600°C in the fluidized bed reactor for a residue of the oil industry

7. Conclusions

An experimental procedure has been proposed to investigate at a lab-scale the potential of biomasses as fuels for pyrolysis and combustion processes. The experimental work coupled physico-chemical characterization tests with pyrolysis under inert and oxidizing conditions and char combustion using different experimental techniques.

Thermogravimetric analysis provides useful information on the temperature range in which pyrolysis/combustion of the fuel can be carried out and allows to estimate the rate and kinetics of the reactive processes. Moreover it provides useful information on the effect of inert/oxidative conditions on the products yield. Examples reported in this paper show that the presence of oxygen upon heating favours pyrolysis reactions in many cases, but when biomasses have a high content of metals and inorganic matter the presence of oxygen hinders the pyrolitic reactions at low-moderate temperature through formation of oxygen complexes.

Tests of pyrolysis in lab scale reactors show that the composition of the pyrolysis gas and tar are strongly affected by the heating rate and by the presence of even minor concentrations of oxygen. As far as gas composition is concerned, slow heating under rigorously inert conditions produces mainly methane and minor amounts of hydrogen, methane, propane, ethylene, CO, CO_2. When heating is carried out in an even mild oxidizing atmosphere the gas produced contains mainly CO_2 and CH_4 and modest amounts of alkanes and alkenes of higher order. As far as tar is concerned, both fast heating and the presence of oxygen increase the low boiling point fraction.

Experiments in a fluidized bed reactor allows to estimate the time of pyrolysis and of char combustion under different conditions.

Characterization of the solid products by ICP and XRD allows to investigate the fate of mineral matter and metals. The examples reported for some metal rich fuels show that metals mainly remain in the solid residue during pyrolysis under rigorously inert conditions (up to 600°C). On the contrary pyrolysis under oxidizing conditions and char combustion at temperatures in excess of 800°C produce the oxidation and loss of selected volatile metals, most likely in their oxidized forms. This result has severe environmental implications and needs to be taken into account in process design.

8. Acknowledgments

Several people contributed to the work and are gratefully acknowledged, in particular Mr Vitale Stanzione for ICP and GC analysis, Dr Paola Ammendola and Dr Giovanna Ruoppolo for pyrolysis experiments in the tubular reactor, Mr Sabato Russo for SEM analysis. Special thanks to Mr Luciano Cortese for the valuable support in several aspects of the experimental work and Dr Riccardo Chirone for guidance and assistance.

9. References

[1] Pedersen K.H, Jensen A.D. Berg M., Olsen L.H , Dam-Johansen K., Fuel Proc.Tech. 90 (2009) 180-185

[2] Senneca O., Chirone R., Salatino P., J. Anal. Appl. Pyrolysis 71 (2004) 959;

[3] O. Senneca, P. Salatino, Combust. Flame 3 (2006) 578

[4] Braguglia C.M., Marani D., Mininni G., Mescia P., Bemporad E., Carassiti F. Water, Air, and Soil Pollution 158 (2004) 193-205

[5] Tillman D.A., Harding N.S., Fuels of Opportunity, Characteristics and Uses in Combustion Systems (2004) 29-87

[6] Struckmann P., Dieckmann H.-J., Brandenstein J., Ochlast M,. VGB Power Tech 84 (2004) 72-76

[7] Day M., Cooney J.D.,Touchette-Barrette C., Sheehan S.E., Fuel Proc.Tech. 63 (2000) 29–44

[8] Nnorom, I. et Al., 2007, Resources, Conservation and Recycling 52 (2008) 5

[9] Afzal A., Chelme-Ayala P., El-Din A.G., El-Din M.G., Water Environment Research (2008) 1397-1415

[10] Senneca O., Fuel 87 (2008) 3262 - 3270
[11] Monte M.C., Fuente E., Blanco A. , Negro C., Waste Management 29 (2009) 293–308
[12] H.L. Friedman, J. Polym. Sci. C6 (1964) 183.
[13] T. Ozawa, J. Therm. Anal. 31 (1986) 547.
[14] H.E. Kissinger, Anal. Chem. 29 (1957) 1702.
[15] T. Ozawa, J. Therm. Anal. 2 (1970) 301.
[16] T. Ozawa, Bull. Chem. Soc. Jpn. 38 (1965) 1881.
[17] J.H. Flynn, L.A. Wall, J. Polym. Sci B4 (1966)
[18] Senneca O, Fuel Processing Technology 88 (2007) 87-97
[19] Salatino P., Senneca O., Masi S., Gasification of a coal char by oxygen and carbon dioxide, Carbon, 36 (1998) 443

17

Estimating Nonharvested Crop Residue Cover Dynamics Using Remote Sensing

V.P. Obade[1], D.E. Clay[1], C.G. Carlson[1],
K. Dalsted[1], B. Wylie[2], C. Ren[1] and S.A. Clay[1]
[1]*South Dakota State University*
[2]*United States Geological Survey (EROS), Sioux Falls*
United States of America

1. Introduction

Non harvested above and below ground carbon must be continuously replaced to maintain the soil resilience and adaptability. The soil organic carbon (SOC) maintenance requirement is the amount of non-harvested carbon (NHC) that must be added to maintain the SOC content at the current level (NHC_m) (Mamani-Pati et al., 2010; Mamani-Pati et al., 2009). To understand the maintenance concept a basic understanding of the carbon cycle is needed (Mamani-Pati et al., 2009). The carbon cycle is driven by photosynthesis that produces organic biomass which when returned to soil can either be respired by the soil biota or contribute to the SOC. The rates that non-harvested biomass is converted from fresh biomass to SOC and SOC is converted to CO_2 are functions of many factors including, management, climate, and biomass composition. First order rate mineralization constants for nonharvested carbon (k_{NHC}) and SOC (k_{SOC}) can be used to calculate half lives and residence times. Carbon turnover calculations are based on two equations,

$$\frac{dSOC}{dt} = k_{NHC}\left[NHC_a - NHC_m\right] \tag{1}$$

$$k_{SOC} \times SOC_e = k_{NHC} \times NHC_m \tag{2}$$

In these equations, SOC is soil organic C, NHC_a is the non-harvested carbon returned to soil, NHC_m is the nonharvested carbon maintenance requirement, k_{soc} is the first order rate constant for the conversion of SOC to CO_2, and k_{NHC} is the first order rate constant for the conversion of NHC to SOC (Clay et al., 2006). These equations state that the temporal change in SOC (dSOC/dt) is equal to the non-harvested carbon mineralization rate constant (k_{NHC}) times the difference between the amounts of carbon added to the soil (NHC_a) and the maintenance requirement (NHC_m) and that at the SOC equilibrium point (SOC_e), the rate that non-harvested C (NHC) is converted into SOC ($k_{NHC} \times NHC_m$) is equal to the rate that SOC is mineralized into CO_2 ($k_{SOC} \times SOC_e$). Through algebraic manipulation, these equations can be combined to produce the equation,

$$\frac{NHC_a}{SOC_e} = \frac{k_{SOC}}{k_{NHC}} + \frac{dSOC}{dt}\left[\frac{1}{k_{NHC} \times SOC_e}\right] \tag{3}$$

When fit to a zero order equation, the y-intercept and slopes are $\frac{k_{SOC}}{k_{NHC}}$ and $\frac{1}{k_{NHC} \times SOC_e}$, respectively.

Based on this equation, NHC_m, k_{NHC}, and k_{SOC} can be calculated using the equations, $NHC_m = b \times SOC_e$; $k_{NHC} = 1/(m \times SOC_e)$; and $k_{SOC} = b/(m \times SOC_e)$. This approach assumed that above and below ground biomass make equal contributions to SOC; that the amount of below ground biomass is known; and NHC is known and that initial (SOC_e) and final (SOC_{final}) SOC values are near the equilibrium point. Advantages with this approach are that k_{SOC} and k_{NHC} are calculated directly from the data and the assumptions needed for these calculations can be tested. A disadvantage with this solution is that surface and subsurface NHC must be measured or estimated. Remote sensing may provide the information needed to calculate surface NHC, through estimating the spatial variation of crop residues which are a major source of NHC.

Traditionally crop residue cover estimates have relied on visual estimation through road side surveys, line-point transect or photographic methods (CTIC, 2004; McNairn and Protz, 1993; Serbin et al., 2009 a). However, such ground-based survey methods tend to be time-consuming and expensive and therefore are inadequate for crop residue quantification over large areas (Daughtry et al., 2005; Daughtry et al., 2006). The need to improve these estimates has prompted much research on the extraction of surface residue information from aerial and satellite remote sensing (Bannari et al., 2006; Daughtry et al., 2005; Gelder et al., 2009; Serbin et al., 2009 a & b; Thoma et al., 2004). Previous research has shown crop residues may lack the unique spectral signature of actively growing green vegetation making the discrimination between crop residues and soils difficult (Daughtry et al., 2005). Daughtry and Hunt (2008) reported that dry plant materials have their greatest effect in the short wave infra-red (SWIR) region between 2000 and 2400 nm related to the concentration of ligno-cellulose in dry plant residue.

Other studies have proposed the Cellulose Absorption Index (CAI), the Lignin Cellulose Absorption index (LCA) and the Shortwave Infrared Normalized Difference Residue Index (SINDRI) for estimating field residue coverage (Daughtry et al., 2005; Daughtry et al., 2006; Thoma et al., 2004; Serbin et al., 2009 c). However, neither CAI, LCA nor SINDRI are currently practical for use in spaceborne platforms (Serbin et al., 2009 a). For example, EO-1 Hyperion which was sensitive to the spectral ranges of CAI and LCA (2100 and 2300 nm wavelengths), is past its planned operational lifetime and suffers bad detector lines (USGS, 2007), while the shortwave infrared (SWIR) detector for ASTER satellite failed in April, 2008 (NASA, 2010; Serbin et al., 2009 c). Therefore, indices that utilize the multispectral wavelength ranges (450-1750 nm) appear to be the most viable economical alternative. The objective of this research was to assess if remote sensing can be used to evaluate surface crop residue cover, and the amount of nonharvested biomass returned to soil.

2. Materials and methods

2.1 Study area

A randomized block field experiment was conducted in South Dakota (SD) in the years 2009 and 2010. The coordinates at the site were 44° 32'07"North and 97° 22' 08"West. Soil at the site was a fine-loamy, mixed, superactive, frigid typic calciudoll (Buse). The treatments considered were residue removed (baled) or returned (not baled) with each treatment replicated 36 times. The field was chisel plowed and corn was seeded at the site during the first week of May in 2009 and 2010. The row spacing was 76 cm and the population was 80,000 plants per hectare. Following physiological maturity in October, grain and stover yields were measured. In all plots corn residue was chopped after harvesting. In residue removal plots, stover was baled, and removed. The amount of residue remaining after baling was measured in at 16 locations that were 0.5806 m^2 in size. For these measurements, the stover was collected, dried, and weighed. Approximately 56% (±0.08) of the corn residue was removed by this process. Following residue removal, soil surface coverage was measured using the approach by Wallenhaupt (1993) on 27th November 2009 and 13th November 2010.

2.2 Field measurements and model development

Spectral reflectance measurements of corn residues were collected with a Cropscan handheld multispectral radiometer (Cropscan Inc., Rochester, Minnesota) under clear sky conditions between 10 a.m. and 3 p.m. for all the field sites on 28th November 2009 and 13th November 2010. The Cropscan radiometer measures incoming and reflected light simultaneously. It measures within the following band widths, 440-530 (blue), 520-600 (green), 630-690 (red), 760-900 (near infra red, NIR), 1550-1750 (mid infra red, MIR), for wide (W) bands, and 506-514, 563-573, 505-515, 654-666, 704-716, 755-765, 804-816, 834-846, 867-876, 900-910, 1043-4053 nanometer (nm) for narrow wavelength bands.

The Cropscan radiometer was set at a height of 2 m above ground, so as to approximate a 1 m^2 spatial resolution on the ground. The Cropscan was calibrated by taking five spectral radiance readings on a standard reflectance, white polyester tarp, before beginning the scanning and after the whole field had been scanned. Scanning errors were minimized by following the protocols as reported by Chang et al. (2005). For calculations it is assumed that the irradiance flux density at the top of the radiometer is identical to the target. Reflectance data were corrected for temperature and incident light angles, relative to top of the sensor. Based on measured reflectance information, four wide reflectance bands and four indices derived from the wide spectral bands were calculated (Table 1).

Vegetation Index	Equation (modified)	Reference
Normalized Difference Vegetation Index ($NDVI_w$)	$NDVI_w$= $(R_{830}-R_{660})/(R_{830}+R_{660})$	Rouse et al. 1974
Green Normalized Difference Vegetation Index$_w$ ($GNDVI_w$)	$GNDVI_w$ = $(R_{830}-R_{560})/(R_{830}+R_{560})$	Daughtry et al. 2000 Gitelson and Merzlyak 1996
Normalized Difference Water Index ($NDWI_w$)	$NDWI_w$= $(R_{830}-R_{1650}/(R_{830}+R_{1650})$	Gao 1996
Blue Normalized Difference Vegetation Index ($BNDVI_w$)	$BNDVI_w$= $(R_{830}-R_{485})/(R_{830}+R_{485})$	Hancock and Dougherty 2007

Table 1. Spectral band combinations (indices)

2.3 Statistical analysis

Proc Mixed available within the Statistical Analysis System (SAS Institute, North Carolina) software, was used to determine reflectance differences in the residue removed and returned plots. Correlation (r) coefficients between reflectance values and weights of stover returned and % surface residue cover were determined. Finally, graphs of percent residue cover versus spectral band and index for the models with the highest correlations were compared.

3. Results and discussion

In 2009, 28.8 % of the soil was covered with residue in the removed (baled) plots, while in 2010, 54% of the soil was covered with residue (Table 2). In the residue returned (not baled) plots, the surface cover was 100 and 70%, in 2009 and 2010, respectively. The residue removal plots (28.8% cover) in 2009 had the lowest reflectance in the green, red, and NIR bands, while the residue returned plots in 2010 had the highest reflectance in the green, red, NIR, and MIR bands. These results imply that corn residues have a relative high albedo, compared to soil. Slightly different results would be expected in soybean (*Glycine max*) where Chang et al. (2004) did not detect reflectance differences between bare and soybean residue covered soil.

Year	Residue	Percent Cover	Weight Mg/ha	Blue W.	Green W.	Red W.	NIR W.	MIR W.	NDVIw	GNDVIw	BNDVIw	NDWIw
2009	Removed	28.8 d	3.7a	4.60 c	6.50 c	8.84 c	13.75 d	19.50 b	0.22b	0.36b	0.50b	-0.0035b
2009	Returned	100 a	7.1b	7.72 a	11.10 ab	15.60 b	23.05 c	24.02 a	0.20c	0.35c	0.50b	0.026a
2010	Removed	54.2 c	2.7c	6.60 b	11.22 b	16.53 b	26.6 b	26.61 a	0.24a	0.41a	0.60a	-0.15c
2010	Returned	70.0 b	5.5d	7.18 a	12.28 a	18.16 a	28.91 a	27.30 a	0.23ab	0.41a	0.60a	0.02ab
p-value		0.0001	0.001	0.0001	0.0001	0.0001	0.0001	0.0005	0.0001	0.0047	0.1691	0.0001
2009		64.4	10.9	6.2	8.77	12.2	18.4	21.76	0.21b	0.36	0.50	-0.08
2010		62.1	7.1	6.9	11.75	17.3	27.74	26.96	0.23a	0.41	0.60	0.011
p-value		0.464	0.001	0.14	0.0010	0.0002	0.0001	0.0368	0.013	0.0001	0.0001	0.0003

*Values within a column that have different letters are significantly different at the 0.05 probability level.

Table 2. Variation in residue cover over several wavelengths reflected from corn residues on the ground near Badger site, SD in the years 2009 and 2010.

	Blue	Green	Red	NIR	MIR
			r		
Residue returned (ton/ha)	0.39	0.30	0.27	0.22	0.002
% residue cover	0.61	0.56	0.53	0.48	0.24
	$NDVI_w$	$GNDVI_w$	$NDWI_w$	$BNDVI_w$	
Residue returned (ton/ha)	-0.35	-0.24	0.35	-0.19	
% residue cover	-0.34	-0.15	0.47	0.01	

Table 3. The correlation between the amounts of residue returned in 2009 and 2010 to the soil and the ground cover with surface reflectance. r values greater than 0.174 are significant at the 0.05 level.

The correlation coefficient between the residue returned in ton/ha and percent residue cover with surface reflectance are shown in Table 3. The correlation coefficients between residue returned and reflectance ranged from 0.002 in the MIR band to 0.39 in the blue band. For the % surface residue cover, higher r values were observed. These results suggest that surface reflectance measurements were better at predicting the crop residue coverage than residue amount. The highest r value between % ground cover and reflectance was observed for the blue band. The different bands previously have been reported for different uses (http://landsat.usgs.gov/best_spectral_bands_to_use.php). The blue band is useful for distinguishing soil from vegetation, while green is useful for assessing plant vigor. The NIR (770-900 nm) and short-wave infrared (1550 - 1750 nm) discriminates biomass content from soil moisture content. Although blue has a high correlation with surface residue cover, atmospheric scattering may reduce its suitability for space-based sensors (Lillesand and Kiefer, 2000; Wang et al., 2010).

The amount of residue retained on the soil was correlated to NDVI, GNDVI, and NDWI, while the percent coverage was correlated to NDVI and NDWI. Of the indices determined, $NDWI_w$ had the highest r value with percent residue cover followed by $NDVI_w$, $GNDVI_w$, and $BNDVI_w$ respectively. It is important to note, that the results are limited by the boundary conditions of the experiment. Although only the percent residue cover and residue weights versus the reflectance were analyzed, other factors such as soil cover, color or moisture could have impacted the reflectance values (Barnes et al. 2003; Daughtry et al., 2005; Daughtry et al., 2006; Pacheco and McNairn, 2010; Thoma et al., 2004).

A comparison of the reflectance data across years for the blue band suggests that a zero order equation ($y = 4.31 + 0.036x$; $r^2 = 0.38$) could explain the relationship between reflectance and surface coverage (Fig. 1). Slightly different results were observed for the $NDWI_w$ indices where a second order quadratic equation ($y = -0.22 + 0.005x - 0.0000263\ x^2$; $R^2 = 0.26$) was used to describe the relationship with surface coverage. The graph of % residue cover versus NDWIw flattens after the 60 % residue cover which implies that $NDWI_w$ may saturate with increasing coverage. A limitation of this study is that only one site was analyzed, therefore any models generated would be suitable for the specific site and only after fall harvest. Other errors can occur when extrapolating plot measurements data to the whole field coverage. In future, research that confirms the finding for other sites and harvesting approach needs to be conducted.

4. Conclusion

The main objective of this study was to investigate the potential of remote sensing to assess residue coverage. The research showed that surface reflectance was more closely correlated with percent cover than the amount of residue returned. Of the spectral band widths measured, reflectance in the blue range provided the most consistent results across the two years. $NDWI_w$ had a higher correlation with residue returned and % cover than $NDVI_w$, $GNDVI_w$, or $BNDVI_w$. Future studies should not only consider more field sites, but incorporate factors such as the decomposition rates of residues on spectral reflectance and harvesting approaches (see Daughtry et al. 2010), so as to develop an accurate and standard approach for mapping residue cover in real time over large geographic areas.

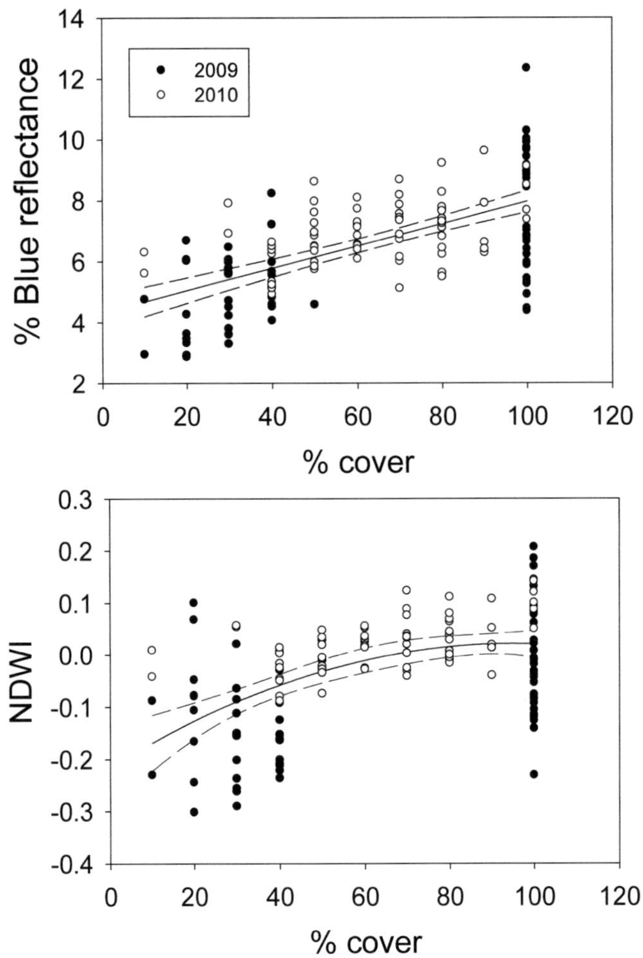

Fig. 1. Percent residue cover versus spectral bands (top) and $NDWI_w$ index (below)

5. Acknowledgements

Funding for this project was provided by NASA, South Dakota Corn Utilization Council, SD 2010 initiative, SD Soybean Research and Promotion Council.

6. References

Bannari, A., Pacheco A., Staenz, K., McNairn, H., Omari, K. 2006. Estimating and mapping crop residue cover on agricultural lands using hyperspectral and IKONOS data. *Remote Sensing of Environment* 104: 447-459.

Barnes, E. M., Sudduth, K.A., Hummel, J. W., Lesch, S. M., Corwin, D. L., Yang, C., Daughtry, C.T. and Bausch, W. C. 2003. Remote- and ground-based sensor techniques to map soil properties. *Photogrammetric Engineering & Remote Sensing*. Vol. 69, No. 6: 619–630.

Chang J., Clay S., Clay D., Aaron D., Helder D., Dalsted K. 2005. 'Clouds influence precision and accuracy of ground-based spectroradiometers'. *Communications in Soil Science and Plant Analysis* 36: 1799-1807.

Chang, J., Clay, S.A.and Clay, D.E. 2004. Detecting weed free and weed infested areas of a soybean (Glycine max) field using NIR reflectance data. Weed Sci. 52:642-648.

Clay, D.E., C.G. Carlson, S.A. Clay, C. Reese, Z. Liu, and M.M. Ellsbury. 2006. Theoretical Derivation of New Stable and Non-isotopic Approaches for Assessing Soil Organic C Turnover. *Agron. J.* 98:443-450.

CTIC. 2004. National survey of conservation tillage practices. Conservation Technology Information Center, CTIC, West Lafayette, IN. http://ctic.purdue.edu/ctic.html (accessed on 26th September 2009).

Daughtry, C. T., Doraiswamy, P. C., Hunt Jr., E. R., Stern, A.J., McMurtrey, J. E., Prueger, J.H. 2006. Remote Sensing of crop residue cover and soil tillage intensity. *Soil and Tillage Research* 91: 101-108.

Daughtry, C. T., Hunt Jr., E. R, Doraiswamy, P. C., McMurtrey, J. E. 2005. Remote sensing the spatial distribution of crop residues. *Agronomy Journal* 97: 864–871.

Daughtry, C. T., Hunt Jr., E.R. 2008. Mitigating the effects of soil and residue water contents on remotely sensed estimates of crop residue cover. *Remote Sensing of Environment* 112: 1647-1657.

Daughtry, C. T., Serbin, G., Reeves III, J. B., Doraiswamy, P. C., Hunt Jr., E.R. 2010. Spectral reflectance of wheat residue during decomposition and remotely sensed estimates of residue cover. *Remote Sensing* 2: 416-431.

Daughtry, C. T., Walthall, C.L., Kim, M.S., Brown de Colstoun, E., McMurtrey, J.E. 2000. Estimating corn leaf chlorophyll concentration from leaf and canopy reflectance. *Remote Sensing of Environment* 74: 229–239.

Gao, B.C. 1996. NDWI—a normalized difference water index for remote sensing of vegetation liquid water from space. *Remote Sensing of Environment* 58: 257–266.

Gelder, B. K., Kaleita, A. L., Cruse, R. M. 2009. Estimating mean field residue cover on midwestern soils using satellite imagery. *Agronomy Journal* 101: 635–643.

Gitelson, A., Merzlyak, M. 1996. Signature analysis of leaf reflectance spectra: algorithm development for remote sensing of chlorophyll. *J. Plant Physiol.* 148: 494-500.

Hancock, D.W., Dougherty, C.T. 2007. Relationships between blue- and red-based vegetation indices and leaf area and yield of alfalfa. *Crop Science* 47: 2547-2556.

Lillesand, T., and Kiefer, R. 2000. *Remote Sensing and Image Interpretation*. New York: John Wiley and Sons, Inc. ISBN 0-471-25515-7.

Mamani-Pati, F., D.E. Clay, C.G. Carlson, S.A. Clay, G. Reicks, and K. Kim. 2010. Nitrogen Rate, Landscape Position, and Harvest Corn Stover Impacts on Energy Gains and carbon budgets of Corn Grown in South Dakota. Agron. J. 102:1535-1541.

Mamani Pati, E.M., D.E. Clay, C.G. Carlson, and S.A. Clay. 2009. Calculating soil organic carbon maintenance using stable and isotopic approaches: A review. P. 189-216. *In E.* Lichtfouse (ed.). Sustainable Agricultural Reviews: Sociology, Organic Farming, Climate Change and Soil Science, Springer.

McNairn, H., Protz, R. 1993. Mapping corn residue cover on agricultural fields in Oxford county, Ontario, using thematic mapper. *Canadian Journal of Remote Sensing* 19: 152-159.

NASA 2010. NASA jet propulsion laboratory SWIR-ASTER user advisory. http://asterweb.jpl.nasa.gov/swir-alert.asp (accessed online 14th August, 2010)

Pacheco, A. and McNairn, H. 2010. Evaluating multispectral remote sensing and spectral unmixing analysis for crop residue mapping, *Remote Sensing of Environment*, 114: 2219-2228.

Rouse, J., Haas, R., Schell, J., Deering, D., and Harlan, J.C. 1974. Monitoring the vernal advancements and retrogradation of natural vegetation. NASA/GSFC, Greenbelt, MD.

Serbin G., Craig T., Daughtry C.S.T., Hunt Jr., E. R., Brown, D.I., McCarty, G.W. 2009a. Effect of soil spectral properties on remote sensing of crop residue cover. *Soil Science of America Journal* 73: 1545-1558.

Serbin G., Daughtry, C.S.T., Hunt Jr., R., Reeves J., Brown D.I. 2009b. Effect of soil composition and mineralogy on remote sensing of crop residue cover. *Remote Sensing of Environment* 113: 224-238.

Serbin, G., Hunt Jr., E. R., Daughtry, C.S.T., McCarty, G. W., Doraiswamy P. C. 2009c. An improved ASTER index for remote sensing of crop residue. *Remote Sensing* 2009 1: 971-991.

Thoma, D., Gupta, C., Bauer, E. 2004. Evaluation of optical remote sensing models for crop residue cover assessment. *Journal of Soil and Water Conservation* 59: 224-233.

USGS. 2007. Earth Observing 1 (EO-1), Sensors-Hyperion. eo1.usgs.gov/hyperion.php (Accessed 30th May, 2010). EROS data center, USGS, Sioux Falls, SD.

Wang Y., Lyapustin, A. I., Privette, J. L., Cook, R. B., SanthanaVannan, S. K., Vermote, E. F., Schaaf, C.L. 2010. Assessment of biases in MODIS surface reflectance due to Lambertian approximation. *Remote Sensing of Environment* 114: 2791–2801.

Wollenhaupt, N. 1993. Estimating residue : line transect method. G1570. MU Extension. Available at http://extension.missouri.edu/publications/DisplayPub.aspx?P=G1570

18

Activated Carbon from Waste Biomass

Elisabeth Schröder[1], Klaus Thomauske[1], Benjamin Oechsler[1], Sabrina Herberger[1], Sabine Baur[1] and Andreas Hornung[2]
[1]*Institute for Nuclear and Energy Technologies, Karlsruhe Institute of Technology (KIT)*
[2]*European Bioenergy Research Institute (EBRI), Aston University, Birmingham*
[1]*Germany*
[2]*United Kingdom*

1. Introduction

As a result of environmental requirements in many countries and new areas of application the demand on activated carbon is still growing. Due to the unavailability of the main basic materials like hard coal, wood or coconut shells in many countries other biomass matters were tested for their appropriateness of activated carbon production.
The objective of this experimental work is the conversion of waste biomass into activated carbon. Waste biomass like straw matters, olive stones, nut shells, coffee grounds and spent grain is converted thermally in two steps. First the biomass undergoes a pyrolysis process at 500°C–600°C in nitrogen atmosphere. The gaseous and liquid pyrolysis products can be used energetically either for heating the facilities or for electricity production.
Second, the solid residue, the char, is treated in an activation process at 800°C–1000°C in steam atmosphere in order to enhance the char surface area which was analyzed by standard BET method. The increase of surface area depends on the type of biomass and on the activation parameters. The production methods were investigated in lab-scale facilities whereas a pilot scale reactor was designed for the transformation of the discontinuous activation process to a continuous production process.
The use of agricultural by-products for activated carbon production as well as the influence of ash content, pyrolysis and activation conditions on the activated carbon quality is investigated by many authors. The high ash content of rice straw makes it difficult to achieve a sufficiently high surface area (Ahmedna et al., 2000). The influence of a one step and a two step thermal treatment of rice straw in CO_2 atmosphere is discussed in (Yun et al., 2001). The two step treatment leads to higher surface areas than the one step treatment which correspond to the own results. Higher temperatures of physical activation in CO_2 atmosphere leads to pore widening which causes an increase of mesopores. Physical activation by the use of an oxidizing gas like steam or CO_2 results in carbons with low surface area whereas chemical activation enhances the carbon surface area (Ahmedna et al., 2000). Chemical activation of rice husks and rice straw is investigated in (Guo et al., 2002; Oh & Park, 2002). The impregnation of rice precursors with KOH or NaOH enhances the surface area. In addition the activation temperature can be lowered. Washing rice straw with alkaline solutions like NaOH allows to reduce the ash content as shown in Table 1 and (Huang et al., 2001). Carbonisation and activation of pretreated rice straw leads to higher surface areas than of non-treated straw

matters but only in a certain range of washing time and temperature due to the effect that lignin and hemicelluloses are dissolved as well which leads to the reduction of straw carbon content (Finch & Redlick 1969; Sun et al. 2001). Ash extraction of straw matters is also discussed in (Di Blasi et al., 2000; Jensen et al., 2001a, 2001b). Activated carbons from olive stones and other waste biomass matters are given in (Daifullah & Girgis, 2003). High porosity can be attained by the use of phosphoric acid prior to heat treatment. Olive stones as precursor are also investigated in (El-Sheikh et al., 2004). They point out the microporous structure of their carbons which were activated in both steam and CO_2 atmosphere. Pretreatment of olive stones with hydrogen peroxide has a negative effect on porosity and surface area. The influence of gas atmosphere on the formation of mesopores in olive stone chars is investigated in (Gonzalez, 1994; Molina-Sabio, 1996). CO_2 activation leads to larger micropore volume than steam activation. Here, pore widening is the predominant effect. Compared to CO_2 activation chemical activation of olive stones with of ZnCl leads to higher surface areas with a high amount of micropores (Lopez-Gonzalez, 1980). Highly microporous carbons with high surface areas are produced by chemical activation of hazelnut, walnut and almond shells and of apricot stones (Aygün, et al., 2003). Pistachio shells and fir wood were investigated in (Wu et al., 2005) by both physical and chemical activation. Chemical activation and the influence of KOH and NaOH on the formation of micropores of loquat stones is reported in (Sütcü & Demiral, 2009). High surface areas are attained with KOH and an increase of chemical agent leads to an increase of surface area. The influence of pyrolysis conditions on pore generation is investigated by pyrolysis of oil palm shells under both, nitrogen and vacuum (Qipeng & Aik, 2008). It is shown that vacuum pyrolysis avoids pore blocking which results in higher surface areas. The effect of binders and pressing conditions on the production of granulated activated carbons are worked out in (Ahmedna et al., 2000; Pendyal et al., 1999). Straw matters and binders from agricultural byproducts like molasses from sugarcane and sugar beet, corn syrup and coal tar were mixed and pressed prior to pyrolysis and CO_2 activation. Molasses as binder leads to lower hardness and higher ash content of the activated carbons than corn syrup or coal tar.

Also chemical activation leads to highly microporous activated carbons with high surface areas this work considers steam activation which is regarded be a low cost method for technical use.

2. Experimental method of biomass pyrolysis and char activation

The experiments on pyrolysis and activation of waste biomass matters were run in lab-scale facilities. The advantage of these small-scale equipments is that the experiments could be run very quickly without long heat-up times and with one operating person. Only small amounts of biomass were needed and the operation conditions could be changed quite easily. Not many efforts had to be made in gas cleaning procedure due to the low exhaust gas flow. The screening test to figure out the optimal char residence time in the activation facility was a one or two day work with an output of 6 – 10 data points. The description of the lab-scale experiments is given in detail for both, pyrolysis and activation activities.

2.1 Biomass properties

For the generation of activated carbon from waste biomass more than 12 different waste biomass matters were investigated. The properties of some of the investigated types of biomass are given in Table 1.

	C	H	O	N	S	Cl	Ash*	H_2O
Rice straw untreated	39.6	4.6	36.4	0.7	0.1	0.2	18.3	8
Rice straw pretreated	42.4	5.9	n.m	0.76	n.m	n.m	3.6	None
Olive stones	48	5.6	n.m	<1	n.m	n.m	5	4
Wheat straw	44.1	6	44.9	0.5	0.2	0.7	7.9	9.8
Wheat straw pellets	43.1	5.9	45.5	1	n.m	n.m	~8	6.5
Walnut shells	50.7	6	n.m	n.m	n.m	n.m	0.9	10.7
Pistachio shells	43.7	5.9	n.m	n.m	n.m	n.m	0.8	Dry

Table 1. Elemental analysis of different types of biomass based on dry matters (wt%). *Appendix C.7 Alkali Deposit Investigation Samples Alkali Deposits Found in Biomass Power Plants: A Preliminary Investigation of Their Extent and Nature National Renewable Energy Laboratory Subcontract TZ-2-11226-1; n.m.: not measured

2.2 Lab-scale pyrolysis

The pyrolysis experiments were run in a "pocket"-reactor which was originally designed for fast pyrolysis experiments and which was reconverted to slow pyrolysis. Heating of biomass at low heating rates of 5-10 K/min was considered to be better than fast heating rates with respect to activated carbon production. A scheme of the reactor is shown in Fig. 1.

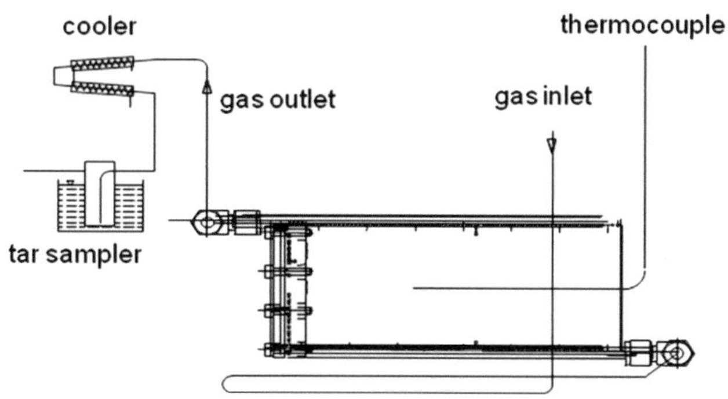

Fig. 1. Scheme of the pyrolysis reactor. Four pockets are connected in parallel and wrapped round with an electric heater. The width of the pockets was 5 mm.

The pockets altogether were filled with about 100 g of biomass. The feed was heated by a flow of hot nitrogen and additionally by electric heaters which were fixed to the walls of each pocket. The pyrolysis temperature was varied but it had only a marginal influence on the quality of the activated carbon because the biomass was not completely devolatilized after pyrolysis. The reason is that activation took place at higher temperature than pyrolysis therefore the entire devolatilization had been realized in the activation step. The disadvantage of incomplete pyrolysis is that some oils which are produced in the activation step require an additional cooling and filter system. The primary pyrolysis gases were cooled in a gas cooler to 5 °C and the oils were collected in order to use them as binder material for the production of granulated activated carbon. After the run of the experiments the char was taken out of the pockets and the mass balance was established.
The total amounts of the pyrolysis products char, tar and gas of the investigated biomass matters are given in Table 2:

Biomass	Char [wt% dm]	Tar [wt% dm]	Gas [wt% dm]
Rice straw	27	40	33
Rice straw washed*	19	30-40	50-40
Wheat straw	28	22	50
Wheat straw pellets	32	33	35
Olive stones crashed	30	49	21
Pistachio shells	29	36	35
Walnut shells	31	29	40
Coconut shells	33	40	27
Coffee grounds	23	53	24
Spent grain	29	20	51
Beech wood (525 °C)+	24	46	30
Coconut press residue	27	51	22
Rape seed	17	63	20

Table 2. Yields of pyrolysis products based on dry biomass matter. The pyrolysis temperature was 600 °C, the heating rate amounted to 10 K/min. * Based on washed and dried straw. + Pyrolysis temperature was 525°C.

The influence of heating rate on the pyrolysis product yields is shown in Table 3. The pyrolysis temperature was set to 600 °C for some biomass matters whereas the heating rate amounted to 30 K/min.

Biomass	Char [wt% dm]	Tar [wt% dm]	Gas [wt% dm]
Rice straw washed	24	36	40
Wheat straw pellets	31	25	44
Pistachio shells	24	54	22

Table 3. Yields of pyrolysis products. The pyrolysis was run at 600 °C, the heating rate amounted to 30 K/min.

As shown from Tables 2 and 3 the tar yield increases if the heating rate is enhanced whereas the char yield slightly decreases. From the aspect of using the tars/oils for energy production in a combined heat and power plant the higher heating rate is more reasonable. The influence of pyrolysis heating rate on the surface area of activated carbon is marginal in this range. A negative effect on the activated carbon quality can be detected at heating rates of more than 250 K/min. For optimization reasons, the amount and quality of the liquid pyrolysis products may be a decision criterion for higher heating rates.

2.3 Lab-scale activation

The activation experiments were run in a reaction tube which was installed in an oven. The scheme of the activation lab-scale facility is shown in Fig. 2.

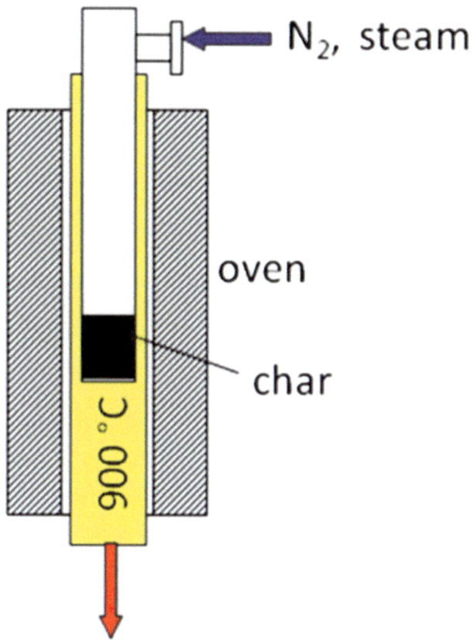

Fig. 2. Scheme of the activation reactor. The reaction tube can be passed through by steam flow. The case where the char is inserted has a porous bottom and can be removed from the tube.

The activation reactor consists of a tube furnace which can be heated to 1100°C. Inside of the furnace a tube with a small case at the bottom is inserted. The case contains the char and has a porous bottom to ensure, that the incoming gas (nitrogen or steam) flows through the char bed. The tube can be taken out of the oven. In the beginning of the experiment 5–10 g of char were inserted into the case with the porous bottom. Afterwards the case was fixed to the tube. The tube was then inserted into the hot furnace and the char was kept under nitrogen atmosphere. When the desired char temperature was reached the nitrogen flow was substituted by steam flow. After some minutes of reacting time, the steam flow was switched off, the tube was taken out of the reactor and cooled to ambient temperature under nitrogen atmosphere. The char mass was recorded and a sample of char was taken out of the case for surface analysis. The remaining char was again inserted into the oven for the next time interval. In this way the surface area of the char could be recorded as function of the conversion rate, i.e. actual char mass/initial char mass.
In the hot steam atmosphere the char got partially oxidized which lead to the loss of char mass and the production of gaseous products like H_2, CO and CO_2. Higher amounts of gaseous long-chain hydrocarbons were produced during the heat-up interval of the char as a result of incomplete pyrolysis at 600 °C. These gases may be of interest in terms of energetic utilization in order to rise the economy of the activated carbon production chain. One way of enhancing the calorific value of the exhaust gases may be a catalytic reforming process as reported in (Hornung et al., 2009a; Hornung et al., 2009b).
As a result of partial oxidation under steam atmosphere, the surface area of the char increases which is shown in Fig. 3-14. The surface area created by the chemical reactions in the steam atmosphere reaches a maximum. Higher char conversion leads to diminishing surface areas due to the lack of carbon. In the final stage, only ash remains.
Some of the char yields which remained at maximum surface area are given in Table 4 for rice straw and olive stones.

Time [min]	Rice straw [wt%] Act. Temp.: 800 °C	Olive stones [wt%] Act. Temp.: 750 °C
30	55	70
45	50	60
60	45	50
90	40	30

Table 4. Char yield as function of activation time for different biomass matters based on the dry initial char mass.

2.4 Surface measurement – BET method

The surface area of pyrolysis char and activated carbon is measured by standard BET-method (Bunauer, Emmett, Teller) with the automatically operating measurement technique NOVA 4000e from Quantachrome Partikelmesstechnik GmbH. The char is exposed in nitrogen atmosphere at the boiling temperature of liquid nitrogen. The amount of nitrogen molecules which are adsorbed in a monolayer on the particles´ surface specify the surface area. In addition pore size analysis and pore volume measurements are made with this technology (Klank, 2006).

2.5 Activation results

The following diagrams show the BET surface area as function of conversion rate, i.e. loss of char mass resulting from steam activation. The values are based on dry initial char mass. The initial char was produced in the lab-scale pyrolysis reactor by the use of various biomass matters. As shown from the diagrams the surface area increases with increasing conversion rate. At conversion rates of more than 80 wt% the surface area diminishes due to the lack of carbon.

Fig. 3 and 4 show the influence of conversion rate on the formation of surface area and the influence of activation temperature on activation time. The higher the activation temperature the lower the resulting activation time for the accessibility of a high surface areas. This example is given for crashed olive stones, but can be observed at all the other investigated biomass matters. Fig. 5-14 give a summary of the biomass type investigation for the applicability of activated carbon production.

From Fig. 3 to 14 it is shown that any kind of nut shell is appropriate for activated carbon production. Straw materials end up with surface areas around 800 m^2/g which is the minimum value that commercially available activated carbons provide.

Activated carbon from rice straw with sufficient quality can only be attained if the straw matter is washed in alkaline solution like NaOH prior to the thermal treatment in order to extract the inorganic compounds (Finch, 1969). Intermediate surface areas can be attained with olive stones, spent grain, coffee grounds and sunflower shells. Due to the low feedstock price activated carbon which is made from these materials seems to have the most economic perspective.

The residence time of the biomass in the pyrolysis reactor averaged 1 hour at a heating rate of 10 K/min. A rotary kiln reactor which is described in (Hornung et al., 2005; Hornung & Seifert, 2006) was tested for pyrolysis of wheat straw pellets and rape seeds. Here the pyrolysis was run at heating rates of 30 K/min.

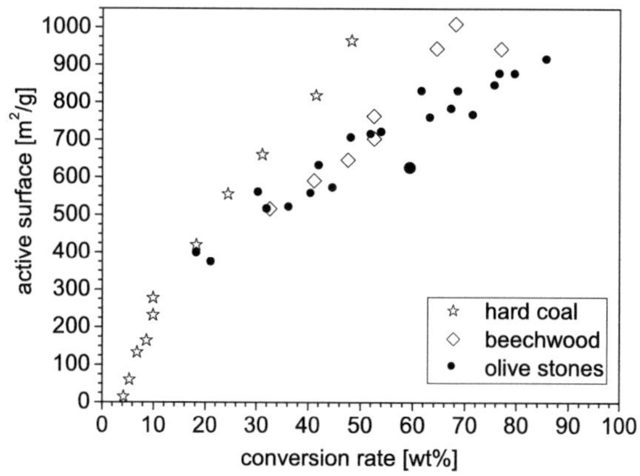

Fig. 3. Active surface of crashed olive stones compared with prevalent raw materials.

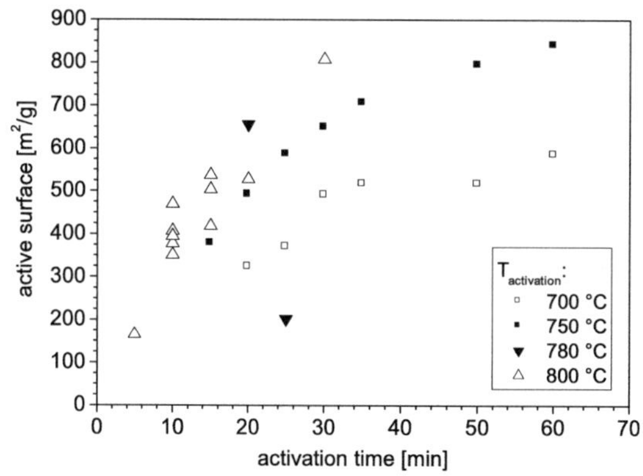

Fig. 4. Influence of activation temperature on activation time in the case of crashed olive stones.

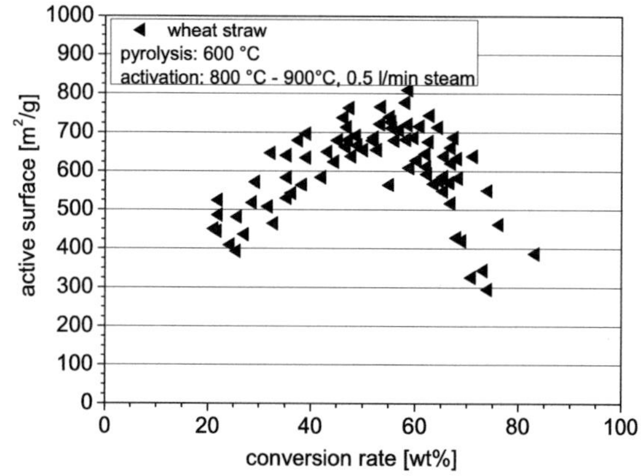

Fig. 5. Wheat straw

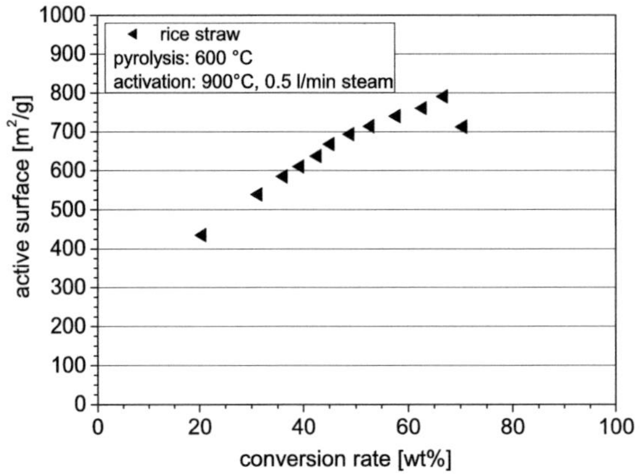

Fig. 6. Washed rice straw

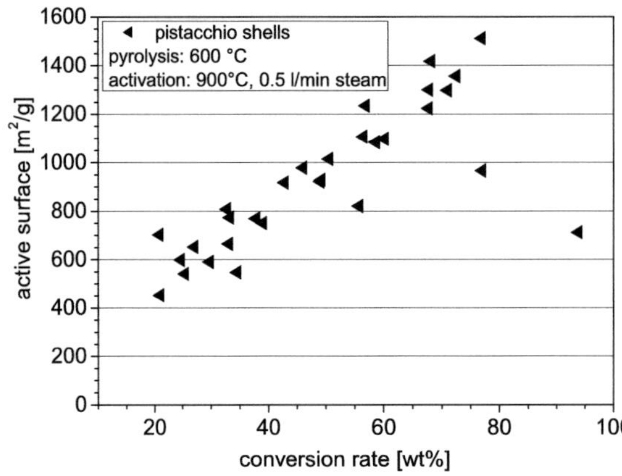

Fig. 7. Pistachio shells

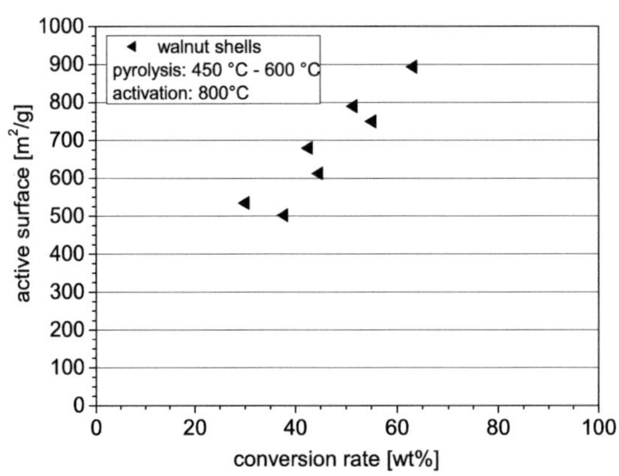

Fig. 8. Walnut shells. The steam flow was 0,5 l/min.

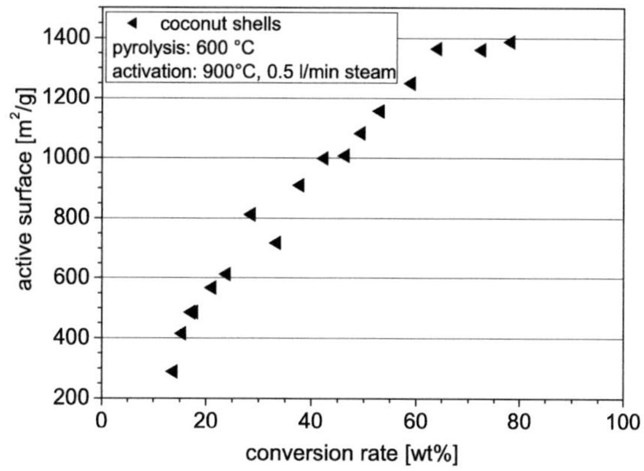

Fig. 9. Coconut shells

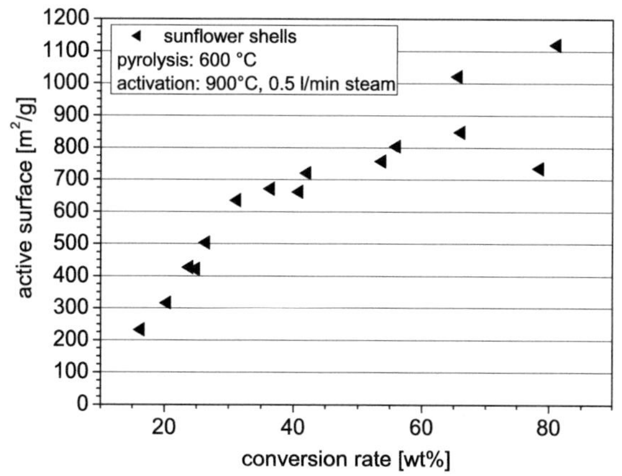

Fig. 10. Sunflower shells

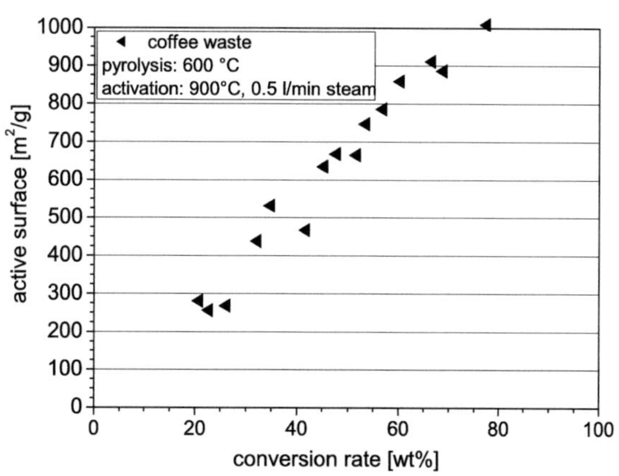

Fig. 11. Coffee waste

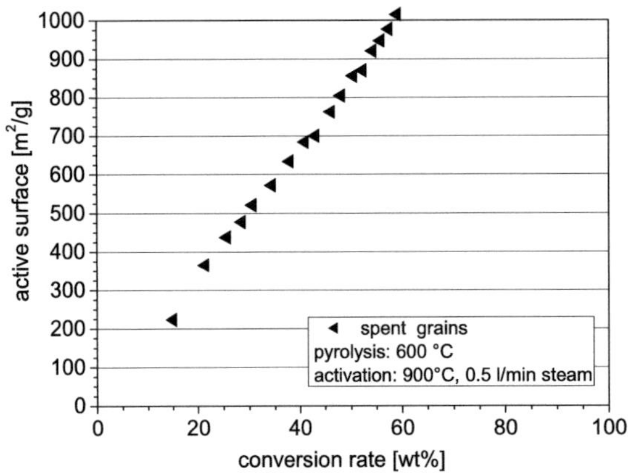

Fig. 12. Spent grain

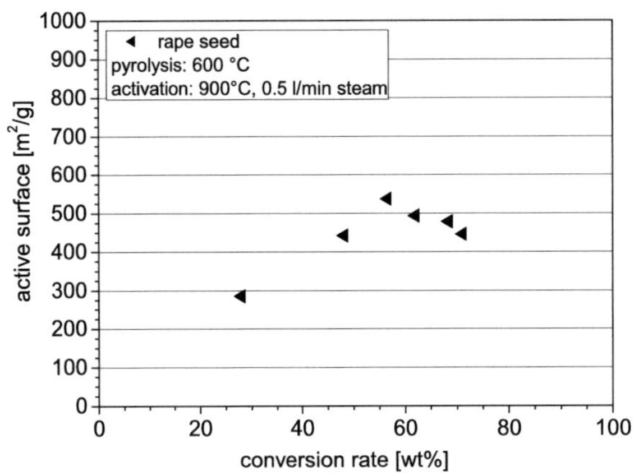

Fig. 13. Rape seed

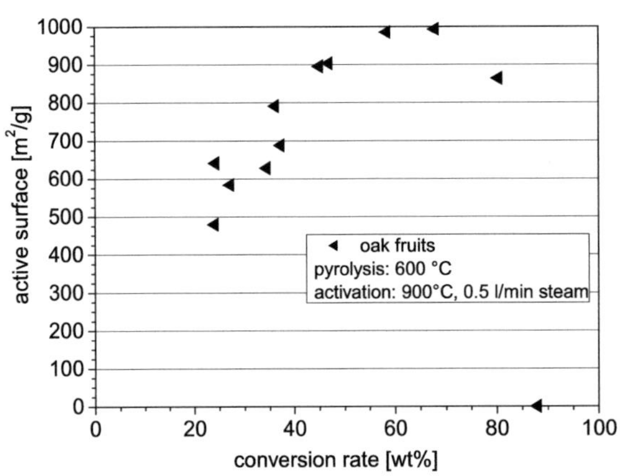

Fig. 14. Oak fruit

Within this heating range the influence of heating rate on the activated carbon quality is negligible. Lower residence times i.e. 10 - 20 min should be chosen for economic reasons. For this the use of the rotary kiln reactor (Hornung et al., 2005, 2006) is suitable.
The residence time of char in the activation step is given as function of conversion rate in following diagrams, Fig. 15 and 16.

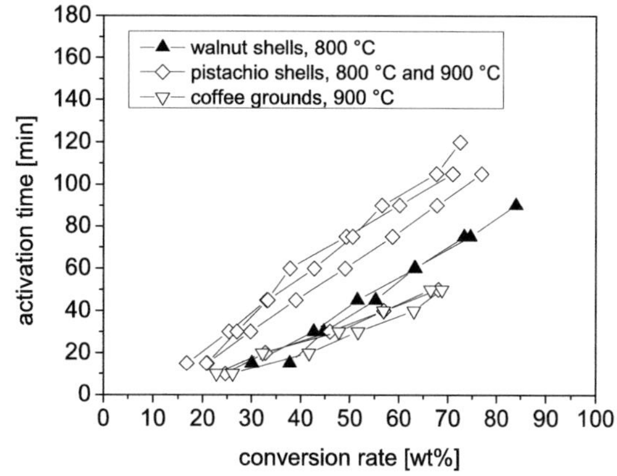

Fig. 15. Activation time as function of conversion rate.

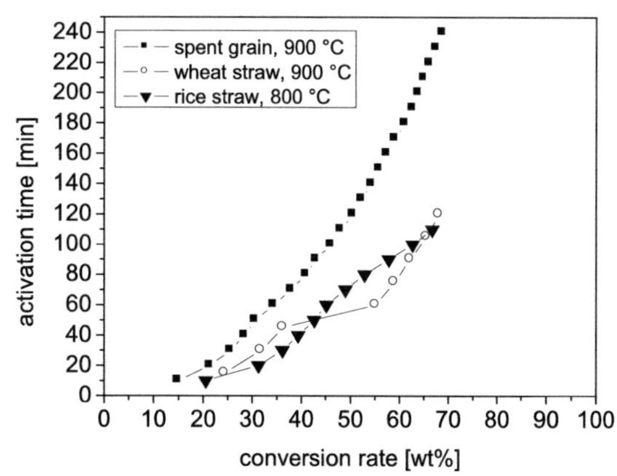

Fig. 16. Activation time as function of conversion rate.

In Fig. 15 the values of walnut shells and pistachio shells belong to 800°C activation temperature except the lower pistachio values which correspond to the activation temperature of 900°C. The activation time was varying from experiment to experiment. The reason for this might have been local effects due to inhomogeneous flow through of the small fixed bed. But nevertheless, experiments with wheat straw pellets exhibits that the char residence time needs to be in the range of 60 - 75 min. These results in combination with the lab-scale pyrolysis experiments are helpful to determine the production parameters of activated carbon from a special type of biomass in a continuous production process.

2.6 Pore size distribution

The pore size distribution of the investigated activated carbons were calculated by DFT method (Density Functional Theory) from the corresponding adsorption isotherms in Figure 17. DFT method (Evans et al., 1986) allows for the determination of the micro- and mesopore volume. The investigated carbons which were used for the isotherm measurements, Fig. 17, were high surface area carbons. As shown from Fig. 17 activated carbons from coconut shells, wheat straw, olive stones and walnut shells follow the same type of isotherm which was detected to be of IUPAC classification type I, indicating the presence of micropores (Sing et al., 1985; Klank, 2006). The hysteresis loops follow type H4 which refers to the presence of mesopores with a predominance of narrow slit-like pores.

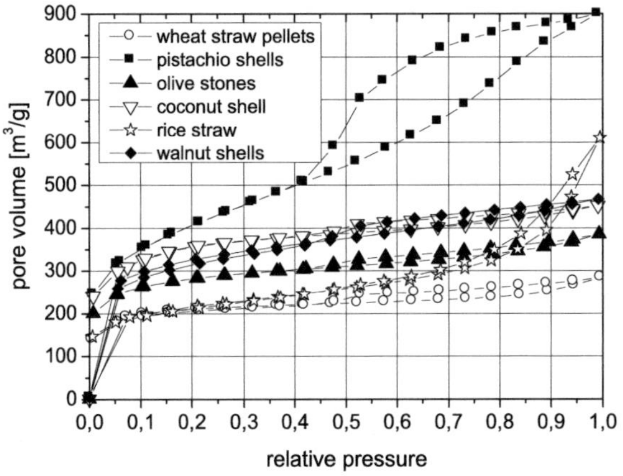

Fig. 17. Adsorption isotherms of nitrogen on various activated carbons.

Activated carbons from rice straw offer as well a high microporosity but at high pressure ratio, there is a steep increase in pore volume which gives rise to the presence of meso- and macropores. The adsorption isotherm of pistachio shells is different from the other curves and tends to type IV. There is a big hysteresis loop visible which can be regarded as hysteresis type H2 and gives rise to complex pore structure and network effects. Due to its resemblance to H1-hysteresis the pore size calculations were based on spherical pores.

The pore size diagrams of Fig. 18 a-d are similar and exhibit a predominant pore diameter of 40 Å, a high amount of micropores but nearly no macropores except the curves of pistachio shells and rice straw. The first one has a sharp peek in the range between 50 and 60 Å indicating a high amount of meso-/macropores. The micropore volume of pistachio shell char is lower compared to the other activated carbon types. The light slope of the curve at higher pore diameters gives rise to a high amount of macropores. For this activated carbons from pistachio shells are predominated by meso- and macropores. Activated carbons from rice straw are predominated by micropores but meso- and macropores are present as well. The shape of the isotherms indicates that all investigated biomass based activated carbons are characterized by a high amount of micro- and mesopores. Macropores are only present in activated carbons from pistachio shells and rice straw.

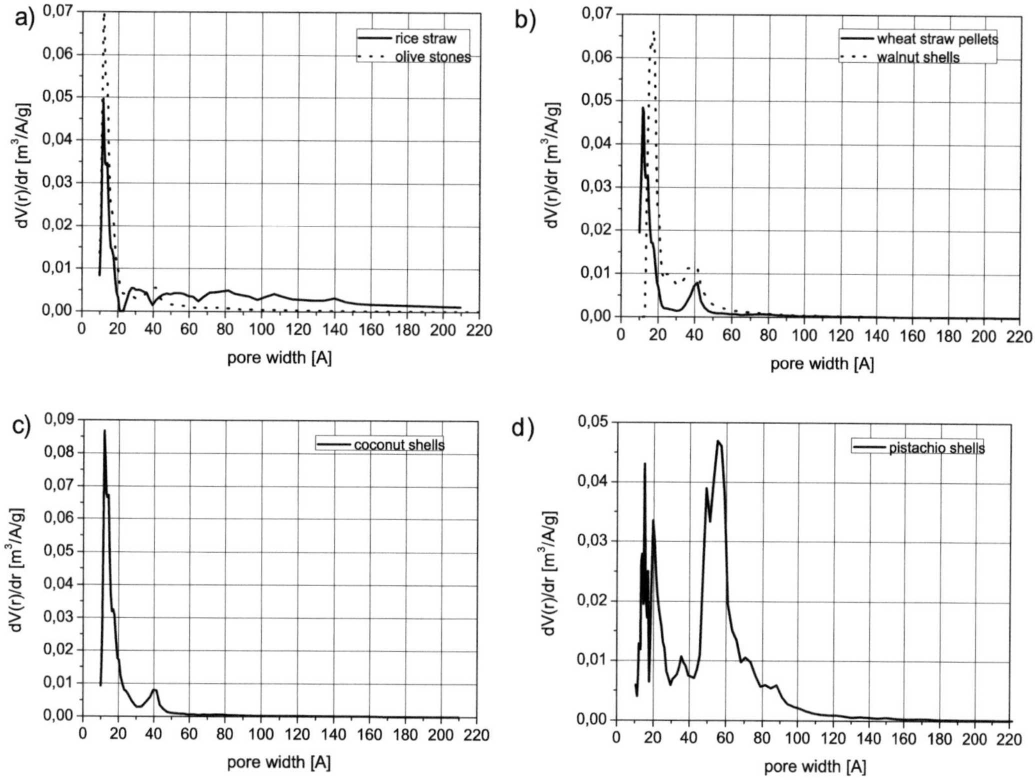

Fig. 18. a-d: Pore size distribution of various activated carbons by use of DFT-method. Activated carbons from pistachio shells are based on a cylindrical pore model whereas the others are calculated on a slit-like model.

3. Generation of granular activated carbon

Dependent on the application of activated carbon, the material has to be granulated for better handling purposes. For this pelletizing tests of char powder were made in order find out the pelletizing conditions. The chars which came out of the pyrolysis reactor were in nearly the same shape than the biomass was before. For this the chars had to be milled to particle sizes of 40 - 280 µm. To form stable pellets the use of a binder is necessary. State of the art is the use of molasses as binder material prior to pyrolysis (Pendyal et al., 1999). For economic reasons and from the aspect of using the high viscous pyrolysis tars for energetic applications the biomass pyrolysis tars were tested as binder material. The scheme of the pellet production is shown in Fig. 19.

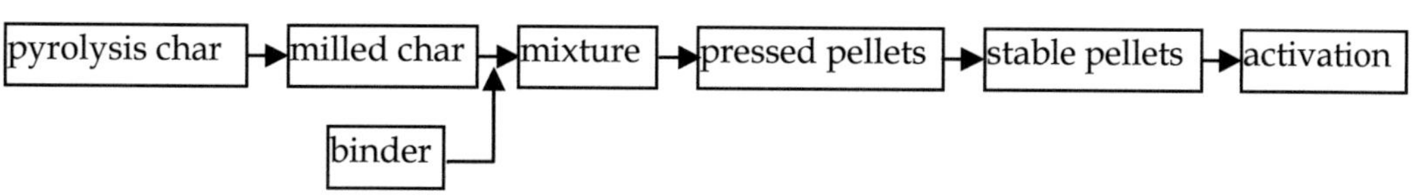

Fig. 19. Scheme of the pelletizing method.

The pelletizing procedure is implemented in between the pyrolysis and the activation step. During activation the binder was decomposed. For this many tar components passed into the gas phase during activation combined with the gaseous products from the char–steam reactions. This lead to an energy rich exhaust gas. Measurements of the exhaust gas composition were made in the pilot-scale rotary kiln reactor and are given in Section 4. For further applications and for economic reasons the possibility of using the exhaust gas energetically p.e. in a gas engine should be taken into account. This procedure allows for further use of pyrolysis tars which are too viscous in order to use them directly as a fuel in an engine. Especially for the energy rich tars from the pyrolysis of straw matters which are difficult to handle this procedure may be an option for further use.

The mixtures of chars and binder had to be put into a shape which was stable enough, to overcome the activation process. Test by extruding the mixtures were not successful. The extruder was either blocked or the pellets were unstable and melt after extruding. For this the concept of using a static press arose. A small lab-scale pressing unit was designed and build in order to test the pressing conditions. With the lab-scale pressing unit the pelletizing conditions were worked out. Several binders were tested and the char/binder ratios were varied. Furthermore the pressing conditions and the pressing temperatures were investigated.

The following Table 5 gives an overview of the pelletizing combinations.

Char	Binder Pyrolysis oils from	Pressing conditions	Temperature
wheat straw rice straw pistachio shells olive stones coffee grounds mixtures of wheat straw and pistachio shells mixtures of rice straw and pistachio shells	coconut press cake coffee ground wheat straw	150 bar – 350 bar	200 °C after pressing 200 °C while pressing

Table 5. List of pelletizing conditions. Combinations of these conditions were investigated.

The results of the pelletizing investigations can be summarized as follows. After activation the binder had passed into the gas phase. Therefore the loss of pellet mass was higher than the loss of unpelletized char mass at the same activation time. The surface area of the pellets corresponds to the surface area of the unpelletized char. With respect to surface area the influence of the binder is marginal.

Further, mixtures with different chars, i.e. chars with good and bad quality like rice straw and pistachio shells, lead to activated carbons with intermediate surface areas. This allows for the enhancement of surface areas from chars which are not of sufficient quality for activated carbon production.

The following pictures show the influence of the binder on the formation of surface area, Fig. 20, and the influence of char mixing, Fig. 21, for wheat straw carbons. Fig. 22 and 23 present the same effects of pelletizing and char mixing by the use of rice straw.

In Fig. 20 it is shown that the surface area of the pelletized char is similar to the unpelletized char but the values are shifted to higher conversion rates due to the fact that the binder evaporates and/or reacts with the steam atmosphere. Fig. 21 and 23 demonstrate that the surface area is shifted to higher numbers when the wheat/rice straw char is mixed with char from pistachio shells.

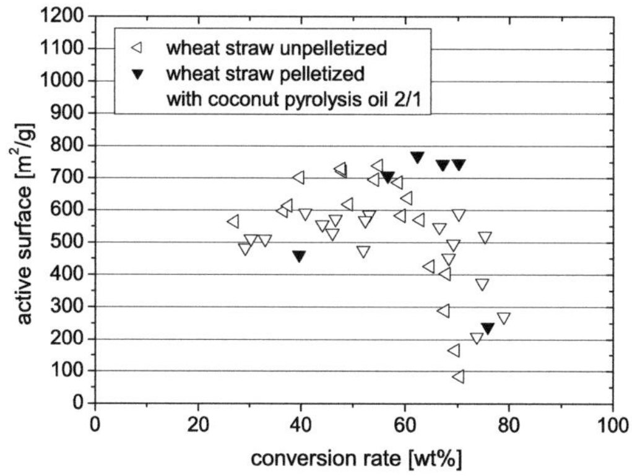

Fig. 20. Influence of binder on the formation of active surface during activation of wheat straw.

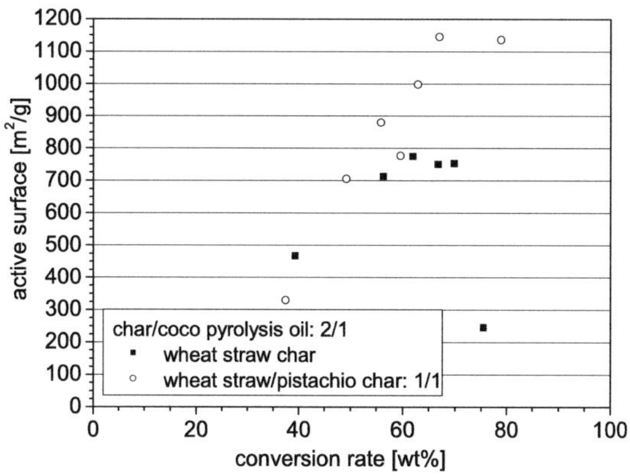

Fig. 21. Influence of char mixing on surface formation during activation in the case of wheat straw.

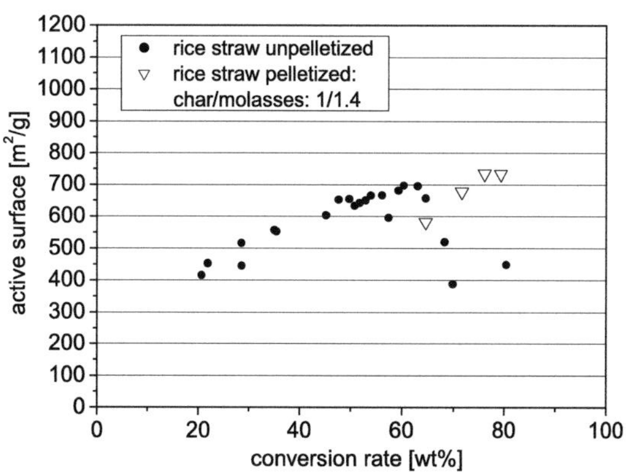

Fig. 22. Influence of binder on the formation of active surface during activation of rice straw.

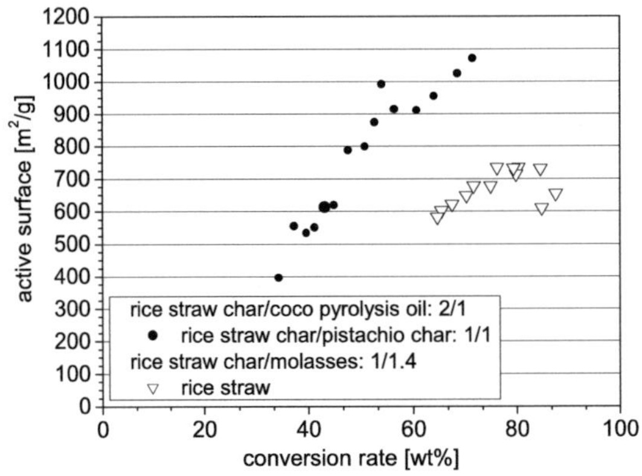

Fig. 23. Influence of char mixing on surface formation during activation in the case of rice straw.

The stability of the pellets, 4 mm in diameter and 20 mm long, was tested by the use of different char/binder ratios and pressing conditions. For this some of the pellets were disposed between the dies of a pressing unit. The break force of the pellets was recorded. These values are given in Tables 6 and 7. Best hardness was attained at a char/binder ratio of 2/1 and 1.5/1. To form the pellets, the char/binder mixtures were pressed at 20 bar and afterwards the matrix was heated in an oven for 2 hours at 200°C. Subsequently the hot matrix was put under the press again to form the pellets at 200 bar. After cooling to ambient temperature the pellets were taken out of the matrix. Another possibility of pellet formation was cold pressing at 200 bar and subsequent heating at 200 °C. As shown from Tables 6 and 7 heating and subsequent pressing leads to higher hardness of the pellets. A higher pressure did not enhance the hardness significantly.
The binders which lead to stable pellets were pyrolysis oil from coconut press residues, tars from wheat straw pellets and tars from coffee ground pyrolysis. Especially wheat straw and coffee ground tars are low in water content and of sticky consistency. Therefore they were regarded to offer good bonding conditions as reported in (Fütterer, 2008).

Binder: Pyrolysis oil	**Char/Binder ratio**	**Pressing temperature**	**Pressure [bar]**	**Force [N]**
Coffee waste	**2/1**	**cold**	**250**	**136**
Spent grain	2/1	200 °C	200	50.9
Coconut press residue	**1/2**	**cold**	**250**	**103**
	2/1	cold	100	50,6
			330	48.5
			200	34.2
			250	**88.6**
Wheat straw	2/1	cold	200	48
	1.5/1	**200 °C**	**200**	**155**
	2/1	200 °C	200	65.9
				44.5
				49.5

Table 6. Break strength of pellets made from wheat straw char. Bold: best combinations.

Binder: Pyrolysis oil	Char/Binder ratio	Pressing temperature	Pressure [bar]	Force [N]
Coconut press residue	1.5 / 1	cold	250	33.7
			350	73.6
		200 °C	200	38.7
Wheat straw	**1.5 / 1**	**200 °C**	**200**	**205**

Table 7. Break strength of pellets made from olive stone char. Bold: best combinations.

The stability of the pellets was not only influenced by the type of binder but as well by the type of biomass. Pellets from olive stone chars were very hard to form, due to the melting effects after pressing. Stable pellets could only be attained by the use of wheat straw tar as binder.

4. Rotary kiln reactor for char activation

The advantage of the lab-scale pyrolysis and activation facilities is the easy way of handling and the short heat-up times. Many experiments can be made in a short time interval. Unfortunately the possibility of treating larger amounts of biomass is not given. Likewise these facilities do not serve for an up-scale to an industrial production process neither for biomass pyrolysis nor for char activation. For this a new concept of an activated carbon production process had to be worked out.

For the pyrolysis step an already existing screw driven rotary kiln reactor (Hornung et al. 2005; Hornung & Seifert, 2006) was used to transfer the lab-scale experiments into a continuous production process. Unfortunately the pyrolysis temperature was limited to 500°C within this reactor. Tests were run with wheat straw pellets, olive stones, coconut press residues, rape seeds and spent grain. The chars were activated in the lab-scale facility. No influence of the chars from lab-scale experiments and rotary kiln pyrolysis was found after the activation step. The surface area of the chars from rotary kiln pyrolysis was similar to the area of the chars from lab-scale pyrolysis. The mass loss during activation was higher when the rotary kiln chars were used due to the lower pyrolysis temperature of 450°C–500°C. The lab-scale pyrolysis was run at 600°C. For this at lot of volatiles were left in the rotary kiln chars. Nevertheless, this type of reactor serves for the pyrolysis of biomass matters with respect of activated carbon production due to the latter heating of the chars to higher temperatures during activation.

The charcoal activation still needed a new upscale concept but some requirements had to be confirmed. First the production process had to be a continuous process with automatically operating feed and discharge systems. Second the char pellets had to be mixed with the steam quite well to ensure that partial char oxidation takes place over the entire particle´s surface. Third the stirring of the particles had to be made softly because the char pellets were not stable enough to withstand high mechanical forces. Forth the residence time of the char inside of the reactor should be well controlled as well as the steam flux. Fifth the reactor should operate at 1000 °C and the possibility of changing the heat system from electrical heating to the use of gas burners should be taken into account.

As a result of these requirements the use of a further rotary kiln reactor seemed to be the most appropriate method for the scale-up of the activation process. To control the residence time of the char in the rotary kiln, it should be equipped with a rotating screw. The temperature control of the char is realized by the installation of five thermocouples along the screw axis. Although the principles of the rotary kiln pyrolysis reactor (Hornung et al. 2005; Hornung & Seifert, 2006) was used for the activation step, a total redesign of this reactor type was necessary in order to run the experiments at higher temperatures.

A sketch of the new, high temperature rotary kiln is shown in Fig. 24. It consists of a tube which is 2 meters long and the outer diameter amounts to 110 mm. The wall thickness is 6 mm. Inside of this tube a screw is located. Both parts consist of heat resistant steel. The tube and the screw can be turned independently from each other. The rotation of the tube insures the particle mixing whereas the rotation of the screw controls the char residence time. The tube is heated electrically by an oven over a length of one meter but it can be changed to gas burner heating if necessary. The axis of the screw is equipped with an electric heater and in the small gap between heater and wall of the screw axis the steam is flowing. Holes in the screw axis assure that the steam enters the reactor room. The steam itself is generated separately by a steam generator. In addition five thermocouples are fixed to the screw to allow for the char temperature control. The rotation speed of the screw is measured and controlled as well as the rotation speed of the tube. Both, the screw and the tube are driven by electric motors. Two valves, one at the feed system and one at the outlet prevent the air from entering the reactor. At the outlet steam, condensed water and the activated char is separated. The activated carbon is cooled to room temperature after leaving the reactor. The heat-up of the rotary kiln to 950°C needs about 3 hours and has to be run carefully due to the thermal expansion of the metal components. The reactor was designed for a char throughput of ~ 1 kg/hour. The valve on the right hand side of the reactor enables the char input. The steam flows through the screw axis and enters the reactor from the right. The steam and the exhaust gases leave the reactor via a small valve which is located close to the activated carbon outlet on the left hand side.

Fig. 24. Sketch of the high temperature rotary kiln reactor for char activation. The operation temperature is 950°C with steam flow and the char throughput amounts to max. 1 kg/h.

Fig. 25 gives an impression of the build-up of the activation rotary kiln reactor.

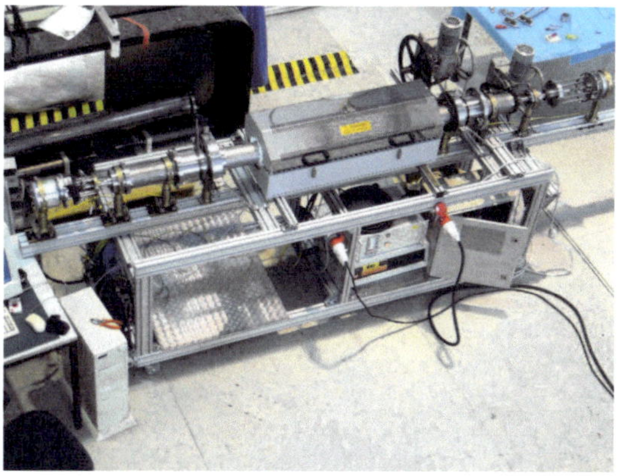

Fig. 25. Photograph of the high temperature rotary kiln reactor for char activation.

To proof whether this reactor is useful for char coal activation batch wise tests were run with char from wheat straw pellets and beech wood cubes. For this 80-100 g of char were inserted into the 950°C hot reactor. The residence time was varied between 40 min and 90 min and the steam flow was adapted to the lab-scale experiments and amounted to 1,7 - 2 m^3/h. After collecting the activated carbon at the reactor outlet, the mass balance was established and the surface area measured. These results were compared with the lab-scale activation results and are given in Fig. 26 and 27. As shown from Fig. 26 and 27 the same or even higher surface areas could be attained with the rotary kiln activation. Only little mass got lost in the reactor as a result of particle destruction. Most of the particles left the reactor in the same shape as they got in but shrinkage due to the chemical reactions could be detected. As expected the particles were not pulverized due to the smooth transport and rotation.

The results are promising and this concept seems to have a good perspective for the activation of the biomass char. This principle allows for the scale-up of the activation step into a continuous production process. For the up-scale of the rotary kiln to a technical plant much attention has to be paid on the heat impact. Inner and outer heating ensures that the steam flux and the char reach the operating temperature.

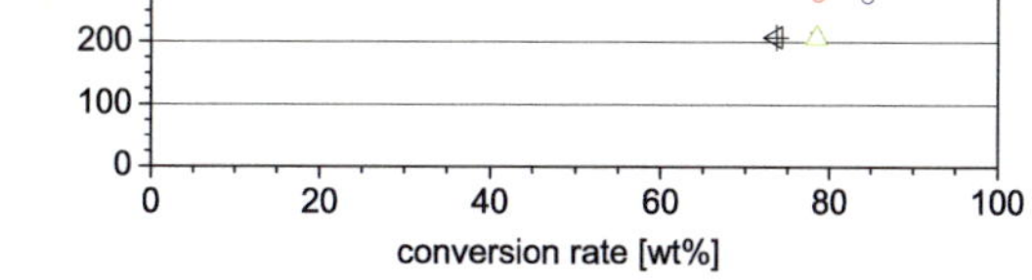

Fig. 26. Comparison of lab-scale and pilot-scale activation in the case of wheat straw pellets. The half-filled pentagons are the pilot scale results of the rotary kiln.

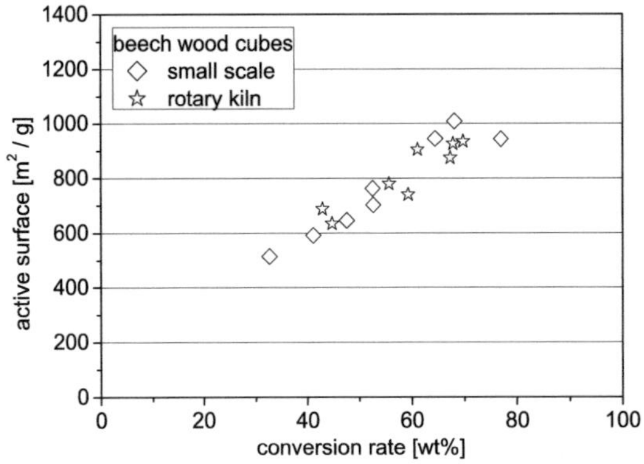

Fig. 27. Comparison of lab-scale and pilot-scale activation in the case of beech wood cubes.

	Experiment 1: 600 g char input				Experiment 2: 600 g char input			
gas component	[vol%] (1)	[wt%] (1)	[vol%] (2)	[wt%] (2)	[vol%] (3)	[wt%] (3)	[vol%] (4)	[wt%] (4)
H_2	52,78	6,87	56,14	7,49	55,56	7,36	58,62	8,07
O_2	0,31	0,65	0,20	0,43	0,01	0,03	0,26	0,58
N_2	1,45	2,62	0,97	1,80	0,05	0,10	1,12	2,13
CO	19,52	35,29	23,27	43,12	23,12	42,52	22,46	42,94
CH_4	10,33	10,70	4,95	5,25	5,97	6,29	3,31	3,63
CO_2	15,18	43,12	14,26	41,52	14,82	42,84	14,18	42,57
C_2H_2	0,01	0,01	0,00	0,00	0,01	0,02	0,00	0,00
C_2H_4	0,41	0,75	0,21	0,38	0,46	0,84	0,05	0,09
H_u [MJ/kg]	17,5		16,15		16,67		15,87	
H_o [MJ/kg]	19,6		18,08		18,85		17,83	
BET [m^2/g]	516		482		474		519	

Table 8. Composition of water free gas atmosphere during steam activation of 600 g wheat straw pellet pyrolysis char. The values are based on the volume resp. mass of water free gas samples. The numbers indicate sampling after 25 min (1), 30 min (2), 37 min (3), 46 min (4).

To proof whether the exhaust gases which were produced during activation of the char in the rotary kiln reactor have the potential of being used energetically, the composition of the gas and steam atmosphere was analyzed by gas chromatography, (Agilent 6890A Plus, packed column CarboxenTM 1000 from Supelco with helium flow of 20 mL/min).

This method required a water free gas sample. For this, the exhaust gas flow was cooled to (-50) °C in several cooling units. An additional filter unit allowed for a water free gasflow.

At the outlet of the cooling section, gas samples were collected at different instants of time. The experiments were run with 600 g of wheat straw pellets and a steam flow of 1,7 - 2 m^3/h. Prior to activation the wheat straw pellets were pyrolysed at 600 °C in the pyrolysis rotary kiln reactor for 20 min. The composition of the water free exhaust gas is documented in (Barth, 2009) and given in Table 8. The experiments were run batch-wise. The reason for it was the better control of the process due to the fact, that the in- and outlet valves did not operate automatically at this instant of time. As shown in Table 8 the calorific value is mainly determined by the gas contents of H_2, CO and by small amounts of CH_4. This gas composition corresponds to a typical synthesis gas which is produced during gasification of hydrocarbons and carbon matters. Behind the cooling unit, the gas flow was measured and amounted to 0.8 m^3/h. Compared to the steam flow of around 2 m^3/h the dilution of the exhaust gas was quite high. Therefore the steam flow should be reduced and its influence on activated carbon quality should be investigated.

5. Conclusion

The generation of activated carbon in a two step process of pyrolysis and steam activation from different waste biomass matters was investigated in both, lab-scale and pilot-scale facilities. The lab-scale experiments provided a database for the production parameters of best quality carbons with high surface areas. The surface measurements were determined by BET method. Activated carbons with high BET surface area can be generated with any kind of nut shells, like pistachio, walnut or coconut. The BET surface amounts to more than 1000 m^2/g. Intermediate values of 800 - 1000 m^2/g can be accomplished with beech wood, olive stones, spent grain, sunflower shells, coffee waste and oak fruits. Straw matters and rape seeds do not serve well for activated carbon production due to their low BET surface of 400–800 m^2/g. Especially rice straw leads to low surface values unless it is not treated with alkaline solvents prior to pyrolysis. The activated carbons are mainly dominated by micro- and mesopores of 40–60 Å. Macropores are as well present in rice straw and pistachio shell carbons.

The composition of the exhaust gases which occur during char activation is determined mainly by H_2, CO, Methane and CO_2. This corresponds to a typical synthesis gas, which occurs during gasification of carbon matters. Due to the high amount of combustible components (50-80 vol%) the dry exhaust gas may serve for energy recovery of the activated carbon production process.

Investigations were made to prove whether pyrolysis tars can be used as binder material for granulated activated carbon production. The pelletizing conditions were worked out and the influence of the binder on the quality and stability of the pellets was tested as well as the influence of char mixing. Heating and pressing of the char/binder mixtures led to stable pellets by the use of pyrolysis oils of coconut press residues, wheat straw and coffee grounds. Mixing of different kinds of chars resulted in intermediate BET surface areas.

Finally a concept for a continuous production process was given. For this a new high temperature rotary kiln reactor was designed which can be heated to 1000 °C. An inner screw allows for a smooth transport of the pelletized material. The char residence time was controlled by the rotation speed of the screw. The experiments showed, that the activated carbons which were produced in the rotary kiln were of same quality than the carbons from the lab-scale facility with respect to surface area. It demonstrates that this type of reactor is suitable for a continuous activated carbon production process.

6. Acknowledgment

We acknowledge support by Deutsche Forschungsgemeinschaft and Open Access Publishing Fund of Karlsruhe Institute of Technology.

7. References

Ahmedna, M., Marshall, W.E. & Rao, R.M. (2000). Production of granular activated carbons from select agricultural by-products and evaluation of their physical, chemical and adsorption properties. *Bioresource Technology,* Vol.71, (2000), pp. 113–123

Aygün, A., Yenisoy-Karakas, S. & Duman, I. (2003). Production of granular activated carbon from fruit stones and nutshells and evaluation of their physical, chemical, and adsorption properties. *Microporous and Mesoporous Materials,* Vol.66, (2003), pp. 189–195.

Barth, N.K. (2009). Herstellung von Aktivkohle aus biogenen Materialien, *Projektarbeit Forschungszentrum Karlsruhe/Duale Hochschule Baden-Württemberg,* 2009

Daifullah, A. A. M. & Girgis B. S. (2003). Impact of surface characteristics of activated carbon on adsorption of BTEX. *Colloids and Surfaces A.: Physiochem. Eng. Aspects,* Vol.214, (2003), pp. 181-193

Di Blasi, C., Branca, C. & D'Errico, G. (2000). Degradation characteristics of straw and washed straw. *Thermochimica Acta,* Vol.364, (2000), pp. 133–142

El-Sheikh, A.H., Newman, A.P., Al-Daffaee, H.K., Phull, S. & Cresswell, N. (2004). Characterization of activated carbon prepared from a single cultivar of Jordanian Olive stones by chemical and physicochemical techniques. *Journal of Analytical and Applied Pyrolysis,* Vol.71, (2004), pp. 151–164.

Evans R., Marconi, U. M. B. & Tarazona, P. (1986). Fluids in narrow pores: Adsorption, capillary condensation, and critical points. *Journal of Chemical Physics,* Vol.84, (1986), 2376-2399

Finch, H.E. & Redlick, R. (1969) *US Patent 3,* 451, 944

Fütterer, L. (2008) Untersuchung von aktivierten Pellets auf die spezifische Oberfläche und von nicht aktivierten Pellets auf die Druckfestigkeit in Abhängigkeit von den Parametern bei der Pelletherstellung, *Projektarbeit Forschungszentrum Karlsruhe/Berufsakademie Mannheim,* 2008

Gonzalez, M. T., Molina-Sabio, M. & Rodriguez-Reinoso F. (1994) Steam activation of olive stone chars, development of porosity. *Carbon,* Vol.32, No.8, (1994), pp. 1407-1413

Gou, Y., Yang, S., Yu, K.,, Zhao, J., Wang, Z. & Xu, H. (2002). The preparation and mechanism studies of rice husk based porous carbon. *Materials Chemistry and Physics,* Vol.74, (2002), pp. 320-223

Hornung, A., Apfelbacher, A., Koch, W., Linek, A., Sagi, S., Schöner, J., Stöhr, J., Seifert, H., Tumiatti, V. & Lenzi F. (2005). Thermochemical conversion of straw - Haloclean® - Intermediate pyrolysis. *Proceedings of the 14th European Biomass Conference and Exhibition,* Paris, October 17 - 21, 2005, p. 913.

Hornung, A. & Seifert, H. (2006) Rotary kiln pyrolysis of polymers containing heteroatoms. In: *Feedstock Recycling and Pyrolysis of Waste Plastics : Converting Waste Plastics into Diesel and Other Fuels*, Scheirs, J. (Ed.), S.549-67, Wiley & Sons Ltd., Chichester, ISBN 0-470-02152-7.

Hornung, U., Schneider, D., Seifert, H., Wiemer, H.-J. & Hornung, A. (2009a). Gekoppelte Pyrolyse und Niedrigtemperatur-Reformierung. *Chemie Ingenieur Technik Special Issue: ProcessNet-Jahrestagung und 27. Jahrestagung der Biotechnologen*, Vol.81, No.8, (August, 2009), pp. 1215–1216

Hornung, U., Schneider, D., Hornung, A., Tumiatti, V. & Seifert, H. (2009b). Sequential pyrolysis and catalytic low temperature reforming of wheat straw. *Journal of Analytical and Applied Pyrolysis*, Vol.85, (2009), pp. 145-50

Huang, S., Jing, S., Wang, J., Wang, Z. & Jin, Y. (2001). Silica white obtained from rice husk in a fluidized bed. *Powder Technology*, Vol.117, No.3, (2001), pp. 232–238.

Jensen, P.A., Sander, B. & Dam-Johansen, K. (2001a). Pretreatment of straw for power production by pyrolysis and char wash. *Biomass Bioenergy*, Vol.20, (2001), pp. 431–446

Jensen, P.A., Sander, B. & Dam-Johansen, K. (2001b). Removal of K and Cl by leaching of straw char. *Biomass and Bioenergy*, Vol.20, (2001), pp. 447–457

Klank, D. (2006). Surface and Pores/particle size and shape/Zeta potential. *Seminar at University of Zürich*, Zürich, October 17-19, 2006

Lopez-Gonzalez, J. de D., Martinez-Vilchez, F. & Rodriguez-Reinoso, F. (1980). Preparation and characterization of active carbons from olive stones. *Carbon*, Vol.18, (1980), pp. 413-418

Molina-Sabio, M., Gonzales, M. T., Rodrigues-Reinoso, F. & Sepulveda-Escribano, A. (1996). Effect of steam and carbon dioxid activation in the micropore size distribution of activated carbon. *Carbon*, Vol.34, No.4, (1996), pp. 505-509

Oh, G.H. & Park, Ch. R. (2002). Preparation and Characteristics of rice-straw-based porous carbons with high adsorption capacity. *Fuel*, Vol.81, (2002), pp 327–336.

Pendyal, B., Johns, M. M., Marshall, W. E., Ahmedna, M. & Rao, R. M. (1999). The effect of binders and agricultural by-products on physical and chemical properties of granular activated carbons. *Bioresource Technology*, Vol.68, (1999), pp. 247-254

Qipeng, J & Aik Chong, L. (2008). Effects of pyrolysis conditions on the physical characteristics of oil-palm-shell activated carbons used in aqueous phase phenol adsorption. *Journal of Analytical and Applied Pyrolysis*, Vol.83, (2008), pp. 175–179

Sing, K. S. W., Everett, D. H., Haul, R. A. W., Moscou, L., Pierotti, R. A., Rouquerol, J. & Siemieniewska, T. (1985). Reporting Physisorption Data for gas/solid Systems with Special Reference to the Determination of Surface Area and Porosity. *Pure & Applied Chemistry*, Vol.57, No.4, (1985), pp. 603—619

Sütcü H. & Demiral H. (2009). Production of granular activated carbons from loquat stones by chemical activation. *Journal of Analytical and Applied Pyrolysis*, Vol.84, (2009), pp. 47–52

Sun, R., Thomkinson, J., Mao, F.C. & Sun, X.F. J. (2001). Physiochemical Characterization of Lignins from rice straw by hydrogen peroxide treatment. *Journal of Applied Polymer Science,* Vol.79, (2001), pp. 719–732.

Wu, F.-C., Tseng, R.-L., Hu, C.-C. & Wang, C.-C. (2005). Effects of pore structure and electrolyte on the capacitive characteristics of steam- and KOH-activated carbons for supercapacitors. *Journal of Power Sources,* Vol.144, (2005), pp. 302–309.

Yun, Ch. H., Park, Y. H. & Park, Ch. R. (2001). Effects of pre-carbonization on porosity development of activated carbons from rice straw. *Carbon,* Vol.39, (2001), pp. 559–567.

Part 6

Fuel Production

19

Ethanol and Hydrogen Production with Thermophilic Bacteria from Sugars and Complex Biomass

Maney Sveinsdottir,
Margret Audur Sigurbjornsdottir and Johann Orlygsson
University of Akureyri, Borgir, Nordurslod, Akureyri
Iceland

1. Introduction

The increase in carbon dioxide (CO_2) emissions has clearly much more profound effects on global climate than earlier anticipated. The main source of CO_2 is by combustion of fossil fuel but its concentration has increased from 355 ppm in 1990 to 391 ppm in 2011 (Mauna Loa Observatory: NOAA-ASRL, 2011). Production of biofuels from biomass has emerged as a realistic possibility to reduce fossil fuel use and scientists have increasingly searched for new economically feasible ways to produce biofuels. The term biofuel is defined as fuel produced from biomass that has been cultivated for a very short time; the opposite of fuel that is derived from fossil fuel biomass (Demirbas, 2009). Plants and autotrophic microorganisms fix gaseous CO_2 into volatile (sugars) and solid compounds (lignocellulose, starch) during growth. These compounds can thereafter be converted to biofuels which, by combustion, releases CO_2 back to atmosphere. This simplified way of carbon flow is not completely true, because growing, cultivating, harvesting and process conversion to biofuels will, in almost all cases, add more CO_2 to atmosphere although less as compared to fossil fuels.
There are several types of biofuels produced and used worldwide today. The most common are methane, ethanol (EtOH) and biodiesel but also, to a lesser extent, hydrogen (H_2), butanol and propanol. There are also several methods to produce biofuels, ranging from direct oil extraction from fat-rich plants or animal fat (biodiesel) to complex fermentations of various types of carbohydrate rich biomass (H_2, EtOH, butanol). Fermentation processes can be performed by both bacteria and yeasts. This overview mainly focuses on the production of EtOH and H_2 from biomass with thermophilic bacteria.

2. Production of EtOH and H_2 from biomass

EtOH as a vehicle fuel originated in 1908 when Henry Ford's famous car, Ford Model T was running on gasoline and EtOH or a combination of both (Gottemoeller & Gottemoeller, 2007). Biomass was however not used as a source for EtOH production until in the early thirties of the 20th century when Brazil started to extract sugar from sugarcane for EtOH production. During the World War II, EtOH production peaked at 77 million liters in Brazil (mixed to gasoline at 42%) (Nardon & Aten, 2008). After the war, cheap oil outcompeted the use of EtOH and it was not until the oil crisis in the mid 70's that interest in EtOH rose again. The program "Pro-Alcool" was launched in 1975 to favour EtOH production from sugarcane. In US, there has been a steady increase in EtOH production from starch based plant material, e.g. corn, since the late 1970's (Nass et al., 2007). Perhaps the main reason for the increase in EtOH production is the discovery that methyl *tert*-butyl ether (MTBE), earlier used in gasoline as an additive, was contaminating groundwater, leading to search for alternative and more environmentally friendly source (Vedenov & Wetzsstein, 2008). Today, US and Brazil produce more than 65.3 billion liters of EtOH which corresponds for 89% of the world production (Renewable Fuel Association, 2010).

Production of EtOH from lignocellulose rich biomass has recently been focused upon. The main reason is the fact that EtOH production from starch and sugar based biomasses is in direct competition with food and feed production. This has been criticized extensively lately, because of the resulting rise in the prizes of food and feed products (Cha & Bae, 2011). Production of EtOH from sugars and starch is called first generation production, opposite to second generation production where lignocellulosic biomass is used. Lignocellulose is composed of complex biopolymers (lignin, cellulose and hemicelluloses) that are tightly bound together in plants. The composition of these polymers varies in different plants (cellulose, 36-61%; hemicellulose, 13-39%; lignin 6-29%) (Olsson & Hahn-Hagerdal, 1996). Of these polymers, only cellulose and hemicelluloses can be used for EtOH production. However, before fermentation, the polymers need to be separated by physiological, chemical or biological methods (Alvira et al., 2010). The most common method is to use chemical pretreatment, either weak acids or bases but many other methods are known and used today (see Alvira et al., 2010 and references therein). This extra pretreatment step has been one of the major factors for the fact that EtOH production from complex biomass has not been commercialized to any extent yet compared to first generation ethanol production. Also, after hydrolysis, expensive enzymes are needed to convert the polymers to monosugars which can only then be fermented to EtOH. Conventionally, most of the EtOH produced today is first generation EtOH but lately, especially after US launched their large scale investment programs (US Department of Energy, 2007), second generation of EtOH seems to becoming a reality within the next few years or decades.

The sugars available for fermentation after the pretreatment and hydrolysis of biomass (when needed) can be either homogenous like sucrose and glucose from sugarcane, and starch, respectively or heterogeneous when originating from lignocellulosic biomass. Thus, the main bulk of biomass used for EtOH production today are two types of sugars, the disaccharide sucrose and the monosugar glucose, both of whom can easily be fermented to EtOH by the traditional baker's yeast, *Saccharomyces cerevisae*. This microorganisms has many advantages over other known EtOH producing microorganisms. The most important are high EtOH yields (>1.9 mol EtOH/mol hexose), EtOH tolerance (> 12%), high robustness and high resistance to toxic inhibitors. However, the wild type yeast does not degrade any pentoses (Jeffries, 2006). The use of genetic engineering to express foreign genes associated with xylose and arabinose catabolism have been done with some success (van Maris et al., 2007) and a new industrial strain with xylose and arabinose genes was recently described (Sanchez et al., 2010). Also, no yeast has been reported to have cellulase or hemicellulase activity. The mesophilic bacterium *Zymomonas mobilis* is a highly efficient EtOH producer. The bacterium is homoethanolgenic, tolerates up to 12% EtOH and grows 2.5 times faster compared to yeasts (Rogers et al., 1982). The bacterium utilizes the Entner-

Doudoroff pathway with slightly higher EtOH yields than yeasts but lacks the pentose degrading enzymes. Many attempts have however been made to insert arabinose and xylose degrading genes in this bacterium (Deanda et al., 1996; Zhang et al., 1995). The company DuPont has recently started to use a genetically engineered *Z. mobilis* for cellulosic EtOH production (DuPont Danisko Cellulosic Ethanol LLC, 2011).

Especially, the lack of being able to utilize arabinose and xylose, both major components in the hemicellulosic fraction of lignocelluloses, has lead to increased interest in using other bacteria with broader substrate spectrum. Bacteria often possess this ability and are capable of degrading pentoses, hexoses, disaccharides and in some cases even polymers like cellulose, pectin and xylans (Lee et al., 1993; Rainey et al., 1994). The main drawback of using such bacteria is their lower EtOH tolerance and lower yields because of production of other fermentation end products like acetate, butyrate, lactate and alanine (Baskaran et al., 1995; Klapatch et al., 1994; Taylor et al. 2008). Additionally, most bacteria seem to tolerate much lower substrate concentrations although the use of fed batch or continuous culture may minimize that problem. On the opposite however, many bacteria show good EtOH production rates. The use of thermophilic microorganisms has especially gained increased interest recently. The main reasons are, as previously mentioned, high growth rates but also less contamination risk as well as using bacteria that can grow at temperatures where "self distillation" is possible, thus eliminating low EtOH tolerance and high substrate concentration problems. Also, the possibilty to use bacteria with the capacity to hydrolyze lignocellulosic biomass and ferment the resulting sugars to EtOH simultaneously is a promising method for EtOH production.

The production of H_2 is possible in several ways but today the main source of H_2 is from fossil fuels and, to a lesser extent, by electrolysis from water. H_2 is an interesting energy carrier and its combustion, opposite to carbon fuels, does not lead to emission of CO_2. Biological production of H_2 is possible through photosynthetic or fermentative processes (Levin et al., 2004; Rupprecht et al., 2006). This chapter will focus on biological H_2 production by dark fermentation by thermophilic bacteria only. Fermentative production of H_2 has been known for a long time and has the advantage over photosynthetic processes of simple operation and high production rates (Chong et al., 2009). Also, many types of organic material, e.g. wastes, can be used as substrates. Thus, its production possesses the use of waste for the production of renewable energy. Fermentative hydrogen production has though not been commercialized yet but several pilot scale plants have been started (Lee & Chung, 2010; Lin et al., 2010).

3. Physiology of thermophilic EtOH and H_2 producing bacteria

Thermophilic bacteria can degrade many carbohydrates and produce various end products, among them both EtOH and H_2. Figure 1 shows the carbon flow from glucose by fermentation by the use of Embden-Meyerhof pathway (EMP). The majority of microorganisms degrade hexoses through this pathway or the Entner-Douderoff pathway (ED). The degradation of glucose with EMP generates two NADH, two pyruvates, the key intermediate in most organisms, together with the formation of two ATP by substrate level phosphorylation. The ED pathway, however, is more restricted to Gram-negative bacteria and Archaea and generates only one mol of ATP, which explains its low distribution among anaerobic bacteria. Some bacteria, especially hyperthermophiles, are known to be able to use both pathways simultaneously (Moat et al., 2002; Siebers & Schönheit, 2005).

There are also some variations of the classical EMP among thermophilic microorganisms. Some archaea e.g. *Pyrococcus* and *Thermococcus* use ADP instead of ATP to transfer phosphate groups to hexoses in the preparation steps of the glycolysis. These bacteria also use ferredoxin-dependent glyceraldehyde-3-phosphate ferredoxin oxidoreductase (GAPOR) for converting glyceraldehyde-3-phosphate to 3-phosphoglycerate in one step (Chou et al., 2008). Thermophilic bacteria, however, use the common glyceraldehydes-3-phosphate dehydrogenase (GAPDH) and reduce glyceraldehydes-3-phosphate to 1,3-glycerate which is thereafter converted to 3-phosphoglycerate. Thus, both groups produce two molecules of ATP by substrate level phosphorylation but the archaea "sacrifice" one and use it to together with two molecules of AMP to produce two molecules of ADP, needed for hexose phosphorylation. Consequently, the amount of energy conserved in glucose to acetate conversion is 3.2 instead of the expected 4.0 ATP/glucose (Sapra et al., 2003).

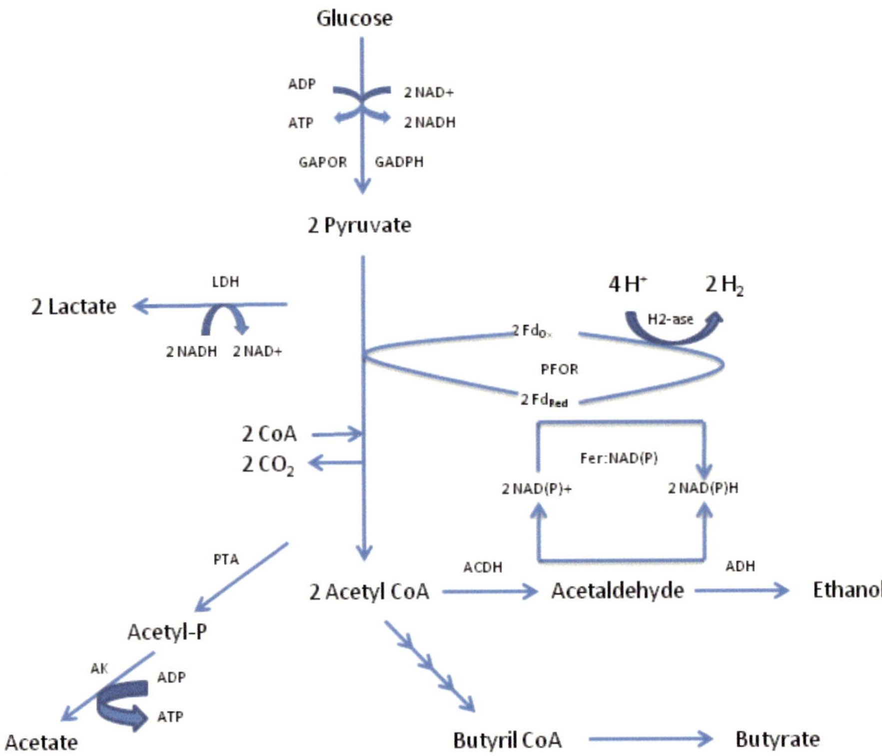

Fig. 1. Simplified scheme of glucose degradation to various end products by strict anaerobic bacteria. Enzyme abbreviations: ACDH, acetaldehyde dehydrogenase; ADH, alcohol dehydrogenase; AK, acetate kinase; Fer:NAD(P), ferredoxin:NAD(P) oxidoreductase; H_2-ase, hydrogenase; LDH, lactate dehydrogenase; PFOR, pyruvate:ferredoxin oxidoreductase; PTA, phosphotransacetylase.

Pyruvate is the end product of glycolysis and can be converted to fermentation products like H_2, EtOH and many more (Fig. 1). The carbon flow depends on the microorganisms involved and the environmental conditions. Pyruvate can e.g. be reduced to lactate by lactate dehydrogenase (LDH) but the most favorable pathway for anaerobic bacteria is to oxidize pyruvate to acetyl-CoA and CO_2 by using pyruvate:ferredoxin oxidoreductase (PFOR) which can be converted to acetate with concomitant ATP synthesis from the acetyl-phosphate intermediate. Acetate is thus the oxidized product but the main advantage for the microorganism is the extra ATP produced. The electrons are transported to reduced ferredoxin which acts as an electron donor for hydrogenases and H_2 is produced as the reduced product. There are mainly two types of hydrogenases; NiFe hydrogenases and the FeFe hydrogenases. Recent overview articles have been published on the subject (Chou et al., 2008; Kengen et al., 2009). Acetyl Coenzyme A can also be converted to acetaldehyde by acetaldehyde dehydrogenase (ACDH) and further to EtOH by alcohol dehydrogenase.

Strict anaerobes can produce H_2 from two major breakpoints during degradation of glucose. Firstly, from a NAD(P)H by GAPDH and from pyruvate ferredoxin oxidoreductase (PFOR) (Jones, 2008). The principal H_2 pathway is through PFOR because of thermodynamics hindrance of reoxidizing NADH (Jones, 2008). It is a well known phenomenon that the low H_2 yields observed by mesophilic and moderate thermophilic bacteria are due to the fact that H_2 production from either ferredoxin or NAD(P)H are thermodynamically unfavorable (Jones, 2008; Hallenbeck, 2009). The redox potential of Fd_{red}/Fe_{ox} couple depends on the microorganism and temperature involved. In nature, high partial pressures of H_2 are relatively uncommon because of the activity of H_2 scavenging microbes, e.g. methanogens or sulfate reducing bacteria (Cord-Ruwisch et al., 1988). This results in a low partial pressure of H_2 which is favorable for a complete oxidation of glucose to acetate and CO_2. At high temperatures, the influence of the partial pressure of H_2 is less on the key enzymes responsible for H_2 production. This is the main reason why extremophilic bacteria have been reported to produce up to 4 moles of H_2 together with 2 moles of acetate in pure cultures and also for the fact that microorganisms growing at lower temperatures direct their end product formation to other reduced products. At lower temperatures, the NADH ferrodoxin oxidoreductase (NOR) that converts NADH to Fd_{red} is strongly inhibited. The E° is – 400 mV for Fd_{red}/Fd_{ox} couple but -320 mV for the $NADH/NAD^+$ couple (Jones, 2008; Hallenbeck, 2009). Therefore, at low temperatures, elevated H_2 concentrations inhibit H_2 evolution at much lower concentrations as compared to extreme temperatures. Mesophilic and moderate thermophilic bacteria respond to this by directing their reducing equivalents to other more favorable electron acceptors and consequently produce reduced products like EtOH, lactate, butyrate and alanine (Fig. 1).

Following are the main stoichiometry equations for the degradation of glucose to various end products by microorganisms with special focus on H_2 and EtOH production.

The amount of H_2 produced depends on the fermentation pathways used and end product formation. For example, if acetic acid is the final product the theoretical yield for one mole of glucose is four moles of H_2:

$$C_6H_{12}O_6 + 4\,H_2O \rightarrow 2CH_3COO^- + 4H_2 + 2HCO_3^- + 4H^+ \quad (1)$$

If on the other hand the final product is butyric acid, the theoretical yield of H_2 is only two moles of H_2 per mole of glucose:

$$C_6H_{12}O_6 + 2\,H_2O \rightarrow CH_3CH_2CH_2COO^- + 2H_2 + 2HCO_3^- + 3H^+ \quad (2)$$

The production of EtOH by *Saccharomyces cerevisae* and *Zymomonas mobilis* occurs according to:

$$C_6H_{12}O_6 + 4\,H_2O \rightarrow 2CH_3COH^- + 2HCO_3^- + 4H^+ \quad (3)$$

Bacteria however, usually produce a mixture of EtOH together with other end products. This results in lower EtOH yields and, in some cases, production of H_2. If lactate is the only end product, no H_2 is formed:

$$C_6H_{12}O_6 + 4\,H_2O \rightarrow Lactate^- + 2HCO_3^- + 4H^+ \quad (4)$$

4. Thermophilic anaerobic bacteria – classification and physiology

In recent years, thermophilic anaerobic bacteria have gained increased attention as potential EtOH and H_2 producing microorganisms. Depending on optimal growth temperatures, thermophilic bacteria can be divided into several categories, e.g. moderate thermophiles (T_{opt} between en 45 to 55°C), true thermophiles (T_{opt} between 55 to 75°C) and extremophiles with optimum temperature above 75°C (Brock, 1986). The ability of thermophiles to live at high temperatures is mainly due to their thermostable proteins; the cell membrane of thermophilic bacteria contains more saturated fatty acids which make it stiffer and more heat resistant as compared to mesophiles (Brock, 1986).

Thermophilic bacteria are capable of adapting to environmental conditions and are able to thrive in geothermal areas although the temperature might be slightly higher than the optimum growth temperature. Geothermal areas offer stability in heat and are thus favorable habitats for thermophilic bacteria (Brock, 1986; Kristjansson & Alfredsson, 1986). Generally, most known thermophilic species are obligate or facultative anaerobes since geothermal areas have low oxygen concentrations (Amend & Shock, 2001). Less variety seems to be of strict anaerobic, heterotrophic thermophilic bacteria (see review of Wagner & Wiegel, 2008 and references therein).

4.1 Thermophilic EtOH and H_2 producing bacteria

There are relatively few genera of thermophiles that include bacteria with good H_2 and EtOH producing capacities. Among good EtOH producers are bacteria that belong to the genera of *Clostridium, Thermoanaerobacter* and *Thermoanaerobacterium* but good H_2 producers are the extremophiles like *Caldicellulosiruptor* and *Thermotoga* and the archaeon *Thermococcus* and *Pyrococcus*. It varies to a great extent how much data is available in literature concerning pure culture studies of individual species on biofuel production. Much data is not on the efficiency of these bacteria to produce H_2 and EtOH but merely on phylogenetic status and basic physiological properties. Also, the data on biofuel production properties from these bacteria on hydrolysates from lignocellulosic biomass is scarce but more is known on yields from monosugars. Below, the discussion will be on the major phylogenetic and physiological characteristics of most of the "good" EtOH and H_2 producing thermophiles known today. Later chapters deal with H_2 and EtOH production rates and yields from both sugars and from complex lignocellulosic biomasses by these bacteria and more.

4.1.1 *Clostridium*

The genus *Clostridium* belongs to the family Clostridiaceae, order Clostridiales, class Clostridia and phylum Firmicutes. These bacteria are spore forming and often present in environments which are rich in plant decaying material. It is thus not surprising that many species are capable of polymer hydrolyzation and this is one of the main reasons for extensive research on biofuel production from complex biomass by these bacteria (Canganella & Wiegel, 1993; Carreira & Ljungdahl, 1993). Several cellulose-degrading enzymes form a structure called cellulosome, located and embedded on the external surface of the cell membrane (Demain et al., 2005). The genus contains a very diverse group of bacteria as shown by a phylogenetic analysis of Collins and co-workers where *Clostridium* species were compared both within species belonging to the genus and to related taxa (Collins, et al., 1994). This investigation and others lead to the conclusion that more than half of the species currently assigned to the genus *Clostridium* are in fact not closely related to the type species *C. butyricum* and should therefore not be included in the newly defined genus *Clostridium*. The genus contains more than 200 validly described species but only about 15 are thermophilic. Two of those thermophilic Clostridia, *C. thermocellum* and *C. thermohydrosulfuricum* (now *Thermoanaerobacter thermohydrosulfuricum*) have attracted the most attention and the cellulosome of *C. thermocellum* has been characterized extensively (Demain et al., 2005). Among other well known thermophilic Clostridia are *C. thermobutyricum* (Wiegel et al., 1989), *C. thermosucciongenes* (Drent et al., 1991) and *C. clariflavum* (Shiratori et al., 2009) and several others.

4.1.2 *Thermoanaerobacterium*

Thermoanaerobacterium together with genus *Thermoanaerobacter* falls within clusters V, VI and VII in phylogenetic interrelationships of *Clostridium* species (Collins et al., 1994). The genus was first described in 1993 when two thermophilic, xylan degrading strains were isolated from Frying Pan Springs in Yellowstone National Park (Lee et al., 1993). They were compared with other xylan degrading bacteria and new taxonomic assignments were proposed thereafter. Today the genus consists of nine validly described species; *T. aciditolerans, T. aotearoense, T. saccharolyticum, T. thermosaccharolyticum, T. thermosulfurigenes, T. xylanolyticum, T. fijiensis, T. polysaccharolyticum and T. zeae* (German Collection of Microorganisms and Cell Cultures and references therein). Most *Thermoanaerobacterium* species have been isolated from hot springs or leachate of waste from canning factories. *Thermoanaerobacterium* species are known for their abilities to convert carbohydrates to various end products like acetate, EtOH, lactate, H_2 and CO_2. Some species have shown promising EtOH and H_2 production capacity but production of mixed end products limit their use (Ren et al., 2008; 2009; 2010; Romano et al., 2010; Sveinsdottir et al., 2010). *T. saccharolyticum* has however been genetically engineered and both acetate and lactate formation has been knocked out (Shaw et al., 2008). According to the description, members of this genus reduce thiosulfate to elemental sulfur while members of *Thermoanaerobacter* reduce thiosulfate to H_2S (Lee et al., 1993).

4.1.3 *Thermoanaerobacter*

Bacteria within this genus were originally classified within the genus *Clostridium* because of close phylogenetic relationship and physiological properties. These bacteria use the classical EMP pathway for sugar degradation and produce EtOH, acetate and lactate as major end products (Lee et al., 1993). Most species have broad substrate range and can degrade both pentoses and hexoses. The genus consists of 24 species (subspecies included) originating from various environments like hot springs and oil fields (Collins et al., 1994; Larsen et al., 1997; Lee et al., 1993; German Collection of Microorganisms and Cell Cultures and references therein). Most species produce EtOH and H_2 as well as lactate, and in some cases alanine as end products. The type species, *Thermoanerobacter ethanolicus* and several other species within the genus has been extensively studied for EtOH production (Fardeau et al., 1996; Georgieva & Ahring, 2007; Georgieva et al., 2008a. b; Lacis & Laword 1988a,b; Lamed & Zeikus, 1980a,b). H_2 production is usually low compared to EtOH by *Thermoanaerobacter* although *Thermoanaerobacter tengcongensis* has been described to produce up to 4 moles of H_2 from one mole of glucose under nitrogen flushed fermentor systems (Soboh et al., 2004).

4.1.4 *Caldicellulosiruptor*

The genus *Caldicellulosiruptor* was first proposed in 1994 by Rainey and co-workers on the basis of physiological characteristics and phylogenetic position of a strain they isolated, *Caldicellulosiruptor saccharolyticus* (Tp8T 6331) (Rainey et al., 1995). Today the genus holds nine different species; *C. acetigenus, C. bescii, C. hydrothermalis, C. kristjanssonii, C. kronotskyensis, C. lactoaceticus, C. obsidiansis, C. owensensis* and *C. saccharolyticus* (German Collection of Microorganisms and Cell Cultures and referenses therein). All species are extremely thermophilic, cellulolytic, non-spore-forming anaerobes that have been isolated from geothermal environments such as hot springs and lake sediments (Rainey et al., 1994; Yang et al., 2010). *Caldicellulosiruptor* species have a relatively broad substrate spectrum capable to utilize e.g. cellulose, cellobiose, xylan and xylose. Extreme thermophiles, have been shown to have superior H_2 production yields and rates compared to mesophiles and produce few other byproduct besides acetate. This makes *Caldicellulosiruptor* species excellent candidates for H_2 production. *C. saccharolyticus* and *C. owensis* have been extensively studied for H_2 production from sugar and hydrolysates from lignocellulosic biomass (Kadar et al., 2004; Vrije et al., 2007; Zeidan & van Niel, 2010).

4.1.5 *Thermotoga*

The genus of *Thermotoga* was first described in 1986 when a unique extremely thermophilic bacteria was isolated from geothermally heated sea floors in Italy and the Azores (Huber et al., 1986). Today, nine different species have been identified; *T. elfii, T. hyphogea, T. lettingae, T. maritima* (type species), *T. naphthophila, T. neapolitana, T. petrophila, T. subterranean* and *T. thermarum* (German Collection of Microorganisms and Cell Cultures and references therein). These species are extremophiles, growing at temperatures that are highest reported for bacteria. All are strictly anaerobic and the cells are rod-shaped with an outer sheethlike structure called toga. (Huber et al., 1986; Jannasch et al., 1988). Most species have been isolated from deep environments, high temperature and pressure environments like oil reservoirs, often rich of sulfur-compounds. Most of them are thus able to reduce either elemental sulfur, thiosulfate or both. Members of *Thermotoga* ferment sugars to mainly acetate, CO_2 and H_2 like *Caldicellulosiruptor* species. Only three species have been reported producing traces of EtOH. Most strains have shown the property of reducing pyruvate to alanine from sugar fermentation and *T. lettingae* produces alanine from methanol (in the presence of elemental sulfur or thiosulfate) (Balk et al., 2002). Other special feature within the genus is the ability of *T. lettingae* to degrade xylan at 90°C and its property of methanol metabolism (Balk et al., 2002). Hydrogen production has been extensively studied for *T. elfi, T. maritima* and *T. neapolitana* (d'Ippolito et al., 2010; Nguyen et al., 2008a,b; van Niel et al., 2002).

4.1.6 Other thermophilic bacteria producing H_2 and EtOH

Apart from the above mentioned genera the capacity to produce EtOH and H_2 has been reported for many other genera. Examples are species within *Caloramator, Caldanaerobacter,*

Caldanerobius and the archaeon *Thermococcus* and *Pyrococcus*. Some species within these genera will be discussed in later chapters.

5. Production of EtOH by thermophilic bacteria

The interest in EtOH production by thermophilic bacteria originates shortly after the oil crisis in the mid 70's of the twentieth century. Earliest reports on EtOH production from sugars include work on *Thermoanaerobacter brockii* and *Clostridium thermocellum* (Ben Bassat et al., 1981; Lamed et al., 1980; Lamed & Zeikus, 1980a, 1980b) but later on other *Thermoanaerobacter* species, e.g. *T. finnii,* (Faredau et al., 1996), *T. thermohydrosulfuricus* (Lovitt et al., 1984; Lovitt et al., 1988), *T. mathrani* (Larsen et al., 1997) and *Thermoanaerobacterium* species (Koskinen et al., 2008a; Sveinsdottir et al., 2009; Zhao et al., 2009, 2010). It was however not until recently that the use of thermophilic bacteria for EtOH production from lignocellulosic biomass arises. The earliest reports on EtOH production of more complex nature are from 1981 on starch (Ben Bassat et al., 1981) and 1988 on avicel (Lamed et al., 1988). The first study on lignocellulosic biomass (hemicellulose fraction of birch- and beechwood) was in 1983 by *Thermoanaerobacter ethanolicus* and several other thermophilic bacteria (Wiegel et al., 1983). Following chapters are divided into two main subchapters; 1) studies of EtOH production from sugars both in batch and continuous cultures with either pure or cocultures of thermophilic bacteria and 2) studies of EtOH production from lignocellulosic biomass by mixed or pure cultures of thermophilic bacteria.

5.1 Production of EtOH from sugars

Although it has been known for a long time that thermophilic bacteria produce EtOH from various carbohydrates it was not until 1980 the first papers appeared in literature with the focus on EtOH production. Earlier investigations include work on *Thermoanaerobacter brockii, Thermoanaerobacter thermohydrosulfuricus* and *Clostridium thermocellum* (Ben Bassat et al., 1981; Lamed & Zeikus, 1980a; 1980b; Lovitt et al., 1984). Ethanol yields by *T. brockii* were only moderate or between 0.38 (Lamed & Zeikus, 1980b) to 0.44 mol EtOH mol glucose^{-1} equivalents (Ben Bassat et al., 1981). In the latter investigation the focus was mostly on the effects of additional acetone and H_2 on end product formation. Much higher yields were later observed by *Thermoanaerobacter thermohydrosulfuricus*, or 0.9 to 1.9 mol EtOH mol glucose^{-1}. (Lovitt et al., 1984; 1988), also with the main focus on the effect of solvents on EtOH production, e.g. EtOH tolerance. *Thermoanerobacter ethanolicus* was described in 1981 (Wiegel & Ljungdahl., 1981) showing extremely good yields of ethanol from glucose (1.9 mol EtOH mol glucose^{-1}). Later this strain has been extensively studied by Lacis and Lawford (Lacis and Lawford 1988a, 1988b, 1989, 1991). Early observation was on high EtOH yields on xylose at low substrate (4.0 g L^{-1}) concentrations. The yields were 1.30 and 1.37 mol EtOH mol xylose^{-1} in batch and continuous cultures, respectively (Lacis & Lawford, 1988a) but only at low substrate concentrations. At higher concentrations (27.5 g L^{-1}) the yields lowered to 0.6 mol EtOH mol xylose^{-1}. Further studies by using xylose limiting continuous cultures, indicated that EtOH yields were more dependent on length of cultivation than upon growth rate and higher yields were presented (1.43 mol mol xylose^{-1}) (Lacis & Lawford, 1988b, 1989). Later data from this strain on glucose showed lower EtOH yields and the direction of the carbon flow was towards lactate formation by increasing substrate concentrations (Lacis & Lawford, 1991). *Thermoanerobacter ethanolicus* JW200 showed also very good EtOH yields from xylose and glucose at low (10 g L^{-1}) substrate concentrations, or

1.45 and 1.95 mol, respectively (Carreira et al., 1982). A mutant strain was later developed (JW200Fe(4)) that showed similar yields but at higher (30 g L^{-1}) substrate concentrations (Carreira et al., 1983). Other investigations on this species on sucrose showed between 1.76 to 3.60 mol EtOH mol sucrose^{-1} with high substrate concentrations (15 to 30 g L^{-1}) (Avci et al., 2006). Recent study on *Thermoanerobacter ethanolicus* strain interestingly shows that the addition of external acetate increases EtOH yields from xylose, glucose and cellobiose (He et al., 2010). EtOH yields on xylose were 1.0 mol EtOH mol glucose^{-1} without any acetate added but increased to 1.17 by adding 150 mM of acetate. Similar increase was observed on glucose, or from 1.16 to 1.34 mol EtOH mol glucose^{-1} without and with added acetate, respectively. It has been suggested that acetate may disrupt energy production through accelerated fermentation (Russel, 1992) which may lead to lower biomass production and higher end product formation. Fardeau et al. (1996) investigated the effect of thiosulfate as electron acceptor on sugar degradation and end product formation by *Thermoanaerobacter finnii*. This strain shows good EtOH yields on xylose or 1.76 mol EtOH mol xylose^{-1} which is actually higher than the theoretical yield (1.67) from this sugar. Yields on glucose were however lower or, 1.45 mol EtOH mol glucose^{-1}. Not surprisingly, the addition of thiosulfate shifted end product formation towards acetate with higher cell yield and lower EtOH production. A study of bacteria isolated from Icelandic hot spring shows that a *Thermoanerobacter* sp. AK33 showed good EtOH yields on monosugars (Sveinsdottir et al., 2009). Glucose and xylose fermentations resulted in 1.5 and 0.8 mol EtOH from one mole of glucose and xylose, respectively. *Thermoanaerobacterium* AK17, isolated from Icelandic hot spring, has been extensively studied for EtOH production (Koskinen et al., 2008a; Orlygsson & Baldursson, 2007; Sveinsdottir et al., 2009). This strain produces 1.5 and 1.1 mol EtOH from one mole of glucose and xylose, respectively. A moderate thermophile, *Paenibacillus* sp. AK25 has also been shown to produce 1.5 mol EtOH mol glucose^{-1} (Sveinsdottir et al., 2009).

One of the main drawbacks for the use of thermophilic bacteria for EtOH production from biomass is their low tolerance towards EtOH. Several studies have been done with *Clostridium thermosaccharolyticum* (Baskaran et al., 1995; Klapatch et al., 1994) and *Thermoanaerobacter* sp. (Georgieva et al., 2008b) to increase EtOH tolerance. The highest EtOH tolerance is by a mutant strain of *Thermoanaerobacter ethanolicus*, or 9% (wt/vol) at 69°C (Carriera & Ljungdahl, 1983) but later studies with JW200 Fe(4), one of its derivatives, show much less tolerance (Hild et al., 2003). Georgieva and co-workers published very high EtOH tolerance (8.3%) for *Thermoanerobacter* BG1L1, a highly efficient xylose degrader in continuous culture studies (Georgieva et al., 2008b). *Thermoanerobacter thermohydrosulfuricus* degrades various pentoses and hexoses as well as starch to high concentrations of EtOH (Ng et al., 1981). By transferring the parent strain (39E) to successively higher concentrations of EtOH, an alcohol tolerant strain (39EA) was obtained (Lovitt et al., 1984). The mutant strain grows at 8% EtOH concentrations (wt/vol) at 45°C but only up to 3.3% at 68°C. The parent strain produces 1.5 mol EtOH mol glucose^{-1} without any addition of EtOH but the yield lowered to 0.6 mol at 1.5% initial EtOH concentrations. The mutant strain showed lower EtOH yields without any addition of EtOH, or 0.9 mol EtOH mol glucose^{-1} but the yields did not decrease to any extent by increasing initial EtOH concentrations up to 4%. Further experiments with the wild type also indicated the role of H_2 production and its influence on EtOH production (Lovitt et al., 1988). Thus, by changing the gas phase from nitrogen to H_2 or carbon monoxide, EtOH yields increased from 1.41 mol EtOH mol glucose^{-1} to 1.60 and 1.90 mol, respectively.

Organisms	Sugar	Cultivation method	Sugar conc. (gL^{-1})	Ethanol yield (mol EtOH mol sugar^{-1})	Temp. (°C)	Reference
T. brockii	Cellobiose	Batch	10.0	0.38	60	Lamed & Zeikus (1980)
T. brockii	Glucose	Batch	5.0	0.44	nd	Ben Bassat et al. (1981)
T. ethanolicus	Glucose	Batch	8.0	1.90	72	Wiegel & Ljungdahl. (1981)
T. ethanolicus	Glucose	Batch	20.0	1.90	68	Carreira et al. (1983)
T. thermohydrosulfuricus	Glucose	Batch	5.0	1.60	60	Lovitt et al. (1984)
T. thermohydrosulfuricus	Glucose	Batch	5.0	0.90	60	Lovitt et al. (1984)
T. thermohydrosulfuricus	Glucose	Batch	10.0	1.40-1.90	60	Lovitt et al. (1988)
T. ethanolicus	Xylose	Batch	4.0-27.5	0.60-1.30	60	Lacis & Lawford (1988a)
T. ethanolicus	Xylose	Con	4.0	1.37	60	Lacis & Lawford (1988a)
T. ethanolicus	Xylose	Con	4.0	1.43	60	Lacis & Lawford (1988b)
T. ethanolicus	Xylose	Con	4.0	1.37	68	Lacis & Lawford (1989)
T. ethanolicus	Xylose	Con	4.0	1.37	67-69	Lacis & Lawford (1991)
T. ethanolicus	Xylose	Con	20.0	1.06	67-69	Lacis & Lawford (1991)
T. finnii	Glucose	Batch	NA	1.45	60	Fardeau et al. (1996)
T. finnii	Xylose	Batch	NA	1.76	60	Fardeau et al. (1996)
C. thermocellum	Cellobiose	Batch	2.6	1.60	60	Knutson et al. (1999)
T. ethanolicus	Xylose	Con	4.0	1.30	69	Hild et al. (2003)
T. ethanolicus	Sucrose	Batch	15-30	1.80-3.60	65	Avci et al. (2006)
T. thermohydrosulfuricus	Sucrose	Batch	15-30	1.10 - 3.00	65	Avci et al. (2006)
Thermoanaerobacter ap 65-2	Sucrose	Batch	15-30	1.30-3.20	65	Avci et al. (2006)
Thermoanaerobacter BG1L1	Xylose	Con	10.0	1.28	70	Georgieva at al. (2008)
Enrichment cultures	Glucose	Batch	18.0	0.10-1.70	50-78	Koskinen et al. (2008)
Coculture	Glucose	Con	12.6-25.2	1.37	60	Koskinen et al. (2008a)
Thermoanaerobacterium AK17	Glucose	Batch	3.6	1.50	60	Sveinsdottir et al. (2009)
Thermoanaerobacterium AK17	Xylose	Batch	3.0	1.10	60	Sveinsdottir et al. (2009)
Thermoanaerobacter Ak33	Glucose	Batch	3.6	1.50	70	Sveinsdottir et al. (2009)
Thermoanaerobacter Ak33	Xylose	Batch	3.0	0.80	70	Sveinsdottir et al. (2009)
Paenibacillus AK25	Glucose	Batch	3.6	1.50	50	Sveinsdottir et al. (2009)
Paenibacillus AK25	Xylose	Batch	3.0	0.90	50	Sveinsdottir et al. (2009)
Mixed culture	Glucose	Batch	5.0	1.53	70	Zhao et al. (2009)
Mixed culture	Xylose	Batch	2.0	1.60	70	Zhao et al. (2010)
Enrichment cultures	Glucose	Batch	9.0	1.34	50-75	Orlygsson et al. (2010)
Enrichment cultures	Xylose	Batch	7.5	1.30	50-75	Orlygsson et al. (2010)
T. ethanolicus	Xylose	Batch	5.0	1.00-1.20	65	He et al. (2010)
T. ethanolicus	Glucose	Batch	5.0	1.20-1.30	65	He et al. (2010)

Table 1. EtOH production from sugars by defined and mixed cultures of thermophilic bacteria. Cultivation was either in batch or continuous (con). EtOH yields as well as substrate concentrations and incubation temperature are also shown.

Recent studies with mixed cultures (batch) were conducted on glucose (Zhao et al., (2009) and xylose (Zhao et al., 2010) where various environmental parameters were optimized for both EtOH and H_2 production. The main bacterial flora, originating from biohydrogen reactor operated at 70°C and fed with xylose and synthetic medium, was identified as various species of *Thermoanaerobacter, Thermoanaerobacterium* and *Caldanaerobacter*. Highest yields observed to be 1.53 and 1.60 mol EtOH mol glucose^{-1} and xylose^{-1} respectively.
Several efforts have recently been made to enrich for new ethanologenic thermoanerobes. Two surveys have been done from Icelandic hot springs where several interesting bacteria were isolated with EtOH yields of > 1.0 mol EtOH from one mol glucose and xylose (Koskinen et al., 2008; Orlygsson et al., 2010).

5.2 Production of EtOH from complex biomass

Production of EtOH from lignocellulosic biomass has gained increased interest in recent years. The type of biomass used has varied to a great extent, e.g. wheat straw, barley straw, hemp, grass, paper and more. Also, the type of pretreatment used is different from one experiment to another. Most data is on biomass pretreated with dilute sulfuric acid or with alkaline pretreatment. The concentration of hydrolysates made from the biomass is also very broad, mostly varying from 0.2 % (w/v) to 15% (w/v). Finally, either pure or mixed cultures are used and either batch or continuous mode. The maximum yield of EtOH from glucose fermentation is 0.51 g EtOH g glucose^{-1}. This corresponds to 2 mol EtOH/mol hexose or 11.1 mM g^{-1}. Considering the complex structure of lignocellulosic biomass, it is not surprising that EtOH yields are usually considerable lower from such substrates (Table 2). Earliest available data on thermophilic bacteria using polymeric biomass originates from studies on *Thermoanaerobacter ethanolicus* and *Clostridium thermocellum* on hemicellulose from birch- and beechwood (Wiegel et al., 1983). These early reports showed promising results but highest yields were observed from the mutant strain *T. ethanolicus*, 4.5 mM g^{-1} xylose equivalent used. Three strains of *Clostridium thermocellum* produced between 1.40 to 2.60 mM EtOH g avicel^{-1} (Lamed et al., 1988). Higher yields (5.0 mM g^{-1} and 5.5 mM g^{-1}) by this bacterium were shown on the same substrate by others (Ahn et al., 1996; Lynd et al., 1989). Rani and co-workers studied EtOH production from both cellulose and lignocellulosic biomass by *C. thermocellum* (Rani et al., 1998). EtOH yields on avicel and Whatman paper was up to 7.2 and 8.0 mM g^{-1} EtOH, respectively. Similar yields were obtained from paddy straw, sorghum stover and corn stubs, pretreated with alkali. The highest yields of EtOH production from cellulosic biomass by *C. thermocellum* are from filter paper, 8.2 mM g^{-1} substrate (Balusu et al., 2004; 2005). In all studies mentioned above with *C. thermocellum* the concentration of cellulose was below 8.0 g L^{-1}. Lin and co-workers recently investigated degradation of napier grass and cellulose (avicel) by *C. thermocellum* and a mixed enrichment culture (Lin et al., 2010). They used from 2.0 to 40.0 g L^{-1} substrate concentrations. The pure culture produced merely 0.72 mM g^{-1} avicel but up to 3.87 mM g^{-1} Napier grass. The mixed culture produced between 0.7-0.9 mM g^{-1} Napier grass and 0.4–5.7 mM g^{-1} avicel. A dramatic decrease in yields was observed by increasing substrate concentrations.

Ahring and co-workers (Ahring et al., 1996) investigated the potential of five thermoanaerobes for EtOH production from the hemicelluloses fraction of wheat straw hydrolysates. Three of the strains produced only minor amounts of EtOH from xylan but *Thermoanaerobacterium saccharolyticum* HG8 and strain A3 produced 6.30 and 5.43 mM g xylan^{-1}, respectively. Strain A3 was further investigated on hydrolysates made from wheat straw, pretreated with wet oxidation. EtOH yields were lower as compared to xylan, or 2.61 mM g wheat straw^{-1} pretreated without oxygen.

Thermoanerobacter mathranii was isolated in 1993 from Hveragerdi in Iceland (Larsen et al., 1997) and has been adapted by Ahring et al., (1996). The strain has been investigated for EtOH production capacity on wet oxidized wheat straw (Ahring et al., 1999). By using very high substrate concentrations (60 g L^{-1}) and wet oxidation with different amounts of sodium carbonate the amount of total sugars released varied from 3.5 to 9.9 g L^{-1}. A fermentation of the strain on undiluted hydrolysate by the strain resulted in the production of approximately 9 mM of EtOH, or 1.3 mM g sugar^{-1}. This strain was also investigated for the effects of inhibitory compounds and hydrolysate concentration on the fermentation of wheat straw hydrolysates (Klinke et al., 2001). The main outcome was that the addition of hydrolysate to a medium containing 4 g L xylose^{-1} did not inhibit EtOH production and it produced 5.5 mM g xylose^{-1}. Increased concentrations of aromatic compounds and hydrolysates however, severely inhibited EtOH production by the strain. Wheat straw hydrolysates have also been investigated by other thermophilic bacteria (Sommer et al., 2004) but with lower EtOH yields.

Organisms	Biomass	Cultivation method	Substr. conc. (gL^{-1})	Ethanol yield (mM g sugar^{-1})	Temp. (°C)	Reference
T. ethanolicus	Wood hydrolysate	Batch	8.0	3.30-4.50	70	Wiegel et al. (1983)
C. thermocellum (3 strains)	Avicel	Batch	20.0	1.40-2.60	60	Lamed et al. (1988)
C. thermocellum	Avicel	Batch	2.5	5.00	60	Lynd et al. (1989)
C. thermocellum	Wood hydrolysate	Batch	4.8	3.10	60	Lynd et al. (1989)
C. thermocellum	Avicel	Con	5.0	5.48	60	Ahn et al. (1996)
C. thermocellum	Avicel	Batch	5.0	3.66	60	Ahn et al. (1996)
C. thermocellum	Whatman paper	Batch	8.0	7.20-8.00	60	Rani et al. (1997)
C. thermocellum	Avicel	Batch	8.0	6.50-7.20	60	Rani et al. (1997)
C. thermocellum	Paddy straw	Batch	8.0	6.10-8.00	60	Rani et al. (1997)
C. thermocellum	Sorghum stover	Batch	8.0	4.80-8.10	60	Rani et al. (1997)
C. thermocellum	Corn stubs	Batch	8.0	4.60-7.80	60	Rani et al. (1997)
Thermophilic strain A3	Xylan	Batch	10.0	5.43	70	Ahring et al. (1996)
T. saccharolyticum	Xylan	Batch	10.0	6.30	60	Ahring et al. (1996)
Thermophilic strain A3	Wheat straw	Batch	60.0 (10.0)*	2.61	70	Ahring et al. (1996)
T. mathranii	Wheat straw	Batch	60.0 (6.7)*	2.61	70	Ahring et al. (1999)
T. mathranii	Wheat straw	Batch	60.0	5.30	70	Klinke et al. (2001)
Several	Wheat straw	Batch	30.0	0.30-0.50	70	Sommer et al. (2004)
Several	Wheat straw	Batch	60.0	0.20-0.40	70	Sommer et al. (2004)
C. thermocellum	Filter paper/Corn steep liq.	Batch	45.0/8.0	8.18	60	Balusu et al. (2005)
T. ethanolicus	Beet molasses	Batch	40.0 (19.5)*	4.81	65	Avci et al. (2006)
T. thermohydrosulfuricus 70-1	Beet molasses	Batch	40.0 (19.5)*	2.95	65	Avci et al. (2006)
Thermoanaerobacter sp. 65-2	Beet molasses	Batch	40.0 (19.5)*	7.25	65	Avci et al. (2006)
Thermoanaerobacter BG1L1	Corn stover	Batch	25.0-150.0	8.50-9.20	70	Georgieva et al. (2007)
Thermoanaerobacter BG1L1	Wheat straw	Batch	30.0-120.0	8.50-9.20	70	Georgieva et al. (2008)
Thermoanaerobacter BG1L1	Corn stover	Con	25.0-150.0	8.50-9.20	70	Georgieva et al. (2008)
Clostridium thermocellum	Avicel	Batch	300-700**	0.70	60	Chinn et al. (2008)
T. ethanolicus	Been card HL	Batch	10.0	1.80	60	Miyazaki et al. (2008)
Clostridium sp.	Been card HL	Batch	10.0	0.85	60	Miyazaki et al. (2008)
Thermoanaerobacterium sp.	Been card HL	Batch	10.0	0.90	60	Miyazaki et al. (2008)
Thermoanaerobacterium AK17	Cellulose	Batch	7.5	5.81	60	Sveinsdottir et al. (2009)
Thermoanaerobacterium AK17	Grass	Batch	7.5	2.91	60	Sveinsdottir et al. (2009)
Thermoanaerobacterium AK17	Paper	Batch	7.5	2.03	60	Sveinsdottir et al. (2009)
Mixed	Napier grass	Batch	2.0-40.0	0.70-0.90	60	Lin et al. (2010)
Mixed	Avicel	Batch	2.0-40.0	0.40-5.70	60	Lin et al. (2010)
C. thermocellum	Napier grass	Batch	2.0-40.0	0.80-3.90	60	Lin et al. (2010)
C. thermocellum	Avicel	Batch	10.0	0.70	60	Lin et al. (2010)
Mixed (*C. thermocellum*)	Banana waste	Batch	10.0-100.0	5.50-9.20	60	Harish et al. (2010)

Table 2. EtOH production from lignocellulosic biomass by defined and mixed cultures of thermophilic bacteria. Cultivation was either in batch or continuous (con). EtOH yields given in mM/g substrate degraded as well as substrate concentrations and incubation temperature are also shown. * = sugar concentration, ** = 30 to 50% as hydrolysate.

Fermentation of beet molasses by three thermophilic *Thermoanaerobacter* species (*T. ethanolicus, Thermoanaerobacter* sp. and *T. thermohydrosulfuricus*) were recently investigated

(Avci et al., 2006). The concentration of sugars were 19.5 g L^{-1} and and fermentation resulted in yields between 3.0 (*T. thermohydrosulfuricus*) and 7.26 mM g^{-1} (*Thermoanaerobacter* sp.). The highest reported EtOH yields reported from complex biomass are by *Thermoanaerobacter* BG1L1 on corn stover and wheat straw (Georgieva & Ahring, 2007; Georgieva et al., 2008a). The biomass was pretreated with acid or wet oxidation and EtOH yields were up to 9.2 mM g^{-1} for biomass hydrolysates.

Studies on *Thermoanaerobacterium* sp and *Clostridium* sp. on been curd refuse hydrolysates were investigated by Miyazaki and co-workers (Miyazaki et al., 2008) with emphasis on cooperation between aerobic cellulose degrading *Geobacillus* with the anaerobes. EtOH yields in this study were relatively low, or between 0.72 to 1.80 mM g substrate^{-1}. Studies on EtOH production by *Thermoanaerobacterium* sp. AK17, isolated from Icelandic hot spring, on various types of lignocellulosic biomass were reported recently (Sveinsdottir et al., 2009). Batch culture studies on 7.5 g L^{-1} of cellulose, grass and newspaper, pretreated with heat and enzymes, showed EtOH yields of 2.0 (paper), 2.91 (grass) to 5.81 (cellulose) mM/g biomass. Optimization experiments were recently done on this strain where EtOH yields on grass and cellulose were increased to 4.0 and 8.6 mM g^{-1}, respectively. The main environmental factors concerning increasing EtOH yields were the use of acid/alkali for pretreatment and by lowering the substrate concentration from 7.5 to 2.5 g L^{-1} (unpublished results).

6. Production of H_2 from thermophilic bacteria

H_2 production from various organic materials by fermentation has been known for a long time. Firstly, the focus was mainly on facultative mesophilic bacteria within the genera of e.g. *Enterobacter, Citrobacter* and strict anaerobes like the typical acetate/butyrate fermentative *Clostridia*. There are numerous publications which focus on mesophilic bacteria that will not be dealt with in this paper. It has not been until relatively recently that H_2 production by thermophiles has gained increased interest and in the past three years there has been an explosion of number of publications within this field of research. Thermophilic bacteria have many advantages as compared to mesophiles concerning H_2 production, however, have remained less studied. High temperatures favor the stoichiometry of H_2 production resulting in higher H_2 yields compared to mesophilic systems (van Groenestijn et al., 2002; van Niel et al., 2003). Furthermore, thermophilic fermentation results in less variety of end products as compared to those of mesophilic fermentation (van Niel et al., 2003). The discussion below is divided into production of H_2 from sugars and from other biomass.

6.1 Production of H_2 from sugars

Pure cultures are, for the most part, used to study effects of environmental factors affecting commercial H_2 production. Several studies on H_2 production on sugars, using pure thermophilic cultures have been reported. The most common are dealing with bacteria belonging to the genera of *Thermoanaerobacterium, Caldicellulosiruptor* and *Thermotoga*. Table 3 summarizes studies using pure cultures for H_2 production from sugars.

Thermotoga neopolitana was first described by Jannasch and co-workers (1988) but earliest data of H_2 production is from 2002 where the bacterium produced 2.0 ml L^{-1} h^{-1} on glucose in batch cultures (van Ootegehem et al., 2002). H_2 production capacity from glucose by this species has since then been investigated in detail by others (Eriksen et al., 2008; d'Ippolito et al., 2008; Nguyen et al., 2008, 2010; Munro et al., 2009) showing yields between 1.84 to 3.85 mol H_2 mol glucose^{-1}. Xylose can also be used by the bacterium with good yields, or 2.20 mol H_2 mol xylose^{-1} (Nguyen et al., 2010b). Most studies reported on H_2 production by *T. neopolitana* have been conducted in batch experiments with relatively low sugar concentrations (5 to 7 g L^{-1}). The only experiment in continuous culture is reported by d'Ippolito et al., (2010) on glucose but very high yields were reported (3.85 mol H_2 mol glucose^{-1}). Other studies on species within the genus have been on *T. elfii* (van Niel et al., 2002) and *T. maritima* (Nguyen et al., 2008; Schröder et al., 1994) with H_2 yields varying from 1.67 to 4.00 (maximum) mol H_2 mol glucose^{-1}.

Organisms	Substrate	Cultivation method	Biomass conc. (g L^{-1})	Volumetric H_2 productivity (mL L^{-1} h^{-1})	H_2 yield (mol H_2 mol glu^{-1} equiv.)	Temp. (°C)	Reference
P. furiosus	Maltose	Con	0.22	5.5-22.0	2.90	98	Schicho et al. (1993)
T. maritima	Glucose	Batch	0.1	6.9	4.00	80	Schroder et al. (1994)
T. elfii	Glucose	Con	10.0	0.6	3.30	65	van Niel et al. (2002)
C. saccharolyticus	Sucrose	Con	10.0	0.6	3.30	70	van Niel et al. (2002)
T. neapolitana	Glucose	Batch	5.0	0.6	N/A	70	Van Ooteghem et al. (2002)
T. tengcongensis	Glucose	Con	4.5	N/A	4.00	75	Soboh et al. (2004)
C. saccharolyticus	Glucose	Batch	1.7	N/A	2.50	70	Kadar et al. (2004)
C. saccharolyticus	Xylose	Batch	1.6	11.3	2.70	70	Kadar et al. (2004)
C. saccharolyticus	Xyl/Glu	Batch	1.0	9.2	2.40	70	Kadar et al. (2004)
C. saccharolyticus	Glucose	Con	4.0	2.5	3.60	70	Vrije et al. (2007)
T. thermosaccharolyticum	sucrose	Batch	20.0	3.0	2.53	60	O-Thong et al. (2008)
T. thermosaccharolyticum	Glucose	Batch	10.0	1.6	2.42	60	Ren et al. (2008)
T. thermosaccharolyticum	Xylose	Batch	10.0	1.6	2.19	60	Ren et al. (2008)
T. neapolitana	Glucose	Batch	5.0	N/A	2.40	80	Eriksen et al. (2008)
T. neapolitana	Glucose	Batch	7.5	N/A	1.84	80	Nguyen et al. (2008a)
T. maritima	Glucose	Batch	7.5	N/A	1.67	80	Nguyen et al. (2008a)
T. neapolitana	Glucose	Batch	2.5	0.1	3.85	77	Munro et al. (2009)
C. thermocellum	Cellobiose	Batch	1.1	N/A	1.73	60	Levin et al. (2006)
C. saccharolyticus	Glucose	Con	10.0	N/A	3.00	70	Willquist et al. (2009)
T. neapolitana	Glucose	Batch	7.0	N/A	3.24	77	Nguyen et al. (2010b)
T. neapolitana	Xylose	Batch	4.0	N/A	2.20	77	Nguyen et al. (2010b)
T. thermosaccharolyticum	Xylose	Batch	12.2	N/A	2.37	60	Cao et al. (2010)
T. neapolitana	Glucose	Con	5.0	6.3	3.85	80	d'Ippolito et al. (2010)
C. owensensis	Glucose	Con	10.0	1.9	3.80	70	Zeidan & van Niel (2010)
C. owensensis	Xylose	Con	10.0	1.4	2.70	70	Zeidan & van Niel (2010)
C. thermolacticum	Lactose	Batch	10.0	N/A	1.80	58	Collet et al. (2003)
Clostridium AK14	Glucose	Batch	3.6	N/A	2.21	50	Almarsdottir et al. (2010)
Clostridium AK14	Xylose	Batch	3.0	N/A	2.55	50	Almarsdottir et al. (2010)

Table 3. H_2 production from sugars by pure cultures of thermophilic bacteria. Cultivation was either in batch or continuous (con). Volumetric H_2 production rates, H_2 yields as well as substrate concentrations and incubation temperature are also shown.

Species belonging to genus *Caldicellulosiruptor* have been intensively investigated for H_2 production. *C. saccharolyticus* grown on sucrose showed good yields in continuous culture, or 6.6 mol H_2 mol sucrose^{-1} (= 3.3 mol H_2 mol hexose^{-1}) (van Niel et al., 2002) and between 2.5 and 3.0 mol H_2 for one mole of xylose and glucose in batch (Kadar et al., 2004; Willquist et al., 2009). Higher yields were observed in continuous culture, or 3.6 as well as high H_2 production rates (Vrije et al., 2007). Recently *C. owensis* has also been shown to be a good H_2 producer both in continuous culture with H_2 yields of 3.8 and 2.7 from glucose and xylose, respectively (Zeidan & van Niel, 2010). Hydrogen production from glucose (4.5 g L^{-1}) in batch by *Thermoanaerobacter tengcongensis* has been investigated (Soboh et al., 2009). The culture was continuously flushed with N_2 to keep the partial pressure of H_2 low. This resulted in higher growth rates but due to high N_2 flushing rates H_2 could not be quantified. However, glucose was almost completely converted to acetate and since no external electron acceptor was added, it was assumed that 4.0 mol H_2 were formed per mol glucose degraded. Other thermophilic bacteria that have been investigated for H_2 production capacity are e.g. *Clostridium* sp. (Almarsdottir et al., 2010 ; Levin et al., 2006), *Thermoanerobacterium saccharolyticum* (Cao et al., 2010; Kadar et al., 2004) and *Pyrococcus furiosus* (Schicho et al., 1993).

In practice it may not be feasible to use pure cultures for H_2 production in large scale production facilities. Therefore, a more attention has recently been upon the use of mixed culture studies for H_2 production, often with sugars as model substrates.

Origin	Substrate	Cultivation method	Biomass conc. (g L^{-1})	Volumetric H_2 productivity (mL L^{-1} h^{-1})	H_2 yield (mol H_2 mol glu^{-1} equiv.)	Temp. (°C)	Reference
Mixed	Glucose	Con	4.9	N/A	2.47	70	Kotsopoulus et al. (2005)
Compost	Lactose	Fed-batch	2.0	N/A	3.70	55	Calli et al. (2008)
Compost	Xylose	Fed-batch	2.0	N/A	1.70	55	Calli et al. (2008)
Natural anaerobic mixed culture	Xylose	Batch	20.0	N/A	0.80	55	Lin et al. (2008)
Anaerobic culture fromhot spring	Glucose	Batch	4.5	N/A	1.16	52	Karadag et al. (2009)
Household solid waste	Xylose	Batch	0.5	N/A	1.62	70	Kongjan et al. (2009)
Household solid waste	Xylose	Con	1.0	2.6	1.61	70	Kongjan et al. (2009)
Cow manure	Glucose	Con	5.0	50.8	3.32	75	Yokoyama et al. (2009)
Mixed	Xylose	Con	6.0	3.4	2.60	70	Zeidan et al. (2010)
Mixed	Glucose	Batch	2.0	N/A	1.58	70	Zhao et al. (2009)
Mixed	Xylose	Batch	2.0	N/A	1.84	70	Zhao et al. (2010)
Sediments-rich samples from hot springs	Glucose	Batch	10.0	N/A	1.71	60	Hniman et al. (2010)
Sediments-rich samples from hot springs	Xylose	Batch	10.0	N/A	1.57	60	Hniman et al. (2010)
Anaerobic culture from hot spring	Glucose	Con	9.0	N/A	1.10	37	Karadag & Puhakka (2010)
Enrichment cultures from hot springs	Glucose	Batch	18.0	N/A	2.10	59	Koskinen et al. (2008a)
Mixed	Glucose	Con	3.6	6.1	0.80	60	Koskinen et al. (2008b)
Enichment culture from hot spring	Glucose	Batch	5.9	N/A	3.20	60	Koskinen et al. (2008c)
Enichment culture from hot spring	Glucose	Con	18.0	N/A	2.74	58	Koskinen et al. (2008c)

Table 4. H_2 production from sugars by mixed cultures of thermophilic bacteria. Cultivation was either in batch or continuous (con). Volumetric H_2 production rates, H_2 yields as well as substrate concentrations and incubation temperature are also shown.

The origin of bacteria used in such studies are from e.g. compost, hot springs, manure or anaerobic digestion systems (Calli et al., 2008; Hniman et al., 2010; Karadag et al., 2009; Karadag & Puhakka, 2010; Lin et al., 2008; Zhao et al., 2009; Zhao et al., 2010). Available data from such experiments are presented in Table 4. Although the yields of H_2 production are usually lower as compared to pure culture studies, very high yields have indeed been obtained. An example of this is from the study of xylose and lactose, fed batch fermentation with bacteria from compost. Yields on lactose were 3.70 mol H_2 mol $lactose^{-1}$ (Calli et al., 2008). Glucose fermentation in continuous culture with bacteria from manure resulted in 3.32 mol H_2 mol $glucose^{-1}$ (Yokoyama et al., 2009). Enrichment culture from Icelandic geothermal hot spring produced H_2 of up to 3.20 mol H_2 mol $glucose^{-1}$ in batch assay (Koskinen et al., 2008c). A continuous culture study showed H_2 yields of 2.74 mol H_2 mol $glucose^{-1}$. The enrichment culture was dominated by strains closely affiliated with *Thermobrachium celere.*

6.2 Production of H_2 from complex biomass

Available data on H_2 production from complex biomass has exploded in the last three years. Complex biomass, such as food waste and lignocellulosic agricultural residues has been used for thermophilic biohydrogen production in both laboratory and pilot scale. The discussion below will be divided according to H_2 production from different types of biomass.

6.2.1 Agricultural wastes and energy crops

Several studies have been done with various corn straw as substrate both in pure (Ivanova et al., 2009) and mixed (Kongjan & Angelidaki, 2010; Kongjan et al., 2010) cultures. Mixed cultures, originating from methanogenic sludge from a potato factory were used in continuous cultures (UASB, CSTR, AF) with hemicellulose rich wheat straw (Kongjan & Angelidaki, 2010). The highest H_2 production yields of 9.5 mmol H_2 g sugar^{-1} (1.70 mol H_2 mol glucose^{-1}) was achieved in the UASB reactor. The reactors were fed with hydrolysates that contained 4.4% (TS), mainly xylose. The hydrolysate prepared with hydrothermal pretreatment was diluted prior to inoculation to 25% (v/v). The main conclusion from this study was that reactor configuration is of great importance for enhancing and stabilizing H_2 production. In another study on this substrate the focus was on the importance of hydrolysate concentrations (Kongjan et al., 2010). High hydrolysate concentrations strongly inhibited H_2 production. Batch culture trials on 5% hydrolysate concentrations showed highest yield or 14.1 mmol H_2 g sugar^{-1} (2.55 mol H_2 mol hexose^{-1} equivalent) but CSTR-reactor that ran on 20% HL showed considerable lower yields or 7.9 mmol H_2 mol sugar^{-1} (1.43 mol H_2 mol glucose^{-1} equivalent). Phylogenetic analysis of the mixed cultures showed presence of *Caldanaerobacter subterraneus, Thermoanaerobacter subterraneus* and *Thermoanaerobacterium thermosaccharolyticum.*

Caldicellulosiruptor saccharolyticus has been used for H_2 production from hemicellulose-rich pine tree wood shavings, maize leaves, wheat straw, sugarcane bagasse and the sweet sorghum bagasse without chemical pretreatment in batch (Ivanova et al., 2009). The highest yields of 3.8 mol H_2 mol glucose $^{-1}$ equivalents was achived from wheat straw hydrolysates. The maize leaves were used both unpretreated and pretreated with cellulase-producing aerobic bacteria, *Bacillus amyloliquefaciens.* The pretreatment greatly improved the H_2 yields. Unpretreated maize leaves yielded 1.53 mol H_2 mol glucose^{-1} while pretreated leaves yielded 3.67 mol H_2 mol glucose^{-1}. Lower yields were obtained from other biomass. *Thermotoga neapolitana* produced 2.3 to 2.7 mmol H_2 g korean rice straw^{-1} (0.41 to 0.49 mol H_2 mol hexose^{-1} equivalent) from untreated and thermally ammonia or dilute sulfuric acid pretreatment, respectively (Nguyen et al., 2010b). Fermentation of hydrolysates from Miscanthus hydrolysates by *Caldicellulosiruptor saccharolyticus* and *Thermotoga elfi,* pretreated by alkali, resulted in 3.4 and 3.2 mol H_2 mol glucose^{-1} equivalent, respectively (de Vrije et al., 2009).

Corn stover and corn stover cornstalk have been investigated for H_2 production capacity by many (Cao et al., 2009; Datar et al., 2007; Liu et al., 2008b; Liu & Cheng, 2010; Ren et al., 2010). Pure culture studies on *Thermoanaerobacterium thermosaccharolyticum* on corn stover hydrolysates showed maximum of 2.7 mol H_2 mol glucose^{-1} equivalent diluted corn stover hydrolysates that contained a mixture of glucose, xylose and arabinose (total sugar concentration, 10 g L^{-1}) (Ren et al., 2010). Pretreatment consisted of mincing with hammer mill, drying and enzymatic hydrolysis. The bacterium showed classical acetate/butyrate fermentation and yields were similar as on equal amounts of pure sugars. Earlier reports on the production capacity of this bacterium on corn stover pretreated with acid showed similar yields, or 2.24 mol H_2 mol glucose^{-1} (Cao et al., 2009). From a study of Liu and Cheng (2010), corn stover was pretreated with microwave assisted strategy and the resulting biomass hydrolysate fermented with mixed thermophilic microflora from a anaerobic digester. H_2 production capacity was however modest, or 1.53 mol H_2 mol glucose^{-1} equivalents

Culture	Feedstock	Cultivation method	Biomass conc. (g L^{-1})	Volumetric H_2 productivity (mL L^{-1} h^{-1})	H_2 yield (mol H_2 mol glu^{-1} equiv.)	Temp. (°C)	Reference
Mixed	Cellulose wastewater	Batch	5.0	ND	0.82	55	Liu et al. (2003)
C. saccharolyticus	Paper sludge	Batch	8.4	91.8	3.70	70	Kadar et al. (2004)
C. thermocellum	Delignified wood fibers	Batch	0.1-4.5	ND	1.00-2.30	60	Levin et al. (2006)
C. thermocellum 27405	Cellulose	Batch	0.1-4.5	ND	0.80-2.00	60	Levin et al. (2006)
C. thermocellum 27405	Whatman paper	Batch	0.1-4.5	ND	0.80-1.90	60	Levin et al. (2006)
Thermotoga neapolitana	Microcrystalline cellulose	Batch	5.0	ND	1.00-2.20	80	Nguyen et al. (2008b)
C. thermocellum	Dried distillers grain	Batch	5.0	5.1	1.27	60	Magnusson et al. (2008)
C. thermocellum	Barley hulls	Batch	5.0	2.0	1.24	60	Magnusson et al. (2008)
C. thermocellum	Cellulose	Batch	1.1	5.1	0.76	60	Magnusson et al. (2008)
C. thermocellum	Contaminated barley hulls	Batch	5.0	5.4	1.18	60	Magnusson et al. (2008)
Coculture	Cellulose	Batch	5.0	ND	1.80	60	Liu et al. (2008b)
T. thermosaccharolyticum	Corn stover	Batch	6.4-12.2	ND	2.24	60	Cao et al. (2009b)
T. thermosaccharolyticum	Miscanthus hydrolysate	Batch	10.0	282.2	3.40	72	Vrije et al. (2009)
Thermotoga neapolitana	Miscanthus hydrolysate	Batch	14.0	275.5	3.20	80	Vrije et al. (2009)
Mixed	Napier grass	Batch	10.0	ND	1.20	55	Lo et al. (2009)
Coculture	Cellulose (filter paper)	Batch	9.0	ND	1.36	55	Geng et al. (2010)
C. saccharolyticus	Wheat straw	Batch	20.0	ND	3.80	70	Ivanova et al. (2009)
C. saccharolyticus	Sweet sorghum plant	Batch	30.0	ND	1.75	70	Ivanova et al. (2009)
C. saccharolyticus	Sugarcane bagasse	Batch	15.0	ND	2.30	70	Ivanova et al. (2009)
C. saccharolyticus	Maize leaves	Batch	8.0	ND	3.67	70	Ivanova et al. (2009)
Mixed	Oil palm trunk hydrolysate	Batch	10.0	ND	1.94	60	Hniman et al. (2010)
Mixed	Corn stover	Batch	13.3	ND	1.53	55	Liu & Cheng (2010)
Clostridium AK14	Cellulose	Batch	5.0	ND	1.10-1.20	50	Almarsdottir et al. (2010)
Clostridium AK14	Hemp stem	Batch	5.0	ND	0.60-0.70	50	Almarsdottir et al. (2010)
Clostridium AK14	Hemp leaf	Batch	5.0	ND	0.20-0.40	50	Almarsdottir et al. (2010)
Clostridium AK14	Grass	Batch	5.0	ND	0.80-0.90	50	Almarsdottir et al. (2010)
Clostridium AK14	Paper	Batch	5.0	ND	0.10-0.40	50	Almarsdottir et al. (2010)
Clostridium AK14	Barley straw	Batch	5.0	ND	0.70-0.80	50	Almarsdottir et al. (2010)
Mixed	Wheat straw	Con	3.9*	34.2	1.70	70	Kongjan & Angelidaki (2010)
Mixed	Wheat straw	Con	3.9*	10.1	1.51	70	Kongjan & Angelidaki (2010)
Mixed	Wheat straw	Con	3.9*	20.6	1.00	70	Kongjan & Angelidaki (2010)
Mixed	Wheat straw	Batch	0.8-3.9*	ND	1.20-2.60	70	Kongjan et al. (2010)
Mixed	Wheat straw	Con	3.1*	7.7	1.42	70	Kongjan et al. (2010)
Mixed	Wheat straw	Batch	50.0	ND	2.54	70	Kongjan et al. (2010)
T. thermosaccharolyticum W16	Corn stover	Batch	10*	250.9	2.70	60	Ren et al. (2010)
Thermotoga neapolitana	Korean rice straw	Batch	10*	31.8	0.41	75	Nguyen et al. (2010b)
Thermotoga neapolitana	Korean rice straw	Batch	10.0	112.4	0.49	75	Nguyen et al. (2010b)

Table 5. H_2 production from agricultural wastes and energy crops. Cultivation was either in batch or continuous (con). Volumetric H_2 production rates, H_2 yields as well as substrate concentrations and incubation temperature are also shown. * = concentrations of sugars.

A coculture of *Clostridium thermocellum* and *Thermoanaerobacterium thermosaccharolyticum* grown on hydrolysate made from 5 g L^{-1} of corn stalk and corn cob powder (no pretreatment), resulted in 1.80 mol H_2 mol glucose^{-1} (Liu et al., 2008b). *Clostridium* AK14 was used to degrade hemp (both stem and leaf), grass, paper and straw (Almarsdottir et al., 2010). Highest yields were observed on grass pretreated with 0.75% sulfuric acid and enzymes, or 6.23 mol H_2 g VS^{-1}. Pretreatment with either alkali or acid increased H_2 in most cases substantially.

Several studies of H_2 production form cellulose have been conducted (Almarsdottir et al., 2010; Geng et al., 2010; Levin et al., 2006; Liu et al., 2003; Liu et al., 2008b; Nguyen et al., 2008). Various sources of cellulose have been used, e.g. wastewater (Liu et al., 2003) Whatman filter paper (Almarsdottir et al., 2010; Geng et al., 2008), microcrystalline cellulose (Liu et al., 2008b; Nguyen et al., 2008b). Hydrogen yields from these studies (all batch) varied from 0.95 to 2.32 mol H_2 mol glucose^{-1} equivalent. In some studies the focus was on different pretreatment methods used. Studies with pure cultures of *Clostridium* AK14, a moderate thermophilic bacterium showed similar results (1.17 mol H_2 mol glucose^{-1} equivalent) from Whatman paper whether it was only enzymatically pretreated or pretreated with both enzymes and weak acid or alkali (Almarsdottir et al., 2010). Hydrogen production from microcrystalline cellulose by *Thermotoga neapolitana* increased however from 1.59 to 2.20 mol H_2 mol glucose^{-1} equivalent by using ionic liquid pretreatment (Nguyen et al., 2008b). The influence of substrate concentrations on H_2 yields from degradation of cellulosic substrates by *Clostridium thermocellum* were investigated by Levin et al., (2006). Highest yields were observed on delignified wood fibers at 0.1 g L^{-1}, 2.32 mol H_2 mol glucose^{-1} equivalents. At 4.5 g L^{-1} yields dramatically decreased to less than 1 mol H_2 mol glucose^{-1}. Other reports on H_2 production from lignocellulosic biomass presented in Table 5 include studies on paper sludge (Kadar et al., 2004), oil palm trunk hydrolysate (Hniman et al., 2010), Napier grass (Lo et al., 2009) and barley hulls (Magnusson et al., 2008) and are not discussed in detail in this overview.

6.2.2 Starch and mixed biomass

Several studies of H_2 production from starch have been done, both with pure soluble starch and a starch based biomass. Akutsu and co-workers used mixed cultures from five different kinds of sludge as inocula to produce hydrogen from starch in CSTR-reactors without any pretreatment (Akutsu et al., 2008). The highest H_2 production yields (2.30 mol H_2 mol glucose^{-1} equivalent) was obtained with thermophilically digested waste activated sludge as inocula. Phylogenetic analysis showed the presence of *Thermoanaerobacterium* in all reactors. *Janthinobacterium* and aerobic bacteria of the genus *Flavobacterium* were also detected. Two other studies by Akutsu and co-workers focused on the effects of different factors on H_2 production from starch (Akutsu et al., 2009a, 2009b). In the first study (Akutsu et al., 2009b) the effects of substrate concentrations (10-70 g L^{-1}) on H_2 production were investigated in continuous cultures using a mixed culture originating from thermophilic acidogenic sludge treating potato waste. The H_2 yields varied from 1.84 to 2.82 mol H_2 mol glucose^{-1} at 70 and 20 g L^{-1} substrate concentrations, respectively. The maximum H_2 production rate was 182 ml $L^{-1}h^{-1}$. In the other study (Akutsu et al., 2009a), the effects of hydrolic retention time, pH and substrate concentrations were further investigated. Hydrogen production rate was gradually increased from 62 to 167 ml H_2 $L^{-1}h^{-1}$ by lowering the HRT from 40 h to 6h but on the other hand, maximum H_2 yields were obtained at 48 h HRT, or 1.68 mol H_2 mol glucose^{-1} equivalent. Additionally, H_2 production diminished greatly when pH was higher than 6.0 or lower than 4.7 indicating the importance of pH for H_2 production (Akutsu et al., 2009a). Study of starch degradation and H_2 production in repeated batch by extreme mixed cultures, originating from cow manure showed H_2 yields of 1.73 mol H_2 mol glucose^{-1} (Yokoyama et al., 2007). The main emphasis was on the phylogenetic analysis of the microbiological community and presence of various *Caldanaerobacter* species was observed.

Culture	Feedstock	Cultivation method	Biomass conc. (g L^{-1})	Volumetric H_2 productivity (mL L^{-1} h^{-1})	H_2 yield (mol H_2 mol glu^{-1} equiv.)	Temp. (°C)	Reference
T. kodakaraensis	Starch	Con	0.5	6.7	3.30	85	Kanai et al. (2005)
Mixed	Food waste	Con	6.9	ND	2.50-2.80	55	Chu et al. (2008)
Mixed	Soluble starch	Batch*	6.3	ND	1.73	75	Yokoyama et al. (2007)
Mixed	Starch	Con	10.0	42.4-70.8	1.40-2.30	55	Akutsu et al. (2008)
C. saccharolyticus	Sweet sorghum	Batch	2.0	nd	2.63	72	Ivanova et al. (2009)
Mixed	Starch	Con	60.0	nd	1.68	55	Akutsu et al. (2009a)
Mixed	Starch	Con	15.0-70.0	nd	1.84-2.82	55	Akutsu et al. (2009b)
Mixed	Wheat starch	Batch	20.0	7.4	2.40	55	Cakir et al. (2010)
T. neapolitana	Algal starch	Batch	5.0	44.6-227.0	1.80-2.50	75	Nguyen et al. (2010c)
C. saccharolyticus	Carrot pulp	Batch	10.0	351.7	2.80	72	Vrije et al. (2010)
T. neapolitana	Carrot pulp	Batch	10.0	280.0	2.70	80	Vrije et al. (2010)
Mixed	Rice winery wastewater	Con	10.0 (COD)	158.3	2.14	55	Yu et al. (2002b)
Mixed	Food waste	Con	25.0 (sugars)	ND	0.60-1.80	55	Shin et al. (2004)
Mixed	Food waste	Con	14.1 (VSS)	16.7-41.7	1.00-2.40	55	Shin & Youn (2005)
Mixed	POME	Batch	85.0 (COD)	24.2	2.53	60	O-Thong et al. (2008)
Mixed	Household solid waste	Batch	0.5	ND	0.30-2.00	70	Liu et al. (2008a)
Mixed	Household solid waste	Batch*	10.0 (VS)	ND	0.82	70	Liu et al. (2008b)
Mixed	Kitchen waste	Batch	23.7 (VSS)	ND	0.88	55	Lee et al. (2008)
Mixed	Cheese whey (lactose rich)	Con	Variable	12.5 - 329.1	ND	55	Azbar et al. (2009)
Mixed	Cheese whey wastewater	Batch	21.3	ND	1.55	55	Azbar et al. (2009)
Mixed	Pig slurry	Con	45.0 (TS)	3.8	ND	70	Kotsopoulos et al. (2009)
Mixed	Kitchen waste	Con	60.5	66.7	0.23	55	Wang et al. (2009)
Mixed	POME	Con	7.0-8.4 (VSS)	379.2	2.17	60	Prasertsan et al. (2009)
Mixed	Crude Palm Oil + sucrose	Batch	24.0	ND	2.50	55	Ismail et al. (2009)
Mixed	Vegetable kitchen waste	Con*	10.0	41.7	1.70	55	Lee et al. (2010)

Table 6. H_2 production from starch and mixed biomass. Cultivation was either in batch or continuous (con). Volumetric H_2 production rates, hydrogen yields as well as substrate concentrations and incubation temperature are also shown.
* = Repeated batch, ** = Semicontinuous

Cakir and co-workers compared hydrogen production from ground wheat starch under mesophilic (37°C) and thermophilic conditions (55°C) with mixed microflora from a heat-treated anaerobic sludge (Cakir et al., 2010). The starch was pretreated with sulfuric acid and heat in order to convert it to soluble sugars. The highest H_2 yield was 2.40 mol H_2 mol $glucose^{-1}$, obtained under the thermophilic conditions. The hyperthermophilic *Thermotoga neapolitana* was used by to produce hydrogen from green algal biomass (Nguyen et al., 2010c). Starch is a major accumulated constituent of algal biomass and therefore makes a good potential feedstock for both EtOH and H_2 production. Two different pretreatments were used to disrupt the algal cell wall (sonication and MeOH exposure) and two other to improve starch conversion to H_2 (HCl + heat and enzymes). All methods gave good effect on H_2 production but the highest H_2 yield (2.5 mol H_2 mol $glucose^{-1}$) was obtained with enzymatic hydrolysis (Nguyen et al., 2010c).

Many types of different food waste biomass have been used to produce H_2, almost exclusively with mixed cultures from various seed sludge. Lee and co-workers have done two different studies on H_2 production from high vegetable kitchen waste (Lee et al., 2008 and Lee et al., 2010). No pretreatment was used in either study. In the first study, a series of batch fermentation tests were conducted at four different pH levels to observe the effects of pH on the H_2 production. Hydrogen yields from different pH levels were all similar, the highest obtained at pH 7.0 (0.49 mmol H_2 g COD^{-1}) except for pH 5.5 (the lowest pH level), where there was no H_2 production at all (Lee et al., 2008). The main bacteria present belong to the genus *Clostridium*. In the other investigation much higher yields were obtained, or 1.7 mmol H_2 g COD^{-1} and the predominant species was closely affiliated to *Thermoanaerobacterium thermosaccharolyticum* (Lee et al., 2010). Recent study of H_2 production from kitchen waste with mixed cultures from various sources showed good production rates (66.7 ml L^{-1} h^{-1}) but much lower yields (0.23 mol H_2 mol glucose^{-1} equivalent) (Wang et al., 2009). A continuous culture study on H_2 production from food waste by the use of mixed culture originating from anaerobic waste water treatment plant resulted in maximum of 2.8 mol H_2 mol hexose^{-1} (Chu et al., 2008). Other studies with food waste include e.g. continuous culture (CSTR) studies by Shin et al., (2004) and Shin &Youn (2005) at sugar concentration of 25 g L^{-1}. Clearly the effects of substrate concentrations are important but higest yields (1.8 mol H_2 mol hexose^{-1}) were obtained at 8 g VS/L (Shin et al., 2004). Maximum H_2 production rate and yield occurred at 8 g VSL^{-1} d^{-1}, 5 days HRT and pH 5.5 (Shin & Youn, 2005). Hydrogen production from household solid waste by using extreme-thermophilic (70°C) mixed culture resulted in 2 mol H_2 mol hexose^{-1} (Liu et al., 2008a) and 0.82 mol H_2 mol hexose^{-1} (Liu et al., 2008b).

Other studies on various mixed substrates include pig slurry (Kotsopoulous et al., 2009), rice winery wastewater (Yu et al., 2002), palm oil effluent (POME) (Ismail et al., 2010; O'Thong et al., 2008; Prasertsan et al., 2009), and cheese whey (Azbar et al., 2009a, 2009b), and are presented in Table 6. Fewer studies have been done using pure microbial cultures producing H_2 from complex biomass. *Caldicellulosiruptor saccharolyticus* and *Thermotoga neapolitana* showed good H_2 yields from carrot pulp hydrolysate, or 2.8 and 2.7 mol H_2 mol hexose^{-1}, respectively (de Vrije et al., 2010). *Thermococcus kodakaraensis* KOD1 showed very high H_2 yields on starch (3.3 mol H_2 mol hexose^{-1}) in continuous culture in a gas lift fermentor with dilution rate of 0.2 h^{-1} (Kanai et al., 2005).

7. Pros and cons of using thermophiles for biofuel production

The use of thermophilic bacteria for production of H_2 and EtOH has several pros and cons compared to the use of mesophilic bacteria, phototrophic bacteria and yeasts. It is possible to compare the use of different microorganisms by looking at several factors of both practical and economical point of view. Historically, yeasts have been and still are, the microorganisms most widely used for EtOH production from homogenous material like sucrose and glucose. The main reason for this are e.g. very high yields, few end products and high EtOH tolerance. However, wild type yeasts do not have degradation genes for pentose and polymer degradation and genetic engineering studies have not yet delivered stable organisms for large scale production. The main benefits of using bacteria for biofuel production is their broad substrate spectrum and they may therefore be a better choice for EtOH production from more complex biomass e.g. agricultural wastes (Taylor et al., 2008). The main drawback of the use bacteria for biofuel production is their low EtOH tolerance and more diverse end product formation. This is the main reason for no commercialized large scale plants have been built yet. Thermophilic bacteria are often very tolerant towards various environmental extremes. Apart from growing at higher temperatures, often with higher growth rates, many are acid and salt tolerant which may be of importance when various mixed substrates are used. In general bacteria tolerate lower EtOH concentrations as

compared to yeasts and elevated substrate concentrations may inhibit growth. This may possible be solved by either using fed batch or continuous cultures or by „self distillation" of EtOH.
H_2 production by mesophilic bacteria has been known for a long time. The main drawback of using mesophilic bacteria is the fact that H_2 production is inhibited at relatively low partial pressures of H_2 resulting in a change of carbon flow away from acetate (and H_2) towards e.g. EtOH and lactate. Extremophilic bacteria are less phroned towards this inhibition and much higher H_2 concentrations are needed before a change in the carbon flow occurs. H_2 production by photosynthesis has gained increased interest lately but H_2 production rates are much slower as compared to bacteria and a need for large and expensive reactors inhibit its practical use. Additionally, fermentation is not dependent on light and can be runned continuously.
Furfural and hydroxymethylfurfural (HMF) are furan derivatives from pentoses and hexoses, respectively and are among the most potent inhibitory compounds generated from acid hydrolysis of lignocellulosic biomass. Most microorgansisms are more sensitive to furfural than HMF but usually inhibition occurs at concentrations above 1 g L^{-1}. Sensitivity of thermophilic bacteria towards these compounds seem to be similar as compared to yeast (de Vrije et al., 2009; Cao et al., 2010).

8. Genetic engineering of thermophiles – state of the art

The main hindrance of using thermophilic bacteria is low tolerance to EtOH and the production of other end products like acetate and lactate. Several efforts have been done to enhance EtOH tolerance for thermophiles. Most of these studies were performed by mutations and adaptation to increased EtOH concentrations (Lovitt et al., 1984,1988; Georgieva et al., 1988) and has already been discussed. Elimination of catabolic pathways leading to other end products by genetic engineering has only got attention in the past few years.
The first report on genetic engineering on thermophilic bacteria to increase biofuel production is on *Thermoanaerbacterium saccharolyticum* (Desai et al., 2004). The L-lactate dehydrogenase (LDH) was knocked out leading to increased EtOH and acetate production on both glucose and xylose and total elimination of lactate production. The wild type strain produced 8.1 and 1.8 mM of lactate from 5 g L^{-1} of glucose and xylose, respectively. Later study of the same species resulted in elimination of all acid formation and generation of homoethanolic strain. This strain uses pyruvate:ferredoxin oxidoreductase to convert pyruvate to EtOH with electron transfer from ferredoxin to NAD(P) but this is unknown by any other homoethanolgenic microbes who use pyruvate decarboxylase. The strain produces 37g L^{-1} of EtOH which is the highest yields reported so far for a thermophilic anaerobe (Shaw et al., 2008).
Two *Geobacillus thermoglucosidasius* strains producing mixed acids from sugar fermentation with relatively low EtOH yields were recently genetically engineered to increase yields (Cripps et al., 2009). The authors developed an integration vector system that led to the generation of stable gene knockouts but the wild type strains had shown problems of genetic instability. They inactivated lactate dehydrogenase and to deal with the excess carbon flux they upregulated the expression of PDH (pyruvate dehydrogenase) to make it the sole fermentation pathway. One of their mutants (TM242) produced EtOH from glucose at more than 90% of the maximum theoretical yields (Cripps et al., 2009).

A strain of *Thermoanaerobacter mathranii* was genetically engineered to improve the EtOH production (Yao & Mikkelsen, 2010). A strain that had already had the *ldh* gene deleted to eliminate an NADH oxidation pathway (Yao & Mikkelsen, 2010) was used. The results obtained indicated that using a more reduced substrate such as mannitol, shifted the carbon balance towards more reduced end products like EtOH. In order to do that without having to use mannitol as a substrate they expressed an NAD^+-dependent GLDH (glycerol dehydrogenase) in this bacterium.

A possible approach to increase H_2 yields is to convert more of the substrate to H_2 by altering metabolism by genetic engineering. Studies on either maximizing yields of existing pathways or metabolic engineering of new pathways have been published (Hallenbeck & Gosh, 2010). Genetic manipulation and metabolic flux analysis are well developed and have been suggested to be applied to biohydrogen (Hallenbeck & Benemann, 2002; Vignais et al., 2006). However, no study on genetic engineering on thermophilic bacteria considering H_2 production has been published to our knowledge. So far, the main emphasis has been on the mesophilic bacteria *E.coli* and *Clostridium* species.

Fermentative bacteria often possess several different hydrogenases that can operate in either proton reduction or H_2 oxidation (Hallenbeck & Benemann, 2002). Logically, inactivation of H_2 oxidation would increase H_2 yields. This has been shown for *E. coli* where elimination of *hyd*1 and *hyd*2 led to a 37% increase in H_2 yield compared to the wild type strain (Bisaillon et al., 2006).

Studies on metabolically engineering Clostridia to increase H_2 production have been published. One study showed that by decreasing acetate formation by inactivate *ack* in *Clostridium tyrobutyricum*, 1.5-fold enhancement in H_2 production was observed; yields from glucose increased from 1.4 mol H_2-mol glucose^{-1} to 2.2 mol H_2-mol glucose^{-1} (Liu et al., 2006).

9. Conclusion

Many bacteria within the genera *Clostridium, Thermoanaerobacter, Thermoanaerobacterium, Caldicellulosiruptor* and *Thermotoga* are good H_2 and/or EtOH producers. Species within *Clostridium* and *Caldicellulosiruptor* are of special interest because of their ability to degrade cellulose and hemicelluloses. Highest EtOH yields on sugars and lignocelluloses hydrolysates are 1.9 mol EtOH mol glucose^{-1} and 9.2 mM g biomass^{-1} (corn stover and wheat straw) by *Thermoanaerobacter thermohydrosulfuricus* and *Thermoanaerobacter* species, respectively. Highest H_2 yields on sugars and lignocelluloses hydrolysates are 4 mol H_2 mol glucose^{-1} and 3.7 mol H_2 mol glucose^{-1} equivalent (from wheat straw) by *Thermotoga maritima* and *Caldicellulosiruptor saccharolyticus*, respectively. Clearly many bacteria within these genera have great potential for EtOH and hydrogen production, especially from complex lignocellulosic biomass. Recent information in genome studies of thermoanaerobes has led to experiments where *Thermonanaerobacterium* and *Thermoanaerobacter* species have been genetically engineered to make them homoethanolgenic. Thus, the greatest drawback of using thermophilic bacteria for biofuel production, their mixed end product formation, can be eliminated but it remains to see if these strains will be stable for upscaling processes.

10. Acknowledgement

This work was sponsored by the Nordic Energy Research fund (BioH2; 06-Hydr-C13), The Icelandic Research fund (BioEthanol; 081303408), The Technological Development and Innovation Fund (BioFuel; RAN091016-2376).

11. References

Ahn, H.J. & Lynd, L.R. 1996. Cellulose degradation and ethanol production by thermophilic bacteria using mineral growth medium. *Applied Biochemistry and Biotechnology,* 57: 599-604.

Ahring, B.K.; Jensen, K.; Nielsen, P.; Bjerre, A. B. & Schmidt, A.S. 1996. Pretreatment of wheat straw and conversion of xylose and xylan to ethanol by thermophilic anaerobic bacteria. *Bioresource Technology,* 58: 107-113.

Ahring, B.K.; Licht, D.; Schmidt, A.S.; Sommer, P. & Thomsen, A.B. 1999. Production of ethanol from wet oxidised wheat straw by *Thermoanaerobacter mathranii. Bioresource Technology,* 68: 3-9.

Akutsu, Y.; Lee, D.-Y.; Chi, Y.-Z.; Li, Y.-Y.; Harada, H. & Yu, H.-Q. 2009a. Thermophilic fermentative hydrogen production from starch-wastewater with bio-granules. *International Journal of Hydrogen Energy,* 34: 5061-5071.

Akutsu, Y.; Li, Y.; Harada, H. & Yu, H. 2009b. Effects of temperature and substrate concentration on biological hydrogen production from starch. *International Journal of Hydrogen Energy,* 34: 2558-2566.

Akutsu, Y.; Li, Y.; Tandukar, M.; Kubota, K. & Harada, H. 2008. Effects of seed sludge on fermentative characteristics and microbial community structures in thermophilic hydrogen fermentation of starch. *International Journal of Hydrogen Energy,* 33: 6541-6548.

Almarsdottir, A.R.; Taraceviz, A.; Gunnarsson, I. & Orlygsson, J. 2010. Hydrogen production from sugars and complex biomass by *Clostridium* species, AK14, isolated from Icelandic hot spring. *Icelandic Agricultural Sciences,* 23: 61-71.

Alvira, P.; Tomas-Pejo, E.; Ballesteros, M. & Negro, M. J. 2010. Pretreatment technologies for an efficient bioethanol production process based on enzymatic hydrolysis: A review. *Bioresource Technology,* 101: 4851-4861.

Amend & Shock. 2001. Energetics of overall metabolic reactions of thermophilic and hyperthermophilic Archea and Bacteria. *FEMS Microbiology Reviews,*25: 175-243.

Avci, A. & Donmez, S. 2006. Effect of zinc on ethanol production by two thermoanaerobacter strains. *Process Biochemistry,* 41: 984-989.

Azbar, N.; Dokgoez, F.T.; Keskin, T.; Eltem, R.; Korkmaz, K.S.; Gezgin, Y.; Akbal, Z.; Oencel, S.; Dalay, M.C.; Goenen, C. & Tutuk, F. 2009a. Comparative evaluation of bio-hydrogen production from cheese whey wastewater under thermophilic and mesophilic anaerobic conditions. *International Journal of Green Energy,* 6: 192-200.

Azbar, N.; Dokgoz, F.T.C.; Keskin, T.; Korkmaz, K.S. & Syed, H.M. 2009b. Continuous fermentative hydrogen production from cheese whey wastewater under thermophilic anaerobic conditions. *International Journal of Hydrogen Energy,* 34: 7441-7447.

Balk, M.; Weijma, J. & Stams, A.J.M. 2002. *Thermotoga lettingae* sp nov., a novel thermophilic, methanol-degrading bacterium isolated from a thermophilic anaerobic reactor. *International Journal of Systematic and Evolutionary Microbiology,* 52: 1361-1368.

Balusu, R.; Paduru, R.M.R.; Seenayya, G. & Reddy, G. 2004. Production of ethanol from cellulosic biomass by *Clostridium thermocellum* SS19 in submerged fermentation - screening of nutrients using plackett-burman design. *Applied Biochemistry and Biotechnology,* 117: 133-141.

Balusu, R.; Paduru, R.R.; Kuravi, S.K.; Seenayya, G. & Reddy, G. 2005. Optimization of critical medium components using response surface methodology for ethanol production from cellulosic biomass by *Clostridium thermocellum* SS19. *Process Biochemistry,* 40: 3025-3030.

Bisaillon, A.; Turcot, J. & Hallenbeck, P.C. 2006. The effect of nutrient limitation on hydrogen production by batch cultures of *Escherichia coli. International Journal of Hydrogen Energy,* 30: 1504-1508.

Baskaran, S., Hogsett, D.A L. & Lynd, L.R. 1994. Ethanol-production using thermophilic bacteria - growth-medium formulation and product tolerances. *Abstracts of Papers of the American Chemical Society, 207,* 172-BTEC.

Ben-Bassat, A.; Lamed, R. & Zeikus, J.G. 1981. Ethanol-production by thermophilic bacteria - metabolic control of end product formation in *Thermoanaerobium brockii. Journal of Bacteriology,* 146: 192-199.

Brock, T.D. 1986. Introduction: an overview of the thermophiles, In: *Thermophiles: General Molecular and Applied Microbiology,* T.D., Brock, (Ed.), pp. 1-16, John Wiley and Sons, ISBN 0471820016, New York, USA.

Cakir, A.; Ozmihci, S. & Kargi, F. 2010. Comparison of bio-hydrogen production from hydrolyzed wheat starch by mesophilic and thermophilic dark fermentation. *International Journal of Hydrogen Energy,* 35: 13214-13218.

Calli, B.; Schoenmaekers, K.; Vanbroekhoven, K. & Diels, L. 2008. Dark fermentative H_2 production from xylose and lactose - Effects of on-line pH control. *International Journal of Hydrogen Energy, 33:* 522-530.

Canganella, F. & Wiegel, J. 1993. The potential of thermophilic Clostridia in biotechnology. In: *The Clostridia and Biotechnology,* D.R. Woods, (Ed.), pp. (391-429), Butterworths Publishers, ISBN 0750690046, Stoneham, MA, USA.

Cao, G.-L.; Ren, N.; Wang, A.; Guo, W.; Yao, J.; Feng, Y. & Xhao, Q. 2010. Statistical optimization of culture condition for enhanced hydrogen production by *Thermoanaerobacterium thermosaccharolyticum* W16. *Bioresource Technology,* 101: 2053-2058.

Cao, G.-L.; Ren, N.; Wang, A.; Lee, D.; Guo, W.; Liu, B.; Feng, Y. & Zhao, Q. 2009. Acid hydrolysis of corn stover for biohydrogen production using *Thermoanaerobacterium thermosaccharolyticum* W16. *International Journal of Hydrogen Energy,* 34: 7182-7188.

Cao, G.-L.; Ren, N-Q.; Wang, A-J.; Guo, W-Q.; Xu, J-F. & Liu, B-F. 2009. Effect of lignocellulose-derived inhibitors on growth and hydrogen production by *Thermoanaerobacterium thermosaccharolyticum* W16. *International Journal of Hydrogen Energy,* 35: 13475-13480.

Carreira, L.H.; Ljungdahl, L. G.; Bryant, F.; Szulcynski, M. & Wiegel, J. 1982. Control of product formation with *Thermoanaerobacter ethanolicus*: enzymology and physiology. In: *Proc. 4th International Symposium on Genetics of Industrial Microorganisms,* pp. 351-355, Kodansha Ltd., Tokyo, Japan,

Carreira, L.H.; Wiegel, J. & Ljungdahl, L.G. 1983. Production of ethanol from bio-polymers by anaerobic, thermophilic, and extreme thermophilic bacteria. I. Regulation of carbohydrate utilization in mutants of *Thermoanaerobacter ethanolicus. Biotechnology and Bioengineering,* Symph.13, pp. 183-191.

Carreira, L.H. & Ljungdahl, L.G. 1993. Production of ethanol from biomass using anaerobic thermophilic bacteria, In: *Liquid fuel developments*, D.L. Wise, (Ed.), pp. 1-28, CRC Press, ISBN 0849360943, Boca Raton, Fl, USA.

Cha, K.S. & Bae, J.H. 2011. Dynamic impacts of high oil prices on the bioethanol and feedstock markets. *Energy policy*, 39: 753-760.

Chinn, M.S.; Nokes, S.E. & Strobel, H.J. 2008. Influence of moisture content and cultivation duration on *Clostridium thermocellum* 27405 end-product formation in solid substrate cultivation on avicel. *Bioresource Technology*, 99: 2664-2671.

Chong, M-L.; Sabaratnam, V.; Shirai, Y. & Hassan, M.A. 2009. Biohydrogen production from biomass and industrial wastes by dark fermentation. *International Journal of Hydrogen Energy*, 34: 3277-3287.

Chou, C-J.; Jenney, Jr. F.E.; Adams, M.W.W. & Kelly, R.M. 2008. Hydrogensis in hyperthermophilic microorganisms: implications fo biofuels. *Metabolic Engineering*, 10: 394-404.

Chu, C.;Li, Y.; Xu, K.; Ebie, Y.; Inamori, Y. & Kong, H. 2008. A pH- and temperature-phased two-stage process for hydrogen and methane production from food waste. *International Journal of Hydrogen Energy*, 33: 4739-4746.

Collet, C.; Schwitzguebel, J.P. & Peringer, P. 2003. Improvement of acetate production from lactose by growing *Clostridium thermolacticum* in mixed batch culture. *Journal of Applied Microbiology*, 95: 824-831.

Collins, M.D.; Lawson, P.A.; Willems, A.; Cordoba, J.J.; Fernandezgarayzabal, J. & Garcia, P. 1994. The phylogeny of the genus *Clostridium* - proposal of 5 new genera and 11 new species combinations. *International Journal of Systematic Bacteriology*, 44: 812-826.

Cord-Ruwisch, R.; Seitz, H. & Conrad, R. 1988. The capacity of hydrogenotrophic anaerobic bacteria to compete for traces of hydrogen depends on the redox potential of the terminal electron acceptor. *Archives of Microbiology*, 149: 350-357.

Cripps, R.E.; Eley, K.; Leak, D.J.; Rudd, B.; Taylor, M.; Todd, M.; Boakes, S.; Martin, S. & Atkinson, T. 2009. Metabolic engineering of *Geobacillus thermoglucosidasius* for high yield ethanol production. *Metabolic Engineering*, 11: 398-408.

Datar, R.; Huang, J.; Maness, P.; Mohagheghi, A.; Czemik, S. & Chornet, E. 2007. Hydrogen production from the fermentation of corn stover biomass pretreated with a steam-explosion process. *International Journal of Hydrogen Energy*, 32: 932-939.

Demain, A. L.; Newcomb, M. & Wu, J. H. 2005. Cellulase, Clostridia, and Ethanol. *Microbiology and molecular biology reviews*, 69: 124-154.

Demirbas, A. 2009. Political, economic and environmental impacts of biofuels: A review. *Applied Energy*, 86: 108-117.

Deanda, K.; Zhang, M.; Eddy, C. & Picataggio, S. 1996. Deveolpment of an arabinose-fermenting *Zymomonas mobilis* strain by metabolic pathway engineering. *Applied and Environmental Microbiology*, 62: 4465-4470.

Desai, S.G.; Guerinot, M.L. & Lynd, L.R. 2004. Cloning of L-lactate dehydrogenase and elimination of lactic acid production via gene knockout in *Thermoanaerobacterium saccharolyticum* JW/SL-YS485. *Applied Microbiology and Biotechnology*, 65: 600-605.

d'Ippolito, G.; Dipasquale, L.; Vella, F.M.; Romano, I.; Gambacorta, A.; Cutignano, A. & Fontana, A. 2010. Hydrogen metabolism in the extreme thermophile *Thermotoga neapolitana*. *International Journal of Hydrogen Energy*, 35: 2290-2295.

Drent, W.J.; Lahpor, G.A.; Wiegant, W.M. & Gottschal, J.C. 1991. Fermentation of inulin by *Clostridium thermosuccinogenes* sp. nov., a thermophilic anaerobic bacterium isolated from various habitats. *Applied and Environmental Microbiology*, 57: 455-462.

DuPont Danisko Cellulosic Ethanol LLC. 2011. Fermentation, In: *Technology*, (April 4, 2011), Available from http://www.ddce.com/technology/fermentation.html

Eriksen, N.T.; Nielsen, T.M. & Iversen, N. 2008. Hydrogen production in anaerobic and microaerobic *Thermotoga neapolitana*. *Biotechnology Letters*, 30: 103-109.

Fardeau, M.L.; Faudon, C.; Cayol, J.L.; Magot, M.; Patel, B.K.C. & Ollivier, B. 1996. Effect of thiosulphate as electron acceptor on glucose and xylose oxidation by *Thermoanaerobacter finnii* and a *Thermoanaerobacter* sp isolated from oil field water. *Research in Microbiology*, 147: 159-165.

Geng, A.; He, Y.; Qian, C.; Yan, X. & Zhou, Z. 2010. Effect of key factors on hydrogen production from cellulose in a co-culture of *Clostridium thermocellum* and *Clostridium thermopalmarium*. *Bioresource Technology*, 101: 4029-4033.

Georgieva, T.I. & Ahring, B.K. 2007. Evaluation of continuous ethanol fermentation of dilute-acid corn stover hydrolysate using thermophilic anaerobic bacterium *Thermoanaerobacter* BG1L1. *Applied Microbiology and Biotechnology*, 77: 61-68.

Georgieva, T.I.; Mikkelsen, M.J. & Ahring, B.K. 2008a. Ethanol production from wet-exploded wheat straw hydrolysate by thermophilic anaerobic bacterium *Thermoanaerobacter* BG1L1 in a continuous immobilized reactor. *Applied Biochemistry and Biotechnology*, 145: 99-110.

Georgieva, T.I.; Mikkelsen, M.J. & Ahring, B.K. 2008b. High ethanol tolerance of the thermophilic anaerobic ethanol producer *Thermoanaerobacter* BG1L1. *Central European Journal of Biology*, 2: 364-377.

German Collection of Microorganisms and Cell Cultures. Available from: http://www.dsmz.de/

Goettemoeller, J., & Goettemoeller. A. 2007. *Sustainable Ethanol: Biofuels, Biorefineries, Cellulosic Biomass, Flex-Fuel Vehicles, and Sustainable Farming for Energy Independence*, Prairie Oak Publishing, ISBN 9780978629304, Maryville, Missouri.

Hallenbeck, P.C. 2009. Fermentative hydrogen production: Principles, progress and prognosis. *International Journal of Hydrogen Energy*, 34: 7379-7389.

Hallenbeck, P.C. & Benemann, J.R. 2002. Biological hydrogen production; fundamentals and limiting processes. *International Journal of Hydrogen Energy*, 27: 1185-1193.

Hallenbeck, P.C. & Ghosh, D. 2010. Improvements in fermentative biological hydrogen production through metabolic engineering. *Journal of Environmental Management*, doi:10.1016/j.jenvman.2010.07.021 (in press).

He, Q.; Lokken, P.M.; Chen, S. & Zhou, J. 2009. Characterization of the impact of acetate and lactate on ethanolic fermentation by *Thermoanaerobacter ethanolicus*. *Bioresource Technology*, 100: 5955-5965.

Hild, H.M.; Stuckey, D.C. & Leak, D.J. 2003. Effect of nutrient limitation on product formation during continuous fermentation of xylose with thermoanaerobacter ethanolicus JW200 fe(7). *Applied Microbiology and Biotechnology*, 60: 679-686.

Hniman, A., O-Thong, S. & Prasertsan, P. 2010. Developing a thermophilic hydrogen-producing microbial consortia from geothermal spring for efficient utilization of xylose and glucose mixed substrates and oil palm trunk hydrolysate. *International Journal of Hydrogen Energy*, In Press.

Huber, R.; Langworthy, T.A.; Konig, H.; Thomm, M.; Woese, C.R. & Sleytr, U.B. 1986. *Thermotoga maritima* sp. nov., represents a new genus of unique extremely thermophilic eubacteria growing up to 90 °C. *Archives of Microbiology,* 144: 324-333.

Ismail, I.; Hassan, M.A.; Rahman, N.A.A. & Soon, C.S. 2010. Thermophilic biohydrogen production from palm oil mill effluent (POME) using suspended mixed culture. *Biomass & Bioenergy,* 34: 42-47.

Ivanova, G.; Rakhely, G. & Kovács, K.L. 2009. Thermophilic biohydrogen production from energy plants by *Caldicellulosiruptor saccharolyticus* and comparison with related studies. *International Journal of Hydrogen Energy,* 34: 3659-3670.

Jannasch, H.W.; Huber, R.; Belkin, S. & Stetter, K.O. 1988. *Thermotoga neapolitana* sp. nov. of the extremely thermophilic, eubacterial genus *Thermotoga. Archives of Microbiology,* 150: 103-104.

Jeffries, T.W. 2006. Engineering yeasts for xylose metabolism. *Current Opinion in Biotechnology,* 17: 320–326

Jones, P. 2008. Improving fermentative biomass-derived H_2-production by engineered microbial metabolism. *International Journal of Hydrogen Energy,* 33: 5122-5130.

Kádár, Z.; de Vrijek, T.; van Noorden, G.E.; Budde, M.A.W.; Szengyel, Z. & Reczey, K. 2004. Yields from glucose, xylose, and paper sludge hydrolysate during hydrogen production by the extreme thermophile *Caldicellulosiruptor saccharolyticus. Applied Biochemistry and Biotechnology,* 113-116: 497-508.

Kanai, T.; Imanaka, H.; Nakajima, A.; Uwamori, K.; Omori, Y.; Fukui, T.; Atomi, H. & Imanaka, T. 2005. Continuous hydrogen production by the hyperthermophilic archaeon, *Thermococcus kodakaraensis* KOD1. *Journal of Biotechnology,* 116: 271-282.

Karadag, D.; Makinen, A.E.; Efimova, E. & Puhakka, J.A. 2009. Thermophilic biohydrogen production by an anaerobic heat treated-hot spring culture. *Bioresource Technology,* 100: 5790-5795.

Karadag, D. & Puhakka, J.A. 2010. Direction of glucose fermentation towards hydrogen or ethanol production through on-line pH control. *International Journal of Hydrogen Energy,* 35: 10245-10251.

Kengen, S.W.M.; Gorrissen, H.P.; Verhaart, M.; van Niel, E.W.J.; Claassen, P.A.M. & Stams, A.J.M. 2009. Biological hydrogen production by anaerobic microorganisms. In: *Biofuels,* W. Soetaert; E.J. Vandamme, (Ed.), pp. 197-221, John Wiley & Sons, ISBN 9780470026748, Chichester, UK.

Klapatch, T.R.; Hogsett, D.A.L.; Baskaran, S.; Pal, S. & Lynd, L.R. 1994. Organism development and characterization for ethanol production using thermophilic bacteria. *Applied Biochemistry and Biotechnology,* 45-46: 209-223.

Klinke, H.B.; Thomsen, A.B. & Ahring, B.K. 2001. Potential inhibitors from wet oxidation of wheat straw and their effect on growth and ethanol production by *Thermoanaerobacter mathranii. Applied Microbiology and Biotechnology,* 57: 631-638.

Knutson, B.L.; Strobel, H.J.; Nokes, S.E.; Dawson, K.A.; Berberich, J.A. & Jones, C.R. 1999. Effect of pressurized solvents on ethanol production by the thermophilic bacterium *Clostridium thermocellum. Journal of Supercritical Fluids,* 16: 149-156.

Kongjan, P. & Angelidaki, I. 2010. Extreme thermophilic biohydrogen production from wheat straw hydrolysate using mixed culture fermentation: Effect of reactor configuration. *Bioresource Technology,* 101: 7789-7796.

Kongjan, P.; Min, B. & Angelidaki, I. 2009. Biohydrogen production from xylose at extreme thermophilic temperatures (70 °C) by mixed culture fermentation. *Water Research,* 43: 1414-1424.

Kongjan, P.; O-Thong, S.; Kotay, M.; Min, B. & Angelidaki, I. 2010. Biohydrogen production from wheat straw hydrolysate by dark fermentation using extreme thermophilic mixed culture. *Biotechnology and Bioengineering,* 105: 899-908.

Koskinen, P.E.P.; Beck, S.R.; Orlygsson, J. & Puhakka, J.A. 2008a. Ethanol and hydrogen production by two thermophilic, anaerobic bacteria isolated from Icelandic geothermal areas. *Biotechnology and Bioengineering,* 101: 679-690.

Koskinen, P.E.P.; Lay, C.; Beck, S.R.; Tolvanen, K.E.S.; Kaksonen, A.H.; Orlygsson, J.; Lin, C.Y. & Puhakka, J.A. 2008b. Bioprospecting thermophilic microorganisms from Icelandic hot springs for hydrogen and ethanol production. *Energy & Fuels,* 22: 134-140.

Koskinen, P.E.P.; Lay, C.; Puhakka, J.A.; Lin, P.; Wu, S.; Orlygsson, J. & Lin C,Y. 2008c. High-efficiency hydrogen production by an anaerobic, thermophilic enrichment culture from an Icelandic hot spring. *Biotechnology and Bioengineering,* 101: 665-678.

Kotsopoulos, T.A.; Zeng, R. J. & Angelidaki, I. 2006. Biohydrogen production in granular up-flow anaerobic sludge blanket (UASB) reactors with mixed cultures under hyper-thermophilic temperature (70 °C). *Biotechnology and Bioengineering,* 94: 296-302.

Kotsopoulos, T.A.; Fotidis, I.A.; Tsolakis, N. & Martzopoulos, G.G. 2009. Biohydrogen production from pig slurry in a CSTR reactor system with mixed cultures under hyper-thermophilic temperature (70 °C). *Biomass & Bioenergy,* 33: 1168-1174.

Kristjansson, J.K. & Alfredsson, G.A. 1986. Life in Icelandic Hot Springs, *Natturufraedingurinn,* 56: 49-68.

Lacis, L.S. & Lawford, H.G. 1991. *Thermoanaerobacter ethanolicus* growth and product yield from elevated levels of xylose or glucose in continuous cultures. *Applied and Environmental Microbiology,* 57: 579-585.

Lacis, L.S. & Lawford, H.G. 1989. Analysis of the variation in ethanol yield from glucose or xylose with continuously grown *Thermoanaerobacter ethanolicus. Applied Biochemistry and Biotechnology,* 20-21: 479-490.

Lacis, L.S. & Lawford, H.G. 1988a. Effect of growth-rate on ethanol-production by *Thermoanaerobacter ethanolicus* in glucose-limited or xylose-limited continuous culture. *Biotechnology Letters,* 10: 603-608.

Lacis, L.S. & Lawford, H.G. 1988b. Ethanol-production from xylose by *Thermoanaerobacter ethanolicus* in batch and continuous culture. *Archives of Microbiology,* 150: 48-55.

Lamed, R., Su, T.M. & Brennan, M.J. 1980. Effect of stirring on ethanol-production by *Clostridium thermocellum. Abstracts of Papers of the American Chemical Society, 180*(AUG), 44-MICR.

Lamed, R. & Zeikus, J.G. 1980a. Ethanol-production by thermophilic bacteria - relationship between fermentation product yields of and catabolic enzyme-activities in *Clostridium thermocellum* and *Thermoanaerobium brockii. Journal of Bacteriology,* 144: 569-578.

Lamed, R. & Zeikus, J.G. 1980b. Glucose fermentation pathway of *Thermoanaerobium brockii. Journal of Bacteriology,* 141: 1251-1257.

Lamed, R. J.; Lobos, J.H. & Su, T.M. 1988. Effects of stirring and hydrogen on fermentation end products of *Clostridium thermocellum*. *Applied Environmental Microbiology,* 54: 1216-1221.

Larsen, L.; Nielsen, P. & Ahring, B.K. 1997. *Thermoanaerobacter mathranii* sp nov, an ethanol-producing, extremely thermophilic anaerobic bacterium from a hot spring in Iceland. *Archives of Microbiology,* 168: 114-119.

Lee, Y-W. & Chung, J. 2010. Bioproduction of hydrogen from food waste by pilot-scale combined hydrogen/methane. *International Journal of Hydrogen,* 35: 11746-11755.

Lee, Y.E.; Jain, M.K.; Lee, C.Y.; Lowe, S.E. & Zeikus, J.G. 1993. Taxonomic distinction of saccharolytic thermophilic anaerobes - description of *Thermoanaerobacterium xylanolyticum* gen-nov, sp-nov, and *Thermoanaerobacterium saccharolyticum* gen-nov, sp-nov - reclassification of *Thermoanaerobium brockii, Clostridium thermosulfurogenes,* and *Clostridium thermohydrosulfuricum* E100-69 as *Thermoanaerobacter brockii* comb-nov, *Thermoanaerobacterium thermosulfurigenes* comb-nov, and *Thermoanaerobacter thermohydrosulfuricus* comb-nov, respectively - and transfer of *Clostridium thermohydrosulfuricum* 39E to *Thermoanaerobacter ethanolicus*. *International Journal of Systematic Bacteriology,* 43: 41-51.

Lee, Z.; Li, S.; Kuo, P.; Chen, I.; Tien, Y.; Huang, Y.; Chuang, C-P.; Wong, S-C. & Cheng, S-S. 2010. Thermophilic bio-energy process study on hydrogen fermentation with vegetable kitchen waste. *International Journal of Hydrogen Energy,* 35: 13458-13466.

Lee, Z.; Li, S.; Lin, J.; Wang, Y.; Kuo, P. & Cheng, S. 2008. Effect of pH in fermentation of vegetable kitchen wastes on hydrogen production under a thermophilic condition. *International Journal of Hydrogen Energy,* 33: 5234-5241.

Levin, D.B.; Pitt, L. & Love, M. 2004. Biohydrogen production: prospects and limitations to practical application. *International Journal of Hydrogen Energy,* 29: 173-185.

Levin, D.B.; Islam, R.; Cicek, N. & Sparling, R. 2006. Hydrogen production by *Clostridium thermocellum* 27405 from cellulosic biomass substrates. *International Journal of Hydrogen Energy,* 31: 1496-1503.

Lin, C., Wu, C., Wu, J., & Chang, F. 2008. Effect of cultivation temperature on fermentative hydrogen production from xylose by a mixed culture. *Biomass & Bioenergy,* 32: 1109-1115.

Lin, C-W,; Wu, C-H,; Tran, D-T,; Shih, M-C,; Li, W-H. & Wu C-F. 2010. Mixed culture fermentation from lignocellulosic materials using thermophilic lignocellulose-degrading anaerobes. *Process Biochemistry,* 46: 489-493.

Lin C-Y, Wu S-Y, Lin P-J, Chang J-S, Hung C-H, Lee K-S, Lay C-H, Chu C-Y, Cheng C-H, Chang AC, Wu J-H, Chang F-Y, Yang L-H, Lee C-W. & Lin Y-C. A pilot scale high-rate biohydrogen production system with mixed microflora. *International Journal of Hydrogen Energy,* in press.

Liu, C. & Cheng, X. 2010. Improved hydrogen production via thermophilic fermentation of corn stover by microwave-assisted acid pretreatment. *International Journal of Hydrogen Energy,* 35: 8945-8952.

Liu, D.W.; Zeng, R.J. & Angelidaki, I. 2008. Enrichment and adaptation of extreme-thermophilic (70 °C) hydrogen producing bacteria to organic household solid waste by repeated batch cultivation. *International Journal of Hydrogen Energy,* 33: 6492-6497.

Liu, D.; Min, B. & Angelidaki, I. 2008. Biohydrogen production from household solid waste (HSW) at extreme-thermophilic temperature (70 °C) - influence of pH and acetate concentration. *International Journal of Hydrogen Energy,* 33: 6985-6992.

Liu, H.; Zhang, T. & Fang, H.H.P. 2003. Thermophilic H-2 production from a cellulose-containing wastewater. *Biotechnology Letters,* 25: 365-369.

Liu, X.; Zhu, Y. & Yang, S.T. 2006. Construction and Characterization of *ack* Deleted Mutant of *Clostridium tyrobutyricum* for Enhanced Butyric Acid and Hydrogen Production. *Biotechnology Pregress,* 22: 1265-1275.

Liu, Y.; Yu, P., Song, X. & Qu, Y. 2008b. Hydrogen production from cellulose by co-culture of *Clostridium thermocellum* JN4 and *Thermoanaerobacterium thermosaccharolyticum* GD17. *International Journal of Hydrogen Energy,* 33: 2927-2933.

Lo, Y. C.; Huang, C.; Fu, T.; Chen, C. & Chang, J. 2009. Fermentative hydrogen production from hydrolyzed cellulosic feedstock prepared with a thermophilic anaerobic bacterial isolate. *International Journal of Hydrogen Energy,* 34: 6189-6200.

Lovitt, R.W.; Longin, R. & Zeikus, J.G. 1984. Ethanol production by thermophilic bacteria: physiological comparison of solvent effects on parent and alcohol-tolerant strains of *Clostridium thermohydrosulfuricum. Applied and Environmental Microbiology,* 48: 171-177.

Lovitt, R.W.; Shen, G.J. & Zeikus, J.G. 1988. Ethanol-production by thermophilic bacteria - biochemical basis for ethanol and hydrogen tolerance in *Clostridium thermohydrosulfuricum. Journal of Bacteriology,* 170: 2809-2815.

Lynd, L.R.; Grethlein, H.E. & Wolkin, R.H. 1989. Fermentation of cellulosic substrates in batch and continuous culture by *Clostridium thermocellum. Applied and Environmental Microbiology,* 55: 3131-3139.

Magnusson, L.; Islam, R.; Sparling, R.; Levin, D. & Cicek, N. 2008. Direct hydrogen production from cellulosic waste materials with a single-step dark fermentation process. *International Journal of Hydrogen Energy,* 33: 5398-5403.

Miyazaki, K.; Irbis, C.; Takada, J. & Matsuura, A. 2008. An ability of isolated strains to efficiently cooperate in ethanolic fermentation of agricultural plant refuse under initially aerobic thermophilic conditions: Oxygen deletion process appended to consolidated bioprocessing (CBP). *Bioresource Technology,* 99: 1768-1775.

Moat, A.G.; Foster, J.W. & Spector, M.P. 2002. Central pathways of carbohydrate metabolism, In: *Microbial Physiology 4th ed.,* A.G. Moat; J.W. Foster & M.P. Spector, (Ed.), pp. 350-367, Wiley-Liss Inc., ISBN 0-471-39483-1, New York, USA.

Munro, S.A.; Zinder, S.H. & Walker, L.P. 2009. The fermentation stoichiometry of *Thermotoga neapolitana* and influence of temperature, oxygen, and pH on hydrogen production. *Biotechnology Progress,* 25: 1035-1042.

Nardon L. & Aten K. 2008. Beyond a better mousetrap: A cultural analysis of the adoption of etahnol in Brazil. *Journal of World Business,* 43: 261-273.

Nass, L.L.P.; Pereira, A.A. & Ellis, D. 2007. Biofuels in Brazil: An overview, *Crop Science,* 47: 2228-2237.

Ng, T.K.; Ben-Bassat, A. & Zeikus, J.G. 1981. Ethanol production by thermophilic bacteria: Fermentation of cellulosic substrates by cocultures of *Clostridium thermocellum* and *Clostridium thermohydrosulfuricum. Applied Environmental Microbiology,* 41: 1337-1343.

Nguyen, T.A.D.; Kim, J.P.; Kim, M.S.; Oh, Y.K. & Sim, S.J. 2008a. Optimization of hydrogen production by hyperthermophilic eubacteria, *Thermotoga maritima* and *Thermotoga neapolitana* in batch fermentation. *International Journal of Hydrogen Energy,* 33: 1483-1488.

Nguyen, T.D.; Han, S.J.; Kim, J.P.; Kim, M.S.; Oh, Y.K. & Sim, S.J. 2008b. Hydrogen production by the hyperthermophilic eubacterium, *Thermotoga neapolitana,* using cellulose pretreated by ionic liquid. *International Journal of Hydrogen Energy,* 33: 5161-5168.

Nguyen, T.D.; Han, S.J.; Kim, J.P.; Kim, M.S. & Sim, S.J. 2010a. Hydrogen production of the hyperthermophilic eubacterium, *Thermotoga neapolitana* under N_2 sparging condition. *Bioresource Technology,* 101: S38-S41.

Nguyen, T.D.; Kim, K.; Kim, M.S. & Sim, S.J. 2010b. Thermophilic hydrogen fermentation from korean rice straw by thermotoga neapolitana. *International Journal of Hydrogen Energy,* 35: 13392-13398.

Nguyen, T.D.; Kim, K-R.; Nguyen, M-T.; Kim, M.S.; Kim, D. & Sim S.J. 2010c. Enhancement of fermentative hydrogen production from green algal biomass by *Thermotoga neaolitana* by various pretreatment methods. *International Journal of Hydrogen Energy,* 35: 13035-13040.

Mauna Loa Observatory: NOAA-ASRL. March 8, 2011. Atmospheric CO_2, In: *CO_2Now.org,* March 10, 2011, Available from http://co2now.org/

Olsson, L. & Hahn-Hagerdal, B. 1996. Fermentation of lignocellulosic hydrolysates for ethanol production. *Enzyme and Micorbial Technology,* 18: 312-331.

Orlygsson, J. & Baldursson, S.R.B. 2007. Phylogenetic and physiological studies of four hydrogen-producing thermoanaerobes from Icelandic geothermal areas. *Icelandic Agricultural Sciences,* 20: 93-105.

Orlygsson, J.; Sigurbjornsdottir, M.A. & Bakken, H.E. 2010. Bioprospecting thermophilic ethanol and hydrogen producing bacteria from hot springs in Iceland. *Icelandic Agricultural Sciences,* 23: 73-85.

O-Thong, S.; Prasertsan, P.; Intrasungkha, N.; Dhamwichukorn, S. & Birkeland, N. 2008. Optimization of simultaneous thermophilic fermentative hydrogen production and COD reduction from palm oil mill effluent by thermoanaerobacterium-rich sludge. *International Journal of Hydrogen Energy,* 33: 1221-1231.

O-Thong, S.; Prasertsan, P.; Karakashev, D. & Angelidaki, I. 2008. Thermophilic fermentative hydrogen production by the newly isolated *Thermoanaerobacterium thermosaccharolyticum* PSU-2. *International Journal of Hydrogen Energy,* 33: 1204-1214.

Prasertsan, P.; O-Thong, S. & Birkeland, N. 2009. Optimization and microbial community analysis for production of biohydrogen from palm oil mill effluent by thermophilic fermentative process. *International Journal of Hydrogen Energy,* 34: 7448-7459.

Rainey, F.A.; Donnison, A.M.; Janssen, P.H.; Saul, D.; Rodrigo, A.; Bergquist, P.L.; Daniel, R.M.; Stackebrandt, E. & Morgan, H.W. 1994. Description of *Caldicellulosiruptor-saccharolyticus* gen-nov, sp-nov - an obligately anaerobic, extremely thermophilic, cellulolytic bacterium. *FEMS Microbiology Letters,* 120: 263-266.

Rani, K.S.; Swamy, M.V. & Seenayya, G. 1998. Production of ethanol from various pure and natural cellulosic biomass by *Clostridium thermocellum* strains SS21 and SS22. *Process Biochemistry,* 33: 435-440.

Rani, K. S.; Swamy, M.V. & Seenayya, G. 1997. Increased ethanol production by metabolic modulation of cellulose fermentation in *Clostridium thermocellum*. *Biotechnology Letters*, 19: 819-823.

Ren, N.; Cao, G.; Guo, W.; Wang, A.; Zhu, Y.; Liu, B. & Xu, J-F. 2010. Biological hydrogen production from corn stover by moderately thermophile *Thermoanaerobacterium thermosaccharolyticum* W16. *International Journal of Hydrogen Energy*, 35: 2708-2712.

Ren, N.; Cao, G.; Wang, A.; Lee, D.; Guo, W. & Zhu, Y. 2008. Dark fermentation of xylose and glucose mix using isolated *Thermoanaerobacterium thermosaccharolyticum* W16. *International Journal of Hydrogen Energy*, 33: 6124-6132.

Ren, N.; Wang, A.; Cao, G.; Xu, J. & Gao, L. 2009. Bioconversion of lignocellulosic biomass to hydrogen: Potential and challenges. *Biotechnology Advances*, 27: 1051-1060.

Renewable Fuels Association. 2010. Annual Industry Outlook, In: *Climate of Opportunity*, (April 2, 2011), Available from
http://www.ethanolrfa.org/pages/annual- industry-outlook

Rogers, P.L.; Lee., K.J.; Skotnicki, M.L. & Tribe, D.E. 1982. Ethanol production by *Zymomonas Mobilis*. *Advances in Biochemical Engineering*, 23: 37-84.

Romano, I.; Dipasquale, L.; Orlando, P.; Lama, L.; d'Ippolito, G.; Pascual, J. & Gambacorta, A. 2010. *Thermoanaerobacterium thermostercus* sp nov., a new anaerobic thermophilic hydrogen-producing bacterium from buffalo-dung. *Extremophiles*, 14: 233-240.

Rupprecht, J.; Hankamer, B.; Mussgnug, J. H.; Ananyev, G.; Dismukes, C. & Kruse, O. 2006. Perspectives and advances of biological H_2 production in microorganisms. *Applied Microbiology and Biotechnology*, 72: 442-449.

Russel, J.B. 1992. Another explanation for the toxicity of fermentation of acids at low pH: anion accumulation versus uncoupling. *Journal of Applied Bacteriology*, 73: 363-370.

Sanchez, R.G.; Karhumaa, K.; Fonseca, C.; Nogue, V.S.; Almeida, J.R.M.; Larsson, C.U.; Bengtsson, O.; Bettinga, M.; Hahn-Hagerdal, B. & Gorwa-Grauslund, M.F. 2010. Improved xylose and arabinose utilization by an industrial recombinant *Saccharomyces cerevisae* strain using evolutionary engineering. *Biotechnology for Biofuels*, 3: 1-11.

Schroder, C.; Selig, M. & Schonheit, P. 1994. Glucose fermentation to acetate, CO_2 and H_2 in the anaerobic hyperthermophilic eubacterium *Thermotoga maritima* - involvement of the Embden-Meyerhof pathway. *Archives of Microbiology*, 161: 460-470.

Schicho, R. N.; Ma, K.; Adams, M.W.W. & Kelly, R. M. 1993. Bioenergetics of sulfur-reduction in the hyperthermophilic archaeon *Pyrococcus furiosus*, *J. Bacteriol.* 175: 1823-1830.

Sapra, R.; Bagraman, K. & Adams, M.W.W. 2003. A simple energy-conserving system: Proton reduction to proton translocation. *Proceedings of the National Academy of Sciences of the United States of America*. 100: 7545-7550.

Shaw, A.J.; Podkaminer, K.K.; Desai, S.G.; Bardsley, J.S.; Rogers, S.R.; Thorne, P.G.; Hogsett, D.A. & Lynd, L.R. 2008. Metabolic engineering of a thermophilic bacterium to produce ethanol at high yield. *Proceedings of the National Academy of Sciences of the United States of America*. 105: 13769-13774.

Shin, H. S. & Youn, J. H. 2005. Conversion of food waste into hydrogen by thermophilic acidogenesis. *Biodegradation*, 16: 33-44.

Shin, H.S.; Youn, J.H. & Kim, S.H. 2004. Hydrogen production from food waste in anaerobic mesophilic and thermophilic acidogenesis. *International Journal of Hydrogen Energy,* 29: 1355-1363.

Shiratori, H.; Sasaya, K.; Ohiwa, H.; Ikeno, H.; Ayame, S.; Kataoka, N.; Miya, A.; Beppu, T. & Ueda, K 2009. *Clostridium clariflavum* sp nov and *Clostridium caenicola* sp nov., moderately thermophilic, cellulose-/cellobiose-digesting bacteria isolated from methanogenic sludge. *International Journal of Systematic and Evolutionary Microbiology,* 59: 1764-1770.

Siebers, B. & Schönheit, P. 2005. Unusual pathways and enzymes of central carbohydrate metabolism in Archaea, *Current Opinion in Microbiology,* 8: 695-705.

Soboh, B.; Linder, D. & Hedderich, R. 2004. A multisubunit membrane-bound [NiFe] hydrogenase and an NADH-dependent Fe-only hydrogenase in the fermenting bacterium *Thermoanaerobacter tengcongensis, Microbiology,* 150: 2451-2461.

Sommer, P.; Georgieva, T. & Ahring, B.K. 2004. Potential for using thermophilic anaerobic bacteria for bioethanol production from hemicellulose. *Biochemical Society Transactions,* 32: 283-289.

Sveinsdottir, M.; Baldursson, S.R.B. & Orlygsson, J. 2009. Ethanol production from monosugars and lignocellulosic biomass by thermophilic bacteria isolated from Icelandic hot springs. *Icelandic Agricultural Sciences,* 22: 45-58.

US Department of Energy. February 28, 2007. Biorefinery Grant Announcement, In: *US Department of Energy; News Archive,* (April 3, 2001), Available from http://www.energy.gov/news/archives/4828.htm

Taylor, M.P.; Eley, K.L,; Martin, S,; Tuffin, M,; Burton, S.G. & Cowan, D.A. (2008). Thermophilic ethanologenesis: future prospects for second-generation bioethanol production. *Trends in Biotechnology,* 27: 398-405.

van Groenestijn, J.W.; Hazewinkel, J.H.O.; Nienoord, M. & Bussmann, P.J.T. 2002. Energy aspects of biological hydrogen production in high rate bioreactors operated in the thermophilic temperature range. *International Journal of Hydrogen Energy,* 27: 1141-1147.

van Maris, A.J.; Winkler, A.A.; Kuyper, M.; de Laat, W.T.; van Dijken, J.P. & Pronk, J.T. 2007. Development of efficient xylose fermentation in *Saccharomyces cerevisae*: xylose isomerase as a key component. *Adv. Biochem. Eng. Biotechnol.* 108: 179-204.

van Niel, E.W.J.; Budde, M.A.W.; de Haas, G.G.; van der Wal, F.J.; Claasen, P.A.M. & Stams, A.J.M. 2002. Distinctive properties of high hydrogen producing extreme thermophiles, *Caldicellulosiruptor saccharolyticus* and *Thermotoga elfii. International Journal of Hydrogen Energy,* 27: 1391-1398.

van Niel, E.W.J.; Claassen, P.A.M. & Stams, A.J.M. 2003. Substrate and product inhibition of hydrogen production by the extreme thermophile, *Caldicellulosiruptor saccharolyticus. Biotechnology and Bioengineering,* 81: 255-262.

van Ooteghem, S.A.; Beer, S.K. & Yue, P.C. 2002. Hydrogen production by the thermophilic bacterium *Thermotoga neapolitana. Applied Biochemistry and Biotechnology,* 98-100: 177-189.

Vedenov, D. & Wetzstein, M. 2008. Toward an optimal U.S. ethanol fuel subsidy. *Energy Economics,* 30: 2073-2090.

Vignais, P.M.; Magnin, J.-P. & Willison, J.C. 2006. Increasing biohydrogen production by metabolic engineering, *International Journal of Hydrogen Energy,* 31: 1478–1483.

Vrije, G. J. de.; Mars, A.E.; Budde, M.A.W.; Lai, M.H.; Dijkema, C.; Waard, P. de. & Classen, P.A.M. 2007. Glycolytic pathway and hydrogen yield studies of the extreme thermophile *Caldicellulosiruptor saccharolyticus*. *Applied Microbiology and Biotechnology*, 74: 1358-1367.

Vrije, T. de.; Bakker, R.R.; Budde, M.A.W.; Lai, M.H.; Mars, A.E. & Claassen, P.A.M. 2009. Efficient hydrogen production from the lignocellulosic energy crop miscanthus by the extreme thermophilic bacteria *Caldicellulosiruptor saccharolyticus* and *Thermotoga neapolitana*. *Biotechnology for Biofuels*, 2: 12.

Vrije, T. de.; Budde, M.A.W.; Lips, S.J.; Bakker, R.R.; Mars, A.E. & Claassen, P.A.M. 2010. Hydrogen production from carrot pulp by the extreme thermophiles *Caldicellulosiruptor saccharolyticus* and *Thermotoga neapolitana*. *International Journal of Hydrogen Energy*, 35: 13206-13213.

Wagner, I.D. & Wiegel, J. 2008. Diversity of Thermophilic anaerobes. *Incredible Anaerobes: From Physiology to Genomics to Fuels*, 1125: 1-43.

Wang, Y.; Li, S.; Chen, I. & Cheng, S. 2009. Starch hydrolysis characteristics of hydrogen producing sludge in thermophilic hydrogen fermentor fed with kitchen waste. *International Journal of Hydrogen Energy*, 34: 7435-7440.

Wiegel, J. & Lungdahl, L.G. 1981. *Thermoanaerobacter ethanolicus* gen. nov., spec. nov., a new extreme thermophilic, anaerobic bacterium. *Archives of Microbiology*, 128: 343-348.

Wiegel, J.; Carreira, L. H.; Mothershed, C. P. & Puls, J. 1983. Production of ethanol from biopolymers by anaerobic, thermophilic, and extreme thermophilic bacteria. II. *Thermoanaerobacter ethanolicus* JW200 and its mutants in batch cultures and resting cell experiments. *Biotechnology and Bioengineering*, 13: 193-205.

Wiegel, J.; Kuk, S.U. & Kohring, G.W. 1989. *Clostridium thermobutyricum* sp. nov., a moderate thermophile isolated from a cellulolytic culture, that produces butyrate as the major product. *International Journal of Systematic Bacteriology*, 39: 199-204.

Willquist, K.; Claassen, P.A.M. & van Niel, E.W.J. 2009. Evaluation of the influence of CO_2 on hydrogen production by *Caldicellulosiruptor saccharolyticus*. *International Journal of Hydrogen Energy*, 34: 4718-4726.

Yao, S. & Mikkelsen, M.J. 2010. Metabolic engineering to improve ethanol production in *Thermoanaerobacter mathranii*. *Applied Microbiology and Biotechnology*, 88: 199-208.

Yang, S.; Kataeva, I.; Wiegel, J.; Yin, Y.; Dam, P.; Xu, Y.; Westpheling, J. & Adams, M.W.W. 2010. Classification of '*Anaerocellum thermophilum*' strain DSM 6725 as *Caldicellulosiruptor bescii* sp. nov. *International Journal of Systematic and Evolutionary Microbiology*, 60: 2011-2015.

Yokoyama, H.; Moriya, N.; Ohmori, H.; Waki, M.; Ogino, A. & Tanaka, Y. 2007. Community analysis of hydrogen-producing extreme thermophilic anaerobic microflora enriched from cow manure with five substrates. *Applied Microbiology and Biotechnology*, 77: 213-222.

Yokoyama, H.; Ohmori, H.; Waki, M.; Ogino, A. & Tanaka, Y. 2009. Continuous hydrogen production from glucose by using extreme thermophilic anaerobic microflora. *Journal of Bioscience and Bioengineering*, 107: 64-66.

Yu, H. Q.; Zhu, Z. H.; Hu, W. R. & Zhang, H. S. 2002. Hydrogen production from rice winery wastewater in an upflow anaerobic reactor by using mixed anaerobic cultures. *International Journal of Hydrogen Energy*, 27: 1359-1365.

Zeidan, A.A. & van Niel, E.W.J. 2010. A quantitative analysis of hydrogen production efficiency of the extreme thermophile *Caldicellulosiruptor owensensis* OLT. *International Journal of Hydrogen Energy,* 35: 1128-1137.

Zeikus, J.G.; Hegge, P.W. & Anderson, M.A. 1979. *Thermoanaerobium brockii* gen. nov. and sp. nov., a new chemoorganotrophic, caldoactive, anaerobic bacterium. *Archives of Microbiology,* 122: 41-48.

Zhang, M.; Franden, M.A.; Newman, M.; Mcmillan, J.; Finkelstein, M. & Picataggio, S. 1995. Promising ethanologens for xylose fermentation—scientific note. *Appl Biochem Biotechnol,* 51-52: 527–536.

Zhao, C.; Karakashev, D.; Lu, W.; Wang, H. & Angelidaki, I. 2010. Xylose fermentation to biofuels (hydrogen and ethanol) by extreme thermophilic (70 °C) mixed culture. *International Journal of Hydrogen Energy,* 35: 3415-3422.

Zhao, C.; O-Thong, S.; Karakashev, D.; Angelidaki, I.; Lu, W. & Wang, H. 2009. High yield simultaneous hydrogen and ethanol production under extreme-thermophilic (70 °C) mixed culture environment. *International Journal of Hydrogen Energy,* 34: 5657-5665.

20

Analysis of Process Configurations for Bioethanol Production from Microalgal Biomass

Razif Harun[1,2], Boyin Liu[1] and Michael K. Danquah[1]
[1]*Bio Engineering Laboratory, Department of Chemical Engineering, Monash University, Victoria,*
[2]*Department of Chemical and Environmental Engineering, Universiti Putra Malaysia, Serdang,*
[1]*Australia*
[2]*Malaysia*

1. Introduction

Fossil fuel depletion has become a great concern as the world population is increasing at an alarming rate. Current concerns such as global warming, depletion of fossil fuels and increasing price of petroleum-based fuels have forced the search for alternative and cost-effective energy sources with lesser greenhouse gas emissions. Research into the development of renewable and sustainable fuels has recognised bioethanol as a viable alternative to fossil fuels, owing to its low toxicity, biodegradability, and the ability to effectively blend with gasoline without any engine modifications (Harun et al., 2009, 2010a).

The utilization of crops such as sugar cane, sorghum and corn are considered as traditional approaches for bioethanol production (Harun et al., 2010a). The use of such feedstock for bioethanol production competes in the limited agricultural logistics for food production thus escalating the "food versus fuel debate" (Harun et al., 2010b). There has been a considerable interest in the use of microalgal biomass to replace food-based feedstock for renewable transport fuel production. Microalgae are autotrophic photosynthetic organisms considered as the fastest growing plant species known (Wayman, 1996). They can tolerate a wide range of pH and temperature conditions in diverse habitats including freshwater and sea water (Harun et al., 2010b). Microalgae can store considerable amounts of carbohydrates in the form of starch/cellulose, glycogen, hexoses and pentoses that can be converted into fermentable sugars for bioethanol production via fermentation (Wayman, 1996). **Table 1** shows the amount of carbohydrates in various species of microalgae.

Compared to existing edible feedstock, microalgae grow easily with or without soil and offer a very short harvesting cycle (1~10 days) (Harun et al., 2010a). Microalgae also have a high capacity of fixing CO_2 via photosynthesis and other greenhouse gases, resulting in an overall reduction in the net gaseous emissions during the entire life cycle of the fuel (Wayman, 1996). Majority of the work reported in literature relies on straightforward sequential application of the production process involving pre-treatment of the biomass, hydrolysis, fermentation and product recovery. Simultaneous occurrences or a combination of these process steps could hugely impact on the process economics of bioethanol production from microalgae. Different process approaches including Separate Hydrolysis and Fermentation (SHF), Separate Hydrolysis and Co-Fermentation (SHCF), Simultaneous Saccharification and Fermentation (SSF), Simultaneous Saccharification and Co-Fermentation (SSCF), and Consolidated Bioprocessing (CBP).

Algae strains	Carbohydrates (% dry wt)
Scenedesmus obliquus	10–17
Scenedesmus quadricauda	-
Scenedesmus dimorphus	21–52
Chlamydomonas rheinhardii	17
Chlorella vulgaris	12–17
Chlorella pyrenoidosa	26
Spirogyra sp.	33–64
Dunaliella bioculata	4
Dunaliella salina	32
Euglena gracilis	14–18
Prymnesium parvum	25–33
Tetraselmis maculate	15
Porphyridium cruentum	40–57
Spirulina platensis	8–14
Spirulina maxima	13–16
Synechoccus sp.	15
Anabaena cylindrical	25–30

Table 1. Amount of carbohydrates from various species of microalgae on a dry matter basis (%) (Becker, 1994)

2. Pretreatment of biomass

Biomass pretreatment is a crucial step as it breaks down the crystalline structure of cellulose and releases the fermentable sugars so that the hydrolysis of carbohydrate can be achieved more rapidly and with greater yields (Mosier et al., 2005). An appropriate pretreatment process can also prevent the formation of inhibitors to the subsequent hydrolysis and fermentation (Sun & Cheng, 2002). However, the pretreatment process contributes significantly to the cost of production (Alvira et al., 2010). The main methods include physical treatment (such as milling and grinding), thermo-chemical pretreatment (such as steam explosion) and ammonia fibre explosion. Mechanical comminution can be a combination of chipping, milling and grinding. It aims to reduce the particle size of the biomass to attain a larger surface area for enzyme access. The desired final particle size determines the appropriate technique to apply. For example, chipping is used when 10-30mm particle size is required whilst milling and grinding are for more fine particles (0.2-2mm) (Alvira et al., 2010). The higher energy cost of mechanical comminution especially for large-scale applications makes it an unattractive approach for pretreatment (Hendriks & Zeeman, 2009). However, small lab-scale experiments routinely employ mechanical comminution for biomass pretreatment.

3. Hydrolysis

Two main hydrolysis methods are widely used to produce monomeric sugar constituents required for fermentation. These include acid hydrolysis (with dilute and concentrated acids) and enzymatic hydrolysis (Saha et al., 2005). The acid pretreatment process dissolves the hemicellulosic component of the biomass and disassembles the cellulose into fermentable sugars which are accessible to enzymes (Wayman, 1996). The use of concentrated acid is limited owing to higher cost, corrosion of containment material, and the formation of inhibiting compounds (Sun & Cheng, 2002). Dilute sulphuric acid is the most studied acid, and gives high hydrolysis yields (Mosier et al., 2005). It can be applied at 180 °C for a short period of time or at 120°C for 30-90 min in different types of reactors such as plug flow, batch, shrinking-bed and counter-current reactors (Sun & Cheng, 2002). Harun et al., (2010a) investigated bioethanol production under varying conditions of reaction time, temperature, microalgae loading and acid concentration. It was found that the highest bioethanol yield occurred with 10g/L of microalgae, 3% (v/v) of sulphuric acid at 160°C for 15min.

Enzymatic hydrolysis is the utilization of enzymes to release the fermentable sugars from the biomass. The process cost of enzymatic hydrolysis is lower than acid hydrolysis as it avoids containment corrosion and occurs under mild temperatures and pH (Sanchez et al., 2004). There are few literatures on the study of biological pretreatments of microalgal biomass. However, the advantages of biological pretreatment can be extrapolated from studies using lignocellulosic biomass, which also contains cellulosic and hemicellulosic materials. There is a wide range of bacteria and fungi that can produce cellulases for hydrolysis, but fungi are mostly used due to their less severe growth conditions and high growth rates (Sanchez et al., 2004). Several white-rot fungi have been reported to enhance the hydrolysis of lignocellulosic materials, such as *Phanerochaete chrysosporium, Ceriporia lacerata, Cyathus stercolerus, Ceriporiopsis subvermispora, Pycnoporus cinnarbarinus* and *Pleurotus ostreaus* (Kumar & Wyman, 2009). A study on fungal pretreatment of wheat straw for 10 days showed an increase in the release of fermentable sugars and a reduction in the concentration of fermentation inhibitors (Kuhar et al., 2008). A study by Singh et al., (2008) on fungal pretreatment of sugarcane also showed an increased release of sugars.

4. Fermentation

One of the most successful microorganisms for bioethanol production is *Saccharomyces cerevisiae* (Wyman, 1996). Although the wild-type strain has a high bioethanol productivity and very tolerant to high ethanol concentrations and inhibitory compounds, it is unable to ferment pentoses (hemicelluloses) (Hahn-Hagerdal et al., 2007). *Pichia stipitis, Candida shehatae* and *Pachysolan tannophilus* are promising microbes that are capable of fermenting both hexoses and pentoses (Lin & Tanaka, 2006). However, *S. cerevisiae* is still the most commercialized and dominated strains for bioethanol production (Lin & Tanaka, 2006). The disadvantage of *S. cerevisiae* can be overcome by introducing genetic information of xylose reductase and xylitol dehydrogenase (Tomas-Pejo et al., 2008). Although the fermentation can be performed as a batch, fed batch or continuous process, most ethanol production industries use the batch mode (Tomas-Pejo et al., 2008). Similar to other biomass, the overall process flow diagram for bioethanol production from microalgae is shown in **Fig. 1**.

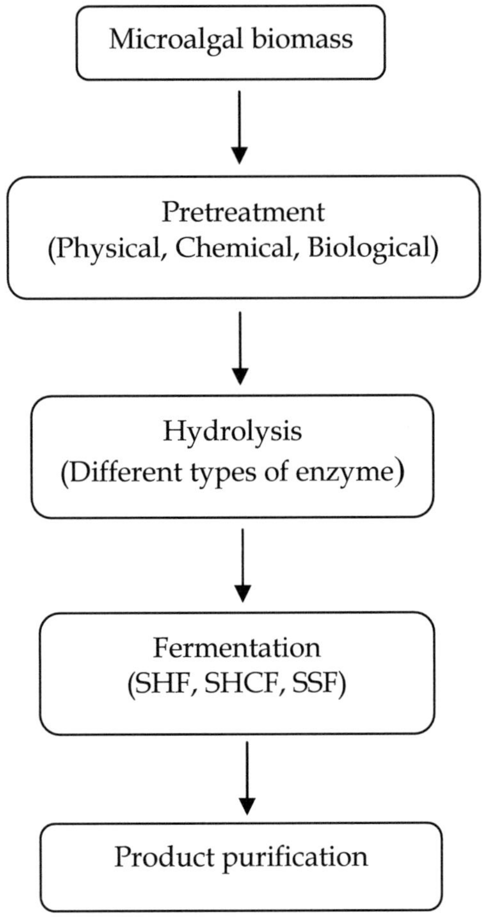

Fig. 1. The overall process flow diagram of bioethanol production from microalgal biomass.

4.1 Separate Hydrolysis and Fermentation (SHF)

In SHF, the enzymatic hydrolysis is performed separately from the fermentation step. Since hydrolysis and fermentation occur in separate vessels, each step can be performed at optimum conditions (Tomas-Pejo et al., 2008). More specifically, it enables enzymes to operate at optimum activities to produce more substrates for yeast fermentation. However, the accumulation of hydrolysis products leads to one of the drawbacks of SHF. Glucose and cellobiose inhibit the activities of the cellulases so the rate of hydrolysis is progressively reduced (Balat et al., 2008).

4.2 Simultaneous Saccharification and Fermentation (SSF)

SSF is an important process strategy for bioethanol production where the enzyme hydrolysis and fermentation are run in the same vessel. In contrast to SHF, the end-product inhibition from cellobiose and glucose in hydrolysis is progressively assimilated by the yeast in the fermentation process. Therefore, compared to SHF, the requirement for enzyme is lower and the bioethanol yield is higher in SSF (Lin & Tanaka, 2006). Furthermore, the higher bioethanol concentration in SSF production also reduces foreign contamination

(Chen & Wang, 2010). Li et al., (2009) also increased the bioethanol yield of SSF by phosphoric acid-acetone pretreatment, which further reduced the inhibitory compounds in the hydrolysis and fermentation with a high solids content (>15% dry matter). Moreover, a fed-batch SSF system was adopted by Li et al., (2009) in order to overcome the problem. In the fed-batch operation, the cellulose suspension after pretreatment and hydrolysis is continuously fed to the bioreactor in order to maintain the liquid viscosity. The fed-batch system turned out to support bioethanol production. Since the hydrolysis and fermentation processes happen at a same temperature, finding an optimal temperature for SSF operation has become the most critical problem.

4.3 Simultaneous Saccharification and Co-Fermentation (SSCF) & Separate Hydrolysis and Co-Fermentation (SHCF)

Microorganisms usually applied for bioethanol production cannot utilize all the sugar sources derived from hydrolysis. For example, the wild-type strain of *S. cerevisiae* is unable to use pentose, and this represents a waste of biomass and reduces the bioethanol yield. To overcome this problem, recombinant yeast or cellulosic enzyme cocktails are introduced during fermentation to convert a wide range of both hexoses and pentoses (Wyman, 1996). Therefore, SSCF can be considered as an improvement to SSF. The hydrolysis and fermentation steps are combined in one vessel for SSCF; hence it has the same characteristics as SSF, such as low cost, short process time, reduced contamination risk and less inhibitory effects (Chandel et al., 2007). A two-step SSCF has been proposed and studied by Jin et al., (2010), where the fermentation time is divided into two equal parts and same conditions were applied as in traditional SSCF. In the two-step SSCF, 4% of total cellulases were used in the first half of the fermentation process, and then the rest of the cellulases were introduced in the second half of the fermentation. The bioethanol yield increased by significantly improving the xylose consumption.

Another similar bioprocess is SHCF, which combines the advantages of SHF and SSCF. The hydrolysis and fermentation processes in SHCF take place in separate vessels so that each step can be performed at its optimal conditions. Besides, since the microbes utilize both pentoses and hexoses effectively in the co-fermentation process in SHCF, the bioethanol yield is higher than SHF. However, there is, to date, few literatures on SHCF operations, but the details can be deduced by referring to SHF and SSCF procedures.

4.4 Consolidated Bioprocessing (CBP)

CBP simultaneously combines biomass hydrolysis, utilization of liberated sugars and fermentation in one bioreactor (Xu et al., 2010). Theoretically, CBP is energy efficient because of reduction of processes and is more cost effective than SSCF (Lynd et al., 2005). However, the crucial problem is to develop an organism to singularly combine all the features during the process. Among all the CBP potential microbes, thermophilic bacteria, such as *Clostridium thermocellum*, are believed feasible as they possess cellulolytic and ethanologenic characteristics under high temperature conditions (Georgieva et al., 2008). Complexes of cellulolytic enzymes contained in *C. thermocellum* known as cellulosome are responsible for cellulose degradation and sugar release. According to the finding from Xu et al., (2010), the temperature of 65°C was used with pH ranging from 6.5-7.4 to compromise between the optimal conditions of the growth of *C. thermocellum* and cellulosome activity. Table 2 shows a summary of the comparison between the different process configurations.

Process	Advantages	Disadvantages
SHF	Hydrolysis and fermentation take place at optimum conditions	Inhibitory effects Increased contamination
SSF	Low quantity of enzyme input High ethanol yield Reduced foreign contamination Less inhibitory effects Lower cost	Either hydrolysis or fermentation can be performed under optimal conditions Difficulty in process control
SHCF	High bioethanol yield Hydrolysis and fermentation take place at optimum conditions	High enzyme load Increased contamination risk Inhibitory effects
SSCF	Shorter process time High bioethanol yield Less contamination risk	High enzyme load Either hydrolysis or fermentation can be perform under optimal conditions
CBP	Cost effective Energy efficient	Lack of suitable organisms Difficulty in process control

Table 2. Comparison of the different fermentation process configurations

5. Experimental work

To further understand the effects of different fermentation approaches on bioethanol production from microalgal biomass, experimental work was designed based on variations of some key process conditions such as the type of substrate, amount of biomass loading and the type of enzymes in order to investigate their influence on the production process. The details of the process are shown in **Fig. 2.**

5.1 Strain and cultivation

Chlorococcum sp. was grown in an outdoor bioreactor (100 L) located in Monash University, Victoria, Australia. The carbohydrate composition of the microalgae strain is shown in Table 3. The microscopic image of the strain is shown in **Fig. 3**. It was composed of 150.0 mg/L $NaNO_3$, 22.7 mg/L $Na_2SiO_3.5H_2O$, 11.3 mg/L $NaH_2PO_4.2H_2O$, 9.0 mg/L $C_6H_8O_7.xFe$, 9.0 mg/L $C_6H_8O_7$, 0.360 mg/L $MnCl_2.4H_2O$, 0.044 mg/L $ZnSO_4.7H_2O$, 0.022 mg/L $CoCl_2.6H_2O$, 0.020 mg/L $CuSO_4.5H_2O$, 0.013 mg/L $Na_2MoO_4.2H_2O$, trace Vitamin B_{12}, Biotin, and Thiamine. Modified F growth medium in synthetic seawater was used for cultivation. The bioreactor was aerated with compressed air to provide the needed CO_2, while other cultivation parameters, such as reactor temperature and illumination level, were not controlled due to its outdoor location. The microalgal culture was dewatered by centrifugation (Heraeus, multifuge 3S-R, Germany) and dried overnight at 60°C in an oven (Model 400, Memmert, Germany). The dried biomass was homogenized by grinding in a laboratory disc miller (N.V Tema, Germany).

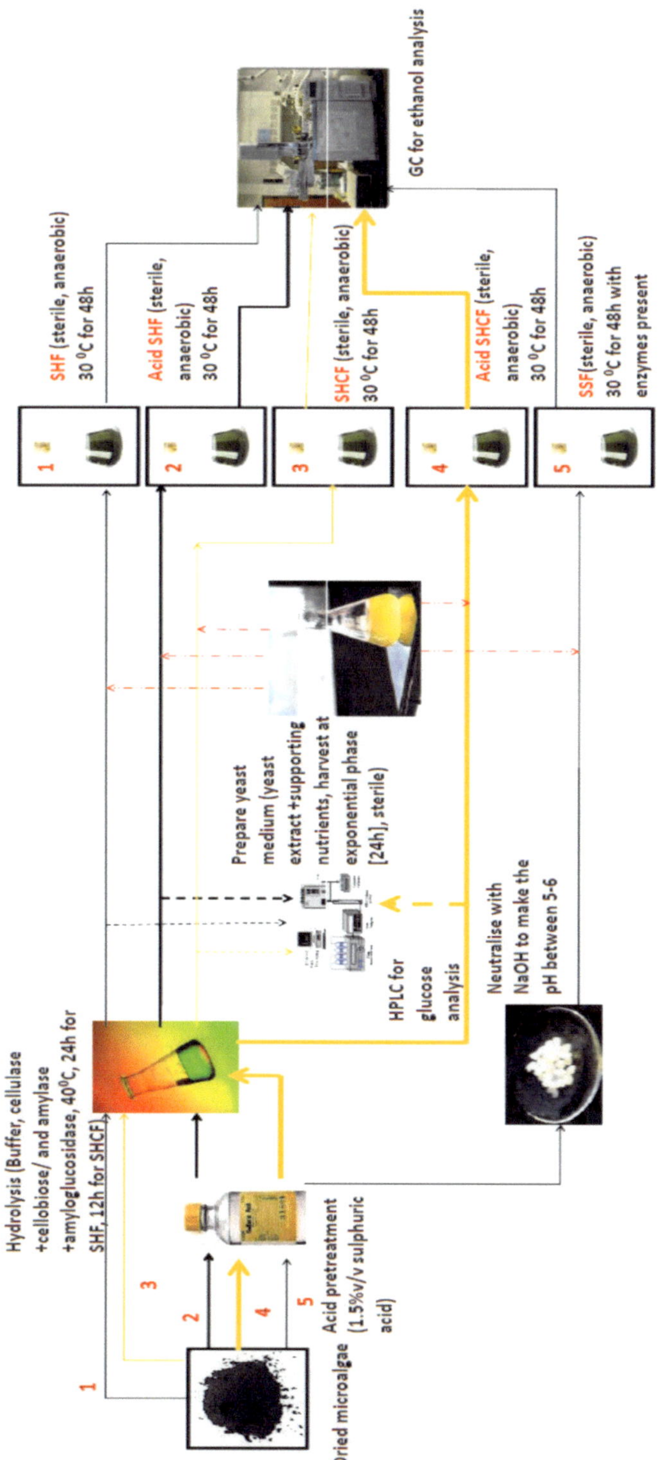

Fig. 2. A flow chart for the experimental procedure.

Component	Composition (%, w/w)
Total carbohydrate	32.52
Xylose	9.54
Mannose	4.87
Glucose	15.22
Galactose	2.89
Starch	11.32
Others	56.16

Table 3. Composition of *Chloroccum sp.* [2]

5.2 Enzymes

The enzymes used in this study were cellulase from Trichoderma reesei (ATCC 26921), cellobiase from Aspergillus niger (Novozyme 188) and α-Amylase from *Bacillus licheniformis*, purchased from Sigma Aldrich, Australia. The activity of cellulase measured as 1.0 unit per mg solid means that one unit of cellulase liberates 1.0 μmole of glucose from cellulose in 1 hour at pH 5.0. Activities of cellobiase and α-amylase were 250 units/mg and 500 units/mg respectively.

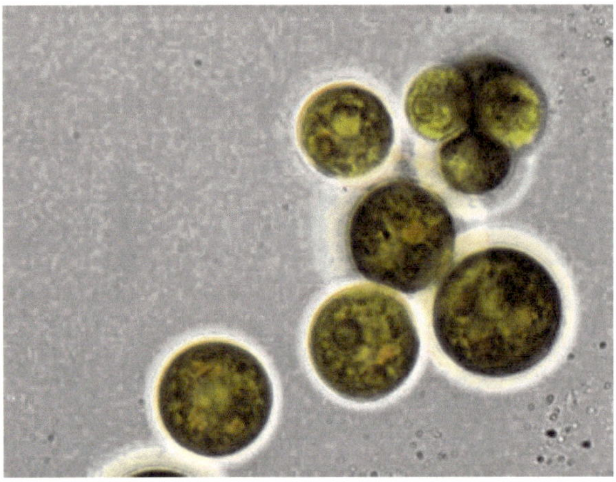

Fig. 3. The microscopic image of the microalgal cells before pretreatment. The images were taken at 40× magnification. The images show that microalgal cells have intact cell walls, thus pretreatment is required to rupture the cell wall to release fermentable sugars (Harun et al., 2010a)

5.3 Fermentation process

5.3.1 Separate Hydrolysis and Fermentation (SHF)

Two types of microalgal substrate were used in this study, acid pre-treated and untreated dried biomass. The acid pre-treated microalgal biomass was obtained after 1% (v/v) sulphuric acid exposure at 140°C for 30 min. The initial amounts of microalgae were varied from 25-100 g/l with a constant mass of 20 mg cellulase for hydrolysis. The enzyme-microalgal biomass mixtures were transferred into shake flasks containing 10 mM of 100 mL sodium acetate buffer solution and incubated (LH Fermentation Ltd., Buckinghamshire, England) at 40 °C, and pH of 4.8 for 24 h. Samples were taken after every 5h and

immediately immersed in a hot water bath at temperature ~90 °C for 10 min in order to stop the enzymatic activity. The samples were then stored in a freezer at -75 °C (Ultraflow freezer, Plymouth, USA) until further analysis.
For the fermentation process, *Saccharomyces cerevisiae*, purchased from Lalvin, Winequip Products Pty Ltd. (Victoria, Australia), was used for bioethanol production. The culture was prepared by dissolving 5.0 g of dry yeast powder in 50 ml sterile warm water (~40 °C) and the pH was adjusted to 7 by 1M NaOH. The yeast was cultured in YDP medium with composition in g/L given as follows: 10 yeast extract, 20 peptone, and 20 glucose. The yeast was harvested after 24h, washed to remove the sugars and then transferred into 500 mL Erlenmeyer flasks containing 100mL of the sugar-containing liquid medium obtained after the hydrolysis process. The flasks were tightly sealed and nitrogen gas was bubbled through to create an oxygen-free environment for bioethanol production. The flasks were incubated at 30 °C under 200 rpm shaking. The pH was maintained at 7 by adding 1M NaOH solution. The fermentation process continued for 50 h and samples for analysis were taken after every 4h.

5.3.2 Separate Hydrolysis and Co-Fermentation (SHCF)

The procedures involved in hydrolysis and fermentation were conducted similarly to the SHF experiment, but the duration of hydrolysis was reduced to 12 h.

5.3.3 Simultaneous Saccharification and Fermentation (SSF)

In the SSF experiment, different concentrations of microalgal biomass within the range 0.2-1.6% w/w were applied. The biomass was diluted using 1.5% w/w sulphuric acid and the slurry was autoclaved at 121 °C for 30min and then transferred into 500 ml Erlmenyer flasks. Cellulase, cellobiase and yeast were aseptically added at 5% (w/w of microalgal biomass). The nutrients mixture, 5 g/L yeast extract, 2 g/L Ammonium chloride (NH_4Cl), 1 g/L Potassium phosphate (KH_2PO_4), and 0.3 g/L Magnesium sulphate ($MgSO_4$), were added to the solution .The flasks were placed in an incubator at 30°C and 200 rpm for 50 hrs. 5 mL sample was taken after every 5 hours from each flask for analytical monitoring. α-amylase (5% w/w of microalgal biomass) was added to the solution in the second set of experiment in order to hydrolyse the starch present.

5.4 Analytical procedures

5.4.1 Quantification of simple sugars

Glucose concentration over time during the fermentation process was analysed using high pressure liquid chromatography (HPLC). The mobile phase used was a mixture of acetonitrile and water (85:15) at a flow rate of 1 mL/min. 30 μL sample was injected at 50 °C. The sample was filtered through a 13mm membrane filter prior to injection. The glucose concentration was evaluated using a calibration curve generated from a HPLC-grade glucose.

5.4.2 Quantification of bioethanol concentration

The bioethanol concentration was analysed using gas chromatography (GC) (Model 7890A, Agilent, CA). The GC consists of an auto sampler, flame ion detector (FID) and HP-FFAP column, 50 m x 0.20 mm x 0.33μm. The injector, detector and oven temperatures were maintained at 150, 200 and 120°C respectively. Nitrogen gas was used as the carrier gas. The bioethanol concentration was quantified using a calibration curve prepared by injecting different concentrations of ethanol standard (0.1-10%v/v).

6. Results and discussion

As shown in **Fig. 4,** it was generally observed that a lower concentration of biomass resulted in a higher bioethanol yield (g bioethanol/g biomass). This is partly due to enhanced interactions between available enzymes and the microalgal biomass. Also, high biomass concentrations could result in the production of inhibitors and toxins during the hydrolysis process, and this could retard the enzymatic activity. Therefore, the bigger inhibitory effect offset the advantage of biomass loading. Although SSF process does not have a separate hydrolysis stage, the enzymatic activity still affects the hydrolysis reaction which happened simultaneously with the fermentation.

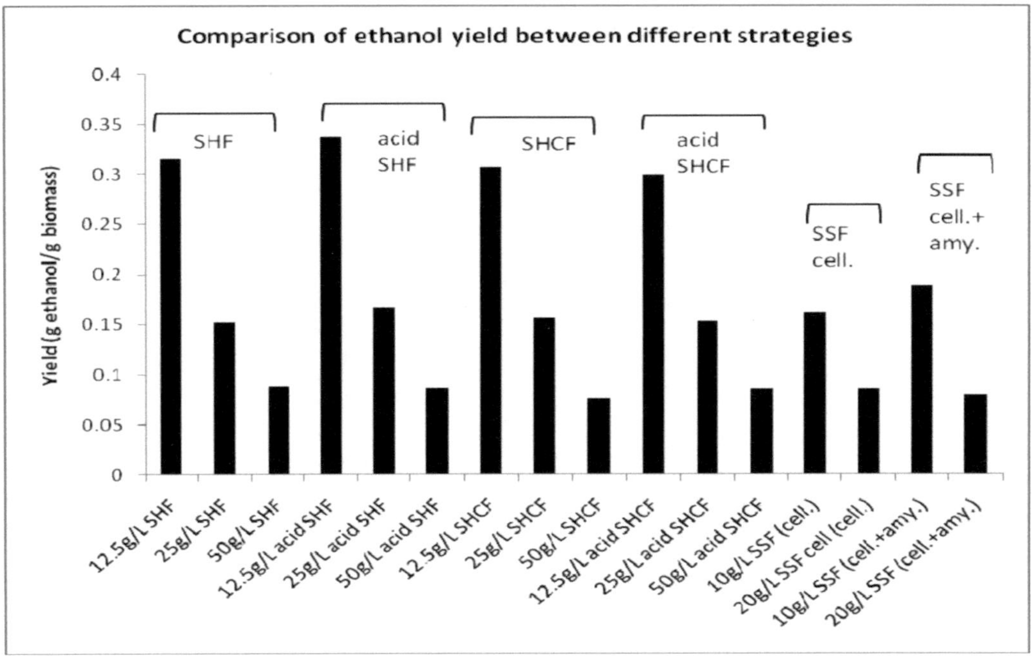

Fig. 4. Comparison of bioethanol yields for different process systems. All the fermentation processes were carried out at 30 °C for 48h, with the pH maintained around 5-6 and a shaking speed of 200rpm

*cell. = cellulose+cellobiose; amy.= amylase+amyloglucosidase

*g/L is the concentration of biomass

Overall, SHF generated the highest bioethanol yield and SSF the lowest, as shown in **Fig. 4**. It has been reported that SSF produces the highest bioethanol yield as it aims to reduce the contamination to the yeast by combining hydrolysis and fermentation into one step. However, due to inherent contamination during the anaerobic treatment, SSF did not achieve the desired yield. Furthermore, since hydrolysis and fermentation works at different temperatures, finding an optimal temperature for SSF is also a critical problem. Hence the unexpected lower yield from SSF could also be attributed to un-optimised temperature conditions. Another factor of influence is pH. A slight change in pH from the optimum valu could cause a significant change in fermentation yields because yeast and the enzyme are sensitive to pH. The performance of SHCF is slightly lower than SHF, possibly due to the shorter hydrolysis time. Therefore, theoretically, there was less monomeric sugar available for fermentation. Acid SHF process gave the highest yield among all the processes because the acid pretreatment step further disrupted the cell wall of the microalgae to release more fermentable sugars.

The SSF process with amylase and amyloglucosidase showed a significantly higher yield than the SSF process with only cellulase and cellobiase (**Fig. 4**). This is due to the hydrolysis of starch in the microalgae, which is 10% of the total mass, and the released sugars used for fermentation. However, the cost of enzymes will be a key parameter to consider against the increased bioethanol yield in a large-scale application.

Fig. 5 shows the glucose yield kinetics during the hydrolysis stage of SHF and SHCF. Glucose production progressively increased with time. Generally, SHF achieved higher yields than SHCF. This shows that the hydrolysis process requires an optimised duration at which the enzymatic activity is maximal. However, the main purpose of SHCF is co-fermentation, where a genetically engineered strain is used so that both hexoses and pentose can be utilised, but in this study only the hydrolysis stage of SHCF was investigated. The glucose yield profile is also consistent with the bioethanol yield profile. Hydrolysate viscosity also has an influence on glucose yield; lower viscosity of the hydrolysate produces higher glucose yield due to enhanced molecular interactions between the enzyme and the substrate. Since SHF was performed for 24 hours and more cellulose and starch were hydrolyzed, the viscosity of SHF hydrolysate was lower.

Acid SHF process is supposed to produce more glucose because the acid pretreatment step disrupts the microalgal cell wall to release more fermentable sugars. However, Figure 6 shows the opposite. A possible reason is that the high acid concentration damaged the structure of some sugars, making them unavailable for fermentation. More work is required to explore the influence of high acid concentration on the released sugar.

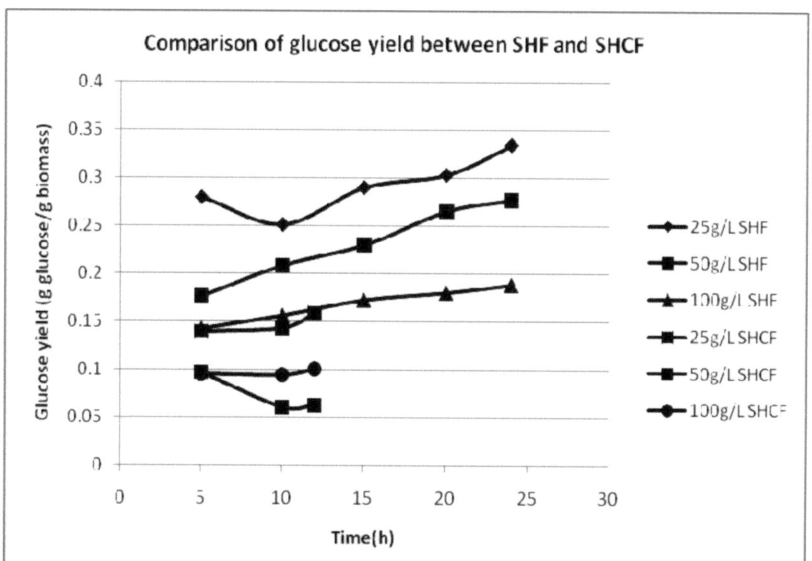

Fig. 5. Comparison of glucose yield between SHF and SHCF. All the hydrolysis processes were carried out at 40 °C with a shaking speed of 200rpm. The pH is maintained at 5 by sodium acetate/acetic acid buffer. SHF was run for 24h while SHCF was only performed for 12h.

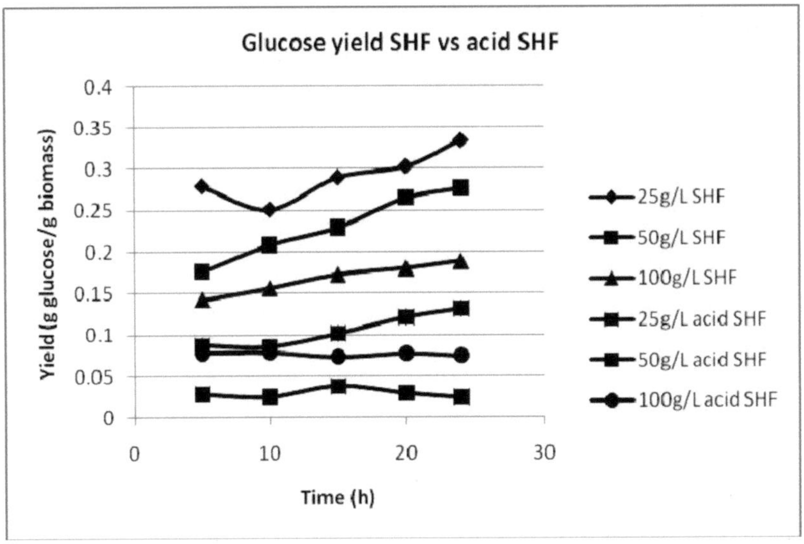

Fig. 6. Comparison of glucose yield between SHF and acid SHF. All the hydrolysis processes were carried out at 40 °C for 24h with a shaking speed of 200rpm. The pH is maintained at 5 by sodium acetate/acetic acid buffer.

7. Conclusions

From the experimental results, it is concluded that acid SHF gave the best performance in terms of bioethanol production. However, there is a potential to improve on the yield of SSF. This could be done by reducing contaminations, applying optimum process conditions, and using genetic engineered yeast strains which can convert pentoses into bioethanol. In industrial applications, the cost of feedstock and cellulolytic enzymes are the two major parameters that contribute to the cost of production. About 40-60% of the total production cost is from raw materials. An integrated approach could improve the production economics; hence it is the main industrial option. Continuous fermentation often gives a higher productivity than batch fermentation. Besides, it reduces inhibitory effects and offers ease of control. Continuous process also keeps the microbes at the exponential phase hence raises the productivity significantly with shorter processing time.

8. Acknowledgements

This work has been supported by the Department of Chemical Engineering, Monash University, Australia and the Ministry of Higher Education, Malaysia.

9. References

Alvira, P.; Tomás-Pejó, E.; Ballesteros, M. & Negro, MJ. (2010). Pretreatment technologies for an efficient bioethanol production process based on enzymatic hydrolysis: a review. *Bioresource Technology*, Vol. 101, No.13, (July 2010), pp. 4851-4861, ISSN 09608524

Balat, M.; Balat, H. & Oz, C. (2008). Progress in bioethanol processing. *Progress in Energy and Combustion Science*, Vol.34, No. 5, (October 2008), pp. 551–73, ISSN 03601285

Becker, EW. (1994). *Microalgae: Biotechnology and Microbiology*, Cambridge University Press, ISBN 0521350204, New York, USA

Chandel, AK.; Chan, ES.; Rudravaram, R.; Narasu, ML.; Rao, LV. & Ravindra, P. (2007). Economics and environmental impact of bioethanol production technologies: an appraisal. *Biotechnology and Molecular Biology Review*, Vol.2, No.1, (February, 2007), pp. 14-32, ISSN 1538-2273

Chen, ML. & Wang FS. (2010). Optimization of a fed-batch simultaneous saccharification and co-fermentation process from lignocellulose to ethanol. *Industrial & Engineering Chemistry Research*, Vol. 49, No.12, (Jun 2010), pp. 5775-5585, ISSN 08885885

Georgieva, TI.; Mikkelsen, MJ. & Ahring, BK. (2008). Ethanol production from wet exploded Wheat straw hydrolysate by thermophilic anaerobic bacterium Thermoanaerobacter BG1L1 in a continuous immobilized reactor. *Applied Biochemistry and Biotechnology*, Vol.145, No. 1-3, (March 2008), pp. 99–110, ISSN 02732289

Harun, R. ; Danquah, MK. & Forde, GM. (2009). Microalgal biomass as a fermentation feedstock for bioethanol production. *Journal of Chemical Technology & Biotechnology*, Vol. 85, No.2, (February 2010), pp. 199-203, ISSN 0268-2575

Harun, R. & Danquah, MK. (2010a). Influence of acid pre-treatment on microalgal biomass for bioethanol production. *Process Biochemistry*, Vol.46, No.1, (January 2011), pp. 304-309, ISSN 13595113

Harun, R.; Singh, M.; Forde GM. & Danquah MK. (2010b). Bioprocess engineering of microalgae to produce a variety of consumer products. *Renewable and Sustainable Energy Reviews*, Vol. 14, No.3, (April 2010), pp. 1037-1047, ISSN 13640321

Hahn-Hagerdal, B.; Karhumaa, K.; Fonseca, C.; Spencer-Martins, I. & Gorwa-Grauslund, M. (2007). Towards industrial pentose-fermenting yeast strains. *Applied Microbiology and Biotechnology*, Vol.74, No.5, (April 2007), pp. 937-953, ISSN 01757598

Hendriks, ATWM. & Zeeman, G. (2009). Pretreatments to enhance the digestibility of Lignocellulosic biomass. *Bioresource Technology*, Vol.100, No.1, (January 2009), pp.10-18, ISSN 09608524

Jin, M.; Lau, MW.; Balan, V. & Dale, BE. (2010). Two-step SSCF to convert AFEX-treated switchgrass to ethanol using commercial enzymes and Saccharomyces cerevisiae 424A(LNH-ST). *Bioresource Technology*, Vol. 101, No. 21, (November 2010), pp. 8171–8178, ISSN 09608524

Kuhar, S.; Nair, LM. & Kuhad, RC. (2008). Pretreatment of lignocellulosic material with fungi capable of higher lignin degradation and lower carbohydrate degradation improves substrate acid hydrolysis and eventual conversion to ethanol. *Canadian Journal of Microbiology*,Vol.54, No.4, (April 2008), pp. 305–313, ISSN 00084166

Kumar, R. & Wyman, CE. (2009). Effects of cellulase and xylanase enzymes on the deconstruction of solids from pretreatment of poplar by leading technologies. *Biotechnology Progress*, Vol.25, No.2, (March 2009), pp. 302–314, ISSN 87567938

Li, H.; Kim, NJ.; Jiang, M.; Kang, JW. & Chang, HN. (2009). Simultaneous saccharification and fermentation of lignocellulosic residues pretreated with phosphoric acid-acetone for bioethanol production. *Bioresource Technology*, Vol.100, No.13, (July2009), pp. 3245-51, ISSN 09608524

Lin, Y. & Tanaka, S. (2006). Ethanol fermentation from biomass resources: current state and prospects. *Applied Microbiology and Biotechnology*, Vol.69, No.6, (February 2006), pp.627-642, ISSN 01757598

Lynd, LR.; van Zyl, WH.; McBride, JE. & Laser, M. 2005. Consolidated bioprocessing of cellulosic biomass: an update. *Current Opinion in Biotechnology*, Vol.16, No.5, (October 2005), pp. 577–583, ISSN 09581669

Mosier, N.; Wyman, CE. ; Dale, BD. ; Elander, RT. ; Lee, YY. ; Holtzapple, M. & Ladisch, CM. (2005). Features of promising technologies for pretreatment of lignocellulosic biomass. *Bioresource Technology*, Vol.96, No.6, (April 2005), pp. 673–86, ISSN 09608524

Saha, BC.; Iten, LB.; Cotta, MA. & Wu, YV. (2005). Dilute acid pretreatment, enzymatic saccharification and fermentation of wheat straw to ethanol. *Process Biochemistry*, Vol. 40, No.12, (December 2005), pp. 3693-700, ISSN 13595113

Sanchez, G.; Pilcher, L.; Roslander, C.; Modig, T.; Galbe, M. & Liden, G. (2004). Dilute acid hydrolysis for fermentation of the Bolivian straw material Paja Brava. *Bioresource Technology*, Vol.93, No.3, (July 2004), pp. 249-256, ISSN 09608524

Singh, P.; Suman, A.; Tiwari, P.; Arya, N.; Gaur, A. & Shrivastava, AK. (2008). Biological pretreatment of sugarcane trash for its conversion to fermentable sugars. *World Journal of Microbiology and Biotechnology*, Vol.24, No.5, (May 2008), pp. 667–673, ISSN 09593993

Sun, Y. & Cheng, J. (2002). Hydrolysis of lignocellulosic materials for ethanol production: a review. *Bioresource Technology*, Vol. 83, No.1, (2002), pp. 1–11, ISSN 09608524

Tomas-Pejo, E.; Oliva, JM.; Ballesteros, M. & Olsson, L. (2008). Comparison of SHF and SSF processes from steam-exploded wheat straw for ethanol production by xylose fermenting and robust glucose-fermenting Saccharomyces cerevisiae strains. *Biotechnology and Bioengineering*, Vol.100, No.6, (August 2008), pp. 1122-1131, ISSN 00063592

Wayman, C. (1996). *Handbook of Bioethanol: Production and Utilization*, Taylor & Fracis, ISBN 1-56032- 553-4, Washington, USA

Xu, C.; Qin, Y.; Li, Y.; Ji, Y.; Huang, J.; Song, H. & Xu, J. (2010). Factors influencing cellulosome activity in Consolidated Bioprocessing of cellulosic ethanol. *Bioresource Technology*, Vol.101, No.24, (December 2010), pp. 9560–9, ISSN 09608524

21

Microbial Conversion of Biomass: A Review of Microbial Fuel Cells

Cagil Ozansoy and Ruby Heard
School of Engineering and Science, Victoria University
Melbourne, Victoria,
Australia

1. Introduction

The cleaner generation of energy is a vital concept if we are to ensure the survival of our current lifestyle past the depletion of the Earth's fossil fuel supply. The study described in this chapter investigates the alternative energy producing method of microbial conversion of biomass to produce electrical energy. Currently techniques are being explored to minimise the cost of building and running cells in which microbial conversion takes place. There are several areas where improvements may be made including the physical design of the cell, the catalyst used in the cathode chamber, the membrane material and the ideal low-cost substrate. The study presented in this chapter provides a review of Microbial Fuel Cells (MFCs), and discusses the design and construction of MFCs and presents the results of tests carried out with a number of constructed cells.

There is no doubt that the world's increasing population is rapidly depleting planet's finite energy resources. It is common knowledge today that the techniques developed to produce electricity and run our vehicles have also been leading to temperature rises resulting in climate changes all over the world. Currently, the world consumes approximately 13 Terawatts of power (Chae et al., 2009) every year while a mere 20 % of this is produced from renewable sources (Hopwood & Cohen, 2000). Techniques of harnessing energy from renewable sources are continuously being further developed all over the world, but are still yet to reach the efficiency levels, which would make these techniques comparable to the traditional, yet unsustainable coal-fired generation.

The discovery of a process in which bacteria are utilised in the fermentation of organic substrates to produce electricity can assist with the transition towards more sustainable ways of energy generation. This is the concept of a Microbial Fuel Cell (MFC). Later, it was also discovered that a variation of the MFC design could produce hydrogen instead through electrolysis. This was termed a Microbial Electrolysis Cell (MEC) (Logan, 2008).

Hydrogen is the most abundant element in the universe with more than 9 out of every 10 atoms being hydrogen atoms. Our most precious resource is made mainly of hydrogen. The Sun's gravitational force pulls hydrogen atoms together which releases helium and energy by the process of fusion. This energy is received on earth as the light and heat which sustains life on Earth. Hydrogen is recognised as an attractive energy carrier due to its clean, efficient and renewable nature (Chae et al., 2009) . It has applications in the production of ammonia and methanol, the refining of metals and most recently as a clean fuel for powering vehicles. Hydrogen can be produced in a number of ways including as a by-product during the cracking of crude oil or by way of electrolysis in a diaphragm cell (Knapp, 2002). The majority of hydrogen gas produced today is developed from fossil fuels contributing to the release of carbon dioxide (Logan et al., 2008). An MEC improves on traditional hydrogen production technology by producing hydrogen yields many times greater than fermentation and at greater energy efficiencies than water electrolysis (Call & Logan, 2008; Logan et al., 2008).

Both MFCs and MECs have great potential in the renewable energy trend. As such many researchers across the world are investigating improvements to the two different systems. While the physical cell structures which can be used are the same, it is the reactions inside the cell which differ. In a microbial fuel cell, bacteria attached to the anode oxidise organic material releasing carbon dioxide and protons into the anode chamber solution. Electrons are transferred to the anode itself which then flow through an electrical circuit to the cathode where they are consumed in the reduction of oxygen. Meanwhile protons cross into the cathode chamber via a membrane. A current is therefore produced as there is a flow of electrons. However, in the absence of oxygen in the cathode chamber no current will be produced. This leads to the variations which constitute an electrolysis cell. With the addition of a small voltage between the anode and cathode, protons now become reduced at the cathode and hydrogen gas is produced (Logan et al., 2006; Call & Logan, 2008; Logan & Cheng, 2008). This process is known as electrohydrogenesis (Logan et al., 2008).

In an attempt to increase the power and hydrogen produced by MFCs and MECs respectively, researchers have recognized a number of areas where variations in design can be studied. Numerous papers have been published as to the effects of differing pH, temperature, electron acceptor, electrode surface area, reactor size and electrode and membrane material (Logan et al., 2008). Many carbon and graphite electrode materials from carbon paper to graphite pencil leads have been trialled with the common conception that graphite brushes produce the best results due to their large surface area (Logan, 2008). Reactor designs are limited only by the imagination and many variations of the original two bottle design have emerged including single chamber, cube, cylinder and U-shaped reactors.

A review of MEC technology has identified several research areas that must still be addressed before MECs can be considered a mature hydrogen production method. These observations are:

- More experience is required with real organic feed stocks containing complex organic substrates such as polymeric and particulate substances;
- Novel, more cost-effective chemical and/or biological cathodes need to be developed that show low potential losses and are not platinum-based;
- Membrane pH gradients need to be eliminated, or membranes should not be used in the reactor;
- Methanogenic consumption of the hydrogen product needs to be prevented (in the case of membrane-less MECs and/or MECs with a biocathode); and, most critically,
- A cost-effective, scalable MEC design needs to be developed.

As mentioned above, often cathodes are platinum based (He & Angenent, 2006; Logan & Cheng, 2008). The platinum acts as a catalyst and is necessary to enhance the rate of reduction of oxygen at the cathode (Logan et al., 2008). As platinum is highly expensive and can be poisoned by components of the substrate solution (Logan et al., 2008; You et al., 2009) it is desirable to eliminate it from MFCs and MECs altogether. Several alternatives have been investigated including chemical catholytes, biocathodes and transitional metals (He & Angenent, 2006). Thus far the highest power ever achieved in an MFC has been using a ferricyanide cathode (He & Angenent, 2006). In order to minimise costs and work towards the eradication of platinum potassium permanganate has been used at the cathode.

2. Problem analysis

There are two significant energy related challenges facing the world. The first one is the production of cleaner, renewable energy at high efficiencies and low costs, the other, the consideration of efficient storage techniques for excess or intermittently produced electricity. Although renewable forms of energy have no or little negative environmental impacts they can have negative social, economical and technical problems associated with them. Wind turbines for instance create cleaner and renewable energy but are extremely expensive to set up and maintain. There are also social issues regarding their location as many people find them to be noisy and unattractive. Hydropower is one of the cleanest forms of renewable energy but it too is bound by locations problems. Hydropower requires suitable dam locations and most importantly, large amounts of water, something that many countries including Australia cannot provide. Solar power again is expensive and suffers low efficiencies while also having environmental effects. Solar panels require large amounts of energy to be produced and silicon production has recently led to unsafe toxic waste disposal. A potential solution to many of these problems are the fuel cells. They are highly efficient, reliable, are noiseless, emission-less and run on widely available fuels.

One problem with several forms of renewable energy is that it is produced intermittently, that is power is only produced at certain times not continuously. Such generation techniques include photovoltaic solar power and wind turbines as they both rely on the natural phenomena of sunlight and wind respectively. For these technologies to be viable options to replace less environmentally friendly generation methods the energy produced during generation periods must be able to be stored for use when generation is not possible. For example, solar power cannot be generated at night or in periods of low sunlight levels however the energy produced on sunny days could be stored and then used at these times. Many forms of energy storage exist including mechanical, chemical, thermal, electrochemical, electrical and biological storage. Other forms of clean, renewable energy are unaffected by environmental factors and can produce electricity constantly. These forms include hydropower, geothermal and fuel cells to name a few.

As previously mentioned, some renewable energy generation techniques require the energy to be stored for later use, in fact this is considered one of the biggest challenges affecting the solar power industry (Zyga, 2009). Electric cars are also being held back by the lack of appropriate storage techniques as batteries are heavy and inefficient when compared with fuels like petrol and gasoline (HowStuffWorks.com, 2000). World energy consumption is predicted to rise 44 percent over the next twenty years with green house gas emissions increasing a staggering 39 percent (Finfacts-Team, 2009). Obviously, this will have significant impact on climate change and put pressure on the power industry to cope with the increasing demand. Already high demand puts strain on the industry, not to mention on the customers wallets. Peak energy periods usually incur a higher cost to consumers.

The power grid also experiences problems of over-demand in extreme weather conditions where the use of heaters/air conditioners is extremely high. This generally results in power outages for long periods of time. In the United States 14 percent of the country's power plants and known as "peaking plants," expensive gas turbines that are only operated during rare hours of extremely high demand (Plumer, 2009). This is an instance of where efficient, cost effective storage methods are highly desirable. The use of renewable generation techniques is predicted to increase throughout the world to contribute to the populations growing energy needs. Therefore, there will also be a greater need for energy storage to facilitate the continuous use of energy from intermittent sources.

3. Possible solutions

This section presents a review of different types of fuel cells and a comparative analysis in terms of design and construction difficulty, material expenses and potential efficiencies. The two main cells chosen are Hydrogen and Microbial Fuel Cells. Viable options are considered for all the necessary components including; a membrane, electrodes and a catalyst as well as possibilities for reactor configurations and substrates.
Several solutions to designing a clean energy storage device have also been evaluated for their suitability to store the electrical energy produced by the fuel cell. The energy can be stored in many forms such as kinetic, chemical, electrochemical and electrical. These types of energy storage have all been reviewed as candidates. The required solution must again prove to be of reasonable expense and effort to design, build and implement.

3.1 Fuel cells

A fuel cell is a device that is capable of converting the chemical energy in a fuel and an oxidant into electricity and a clean by product (FCTec, 2010). Fuels can consist of traditional compounds like hydrogen, natural gas, methanol and gasoline while a common oxidant is oxygen. A fuel cell is an electrochemical system like a battery however a fuel cell does not run down or need recharging; as long as a fuel and oxidiser are present the cell will generate electricity.
All fuel cells are constructed and operate in the same general manner. Figure 1 shows a cell that consists of three materials sandwiched together, namely an anode, electrolyte and cathode. The anode and cathode are known as electrodes and are made of some form of conductive material. The anode and cathode are where the oxidation of the fuel and reduction of the oxidant occur respectively. This is facilitated by a catalyst, which can consist of a range of substances. When the fuel is oxidised, electrons and protons are produced. The protons are able to flow through the electrolyte as it is a substance that allows ion exchange while electrons are blocked. Electrons therefore pass through the anode and to the cathode via an external circuit where they react with the protons and oxidant. This results in a by-product (usually water), while the flow of electrons constitutes an electrical current and therefore electricity generation.
Fuel cells are suitable for a wide range of applications. They are particularly useful as power sources in remote locations such as spacecraft, remote weather stations, rural locations, and even have military applications. Currently the use of fuel cells in cogeneration is a rapidly growing area for office buildings and factories. The advantage of homes and businesses using fuels cells is the savings on grid electricity and that excess energy produced can even be sold back to the grid for profit. Future prospects for the technology includes powering road and sea vehicles, providing off-grid power supplies and emergency power, micro fuel cells to implement in small appliances and as portable charging docks.

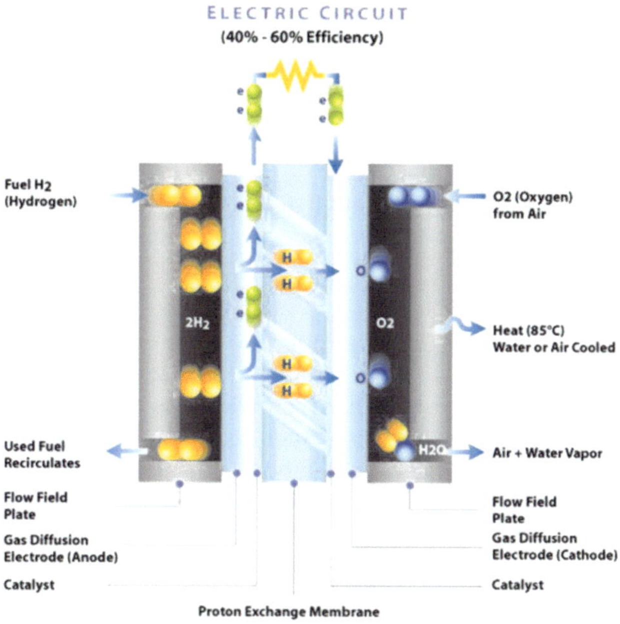

Fig. 1. Basic configuration of a fuel cell (Ballard Power Systems)

3.1.1 Hydrogen fuel cell

The most common form of fuel cell is the hydrogen fuel cell. In this case hydrogen is the fuel and oxygen is the oxidant. Pressurised hydrogen gas enters the anode chamber and is forced through a catalyst (usually platinum) by the pressure. When hydrogen comes in contact with the platinum it is split into two H+ ions and two electrons. Electrons conduct through the anode and to the cathode via an external circuit. Meanwhile in the cathode chamber oxygen gas is also forced into the catalyst where it splits into two strongly negatively charged oxygen atoms. The strong negative oxygen atom charge attracts the positively charged H+ ions through the membrane where they combine with the oxygen atoms and the electrons to form a water molecule (Nice & Strickland, 2010).

Design's can be as simple as using two pieces of platinum coated wire in a glass of water (Field, 2008) however platinum is extremely expensive and the source of fuel also presents difficulties. Water can be used as the source of hydrogen but there must be a way of splitting the atoms to release the hydrogen and oxygen atoms. This process is known as electrolysis and requires energy. Alternatively stored hydrogen and oxygen gas can be pumped into the cell. Efficiencies achieved can be up to 80% (Nice & Strickland, 2010).

3.1.2 Microbial fuel cell

A microbial fuel cell utilises the electron extracting properties of special bacteria attached to the anode to produce electricity. Bacteria attached to the anode oxidise organic material releasing carbon dioxide and protons into the anode chamber solution. Electrons are transferred to the anode itself which then flow through an electrical circuit to the cathode where they are consumed in the reduction of oxygen. Meanwhile protons cross into the cathode chamber via a membrane. A current is therefore produced as there is a flow of electrons.

A range of alternatives are available for all components while a range of organic material can be used as the cell's fuel. One option is to use wastewater in the anode chamber. This has the added benefit of water purification while electricity is produced. There is also no need to deal with hard to store gases such as hydrogen which are also expensive to produce. Organic matter can be sourced simply from household food scraps or garden waste. This makes a microbial fuel cell easy and inexpensive to run. Alternative catalysts are available to platinum which significantly reduces cell costs. Efficiencies achieved are comparable to those achieved with hydrogen fuel cells.

3.2 MFC components

An MFC requires much the same components as any other fuel cell however there are some drastic differences. Most obvious is the requirement for bacteria which is not an addition to any other type of fuel cell. Where most other fuel cells incorporate chemicals to achieve their electricity producing reactions, a microbial fuel cell requires a form of organic matter (substrate) to maintain the bacteria also providing the means of generation. Reactions using oxygen as the terminal electron acceptor are extremely slow so either a catalyst must be used or a terminal electron acceptor with a faster reaction time must replace oxygen in the cathode chamber. These alternatives for these things, along with the more conventional components; membrane and electrodes as well as reactor configuration, are discussed below.

3.2.1 Reactor

Reactor configurations vary greatly from researcher to researcher. The size and shape of a reactor is entirely up to the designer and as no stand out design has been identified the variations continue with the goal of finding a scalable design. To date reactors have been cube shaped, cylindrical, horse shoe shaped, two chamber and single chamber and H-type configured and made of glass and various types of plastic, even buckets. Sizes also vary widely with some reactors having volumes of a few square centimetres and others of up to a square metre. So far researchers have speculated that single chamber reactors may show the most promise but this has not deterred people from using two chamber types. In terms of construction difficulties a single chamber reactor can be the harder of the two options. For this reason the reactor for this research project will be a two chambered cell and to further simplify the design and construction a cube shape will be used. Figure 2 depicts the proposed cell design. The total reactor capacity will be 1.1L, 600mL in the anode chamber and 500mL in the cathode chamber.

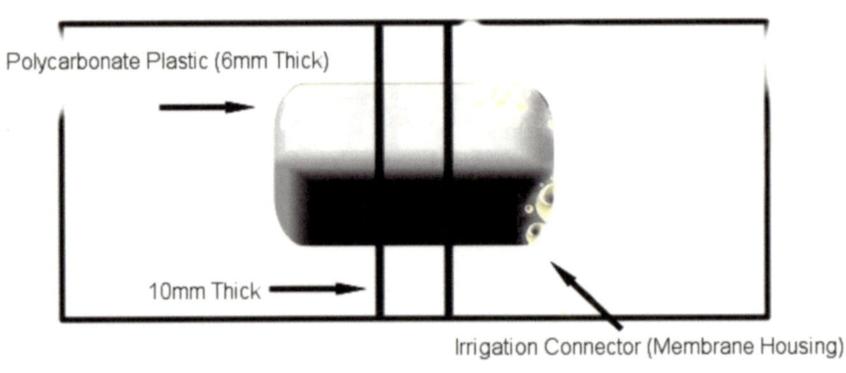

Fig. 2. Basic reactor design

3.2.2 Membrane

A membrane is described as a layer of material which serves as a selective barrier between two mediums that is impermeable to specific particles, molecules or substances. Membranes occur naturally in the bodies of plants and animals and are made artificially for separation purposes in laboratories and industry. Synthetic membranes include anion exchange membranes, cation exchange membranes and ultra filtration membranes. Studies have shown that anion exchange membranes perform better than cation exchange membranes due to a lower resistance (You et al., 2009) . These types of membranes are generally very expensive, have high minimum orders or incur large freight charges as they are only manufactured overseas. An alternative is membrane inclusive water resistant clothing such as Gore-Tex. High quality ski clothing is specially made to contain a membrane within the fabric to repel water. The fabric is suitable for fuel cell applications as it successfully separates the liquids in the two chambers whilst allowing protons to flow from the anode to the cathode chamber. Table 1 contains several membrane alternatives that would be suitable for the microbial fuel cell.

Company	Product	Description	Size (cm)	Price (AUD)	Freight (AUD)	Total (AUD)
FeMa Tech Germany	F-930-EW900 Thickness 30	Ion exchange membrane	20 x 30	$101.64	$73.653	$175.29
Fuel Cell Store	Nafion RE212	Proton exchange membrane	10 x 10	$26.17	$163.89	$190.06
Entrant	Ski Pants	Fabric sourced from pants	20 x 20	Free	$0	$0
Membranes International	AMI-7001S	Anion exchange membrane	15 x 15	Free	$0	$0

Table 1. Membrane solutions

3.2.3 Electrodes

All fuel cells require two electrodes, an anode and a cathode. These electrodes facilitate the transport of electrons through an external circuit hence resulting in electricity. Electrodes can be made of any non corrosive, conductive material. Most commonly used materials include carbon, graphite and steel. Steel has been found to be less effective for use in microbial fuel cells as it is not a porous material and bacteria appear to be unable to attach themselves (Logan, 2008) . Carbon and graphite are both widely available materials which come in many forms. Carbon is available as paper, cloth and foam while graphite comes in the form of rods, granules and brushes. Studies have not shown a distinct favourability towards neither carbon nor graphite so we will assume that performance is very similar and depends more heavily on electrode surface area. All above mentioned products are quite expensive and almost entirely produced overseas so there is also a high freight cost involved. Several companies were found to offer free samples of carbon foams and cloths however only one company does not charge a shipping cost for the sample. For this reason vitreous reticulated carbon foam has been chosen for the electrodes. Foam has an advantage over paper and cloth as it is more porous and has a greater surface area to house more bacteria.

3.2.4 Substrate

A substrate is the substance contained in the anode chamber that is to be oxidised. As mentioned earlier this can include fuels such as hydrogen and gasoline. In a microbial fuel cell the substrate used can be any form of organic matter. Cells have been successfully operated on chocolate (Markusic, 2010) , wine (Danigelis, 2009) , wastewater (Logan, 2008) , acetate (Liu et al., 2005; Sun et al., 2008), glucose (Logan, 2008) and more. Most frequently glucose, wastewater and acetate are used in experiments with the highest results being obtained with acetate (Logan, 2008) .

3.2.5 Catalyst/catholyte

The cathode chamber is where protons and electrons recombine and reduce an electron acceptor. A common electrode acceptor is oxygen due to its abundance in air. When oxygen is used however the reaction is very slow therefore the need for a catalyst arises. Most MFC's use platinum as the catalyst (Logan, 2008) however this is extremely expensive. Due to the expense, which affects the viability of fuel cells, much research is aimed at finding an equally efficient but less expensive catalyst. One option is to use a catholyte to replace oxygen as the terminal electron acceptor. Chemicals such as ferricyanide and potassium permanganate have been used successfully with results comparable to those achieved with platinum (He & Angenent, 2006). These chemicals are far less expensive than platinum however the disadvantage is that they are consumed in the reaction and must be replaced. Research has also been conducted into the use of bacterial cathodes known as biocathodes. The concept of a biocathode is that bacteria are attached to the cathode as they are to the anode. Bacteria then assist the reduction of oxygen without the need for any additional chemicals or substances (He & Angenent, 2006).

3.3 Common storage techniques

Energy storage is not a new development in fact people have been storing energy far before the discovery of electricity. Energy storage can be achieved in many ways; some techniques are simple such as those used in grandfather clocks while others involve complex chemical reactions.

3.3.1 Mechanical (flywheel)

A flywheel is a form of mechanical storage that far predates any battery; in fact it is one of humanities oldest and most familiar technologies. Examples are the potter's wheel which dates back six thousand years and more relevantly today the mechanism regulating the strokes of pistons in a car engine. A flywheel is simply a wheel on an axle which is able to store and regulate energy by continuously spinning. When spun at high speeds a fly wheel becomes a bank for massive amounts of kinetic energy which can be drawn out when required. Fly-wheel based batteries can reach energy densities 3-4 times higher than traditional lead-acid cell batteries. Another advantage of flywheels is that they are able to charge to full capacity in a matter of minutes rather than hours and discharge quickly without damage. They are also unaffected by extreme temperatures and have an impressive efficiency of 85-95% as well as a lifetime in decades (Putnam, 2007). A flywheel may not be appropriate to store the energy produced by the fuel cell in this project due to safety concerns associated with them caused by the high speed rotor. There is a possibility of the rotor breaking loose and releasing energy is an uncontrolled fashion (Rayner, 2008). They also experience a high current loss.

3.3.2 Electrochemical (battery)

A battery is a form of electrochemical energy storage, storing chemical energy and converting it to electrical energy. Chemicals inside the battery produce electrons which cling to the negative terminal. When the negative terminal is connected to the positive terminal via a conductor, electrons flow from the negative terminal to the positive terminal releasing their energy to a load. A battery is charged by doing the reverse of this action and applying an electrical current to the terminals. The disadvantage of batteries is that they are expensive and also heavy. This causes a significant problem for electric vehicles as weight is an important limiting factor in vehicle speed. One application of fuel cells is as an alternative to gasoline. For this reason batteries are not a suitable option for the storage of power from the fuel cell in this project.

3.3.3 Chemical (hydrogen)

Hydrogen is the most copious element in the universe with more than 9 out of every 10 atoms being hydrogen atoms. In fact our most precious resource is made mainly of hydrogen. The Sun's gravitational force pulls hydrogen atoms together which releases helium and energy by the process of fusion. We receive this energy as the light and heat which sustains life on Earth. Hydrogen is recognised as an attractive energy carrier due to its clean, efficient and renewable nature (Chae et al., 2009) . It has applications in the production of ammonia and methanol, the refining of metals and most recently as a clean fuel for powering vehicles. Hydrogen can be produced in a number of ways including as a by-product during the cracking of crude oil or by way of electrolysis in a diaphragm cell (Knapp, 2002) . The majority of hydrogen gas produced today is developed from fossil fuels contributing to the release of carbon dioxide (Logan et al., 2008). Hydrogen can also be produced by a process called electrolysis using either a hydrogen electrolysis cell or a microbial electrolysis cell. A Microbial Electrolysis Cell improves on traditional hydrogen production technology by producing hydrogen yields many times greater than fermentation and at greater energy efficiencies than water electrolysis (Call & Logan, 2008; Logan et al., 2008).

Hydrogen is a good storage option for this project and a microbial electrolysis cell is a suitable addition because it is closely related to the project fuel cell. The efficiencies of this storage technique are also very high. As such this will be suggested as the most suitable storage system to use.

4. Design and construction

This section outlines the design process, construction and testing of two MFC designs. The design process is explained in detail, construction strategies given, and results presented.

4.1 Design 1

Fuel cell design is an important factor in the success of an MFC/MEC. Single chamber cells have evolved from the original two chamber design in an attempt to eliminate the need for a membrane (Call & Logan, 2008). This is desirable to simplify reactor architecture and reduce the internal resistance caused by the inclusion of a membrane between chambers. It is also possible to lower the internal resistance of a two chamber cell by reducing the physical distance between the anode and cathode and increasing the area of the membrane. This is where the two bottle 'H-type' design falls short (Logan et al., 2008) and where cube and cylinder models show improvements. The prototype constructed in this experiment is a cube design for this reason.

4.1.1 Design and construction

The cell is constructed of a sturdy polycarbonate plastic with the anode chamber holding 600mL, slightly more than the cathode chamber which holds 500mL. The cell is airtight for use in MEC mode having just four valves which allow for helium to be flushed into both chambers, carbon dioxide to be removed from the anode chamber and hydrogen to be extracted from the cathode chamber. Figure 3 is a two dimensional, not to scale representation of the cell before any electrodes are incorporated.

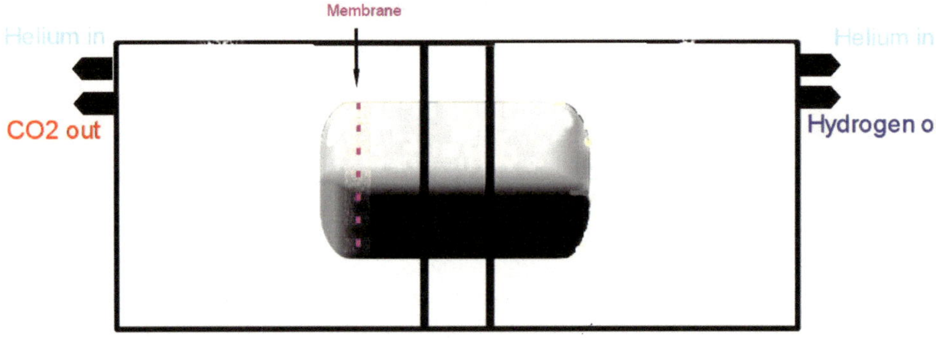

Fig. 3.Two dimensional conceptual model of the fuel cell without electrodes

The original complete reactor design is shown in Figure 4. As shown in the figure, the electrodes used are reticulated vitreous carbon foam. These are suspended from the lid of the cell by metal clips which attach to an external circuit containing a ten ohm resistor outside the cell. The membrane is a common ski clothing material known as Entrant. The material has been designed specifically as wet weather gear as the fabric contains a waterproof membrane. The fabric is suitable for use as a membrane as it will allow protons to pass through while preventing the contents in the two chambers from contaminating one another. Similar materials such as Gore Tex have also been shown to be effective as fuel cell membranes (Blair, 2008).

The membrane is held in place by an irrigation pipe fitting allowing a circular area of 3.5 cm for proton flow. The total cost of essential materials is AUD$57.55 for the MFC setup and $68.59 for the MEC setup (not including the cost of an inert gas for oxygen removal). Figure 5 is a photograph image of the finished reactor.

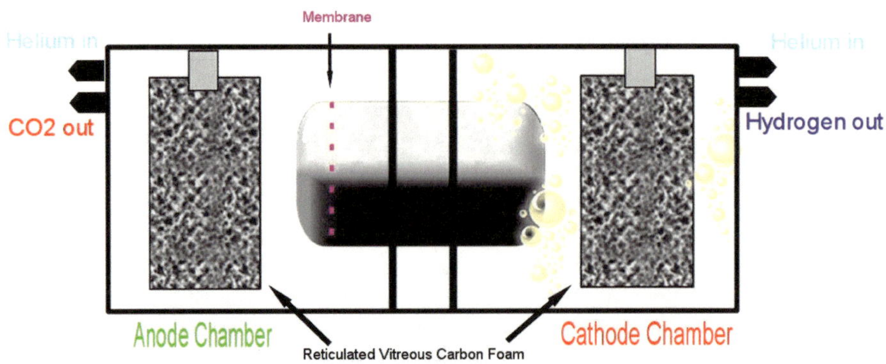

Fig. 4. Two dimensional diagram of complete fuel cell

Fig. 5. Photographic image of the designed and constructed cell

Each chamber contains an inexact buffer solution to stimulate bacteria growth. Due to limited chemical resources where a precisely measured buffer solution would otherwise be used, fertilisers containing the necessary elements were added to each chamber. The anode was inoculated over 12 hours with water from the Maribyrnong River to allow time for the bacteria to attach to the electrode. The anode chamber was then filled with 300mL river water, the buffer ingredients and 300mL vinegar as the substrate. The cathode chamber contained only filtered water and the nutrients. The cell was then left over a period of 9 days to allow for bacterial growth.

4.1.2 Results and discussion

The cell was firstly operated in the MFC mode over a period of 90 hours with potassium permanganate used as the electron acceptor. At the instant of adding the catholyte the system produced 1.7mV. Measurements were taken as frequently as was possible without a system in place for automatic sampling. The recorded voltages and calculated current densities for this period are shown in Figure 6. The cell was not kept under temperature controlled conditions and it was observed that the voltages increased and decreased throughout the days with the fluctuating temperature. To investigate this more closely between the period of 64 and 76 hours from start both the voltage and temperature were recorded every fifteen minutes as shown in Figure 7.

The maximum voltage obtained throughout the cell running time occurred during this period. At 74.5 hours from start a voltage of 28.7mV was observed at a temperature of 39°C. The corresponding current density is 404.8mA/m^2 with power normalised by anode surface area of 11.62mW/m^2 and volumetric power 50.05mW/ m^3. The system failed to produce hydrogen when operated as an MEC. Further experimentation is needed to determine if the reason for this was due to the presence of oxygen despite the cell being flushed with helium or due to the short running time and therefore lack of highly developed bacterial communities.

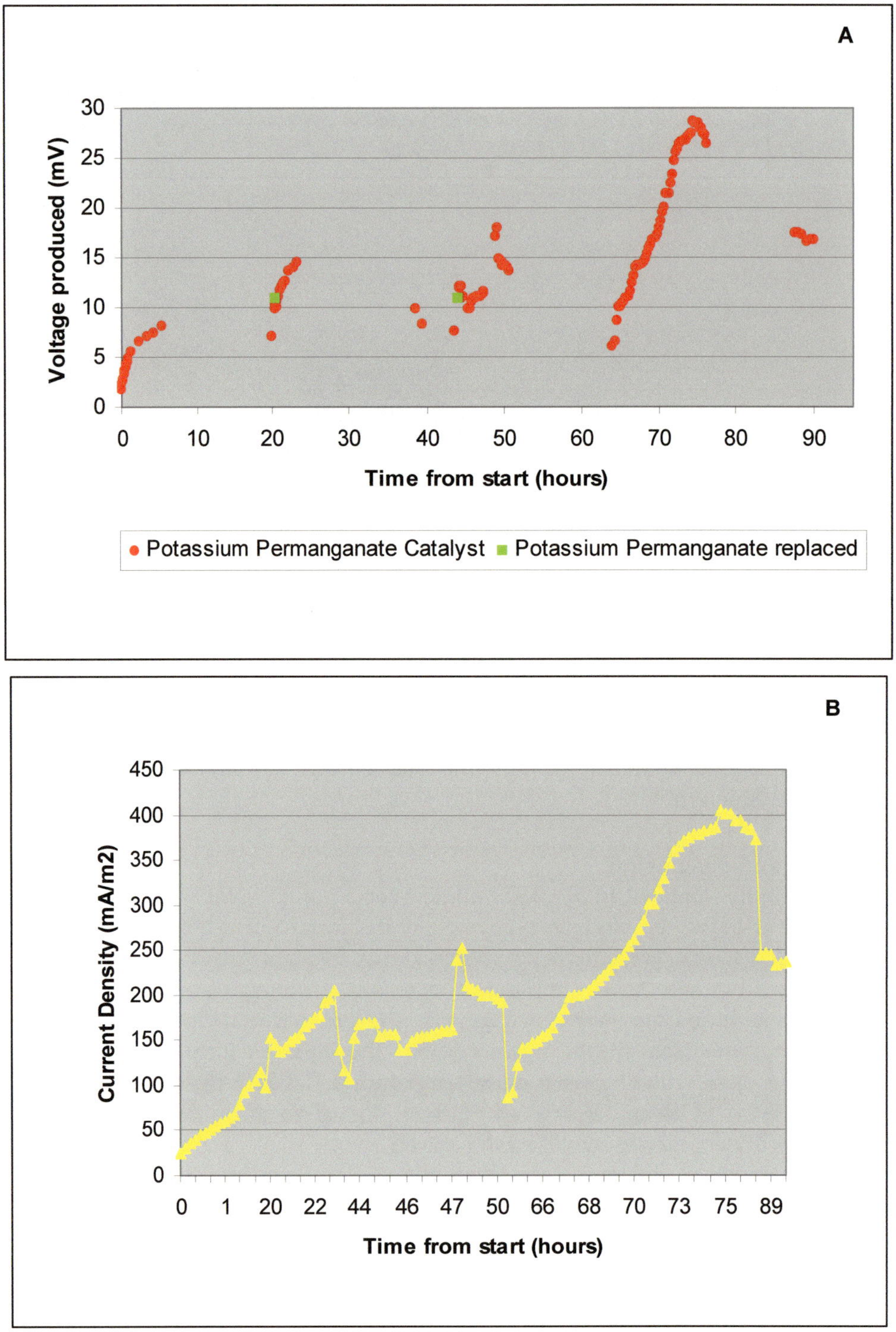

Fig. 6. A) Recorded voltages from start time (Oh) to end time (90h) and B) calculated current densities over the same period.

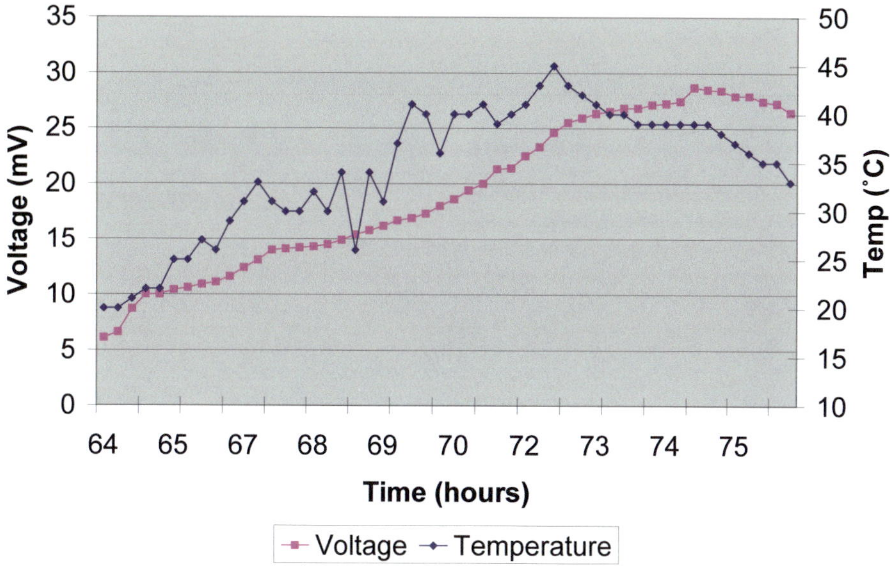

Fig. 7. Effect of varying temperature on voltage produced for the period 64 to 76 hours after start.

4.2 Design 2

Many possibilities could be identified to explain the low performance of the cell. The most obvious ones are the high internal resistance, inadequate bacterial growth and unsuitable/insufficient organic matter present. A combination of these may even have been the case.

4.2.1 Design and construction

In order to overcome the original cells problems several improvements were proposed. These included replacing the Entrant material with an anion exchange membrane, using sewage as both the bacteria source and the organic matter and increasing the amount of electrode material. An illustration of these improvements can be found in Figures 8 and 9. Figure 9 shows a photographic image of the completed reactor.

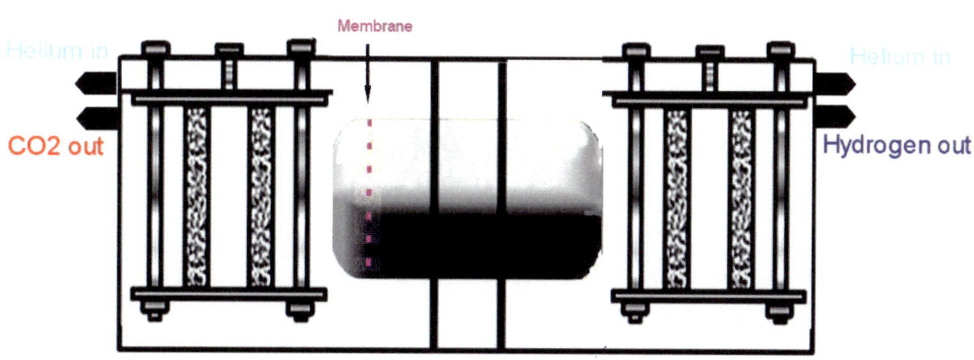

Fig. 8. Two dimensional diagram of the improved cell

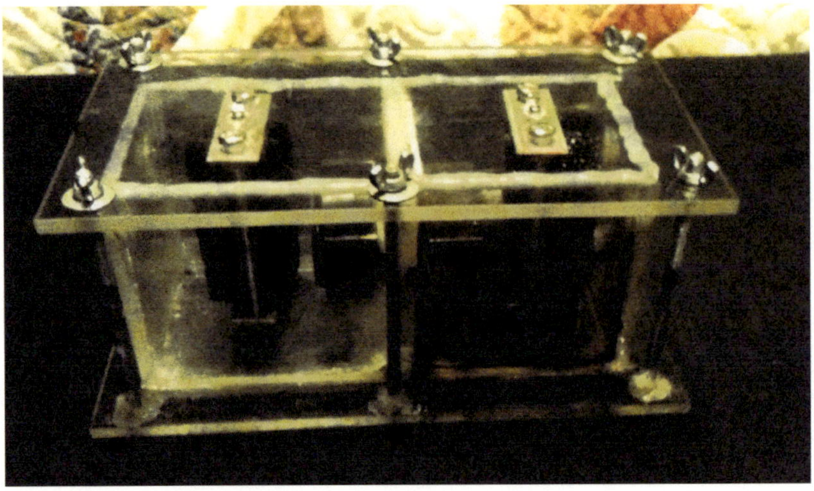

Fig. 9. Photograph of improved cell

As shown in Figures 8 and 9, the single electrodes which were held in place by copper clamps have been replaced by double electrodes held by a stainless steel frame. This adjustment was made for two reasons. Firstly the increased surface area of anode and cathode material will allow for more bacterial growth and more reactions occurring in the cathode chamber. Secondly the copper clamps used previously were highly susceptible to rust causing them to need to be replaced frequently and the chamber contents to be contaminated by rust particles. Due to the constant immersion of the electrodes in water stainless steel was the only viable option to use to hold the electrodes in place and resist corrosion.

The Entrant membrane has been replaced with an anion exchange membrane designed especially for such applications. The new membrane may provide better facilitation for protons than the previous material used. Due to the small amount required a free sample was able to be obtained however larger quantities can be exceedingly expensive. Initially bacteria were sourced from river water which should provide an adequate array of bacteria including many exo-electrogens. However as wastewater is more nutrient rich it was thought that it may also be more bacteria rich. The experiment was carried out in the same manner as the previous trial with the same buffer solution used in the anode and cathode chambers.

4.2.2 Results and discussions

The improved cell was operated over a period of 55 days, almost three times longer than the original cell. At the instant of adding the potassium permanganate the cell produced only 0.4mV as compared with the 1.7mV seen in the previous cell. The initial sharp increase in voltage over the first few hours that was demonstrated by the first cell was not seen in the improved cell. The voltage instead was stagnant around the 0.4mV mark for some time.

The recorded voltages and calculated current densities are shown below in Figures 10 and 11 respectively. The maximum cell voltage observed was 4.73mV which occurred 17 days after start time. The corresponding current density is 66.75mA/m^2 with power normalised by anode surface area of .316mW/m^2 and volumetric power 1.36mW/ m^3.

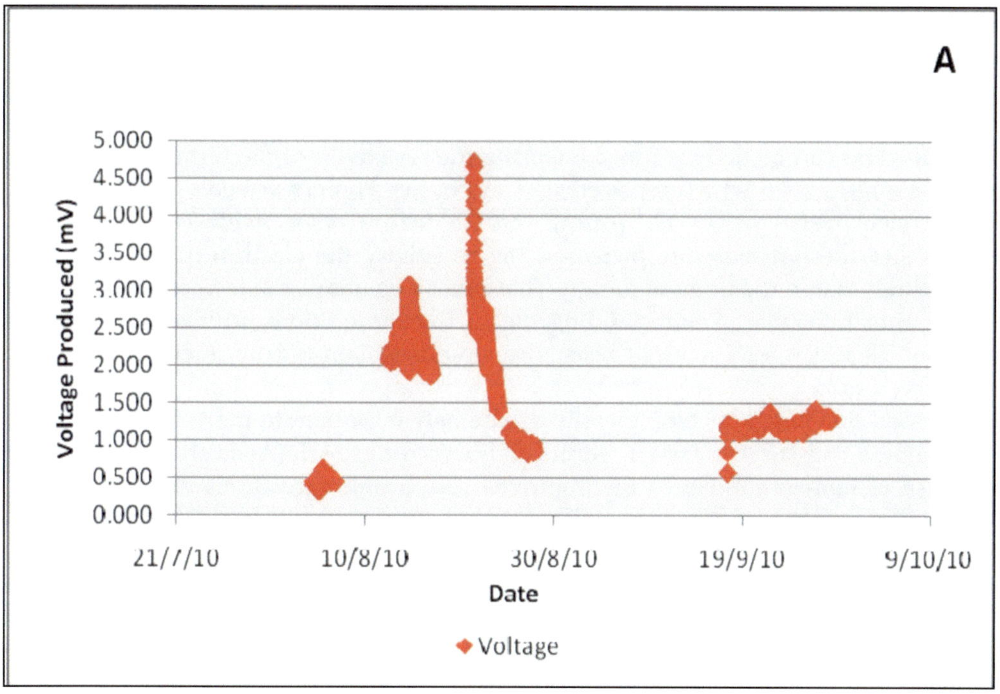

Fig. 10. Recorded voltages from start time (4/8/10) to end time (28/9/10)

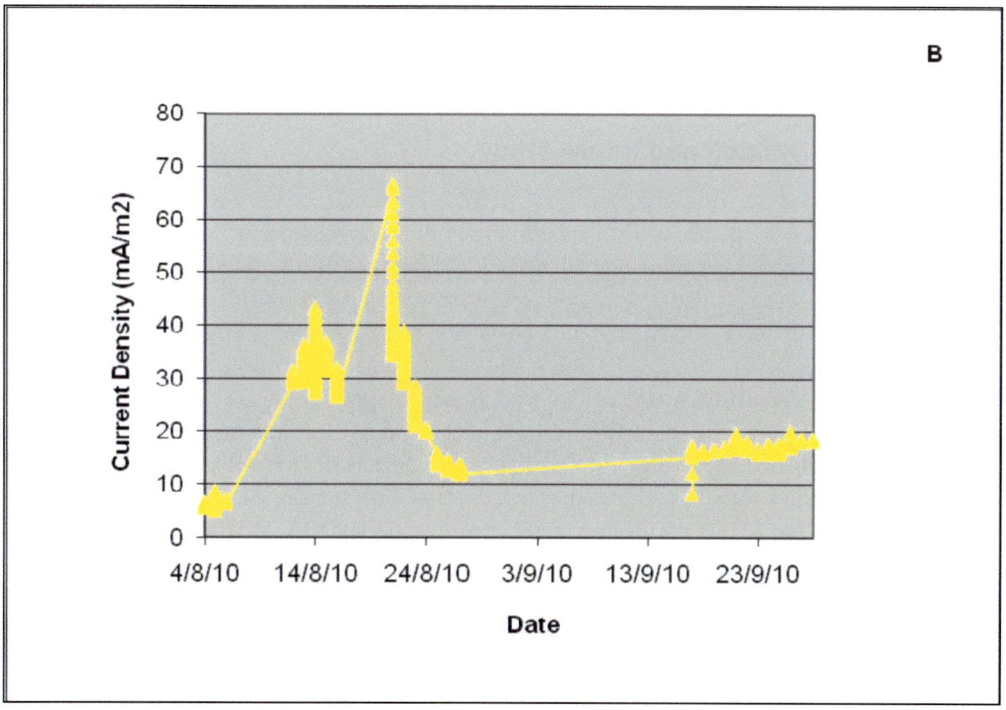

Fig. 11. Calculated current densities over the same period

Rather than improving the performance of the fuel cell, the adjustments made hampered the performance. Time did not permit for a lengthy investigation into the reasons for this however several conclusions can be theorised. It can be concluded that the addition of the stainless steel frame housing the electrodes considerably increased the internal resistance of the cell. This can be deduced by comparing the resistivity of the formerly used copper and the stainless steel which replaced it. The Physics Hypertext book gives copper and stainless steel resistivity of 17.1 (nΩm) and 720 (nΩm) respectively (Alert, 2010). The stainless steel design therefore increased the resistivity the electron travels through by over 40 times that in the original design. This effect was unavoidable as stainless steel was the only metal capable of withstanding the fuel cell conditions without suffering from corrosion. This increased internal resistance would explain a drop in current density as seen in the results.

As discussed previously bacteria growth is extremely important to the voltage produced. It is well known that bacteria favour warm conditions for growth (Answers.com, 2010) . Due to the use of human effluent in the improved cell, health regulations required that it be located in an isolated location. The only available areas were either outside or in a poorly ventilated, cold room. As it was winter at the time of the trial the cell was most often kept at a temperature of around 15 degrees Celsius. The earlier trial however was conducted during summer where the cell was kept at a more suitable 25 - 30 degrees Celsius. These inadequate conditions may have hampered the bacterial growth and decreased the output voltage as a result. The higher temperature may have also contributed to the higher voltage in the cathode chamber. Since heat is a catalyst a higher temperature may have speeded up the reaction in the cathode chamber by causing the protons to have a greater kinetic energy and collide with a higher number of electrons. The heat related effects were demonstrated in the previous trial where the temperature and voltage were recorded and compared over a period of time.

5. Conclusions, outlook and future study

The study presented in this chapter has mainly provided a review of the concept of microbial conversion of biomass into usable energy. An overview of Microbial Fuel Cells (MFCs) has been given, and their significance has been outlined. MFC and MEC systems are recognised as energy production systems with great potentials. The world's need for electricity and fuel is ever increasing and so is the need for clean, renewable methods to produce these things. Microbial cells have the advantage of running off widely available sources of energy including waste water and food scraps. These are things produced by every home around the globe including remote areas and in developing countries.

The study also discussed the design and construction details of two MFCs and presented the results of tests carried out with the constructed cells. The results demonstrate that further work is needed before MFCs could become a commercial success. Reducing the expenses of building and operating an MFC/MEC system as well as increasing cell efficiencies are ongoing issues for the technology. Further research is also required in many areas particularly in the area of catalysts as the cathode with the inclusion of the catalyst has been found to account for almost fifty percent of the cost of an MFC/MEC (Call et al., 2009). Future studies should focus on the incorporation of bacteria on the cathode to replace the current techniques.

Optimal methods for producing a biocathode have not been thoroughly investigated (Logan et al., 2008) however it has been shown that an effective way to produce a biocathode is to reverse the polarity of an MEC (Rozendal et al., 2008; Jeremiasse et al., 2009).
Membrane's also increase the expense of a system so architectures which do not require a membrane are obviously more desirable although they too have their drawbacks. Membrane fabric offers an inexpensive alternative to cation and anion exchange membranes but further investigation as to their comparative performance must be undertaken. Nevertheless MFC and MEC technology is relatively new but advancements to the technology have been rapid and should continue at this pace resulting in higher efficiencies, lower costs and a scalable reactor design in the not too distant future.

6. References

Alert, G. (2010). Electric Resistance. Available from: <http://physics.info/electric-resistance/>.

Answers.com. (2010). What conditions promote bacterial growth? , Available from: <http://wiki.answers.com/Q/What_conditions_promote_bacterial_growth>.

Ballard Power Systems. (24 January 2008). How Fuel Cells Work. Available from: <http://www.ballard.com/_pvw0D3492BE/About_Ballard/Resources/How_Fuel_Cells_Work.htm>.

Blair, S. (2008). Monash fuels the next generation of Hybrid cars. 17 January 2010, Available from: <http://www.monash.edu.au/news/newsline/story/1310>.

Call, D. and B. E. Logan (2008). "Hydrogen Production in a Single Chamber Microbial Electrolysis Cell Lacking a Membrane."

Call, D. and B. E. Logan (2008). "Hydrogen Production in a Single Chamber Microbial Electrolysis Cell Lacking a Membrane." *Environ. Sci. Technol.* Vol. 42, No.9, pp. 3401–3406.

Call, D. F., M. D. Merrill and B. E. Logan (2009). "High Surface Area Stainless Steel Brushes as Cathodes in Microbial Electrolysis Cells." *Environ. Sci. Technol.* Vol. 43, No.6, pp. 2179–2183.

Chae, K., M. Jinchoi, K. Kim, F. F. Ajayi, I. Chang and I. S. Kim (2009). "A Solar-Powered Microbial Electrolysis Cell with a Platinum Catalyst-Free Cathode To Produce Hydrogen." *Environ. Sci. Technol.* Vol. 43, No.24, pp. 9525–9530.

Danigelis, A. (2009). What a Great Vintage of Hydrogen You Have. Available from: <http://news.discovery.com/tech/what-a-great-vintage-of-hydrogen-you-have.html>.

FCTec. (2010). Fuel Cell Basics. Available from: <http://www.fctec.com/fctec_basics.asp>.

Field, S. Q. (2008). Chapter 3: Electrochemistry. Available from: <http://scitoys.com/scitoys/scitoys/echem/fuel_cell/fuel_cell.html>.

Finfacts-Team. (2009). International Energy Outlook 2009: World energy consumption projected to rise 44% from 2006 to 2030; Carbon emissions to jump more than 39% without new policies. 21 March 2010, Available from: <http://www.finfacts.ie/irishfinancenews/article_1016796.shtml>.

He, Z. and L. T. Angenent (2006). "Application of Bacterial Biocathodes in Microbial Fuel Cells." *Electroanalysis.* Vol. 18, No.19-20.

Hopwood, N. and J. Cohen. (2000). Greenhouse Gases and Society. 13 February 2010, Available from: <http://www.umich.edu/~gs265/society/greenhouse.htm>.

HowStuffWorks.com. (2000). What are all the different ways to store energy besides using rechargeable batteries? 21 March 2010, Available from: <http://science.howstuffworks.com/question247.htm#>.

Jeremiasse, A. W., H. V. M. Hamelers and C. J. N. Buisman (2009). "Microbial electrolysis cell with a microbial biocathode." *Elsevier: Bioelectrochemistry,* Vol. 78, No.1, pp. 39-43.

Knapp, B. (2002). *Hydrogen and the Noble Gases.* Atlantic Europe Publishing Company Limited, 0717275736.

Liu, H., S. Grot and B. E. Logan (2005). "Electrochemically Assisted Microbial Production of Hydrogen from Acetate." *Environmental Science & Technology.* Vol. 39, No.11, pp. 4317-4320.

Logan, B. E. (2008). *Microbial Fuel Cells.* Wiley, 0470239484.

Logan, B. E., D. Call, S. Cheng, H. Hamelers, T. Sleutels, A. Jeremiasse and R. Rozendal (2008). "Microbial Electrolysis Cells for High Yield Hydrogen Gas Production from Organic Matter." *Environ. Sci. Technol.* Vol. 42, No.23, pp. 8630–8640.

Logan, B. E. and S. Cheng (2008). "Evaluation of catalysts and membranes for high yield biohydrogen production via electrohydrogenesis in microbial electrolysis cells (MECs)." *Water Science & Technology* Vol. 59, No.10.

Logan, B. E., B. Hamelers, R. Rozendal, U. Schroder, J. Keller, S. Freguia, P. Aelterman, W. Verstraete and K. Rabaey (2006). "Microbial Fuel Cells: Methodology and Technology." *Environmental Science & Technology.* Vol. 40, No.17, pp. 5181-5192.

Markusic, M. (2010). How is Renewable Energy from Chocolate Possible? . Available from: <http://www.brighthub.com/environment/renewable-energy/articles/37144.aspx>.

Nice, K. and J. Strickland. (2010). How fuel cells work. Available from: <http://auto.howstuffworks.com/fuel-efficiency/alternative-fuels/fuel-cell5.htm>.

Plumer, B. (2009). Why Energy Storage -- Yes, Energy Storage -- Needs Love. 21 March 2010, Available from: <http://www.tnr.com/blog/the-vine/why-energy-storage-yes-energy-storage-needs-love>.

Putnam, S. C. (2007). The Mechanical Battery. 12 January 2010, Available from: <http://www.damninteresting.com/the-mechanical-battery>.

Rayner, J. (2008). Flywheel energy storage. 20 February 2010, Available from: <http://www.upei.ca/~physics/p261/projects/flywheel1/flywheel1.htm>.

Rozendal , R. A., A. W. Jeremiasse, H. V. M. Hamelers and C. J. N. Buisman (2008). "Hydrogen Production with a Microbial Biocathode." *Environ. Sci. Technol.* Vol. 42, pp. 629-634.

Sun, M., G.-P. Sheng, L. Zhang, C.-R. Xia, Z.-X. Mu, X.-W. Liu, H.-L. Wang, H.-Q. Yu, R. Qi, T. Yu and M. Yang (2008). "An MEC-MFC-Coupled System for Biohydrogen Production from Acetate." *Environmental Science & Technology.* Vol. 42, No.21, pp. 8095-8100.

You, S. J., N. Q. Ren, Q. L. Zhao, J. Y. Wang and F. L. Yang (2009). "Power Generation and Electrochemical Analysis of Biocathode Microbial Fuel Cell Using Graphite Fibre Brush as Cathode Material." *Fuel Cells.* , No.5, pp. 588–596.

Zyga, L. (2009). Liquid Battery Offers Promising Solar Energy Storage Technique. 21 March 2010, Available from: <http://www.physorg.com/news155569564.html>.

Part 7

Bio-Economic

Methods for Structural and Parametric Synthesis of Bio-Economic Models

Darya V. Filatova
Jan Kochanowski University in Kielce, Kielce, Poland

1. Introduction

The communities of animals and plants are some examples of biological systems where population dynamics $\mathbf{X}(t) \in \mathrm{R}^d$, $t \in [t_0, t_1]$, depends not only on environmental variability and internal transformations, but also on human control factor $\mathbf{u}(t) \in \mathrm{R}^r$, $t \in [t_0, t_1]$. (We will call also $\mathbf{X}(t)$ a phase variable and $\mathbf{u}(t)$ a control variable.)

Since many biological systems have to stay in some dynamic equilibrium, the control representing the action, decision, or policy of the decision-makers can only correct the system development with moderate effects on its natural behavior. In this case controllers are interested in the selection of a nonaticipative decision among the ones satisfying all the changes of existence conditions and inter-connections among different communities.

Suppose that the controller is interested in renewable resource management (fishery or forestry) over the planning horizon $[t_0, t_1]$.

The goal of this task is maximization with respect to $\mathbf{u}(t) \in U(t)$ ($U(t)$ is a given metric space, $t \in [t_0, t_1]$) of the functional

$$\mathrm{J}(\mathbf{X}(\cdot), \mathbf{u}(\cdot)) = \sup_{u \in U} \left\{ \int_{t_0}^{t_1} f(t, \mathbf{X}(t), \mathbf{u}(t)) dt + \psi_0(\mathbf{X}(t_0), \mathbf{X}(t_1)) \right\} \tag{1.1}$$

subjected to the constraints

$$d\mathbf{X}(t) = \tilde{\mathrm{T}}(t, \mathbf{X}(t), \mathbf{u}(t)) dt, \tag{1.2}$$

$$\psi_1(\mathbf{X}(t_0), \mathbf{X}(t_1)) \le 0, \tag{1.3}$$

$$\psi_2(\mathbf{X}(t_0), \mathbf{X}(t_1)) = 0; \tag{1.4}$$

$$g_i(t, \mathbf{X}(t)) \le 0, \ \forall t \in [t_0, t_1], \ i = 1, \dots, \ell_1, \tag{1.5}$$

$$\phi_j(t, \mathbf{u}(t)) \le 0, \ j = 1, \dots, \ell_2, \tag{1.6}$$

where $\tilde{\mathsf{T}}(t,\mathbf{X}(t),\mathbf{u}(t))$ is an operator of a mathematical model of the resource $\mathbf{X}(t)$, $f: \mathsf{R}\times\mathsf{R}^d\times\mathsf{R}^r \to \mathsf{R}$, $\varphi: \mathsf{R}\times\mathsf{R}^d\times\mathsf{R}^r \to \mathsf{R}^d$, $\psi_0: \mathsf{R}^d\times\mathsf{R}^d \to \mathsf{R}$, $\psi_1: \mathsf{R}^d\times\mathsf{R}^d \to \mathsf{R}^{k_1}$, $\psi_2: \mathsf{R}^d\times\mathsf{R}^d \to \mathsf{R}^{k_2}$, $g_i: \mathsf{R}\times\mathsf{R}^d \to \mathsf{R}$ $(1\le i\le \ell_1)$, $\phi_j: \mathsf{R}\times\mathsf{R}^r \to \mathsf{R}$ $(1\le j\le \ell_2)$; d, r, k_1, k_2, ℓ_1, $\ell_2 \in \mathsf{N}$, and $\sup(\cdot)$ is the least upper bound.

The task (1.1) - (1.6) is formulated as an optimal control task [Milyutin & Osmolovski, 1998]. Its solution and accuracy of this solution depend on many different factors, mainly on the successful selection of the object equation. The wrong specification of a model and as a consequence wrong parametric identification can lead to erroneous solution of (1.1) - (1.6). Since the factor of uncertainty is always present at the resource description, it is reasonable to take (1.2) as an stochastic differential equation (SDE).

Let the renewable resource describes a certain population of fish, which natural growth rate depends on different biological parameters. Very often these parameters are evaluating over the time because environmental conditions are not constant. In this case it is reasonable to treat the parameters of SDE as the bivariate functions (or the SDE with time-varying parameters). Since the structure of the SDE is selected, the next step is the construction of the estimation procedure.

It is not easy to describe the bivariate functions by means of certain functional forms. Flexible model does not assume any specific form of the functions. This data-analytic approach called nonparametric regression can be found in statistical literature. However the direct application of the ideas does not bring desired results. The improvements of the identification procedures were presented in [Fan at al. 2003]. The main idea of this work was based on the discretization of the SDE and further approximation of parameter functions by constants at the discretization points. It is clear that the accuracy of the estimates depends on the accuracy of the discretization method. To overcome this problem we propose to consider bivariate functions as an control functions and solve the identification task as an optimal control problem using the maximum principle.

The rest of the paper is organized as follows. The second section is dedicated to the system analysis of a bio-economic models. The third section presents the identification methods based on the ideas [Bastogne at al., 2007], [Hansen & Penland, 2007], [Hurn at al., 2003], [Jang at al., 2003], [McDonald & Sandal, 1998], [Shoji & Ozaki, 1998]. The fourth section shows the solution for the bivariate functions. The paper is ended by the conclusions.

2. System analysis of a bio-economic model

2.1 Problems of structural and parametric synthesis

The mathematical description of any bio-economic model requires taking into account all the elements of the system and all the interrelations among them. Detailed analysis of the "complete" model allows to forecast its behavior and introduce optimal management strategies. Unfortunately, this analysis is mathematically difficult. From one side it is impossible to detect all the elements and nature of their interconnections. From other side there are several sources of uncertainty: the growth, mortality, reproduction rate vary in random manner causing random effects on genetics and age structure of exploited population; the price of the resource depends on economical situation on stock markets, political situation, climate and etc.

Distinctive property of each stochastic object is contained with ambiguous respond on the same input signals. Even for simplest one-dimensional object and for non-stochastic input signal, output signal of stochastic object can't be considered as deterministic one. For output

variable of this objects class scattering growths with increasing of objects "noise properties". For that reason uncertainty in behavior can be explained by noise influence that in addition brings deficiency of a priori information about system.
Therefore, it is impossible to detect a model, whose properties and mathematical description would correspond to the "exact" behavior of the system. Every mathematical model will be only similar with the system. In this case model selection has to be done under two groups of disjoint requirements:

- the main features of the system have to be reflected as precise as possible (the degree of similarity);
- existence of the theoretical methods, which allow to use the model for forecast, optimization, control, etc.

The dynamics of bio-economic system (1.2) can be described by means of its states. Let a union of the values of the phase variable $\mathbf{X}(t) \in \mathsf{R}^d$, $t \in [t_0, t_1]$, denote the state $\mathsf{S}_t \in \mathsf{S}^*$ (S^* is some space) of the system in time t, $\mathsf{S}_{t+\Delta} \in \mathsf{S}^*$ denote the state in time $t+\Delta$ (where Δ is time increment, $\Delta \in \mathsf{R}$, $0 < \Delta < t_1$). The dynamics of the system (1.2) can be written as

$$\mathsf{S}(t+\Delta) - \mathsf{S}(t) = \mathsf{T}\left(\mathsf{S}^*, \left\langle \mathsf{C}(t), \mathsf{G}(t) \right\rangle\right), \tag{2.1}$$

where $\mathsf{T}(t)$ is an operator of a mathematical model, $\mathsf{C}(t) = \{\mathsf{c}_0(t), \mathsf{c}_1(t), ..., \mathsf{c}_k(t)\}$ is a set of k parameters, $\mathsf{G}(t) = \{\mathsf{g}_0(t), \mathsf{g}_1(t), ..., \mathsf{g}_k(t)\}$ is the set of the structural relations of the phase variables of the system.
The formalization of the model (2.1) depends on the selection of:

- the set of phase variables $\mathbf{X}(t) \in \mathsf{R}^{d-n}$, $0 \le n < d$;
- the theoretical method to define S^*;
- the structural relations $\mathsf{G}(t)$ among the selected phase variables;
- the parametric identification method for the estimation of the parameters $\mathsf{C}(t)$.

Even though it is well known task, the problem of the object (2.1) formalization can be solved by a few mathematical tools. Among basic groups of exact and approximated methods, which are used to solve the problem, we can name: exact methods, methods of task simplification, methods of the task linearization, numerical methods, methods of integral transformation, method of infinite series, variation methods, and methods of reduction to the systems of the ordinary differential equations.

2.2 Stochastic differential equation as a bio-economic model

We consider a certain population of fish, whose size at time t is denoted by $X(t) \in \mathsf{R}$, $t \in [t_0, t_1]$. This population has a natural growth rate $\varphi^*(t, X(t))$:

$$\frac{dX(t)}{dt} = \varphi^*(t, X(t)), \quad X(t_0) = X_0, \tag{2.2}$$

where $\varphi^* : \mathsf{R} \times \mathsf{R} \to \mathsf{R}$ is assumed to be a concave function with given properties.
Fish stock $X(t)$ has natural fluctuations and is subjected to many stochastic effects. To take them into account we have to improve the model (2.2) adding some stochastic terms. In fact

the increment $\frac{dX(t)}{dt}$ is not a deterministic, thus the structure of the model (2.2) has to be reorganized.

Let $\left(\Omega,\mathsf{F},\{\mathsf{F}_t\}_{t\geq 0},\mathsf{P}\right)$ be a stochastic basis satisfying the usual conditions. Let the phase variable be $X:\Omega\rightarrow\mathsf{R}$, $\omega\in\Omega$, $X(\omega)=x$ and $\{v(t),t\in[t_0,t_1]\}$ be a continuous stochastic process, defined on $\left(\Omega,\mathsf{F},\{\mathsf{F}_t\}_{t\geq 0},\mathsf{P}\right)$, such that its mean value function $\mathsf{E}\left[v(t)\right]=0$ for every $t\in[t_0,t_1]$ and $v(t_0)=0$. By the intuition we add the term $v(t)$ to (2.2)

$$\frac{dX(t)}{dt}=\varphi^{**}\left(t,X(t),v(t)\right),\ X(t_0)=X_0,\ t\in[t_0,t_1], \tag{2.3}$$

where $\varphi^{**}::\mathsf{R}\times\mathsf{R}\times\mathsf{R}\rightarrow\mathsf{R}$.

The model (2.3) presents a stochastic differential equation, whose solution depends on the selection of the stochastic process $v(t)$ and its properties. Let the process be the Brownian motion process $\{B(t),t\in[t_0,t_1]\}$. This process has many useful theoretical properties, namely

1. the independence and stationarity of the increments $B(t+h)-B(\tau+h)$ for every $t>\tau\in[t_0,t_1]$ and every $h>0$;
2. the mean square continuity of $B(t)$ for every $t\in[t_0,t_1]$;
3. the regularity conditions, i.e. $\mathsf{E}\left[\left|B(t_0)\right|^2\right]<\infty$ and $\mathrm{var}\left[B(t+h)-B(\tau+h)\right]<\infty$ (where $\mathsf{E}[\cdot]$ denotes the expectation operator and $\mathrm{var}[\cdot]$ denotes the variance operator).

We replace the stochastic term $v(t)$ by $B(t)$ and rewrite the model (2.3) as a stochastic differential equation

$$dX(t)=a\left(t,X(t)\right)dt+b\left(t,X(t)\right)dB(t),\ X(t_0)=X_0, \tag{2.4}$$

or as an integral equation

$$X(t)-X(t_0)=\int_{t_0}^{t}a\left(\tau,X(\tau)\right)dt+\int_{t_0}^{t}b\left(\tau,X(\tau)\right)dB(\tau), \tag{2.5}$$

where $a:\mathsf{R}\times\mathsf{R}\rightarrow\mathsf{R}$ and $b:\mathsf{R}\times\mathsf{R}\rightarrow\mathsf{R}$ are Baire functions, $dB(t)$ is an increment of the Brownian motion process. The first integral in (2.5) will be a mean square Riemann integral, whereas the second one will be the Ito stochastic integral. Equation (2.4) is therefore called Ito stochastic differential equation.

Equation (2.4) can be used for the description of the object (1.2). In this case, the optimal control solution of the task (1.1) – (1.6) requires the existence and uniqueness of the solution of the equation (2.4). This solution can be treated in strong or weak sense. The SDE (2.4) has

- a unique strong solution, if any two solutions $X(t)$ and $\tilde{X}(t)$, $t\in[t_0,t_1]$, coincide by all trajectories of the process so that

$$\mathsf{P}\left(\sup_{t\in[t_0,t_1]}\left|X_t-\tilde{X}_t\right|>0\right)=0,$$

- a unique weak solution, if its solutions coincide by probability, but not obligatory by trajectories.

Let us consider the solution of SDE (2.3) only in the strong sense. The following theorem guaranties the existence and uniqueness of the strong solution for (2.4).

Theorem 2.1. *The solution of equation (2.3) $X(t)$ ($t\in[t_0,t_1]$) exists and is unique in strong sense, if the following conditions are held.*

A1 *(Measurability): $a(t,x)$ and $b(t,x)$ are jointly L^2 – measurable in $(t,x)\in[t_0,t_1]\times\mathsf{R}$.*

A2 *(Lipschitz condition): There exists a constant $K>0$ such that*

$$|a(t,x)-a(t,y)|\le K|x-y|$$

and

$$|b(t,x)-b(t,y)|\le K|x-y|$$

for all $t\in[t_0,t_1]$ and $x,y\in\mathsf{R}$.

A3 *(Linear growth bound): There exists a constant $K>0$ such that*

$$|a(t,x)|^2\le K^2\left(1+|x|^2\right)$$

and

$$|b(t,x)|^2\le K^2\left(1+|x|^2\right)$$

for all $t\in[t_0,t_1]$ and $x\in\mathsf{R}$.

A4 *(Initial value): $X(t_0)$ is A_{t_0} -measurable with $\mathsf{E}\left(|X(t_0)|^2\right)<\infty$.*

■

In particular, assumptions (A2) and (A3) ensure that a solution of an SDE does not explode. This is very important to insure the stability of a numerical approximation of $X(t)$, $t\in[t_0,t_1]$, and avoid the “stiffness” and “ill-posed” problems. Among other properties are boundedness and continuity on $t\in[t_0,t_1]$, which guarantee the existence of an adjoint system and partly the optimality of the solution of the task (1.1) – (1.6).

The description of a bio-economic model is not completed since we have not introduce the set of parameters $\mathsf{C}(t)$, $t\in[t_0,t_1]$, in the structure (2.1). Denoting the parameters as $\boldsymbol{\theta}(t)\in\mathsf{R}^q$ we rewrite (2.4)

$$dX(t)=a(t,X(t);\boldsymbol{\theta}(t))dt+b(t,X(t);\boldsymbol{\theta}(t))dB(t),\ X(t_0)=X_0. \tag{2.6}$$

It is clear that the values of the parameters $\boldsymbol{\theta}(t)$ are unknown and have to be evaluate by the proper estimation procedure. If one uses the maximum likelihood estimation method the following can be helpful.

The process $X(t)$, $t\in[t_0,t_1]$, generated by the Ito stochastic differential equation (2.4) is a Markov process and is therefore characterized by the density function $\mathsf{p}(X(t))=\mathsf{p}(t,x)$ for all $t\in[t_0,t_1]$ and the transition probability density function $\mathsf{p}(X(t),X(\tau))=\mathsf{p}_{X(t)|,X(\tau)}(x|y)=\mathsf{p}(t,x;\tau,y)$ for every $t>\tau\in[t_0,t_1]$. Assume the existence of the continuous partial derivatives $\frac{\partial \mathsf{p}}{\partial t}$, $\frac{\partial[\mathsf{p}a(t,x)]}{\partial x}$, $\frac{\partial^2[\mathsf{p}b^2(t,x)]}{\partial x^2}$, where p stands for $\mathsf{p}(t,x)$ and $\mathsf{p}(t,x;\tau,y)$. In this case we can write two equations, namely

a. Kolmogorov's forward equation

$$\frac{\partial \mathsf{p}(t,x;\tau,y)}{\partial t}=-\frac{\partial\left[\mathsf{p}(t,x;\tau,y)a(t,x)\right]}{\partial x}+\frac{1}{2}\frac{\partial^2\left[\mathsf{p}(t,x;\tau,y)b^2(t,x)\right]}{\partial x^2}, \tag{2.7}$$

b. Kolmogorov's backward equation

$$-\frac{\partial \mathsf{p}(t,x;\tau,y)}{\partial \tau}=a(\tau,y)\frac{\partial \mathsf{p}(t,x;\tau,y)}{\partial y}+\frac{1}{2}b^2(\tau,y)\frac{\partial^2 \mathsf{p}(t,x;\tau,y)}{\partial y^2} \tag{2.8}$$

with initial conditions selected in proper manner.

3. Estimation methods for time independent parameters of SDE

3.1 Maximum likelihood method

Consider the stochastic process $X=\{X_t,t\in[t_0,t_1]\}$, which is assumed to be the unique strong solution of the SDE (2.6) with $\boldsymbol{\theta}(t)=\boldsymbol{\theta}$

$$dX(t)=a(t,X(t);\boldsymbol{\theta})dt+b(t,X(t);\boldsymbol{\theta})dB(t),\ X(t_0)=X_0. \tag{3.1}$$

Assume that there are $n\in\mathsf{N}$ real valued observations $Y_0,Y_1,...,Y_n\in\mathsf{R}$ of the process X, which were made at the discretization times $t_0\le\tau_0,\tau_1,...,\tau_n\le t_1$. (We consider, for simplicity, equidistant time discretization with $\tau_n=n\Delta$, where $\Delta=\frac{t_1}{n}\in(0,1)$ and some $n\in\mathsf{N}$). The observations contain information about the parameters $\theta_1,\theta_2,...,\theta_q\in\Theta\subseteq\mathsf{R}$ (where Θ specifies the set of allowable values for the parameters, $q\in\mathsf{N}$) that we wish to estimate. The maximum likelihood estimator $\hat{\boldsymbol{\theta}}=\left(\hat{\theta}_1,\hat{\theta}_2,...,\hat{\theta}_q\right)^{\mathrm{T}}$ is the best estimate of $\boldsymbol{\theta}\in\Theta$, where

$$\mathsf{L}^*=\mathsf{L}\left(\hat{\boldsymbol{\theta}}\right)=\sup_{\boldsymbol{\theta}\in\Theta}\mathsf{L}(\boldsymbol{\theta}), \tag{3.2}$$

here $\sup(\cdot)$ is the least upper bound of $\mathsf{L}(\boldsymbol{\theta})$ over all $\boldsymbol{\theta}\in\Theta$.
This means, $\hat{\boldsymbol{\theta}}$ is the parameter that maximizes the likelihood function with respect to the set of permitted parameter values $\boldsymbol{\theta}\in\Theta$. The hypothesized density in the given parametrized family of probability densities $\mathsf{f}_Y\left(\cdot,\hat{\boldsymbol{\theta}}\right)$ represents the most probable density from the given class of densities having observed $Y_0,Y_1,...,Y_n\in\mathsf{R}$.

Let $\mathsf{p}(t_k, y_k | t_{k-1}, y_{k-1}; \boldsymbol{\theta})$ be the transition probability density of (t_k, y_k) starting from (t_{k-1}, y_{k-1}) given the vector $\boldsymbol{\theta}$, $k \in \{1,2,...,n\}$. The density of the initial state is $\mathsf{p}_0(y_0|\boldsymbol{\theta})$. The joint density corresponds to the likelihood function

$$\mathsf{L}\,(\boldsymbol{\theta}) = \mathsf{p}_0(y_0;\boldsymbol{\theta}) \prod_{k=1}^{n} \mathsf{p}(t_k, y_k | t_{k-1}, y_{k-1}; \boldsymbol{\theta}) \tag{3.3}$$

or in more convenient for the numerical simulation form

$$\mathsf{L}(\boldsymbol{\theta}) = -\ln(\mathsf{L}\,(\boldsymbol{\theta}))\,. \tag{3.4}$$

The task (3.2) for (3.4) can be solved as initial value problem for forward Kolmogorov's equation (2.7) with the initial condition given as $\lim_{t \to \tau} \mathsf{p}_{Y_k|Y_{k-1}}(y_k|y_{k-1}) = \delta(y_k - y_{k-1})$, where $\delta(\cdot)$ stands for a forward difference [Jazwinski, 2007].

Let the SDE (3.1) be the liner SDE

$$dX(t) = \theta_1 X(t)dt + \theta_2 X(t)dB(t),\; X(t_0) = X_0, \tag{3.5}$$

where θ_1 and θ_2 are unknown parameters, which have to be estimated using the observations $Y_0, Y_1, ..., Y_n$.

Taking $Y_0 = X_0$, we rewrite (3.4) as

$$\mathsf{L}(\theta_1, \theta_2) = \sum_{k=1}^{n} \ln \left\{ \frac{\exp\left\{ -\left[\ln\left(\frac{Y_k}{Y_{k-1}}\right) - \left(\theta_1 - \frac{\theta_2^2}{2}\right)\Delta \right]^2 \Big/ 2\theta_2^2\,\Delta \right\}}{Y_k \sqrt{2\pi\theta_2^2\Delta}} \right\}. \tag{3.6}$$

Under suitable conditions, when the true parameter is an interior point of Θ, the maximum likelihood estimate $\hat{\boldsymbol{\theta}}$ can be obtained as a root of the first order conditions

$$\frac{\partial \mathsf{L}\,(\boldsymbol{\theta})}{\partial \theta_i} = 0, \tag{3.7}$$

for all $i \in \{1,2,...,q\}$, where $\frac{\partial}{\partial \theta_i}$ denotes the partial derivative with respect to θ_i.

Using the conditions (3.7) for (3.6) we get the estimates of the parameters

$$\hat{\theta}_1 = \frac{1}{t_1} \ln\left(\frac{Y_n}{Y_0}\right) + \frac{\hat{\theta}_2^{\,2}}{2} \tag{3.8}$$

and

$$\hat{\theta}_2 = \sqrt{\frac{1}{n-1} \sum_{k=1}^{n} \Delta \left[\frac{1}{\Delta} \ln\left(\frac{Y_k}{Y_{k-1}}\right) - \frac{1}{t_1} \ln\left(\frac{Y_n}{Y_0}\right) \right]^2}\,. \tag{3.9}$$

Unfortunately, Kolmogorov's forward equation (2.7) has been solved only in a few simple cases. So, the alternative approach is required to solve the problem (3.4).

3.2 Monte Carlo methods

One difficulty in finding the optimal values $\boldsymbol{\theta}^{opt}$ is that the transition densities are not known. Recall that $\mathsf{p}(t,x;\tau,x^*)$ is a random variable since it depends on values x^* taken on by x_τ. We can take the expectation of (2.7) and, interchanging expectation with differentiation, obtain

$$\frac{\partial \mathsf{p}(t,x)}{\partial t} = -\frac{\partial\left[\mathsf{p}(t,x)a(t,x)\right]}{\partial x} + \frac{1}{2}\frac{\partial^2\left[\mathsf{p}(t,x)b^2(t,x)\right]}{\partial x^2}. \tag{3.10}$$

By the same reasoning as before we rewrite the likelihood function (3.3) as

$$\mathsf{L}(\boldsymbol{\theta}) = -\ln \mathsf{p}_0(y_0;\boldsymbol{\theta}) - \ln\sum_{k=1}^{n}\mathsf{p}(t_k,y_k;\boldsymbol{\theta}) \cdot \tag{3.11}$$

The main problem here is how to find the estimates $\hat{\mathsf{p}}(t_k,y_k;\boldsymbol{\theta})$, $k=1,2,..,n$, since we have only the sequence $(Y_n)_{n\in\{0,1,\dots,n_{t_1}\}}$. One of the ways to solve this task is to use Monte Carlo simulation method. We will consider to two possibilities, namely for one sample data and for the panel data.

3.2.1 One sample data set

Let for the random variable Y_k ($Y_k:\Omega\to\mathsf{R}$, $\omega\in\Omega$, $Y_k(\omega)=y_k$, $k\in\{0,1,\dots,n_{t_1}\}$) the distribution function be denoted as $\mathsf{F}_{Y_k}(y_k)$ and the density function as $\mathsf{f}_{Y_k}(y_k)$. For the simplicity of the reasoning we omit the index n. Both functions are connected as follows

$$\mathrm{f}_Y(y) = \frac{d}{dy}\mathrm{F}_Y(y) \cdot \tag{3.12}$$

The empirical estimate of the distribution function is

$$\hat{\mathsf{F}}_Y(y) \equiv \frac{1}{M}\sum_{j=1}^{M} 1\{\tilde{y}_j \le y\}, \tag{3.13}$$

where $1\{\cdot\}$ is set belonging indicator, M is a number of the realizations $\tilde{y}_j$.
Taking into account (3.12) and (3.13) the estimate for the density function is

$$\hat{\mathsf{f}}_Y(y) = \frac{\hat{\mathsf{F}}_Y(y+h)-\hat{\mathsf{F}}_Y(y)}{h} = \frac{1}{Mh}\sum_{j=1}^{M} 1\{y<\tilde{y}_j \le y+h\}, \tag{3.14}$$

where h is the bandwidth, $h>0$.
Since $\hat{\mathsf{F}}_Y(y)$ is an unbiased estimate of $\mathsf{F}_Y(y)$, we have the unbiased estimate for the density function so that

$$\begin{aligned}\mathsf{E}\left[\hat{\mathsf{f}}_Y(y)\right]=\mathsf{f}_Y(y) &= \mathsf{E}\left[\frac{\hat{\mathsf{F}}_Y(y+h)-\hat{\mathsf{F}}_Y(y)}{h}\right] - \mathsf{f}_Y(y) \\ &= \frac{1}{Mh}\sum_{j=1}^{M} 1\{y<\tilde{y}_j \le y+h\} - \mathsf{f}_Y(y)\end{aligned} \tag{3.15}$$

for $h \to 0$ and $M \to \infty$ and the variance

$$\begin{aligned}
\mathrm{var}\left[\hat{\mathsf{f}}_Y(y)\right] &= \mathrm{var}\left[\frac{1}{Mh}\sum_{j=1}^{M} 1\{y < \tilde{y}_j \le y+h\}\right] \\
&= \frac{1}{Mh^2}\mathrm{var}\left[1\{y < \tilde{y}_j \le y+h\}\right] \\
&= \frac{1}{Mh}\left[\frac{\mathsf{F}_Y(y+h)-\mathsf{F}_Y(y)}{h}\left(1-\left(\mathsf{F}_Y(y+h)-\mathsf{F}_Y(y)\right)\right)\right] \\
&= \frac{\mathsf{f}_Y(y)}{Mh} + \mathrm{O}\left(\frac{1}{M}\right)
\end{aligned} \tag{3.16}$$

is striving to zero for $Mh \to \infty$ and $M \to \infty$. It is clear that quality of the estimate $\mathsf{f}_Y(y)$ depends on values M and h selection. In order $M \to \infty$, it is enough to increase the number of sample paths Y^j, thus we will consider problem of h parameter selection.
Let $h \to 0$ and $Mh \to \infty$ by means of the kernel function the approximation of the mean squared error for the estimate $\hat{\mathsf{f}}_Y(y)$ can be written as

$$MSE\left(\hat{\mathsf{f}}_Y(y)\right) = \frac{1}{Mh}\left\|K(\rho)\right\|_2^2 + \frac{h^4}{4}\left(\mu\left(K(\rho)\right)\right)^2 \left\|\mathsf{f}''_Y(y)\right\|_2^2, \tag{3.17}$$

where $\left\|K(\rho)\right\|_2^2$ and $\mu\left(K(\rho)\right)$ are some constants, depending on kernel function ρ; f''_Y is the second derivative of the function f_Y.
The minimization of (3.17) with respect to h gives following results

$$h^{opt} = \left(\frac{1}{M}\frac{\left\|K(\rho)\right\|_2^2}{\left\|\mathsf{f}''_Y(y)\right\|_2^2\left(\mu\left(K(\rho)\right)\right)^2}\right)^{1/5}, \tag{3.18}$$

where $\mathsf{f}''_Y(y)$ is the only unknown term.
Now the solution of (3.18) depends only on the kernel function selection. For the simplicity we use the parametric identification Epanichnikov kernel function

$$K(\rho) = \frac{3}{4}\left(1-\rho^2\right)1\left(|\rho| \le 1\right), \tag{3.19}$$

where $\rho = \left\{\rho_j = \frac{1}{h}\left(\tilde{y}_j - y\right), j = 1,2,...,M\right\}$.
Substituting (3.19) into (3.18) we get

$$h^{opt} \approx 0.9\sigma_Y M^{-1/5}, \tag{3.20}$$

where the standard deviation of the sample is given as

$$\sigma_Y = \sqrt{\frac{1}{M-1}\left[\sum_{j=1}^{M}\tilde{y}_j^2 - \frac{1}{M}\left(\sum_{j=1}^{M}\tilde{y}_j\right)^2\right]}$$

for $j = 1,2,\dots,M$.

Next, let $\tilde{Y}$ be a discrete time strong approximation of $X = \{X_t, t \in [t_0, t_1]\}$ and the sequences $(\tilde{y}_{kj})_{k\in\{0,1,\dots,n_{t_1}\}, j\in\{1,2,\dots,M\}}$ of values of the numerical approximation at the discretization times $\tau_0, \tau_1, \dots, \tau_{n_{t_1}}$ be computed in an iterative manner. Now for the observations Y any value of $\mathrm{f}_{Y_k}(y_k)$ can be estimated as follows

$$\hat{\mathrm{f}}_{Y_k}(y_k) = \frac{3}{4Mh}\sum_{j=1}^{M}\left(1 - \left(\frac{y_k - \tilde{y}_{kj}}{h}\right)^2\right) 1\{y_k - h \le \tilde{y}_{kj} \le y_k + h\}. \tag{3.21}$$

As far as $h \to 0$, but $h \neq 0$, then estimates (3.21) are biased $\mathrm{E}\left[\ln \hat{\mathrm{f}}_{Y_k}(y_k)\right] - \ln \mathrm{f}_{Y_k}(y_k) \neq 0$.

The value of $\ln \hat{\mathrm{f}}_{Y_k}(y_k)$ has to be corrected for every $k \in \{0,1,\dots,n_{t_1}\}$, namely by repeating the estimation procedure S times for each $\hat{\mathrm{f}}_{Y_k}(y_k)$

$$\ln \mathrm{f}_{Y_k}(y_k) = \ln\left(\mathrm{E}\left[\hat{\mathrm{f}}_{Y_k}(y_k)\right]\right) - \frac{1}{2}\frac{\mathrm{var}\left[\hat{\mathrm{f}}_{Y_k}(y_k)\right]}{\left(\mathrm{E}\left[\hat{\mathrm{f}}_{Y_k}(y_k)\right]\right)^2}, \tag{3.22}$$

where

$$\mathrm{E}\left[\hat{\mathrm{f}}_{Y_k}(y_k)\right] = \frac{1}{S}\sum_{s=1}^{S}\hat{\mathrm{f}}_{Y_k}^{(S)}(y_k)$$

and

$$\mathrm{var}\left[\hat{\mathrm{f}}_{Y_k}(y_k)\right] = \frac{1}{S-1}\sum_{s=1}^{S}\left(\hat{\mathrm{f}}_{Y_k}^{(S)}(y_k) - \mathrm{E}\left[\hat{\mathrm{f}}_{Y_k}(y_k)\right]\right)^2.$$

Since the estimate (3.20) is found on the basis of values of the numerical approximation, the accuracy of the estimation procedure for (3.11) and in consequence of the solution of (3.2). Therefore, the optimization of (3.2) has to be done under constrains. Let us recall the definition.

Definition 3.1. A discrete time approximation Y^{Δ} converges strongly with order $\gamma > 0$ at time t_1 if there exists a positive constant C, which does not depend on Δ, and a $\delta_0 > 0$, such that

$$\varepsilon(\Delta) = \mathsf{E}\left(\left|X_{t_1} - Y_{t_1}^{\Delta}\right|\right) \le C\Delta^{\gamma}, \tag{3.23}$$

for each $\Delta \in (0, \delta_0)$. ■

The order of strong convergence γ is usually know for selected numerical scheme and cant be improved. The only possibility to get better accuracy for the parameters of SDE (3.1) is the chose of the discretization time $\Delta \in (0, \delta_0)$. As it was set before $\Delta = \frac{t_1}{n}$, so this means that $\varepsilon(\Delta) \to 0$ if $n \to \infty$. Taking into account that the generated values $\tilde{y}_{kj}$ depend also on the unknown parameters $\boldsymbol{\theta}$, we set $\hat{f}_{Y_k}(y_k) := \hat{f}_{Y_k}(y_k;\boldsymbol{\theta})$, $k = 0,1,...,n$, and formulate the identification task (3.1) as minimization of the functional

$$J(\boldsymbol{\theta}) = \inf_{\boldsymbol{\theta}\in\Theta}\left\{-\ln\hat{f}_{Y_0}(y_0;\boldsymbol{\theta}) - \sum_{k=1}^{n}\hat{f}_{Y_k}(y_k;\boldsymbol{\theta})\right\} \tag{3.24}$$

subjected to the constraints (3.1) and (3.23).

3.2.2 Panel data sample set

Let us consider a very typical situation for the bio-economic modeling, when the data on the population dynamics $X = \{X_t, t \in [t_0, t_1]\}$ are obtained by different observers, say there are M observers. In this case the parameters of the stochastic differential equation (3.1) can be estimated on the basis of the panel Y_k^j, where $j = 1,2,..,M$ stands for the observer, $k = 0,1,...,n$ refers to the discretization times $t_0 \le \tau_0, \tau_1, ..., \tau_n \le t_1$ and $Y_0 = \mathsf{E}\left[Y_0^j\right]$. It is not difficult to conclude that the hypothesized distribution in the given parametrized family of probability distributions $\mathsf{F}_X\left(\cdot,\hat{\boldsymbol{\theta}}\right)$ represents the most probable distribution from the given class of distributions having observed Y_k^j, $j = 1,2,..,M$, $k = 0,1,...,n$.

We suppose, that for the stochastic process $X = \{X_t, t \in [t_0, t_1]\}$ there exists the equivalent stochastic process $\tilde{X} = \{\tilde{X}_t, t \in [t_0, t_1]\}$, which sample paths w.p.1 are continuous on the interval $[t_0, t_1]$, so that both processes have equivalent distributions, i.e. $\mathsf{F}_X(x) = \mathsf{F}_{\tilde{X}}(\tilde{x})$.

The empirical estimate of $\mathsf{F}_X(x)$ can be found on the basis of Y_k^j as

$$\mathsf{F}_{Y_k}(y_k) = \frac{1}{M}\sum_{j=1}^{M}\mathbb{1}_{(-\infty, y_j]}\left(Y_k^j\right), \tag{3.25}$$

where $j = 1,2,..,M$, $k = 0,1,...,n$.

For the same estimate of $\mathsf{F}_{\tilde{X}}(\tilde{x})$ the generated sample paths are required

$$\hat{\mathsf{F}}_{\tilde{Y}_k}(\tilde{y}_k;\boldsymbol{\theta}) = \frac{1}{N}\sum_{i=1}^{N}\mathbb{1}_{(-\infty, y_i]}\left(\tilde{Y}_k^i\right), \tag{3.26}$$

where N is the number of simulated sample paths of the equivalent stochastic process given by (3.1) with the set of the parameters $\boldsymbol{\theta}$.

Now, the identification task can be solved by means of the testing the hypothesis about the equivalence of the distributions (3.25) and (3.26), using, for example, Kolmogorov-Smirnov's goodness-of-fit test

$$D_{N,M}(\tau_k;\boldsymbol{\theta}) = \sup_{y,\tilde{y}\in\mathsf{R}}\left|\hat{\mathsf{F}}_{Y_k}(y_k) - \hat{\mathsf{F}}_{\tilde{Y}_k}(\tilde{y}_k;\boldsymbol{\theta})\right| \tag{3.27}$$

for all $\tau_k \in [t_0, t_1]$.

The statistic (3.27) has asymptotic null distribution

$$KS\left(D(\tau_k;\boldsymbol{\theta})\right)=\lim_{N,M\to\infty}\mathrm{P}\left(\sqrt{\frac{NM}{N+M}}D_{N,M}(\tau_k;\boldsymbol{\theta})\le D^*\right), \tag{3.28}$$

where D^* is critical value of Kolmogorov's distribution.
The expression (3.28) can be presented also by

$$KS\left(D(\tau_k;\boldsymbol{\theta})\right)=1-2\sum_{\ell=1}^{+\infty}(-1)^{\ell-1}\exp\left[-2\ell^2 D(\tau_k;\boldsymbol{\theta})^2\right]. \tag{3.29}$$

A large value of $D(\tau_k;\boldsymbol{\theta})$, and therefore a small value $KS\left(D(\tau_k;\boldsymbol{\theta})\right)$, indicates that the distributions are not equivalent, whereas small values of $D(\tau_k;\boldsymbol{\theta})$ support that the distributions are equivalent. This fact can be used for the formulation of the identification task for (3.1), that is to say one has to maximize the functional

$$J(\boldsymbol{\theta})=\sup_{\boldsymbol{\theta}\in\Theta}\left\{\sum_{k=0}^{n}KS\left(D(\tau_k;\boldsymbol{\theta})\right)\right\} \tag{3.30}$$

subjected to (3.1) and (3.23).

4. Identification method for the time-varying parameters

4.1 Basic assumptions

Let $(\Omega,\mathsf{F},\{\mathsf{F}_t\}_{t\ge0},\mathsf{P})$ be a complete probability space on which some m - dimensional stochastic process $\mathbf{B}=\left\{\mathbf{B}_t=\left(B_t^1,B_t^2,\ldots,B_t^m,\right)^{\mathrm{T}},t\in[t_0,t_1]\right\}$ is defined such that $\{\mathsf{F}_t\}_{t\ge0}$ is the natural filtration generated by $\mathbf{B}(\cdot)$, augmented by all the P -null sets in F . We suppose that these stochastic processes are independent and replace (2.6) by the following SDE

$$d\mathbf{X}(t)=\mathbf{a}\left(t,\mathbf{X}(t),\boldsymbol{\theta}(t)\right)dt+\mathbf{b}\left(t,\mathbf{X}(t),\boldsymbol{\theta}(t)\right)d\mathbf{B}\ ,\ \mathbf{X}(t_0)=\mathbf{X}_0\in\mathsf{R}^d,\ t\in[t_0,t_1], \tag{4.1}$$

where $\mathbf{a}:[t_0,t_1]\times\mathsf{R}^d\times\Theta\to\mathsf{R}^d$ and $\mathbf{b}:[t_0,t_1]\times\mathsf{R}^d\times\Theta\to\mathsf{R}^{d\times m}$ with Θ being a given metric space, which specifies the set of allowable values for the parameters $\boldsymbol{\theta}$, $\boldsymbol{\theta}(\cdot)$ is the unknown non-random vector of parameters.
The goal is to present the estimation method for the parameters $\boldsymbol{\theta}(\cdot)$ taking into account some properties of the stochastic process, which is assumed to be the unique strong solution of (4.1). For the simplicity in further reasoning we will consider one-dimensional case $(d=m=1)$ of the SDE (4.1) and limit the family of stochastic processes $\mathbf{B}(\cdot)$ to one-dimensional ordinary Brownian motion (fBm). This gives the possibility to rewrite (4.1) as

$$dX(t)=a\left(t,X(t),\boldsymbol{\theta}(t)\right)dt+b\left(t,X(t),\boldsymbol{\theta}(t)\right)dB(t)\ ,\ X(t_0)=X_0\in\mathsf{R},\ t\in[t_0,t_1], \tag{4.2}$$

We point out that although the SDE (4.2) is now assumed to be one-dimensional, results can be extended to m -dimensional case of the SDE (4.1) with the same ideas.

4.2 Estimation principle

There are many possibilities to solve the general optimal control problem (1.1), (1.3) - (1.6), (4.2) with respect to the identification problem of the parameters $\boldsymbol{\theta}$. Since the solution of the object equation (4.2) is a stochastic process, it is reasonable to use stochastic principles as it was done in [Hu at al., 2003]. However, in our case we are not going to solve "pure" optimal control task, because we consider a non-random vector of parameters and thus SDE (4.2) can be converted to an ordinary differential equation (ODE) by means of moment equations.

Let $m_1(t)=\mathsf{E}\left[X(t)\right]$ and $m_2(t)=\mathsf{E}\left[X^2(t)\right]$ be the first and second moments of stochastic process $X(t)$, $t\in\left[t_0,t_1\right]$, generated by the SDE (4.2). Denote a new state variable

$$\mathbf{y}(t)=\left[m_1(t),m_2(t)\right]\in\mathsf{R}^2,$$

where $\mathbf{y}(t_0)=\left[m_1(t_0),m_2(t_0)\right]$ $\left(m_1(t_0)=\mathsf{E}\left[X_{t_0}\right],\ m_2(t_0)=\mathsf{E}\left[X_{t_0}^2\right]\right)$, and describe object dynamics using a system of the ODEs

$$d\mathbf{y}(t)=\varphi\left(t,\mathbf{y}(t),\boldsymbol{\theta}(t)\right)dt \quad \text{a.e. } t\in\left[t_0,t_1\right]. \tag{4.3}$$

In this manner we have the possibility to use the principle maximum in a form, described in [Milyutin & Osmolovskii, 1998] or [Milyutin at al., 2004], to solve the parameter estimation problem. Now we introduce several definitions, which help to construct the estimation method.

Definition 4.1. Any $\boldsymbol{\theta}(\cdot)$ is called a feasible parameters vector $\boldsymbol{\theta}_f(\cdot)$, if

- $\boldsymbol{\theta}(\cdot)\in\mathbf{V}\left[t_0,t_1\right]$, where $\mathbf{V}\left[t_0,t_1\right]\triangleq\left\{\boldsymbol{\theta}:\left[t_0,t_1\right]\to\Theta\,\middle|\,\boldsymbol{\theta}(\cdot)\text{ is measurable}\right\}$;
- $\mathbf{y}(\cdot)$ is the unique solution of the system of the ODEs (4.3) under $\boldsymbol{\theta}(\cdot)$;
- the state constraints (1.3) and (1.4) are satisfied;
- $f\left(t,\mathbf{y}(t),\boldsymbol{\theta}(t)\right)$ belongs to the set of Lebesgue measurable functions such that

$$\int_{t_0}^{t_1}\left|f\left(t,\mathbf{y}(t),\boldsymbol{\theta}(t)\right)\right|dt<\infty. \qquad \blacksquare$$

Definition 4.2. $\hat{\boldsymbol{\theta}}(\cdot)$ is called an optimal estimate of $\boldsymbol{\theta}(\cdot)$, if $\mathsf{J}\left(\hat{\mathbf{y}}(\cdot),\hat{\boldsymbol{\theta}}(\cdot)\right)$ is measurable and there exists $\varepsilon>0$ such that for any $\mathbf{u}_f(\cdot)$ the following inequalities are fulfilled

$$\left\|\mathbf{y}(\cdot)-\hat{\mathbf{y}}(\cdot)\right\|_{\mathsf{C}\left(\left[t_0,t_1\right],\mathsf{R}^2\right)}<\varepsilon,$$

$$\mathsf{J}\left(\mathbf{y}(\cdot),\boldsymbol{\theta}(\cdot)\right)\geq\mathsf{J}\left(\hat{\mathbf{y}}(\cdot),\hat{\boldsymbol{\theta}}(\cdot)\right),$$

where $\mathbf{C}([t_0,t_1],\mathrm{R}^2)$ is the set of all continuous functions, $\hat{\cdot}$ denotes the estimate.

The definitions (4.1) and (4.2) allow us to propose the goal function (goal) as follows

$$\mathtt{J}(\mathbf{y}(\cdot),\boldsymbol{\theta}(\cdot))=\inf_{\boldsymbol{\theta}\in\Theta}\int_{t_0}^{t_1} f(t,\mathbf{y}(t),\boldsymbol{\theta}(t))dt$$

where

$$f(t,\mathbf{y}(t),\boldsymbol{\theta}(t))=\left\|\varphi(t,\mathbf{y},\boldsymbol{\theta})-\varphi(t,\hat{\mathbf{y}},\hat{\boldsymbol{\theta}})\right\|_2^2$$

and $\inf(\cdot)$ is the greatest lower bound. ■

The phase constraints (1.3) and state constraints (1.4) can be defined on the basis of the properties of the stochastic process [Shyryaev, 1998]. As it was said before the Pontryagin's type maximum principle will be used to find the solution to the estimation problem. In this case we introduce the Pontryagin's function

$$\mathtt{H}(t,\mathbf{y}(t),\boldsymbol{\theta}(t),\psi(t))=\psi(t)\varphi(t,\mathbf{y}(t),\boldsymbol{\theta}(t))-\alpha_0 f(t,\mathbf{y}(t),\boldsymbol{\theta}(t)),$$

where $\psi(t)\in(\mathrm{R}^2)'$ is an adjoint function of bounded variation (ψ : $[t_0,t_1]\to\mathrm{R}^2$ is an absolutely continues function), α_0 is a number.

The theorem below, based on Dubovitski-Milyutin method [Milyutin at al., 2004], gives the possibility to find an optimal estimate $\hat{\boldsymbol{\theta}}(\cdot)$ of $\boldsymbol{\theta}(\cdot)$ for SDE (4.2).

Theorem 4.1. Let $\hat{\boldsymbol{\theta}}(\cdot)$ be an optimal estimate of $\boldsymbol{\theta}(\cdot)$ and $(\hat{\mathbf{y}}(\cdot),\hat{\boldsymbol{\theta}}(\cdot))$ be an optimal pair ($\boldsymbol{\theta}(\cdot)\in \mathtt{L}^\infty([t_0,t_1],\mathrm{R}^2)$, $\mathbf{y}(\cdot)\in\mathbf{C}([t_0,t_1],\mathrm{R}^2)$). Then there exist a number α_0, a function of bounded variation $\psi(t)$ (which defines the measure $d\psi$), a function of bounded variation $\lambda(t)$ (which defines the measure $d\lambda$) such that the following conditions hold:

- nontriviality $|\alpha_0|+\|d\lambda\|>0$,
- nonnegativity $\alpha_0\geq 0$, $d\lambda\geq 0$,
- complementary slackness $d\lambda(t)g(t,\mathbf{y}(t))=0$,
- adjoint equation

$$-d\boldsymbol{\psi}(t)=\boldsymbol{\psi}(t)\varphi_{\mathbf{y}}(t,\hat{\mathbf{y}}(t),\hat{\boldsymbol{\theta}}(t))-\alpha_0 f_{\mathbf{y}}(t,\hat{\mathbf{y}}(t),\hat{\boldsymbol{\theta}}(t))-g_{\mathbf{y}}(t,\hat{\mathbf{y}}(t))d\lambda,$$

- transversality condition $\boldsymbol{\psi}(t_1)=0$,
- the local maximum condition

$$\psi(t)\varphi_{\boldsymbol{\theta}}(t,\hat{\mathbf{y}}(t),\hat{\boldsymbol{\theta}}(t))-f_{\boldsymbol{\theta}}(t,\hat{\mathbf{y}}(t),\hat{\boldsymbol{\theta}}(t))=0.$$

■

The proof of the theorem 4.1 is not complicated and can be found in [Milyutin & Osmolovskii, 1998].

5. Conclusions

The stochastic differential equation was considered as the bio-economic model in the task of optimal control of the resource management. Several groups of the parameter estimation methods for the different types of the stochastic differential equation were proposed. First group of the estimation procedures is based on the maximum likelihood method, second one uses principles of Monte Carlo simulations and the last one employs the Pontryagin's type maximum principle. First and second group are very sensitive to the structural selection of the stochastic differential equation, not useful in the case of time-varying parameters or system of stochastic differential equations. However, they can be used for the "first iteration" in the time-varying case. The last method can be easily applied for the mentioned problems. Its scheme, formulated as the theorem, can be used if one is interested in the parametric identification of a system of the ordinary differential equations. In future, the numerical experiments are intended to take place in order to investigate the accuracy of the method.

6. References

Bastogne T.; Thomassin M. & Masse J. 2007. Selection and identification of physical parameters from passive observation. Application to a winding process. *Control Engineering Practice*, 15: 1051 - 1061.

Fan J.; Jiang J.; Zhang Ch. & Zhou Z. 2003. Time-dependent diffusion models for term struture dynamics. *Statistica Sinica*, 13: 965 - 992.

Hansen J.A. & Penland C. 2007. On stochastic parameter estimation using data assimilation. *Physica D*, 230: 88 - 98.

Hu Y.; Øksendal B. & Sulem A. 2003. Optimal consumption and portfolio in a Back Scholes market driven by fractional Brownian motion: Infinite dimensional analysis, *Quantum Probability and Related Topics*, 6 (4): 519 - 536.

Hurn A.S., Lindsay K.A., Martin V.L. 2003. On the efficacy of simulated maximum likelihood for estimating the parameters of stochastic differential equations. Journal of Time Series Analysis, 24: 45 - 63.

Jang M.-J.; Wang J.-S. & Liu Y.-C. 2003. Applying differential transformation method to parameter identification problems. *Applied Mathematics and Computation*, 139: 491 - 502.

Jazwinski A.H. (2007). Stochastic Processes and Filtering Theory. Dover Publiscations Inc.

McDonald A.D. & Sandal L.K. 1999. Estimating the parameters of stochastic differential equations using a criterion function based on the Kolmogorov-Smirnov statistic. *J. Statist. Comput. Simul.*, 64: 235 - 250.

Milyutin A.A. & Osmolovskii N.P. 1998. Calculus of Variations and Optimal Control, *Translations of Mathematical Monographs*, volume 180, American Mathematical Society, Providence.

Milyutin A.A.; Dmitruk A.V. & Osmolovskii N.P. 2004. Maximum Principle in Optimal Control, *Moscow State University, Moscow* (in Russian).

Shoji I. & T. Ozaki 1998. Estimation for the nonlinear stochastic differential equations by a local linearization method, Stochastic Analysis and Applications, 16: 733 - 752, 1998.

Shyryaev A.N. (1998). The Basis of Stochastic Financial Mathematics: Facts, Models (in Russian).